THE GEOLOGICAL MODELLING
OF HYDROCARBON RESERVOIRS
AND OUTCROP ANALOGUES

The Geological Modelling of Hydrocarbon Reservoirs and Outcrop Analogues

EDITED BY STEPHEN S. FLINT
AND IAN D. BRYANT

SPECIAL PUBLICATION NUMBER 15 OF THE
INTERNATIONAL ASSOCIATION OF SEDIMENTOLOGISTS
PUBLISHED BY BLACKWELL SCIENTIFIC PUBLICATIONS
OXFORD LONDON EDINBURGH BOSTON
MELBOURNE PARIS BERLIN VIENNA

Editorial Offices:
Osney Mead, Oxford OX2 0EL
25 John Street, London WC1N 2BL
23 Ainslie Place, Edinburgh EH3 6AJ
238 Main Street, Cambridge
Massachusetts 02142, USA
54 University Street, Carlton
Victoria 3053, Australia

Other Editorial Offices:
Librairie Arnette SA
2, rue Casimir-Delavigne
75006 Paris
France

Blackwell Wissenschafts-Verlag
Meinekestrasse 4
D-1000 Berlin 15
Germany

Blackwell MZV
Feldgasse 13
A-1238 Wien
Austria

First published 1993

Set by Excel Typesetters Company, Hong Kong
Printed and bound in Great Britain
at the Alden Press, Oxford

DISTRIBUTORS

Marston Book Services Ltd
PO Box 87
Oxford OX2 0DT
(*Orders*: Tel: 0865 791155
Fax: 0865 791927
Telex: 837515)

USA
Blackwell Scientific Publications, Inc.
238 Main Street
Cambridge, MA 02142
(*Orders*: Tel: 800 759-6102
617 876-7000)

Canada
Oxford University Press
70 Wynford Drive
Don Mills
Ontario M3C 1J9
(*Orders*: Tel: 416 441-2941)

Australia
Blackwell Scientific Publications Pty Ltd
54 University Street
Carlton, Victoria 3053
(*Orders*: Tel: 03 347-5552)

A catalogue record for this title
is available from the British Library

ISBN 0-632-03392-4

Library of Congress
Cataloging in Publication Data

The Geological modelling of hydrocarbon reservoirs
and Outcrop Analogues/edited by
Stephen S. Flint and Ian D. Bryant.
p. cm.
(Special publication number 15 of the
International Association of Sedimentologists)
Includes bibliographical references
and index.
1. Petroleum—Geology.
2. Natural gas—Geology.
3. Geological modeling.
I. Flint, Stephen S. II. Bryant, Ian D.
III. International Association of Sedimentologists.
IV. Series: Special publication . . .
International Association of Sedimentologists; no. 15.
TN870.5.G3868 1993
553.2′8′011—dc20

Contents

vii Preface

Quantitative Data Collection

3 Quantitative clastic reservoir geological modelling: problems and perspectives
I.D. Bryant & S.S. Flint

21 Alluvial architecture in a sequence stratigraphic framework: a case history from the Upper Cretaceous of southern Utah, USA
K.W. Shanley & P.J. McCabe

57 Sedimentary architecture of field analogues for reservoir information (SAFARI): a case study of the fluvial Escanilla Formation, Spanish Pyrenees
T. Dreyer, L.-M. Falt, T. Høy, R. Knarud, R. Steel & J.-L. Cuevas

81 Quantitative facies analysis of coal-bearing sequences in the Bowen Basin, Australia: applications to reservoir description
A. Falkner & C. Fielding

99 Quantification of turbidite facies in a reservoir-analogous submarine-fan channel sandbody, south-central Pyrenees, Spain
J.D. Schuppers

113 Outcrop studies of shale smears on fault surface
N.G. Lindsay, F.C. Murphy, J.J. Walsh & J. Watterson

125 Applications of the formation micro-scanner to modelling of Palaeozoic reservoirs in Oman
C.G.L. Mercadier & S.E. Livera

143 Predicting reservoir sandbody orientation from dipmeter data: the use of sedimentary dip profiles from outcrop studies
H. Williams & H.F. Soek

157 Permeability patterns in point bar deposits: Tertiary Loranca Basin, central Spain
C.A. Hartkamp-Bakker & M.E. Donselaar

169 Knowledge base development for the estimation of reservoir rock properties in the interwell area: examples from the Texas Gulf coast
F.C.J. Mijnssen, N. Tyler & K.J. Weber

181 The use of 3-D seismic in reservoir geological modelling
K.J. Weber

Modelling Methods

191 Sedimentary flow units in hydrocarbon reservoirs: some shortcomings and a case for high-resolution permeability data
A. Hurst

205 The use of length distributions in geological modelling
G. Geehan & J. Underwood

213 A theoretical study of fluvial sandstone body dimensions
J.S. Bridge & S.D. Mackey

237 Stochastic modelling of fluvial sandstone bodies
J.P.P. Hirst, C.R. Blackstock & S. Tyson

253 A 3-D modelling approach for providing a complex reservoir description for reservoir simulations
G.L. Krum & C.R. Johnson

259 Index

Preface

Recent technological advances and the increasing emphasis on maximizing recovery from existing oil and gas fields has led to an upsurge of interest in reservoir characterization and quantitative modelling of physical rock properties in 3-D inter-well space. A variety of research conferences have been convened over the last 5 years to provide a forum for debate of new ideas and directions in this area of interdisciplinary research; these meetings have, in the USA, included NIPER conferences, SPE symposia and specialist sessions within the AAPG Annual Conventions. Within Europe, similar topics have been covered by specialist meetings of the Geological Society of London and the EAPG. However, much of this debate has been driven by reservoir engineers and computer software engineers keen to take advantage of advances in new methods of formation evaluation and information technology, but to a certain extent unfamiliar with sedimentological concepts. Over the same time period much academic research in applied sedimentology seemed to us to be carried out divorced from these industry developments in subsurface modelling.

As a consequence it was decided to convene 'The Geological Modelling of Hydrocarbon Reservoirs' symposium as part of the 1990 International Sedimentological Congress at Nottingham, UK. The symposium was conceived initially in 1988 when we were both working on 'how to correlate well logs' at the scale of individual reservoir flow units. The object of the symposium was to bring together geologists, mathematicians, engineers and others involved in reservoir geological modelling at an international meeting in order to encourage debate and mutually beneficial exposure to ideas from a variety of disciplines. A total of 41 oral presentations and 24 posters were given in what proved to be a lively meeting.

Discussions at the meeting emphasized the need to select carefully appropriate outcrop analogues when trying to supplement sparse subsurface data with outcrop-derived measurements. Criteria suggested for appropriate analogue selection included tectonic setting, geological age and subsidence rates, while a small number of delegates emphasized the important role of sequence stratigraphy in determining suitable analogues. Other participants stressed the need to make more use of existing subsurface data sets. A general consensus was reached that quantitative modelling of carbonate reservoirs was rather more difficult than clastic reservoirs, due to the less obvious link between depositional facies and reservoir quality. Unfortunately this is reflected by the scarcity of carbonate papers in this volume but many of the techniques discussed herein are valid for carbonate reservoir modelling.

Our approach to the topic of this book is based on the belief that well data alone cannot provide an unambiguous 3-D volume model of the subsurface. Paradoxically 3-D data are widely available in seismic form, but only recently, and still only under favourable circumstances, can seismic data resolve individual reservoir sands. The basic drive of this volume is therefore to illustrate methods of dealing with the resultant uncertainty which haunts the development geologist. The 16 papers, which cover outcrop and subsurface data sets from Europe, the Middle East, North America and Australia, represent a selection of those for which full manuscripts were submitted. They are grouped into sections on quantitative data collection and computer modelling methods. This is perhaps the first IAS volume that highlights such close research links and equal levels of interest between industry and academic earth scientists, geostatisticians and reservoir engineers. Such collaborative studies are surely the way forward in this research area—the answers to practical questions providing a powerful positive feedback circuit to the academic science.

We would like to thank the authors of the papers for their cooperation in meeting our deadlines, despite disruption to certain postal services during the Gulf conflict. The referees were central in helping to make this publication a good reflection of the state of the art in reservoir modelling. These included many of the authors plus J. Brint, T. Cross, O. Dubrule, D. Horowitz, J. Kantorowicz, D. Mann, J. Marshall, J. Martin, C. O'Byrne, A. Paardekam, M. Rider, A. Ryseth, S. Smith and K. Sorbie.

STEPHEN S. FLINT, *Department of Earth Sciences, University of Liverpool, Liverpool, UK*;
IAN D. BRYANT, *Shell Todd Oil Services, New Plymouth, New Zealand*

Quantitative Data Collection

Spec. Publs int. Ass. Sediment. (1993) **15**, 3–20

Quantitative clastic reservoir geological modelling: problems and perspectives

I.D. BRYANT* *and* S.S. FLINT†

**Shell Todd Oil Services Ltd., Private Bag, New Plymouth, New Zealand;*
†Department of Earth Sciences, University of Liverpool, PO Box 147, Liverpool L69 3BX, UK

ABSTRACT

The emerging technology of representing reservoir geological descriptions as three-dimensional (3-D) numerical models provides a method of reservoir characterization that is capable of rigorous testing using independent methods. Many of these models are also capable of integrating the increasingly large volumes of data concerning the reservoir formation that are derived from 3-D seismic surveys, improved wireline logging tools and wells that pass horizontally through the reservoir.

The quality of these models is, however, to a large extent dependent on the quality of the quantitative geological knowledge that is used in their construction. For example, an increasingly large number of data sets concerning the lateral extent of fluvial channel-fill sandbodies as a function of their thicknesses is being amassed, although critical application of these statistics is required. Only sparse data sets are available to constrain the modelling of widths from thicknesses for other types of genetic reservoir sandbodies. It is critical to compare 'like with like' when compiling data sets; thus sequence stratigraphy provides an important framework within which to group geometrical data. Furthermore, the application of both object-based and cell-based modelling procedures also requires that the orientation of the genetic bodies being modelled is known. Whilst new formation imaging tools are increasingly capable of providing stratigraphic dips, the interpretation of sandbody orientation from these data is still, in many instances, problematic. Where well data are particularly sparse with respect to the scale of geological variation being modelled, forward-modelling procedures and sequence stratigraphic concepts may provide a more reliable approach to reservoir modelling. Such models may be based on deterministic or mixed deterministic/probabilistic rather than purely probabilistic techniques.

INTRODUCTION

The techniques of quantitative reservoir geological modelling are not new. Even in the 1960s, when modern facies sedimentology was still in its infancy, detailed quantitative reservoir models were generated by integrating quantitative outcrop data with cores, geophysical wireline logs and pressure measurements (e.g. Dupuy & Pottier, 1963; Montadert, 1963; Grout *et al.*, 1965; Verrien *et al.*, 1967). Quantitative measurements were also made from recent sediments with the aim of understanding better the spatial distribution of reservoir properties in analogous, but less accessible, reservoir formations (e.g. Oomkens, 1967; Weber *et al.*, 1972; Pryor, 1973).

The rapid growth in the number and complexity of reservoir models in recent years has two main causes. The first is the increasing amount of information that new technologies are able to obtain from the reservoir formations, and the second is the development of more powerful and efficient computing systems, which enable much more detail to be incorporated into numerical reservoir descriptions and numerical reservoir simulation models. The increasingly widespread use of 3-D seismic surveying techniques (Fig. 1) has resulted in much more detailed measurement of both structural and depositional features (Nestvold, 1987, 1989; Ruitenberg *et al.*, 1990). Wireline logging tools are

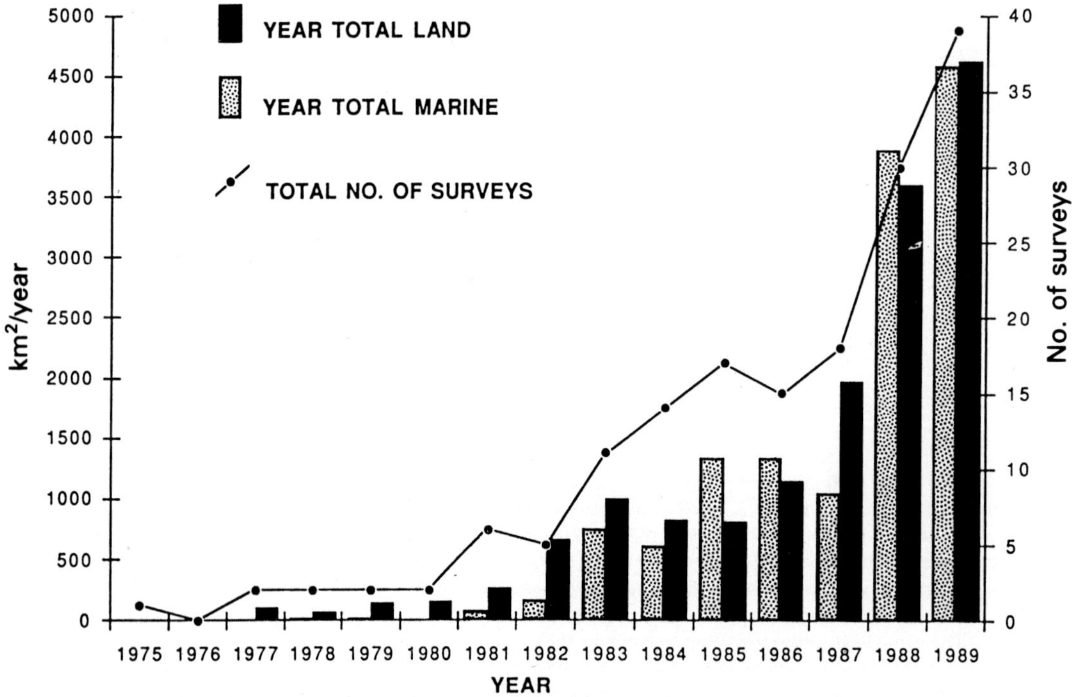

Fig. 1. Histogram to show the increasingly widespread application of 3-D seismic surveying within just one company (Shell). From Nestvold (1989).

able to image the borehole wall such that, even in the absence of core, detailed information concerning the distribution and orientation of sedimentary structures is available (Grace & Pirie, 1986; Serra, 1987; Luthi & Banavar, 1988; Anxionnaz *et al.*, 1990; Mercadier & Livera, this volume). A more recently developed technology—that of drilling wells which pass horizontally through reservoir formations—is beginning to be applied widely and will increasingly provide information concerning lateral geological changes. This wealth of new information is collected at widely differing scales, ranging from the centimetric (e.g. wireline log data) to the kilometric (seismic data). The need to integrate these data into a model that is consistent with all of them has led to the development of a new generation of powerful computer workstations (e.g. Abib *et al.*, 1988; Clemetsen *et al.*, 1990; Gundesø & Egeland, 1990; Wadsley *et al.*, 1990; Bryant *et al.*, 1991).

Despite these technological advances the difficulty remains that the measured data are never sufficient to describe unambiguously the reservoir. This is a consequence of the scale and resolution at which measurements are made. Seismic methods result in a description of the entire reservoir volume but are limited in the vertical (and lateral) resolution of the information that they provide as a consequence of the relatively low frequency content of the seismic data. Well data, on the other hand, may permit a very detailed description of the reservoir but only sample a very small fraction (commonly 10^{-19}) of the total reservoir volume. Furthermore, these data have traditionally come from vertical or near vertical wells, leading to an emphasis on vertical facies analysis. The increasing number of wells that pass horizontally through reservoir formations yield information concerning lateral facies changes and will undoubtedly provide a test for geologists' abilities in lateral facies analysis.

Traditionally, qualitative geological models have been used to extrapolate high-resolution well data into the interwell volume. In some cases outcrop-derived statistics have been used to constrain further this extrapolation (e.g. Knutson *et al.*, 1971; Lorenz *et al.*, 1985; Rudkiewicz & Eschard, 1991). Models generated by these traditional methods are generally presented in the form of selected cross-sections

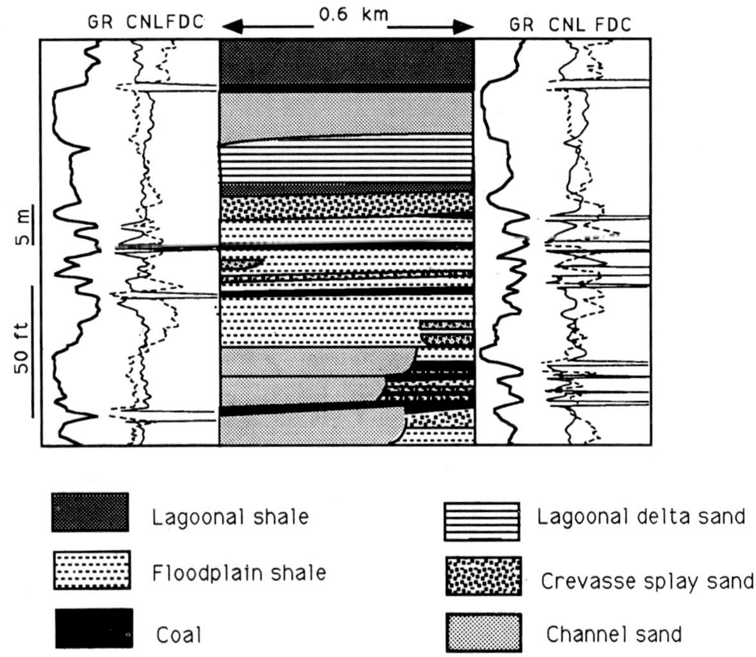

Fig. 2. Traditional correlation of two wells within the Ness Formation of the Brent field, UK North Sea. (GR = Gamma Ray; FDC = Density log; CNL = Compensated Neutron Log). Note that whilst the correlation may be correct, neither the size nor the orientation of the channel-fill reservoir sandbody is defined by this 2-D representation of the reservoir.

and/or maps (Fig. 2). These representations of the reservoir are ambiguous descriptions because they are only defined in two dimensions and consequently are not capable of rigorous testing, e.g. neither the precise location of the termination of the channel-fill sandbody shown in Fig. 2 nor its orientation are defined by this schematic 2-D representation.

The new generation of quantitative reservoir models are usually 3-D, cell-based models, where each cell requires to be allocated attributes on the basis of the geological model (Fig. 3). These models put much greater demands on the geologist than traditional models since the geology requires complete description at every point within the 3-D volume of the reservoir (Haldorsen *et al.*, 1987), e.g. the channel sandbodies shown in Fig. 3 have each been assigned a 'most likely' orientation and width. The model may therefore be subjected to independent testing (e.g. is the rock volume represented by one channel consistent with the volume of produced hydrocarbons?) and is thus intrinsically a more scientific description of the reservoir, even though it is not necessarily any more correct than the 2-D description. These 3-D models may then be used to simulate numerically fluid flow through the reservoir with time. These simulations in turn provide a further method of testing the validity of the geological model and may help to distinguish between alternative models that are equally probable from a geological perspective. Whilst these 3-D numerical reservoir descriptions represent a significant advance over traditional methods, they do, however, require more quantitative geological information in order to be constructed in a geologically realistic manner.

METHODOLOGY

One commonly applied methodology for modelling clastic reservoirs constitutes:

1 definition of the space occupied by the reservoir interval;

2 recognition of geological units within this space;

3 assignment of geometries to the units;

4 arrangement of these units within the defined space (architecture);

5 assignment of reservoir properties to the units.

We will discuss each of these steps in turn whilst at the same time reviewing:

(a) what quantitative data are required to generate these reservoir descriptions;

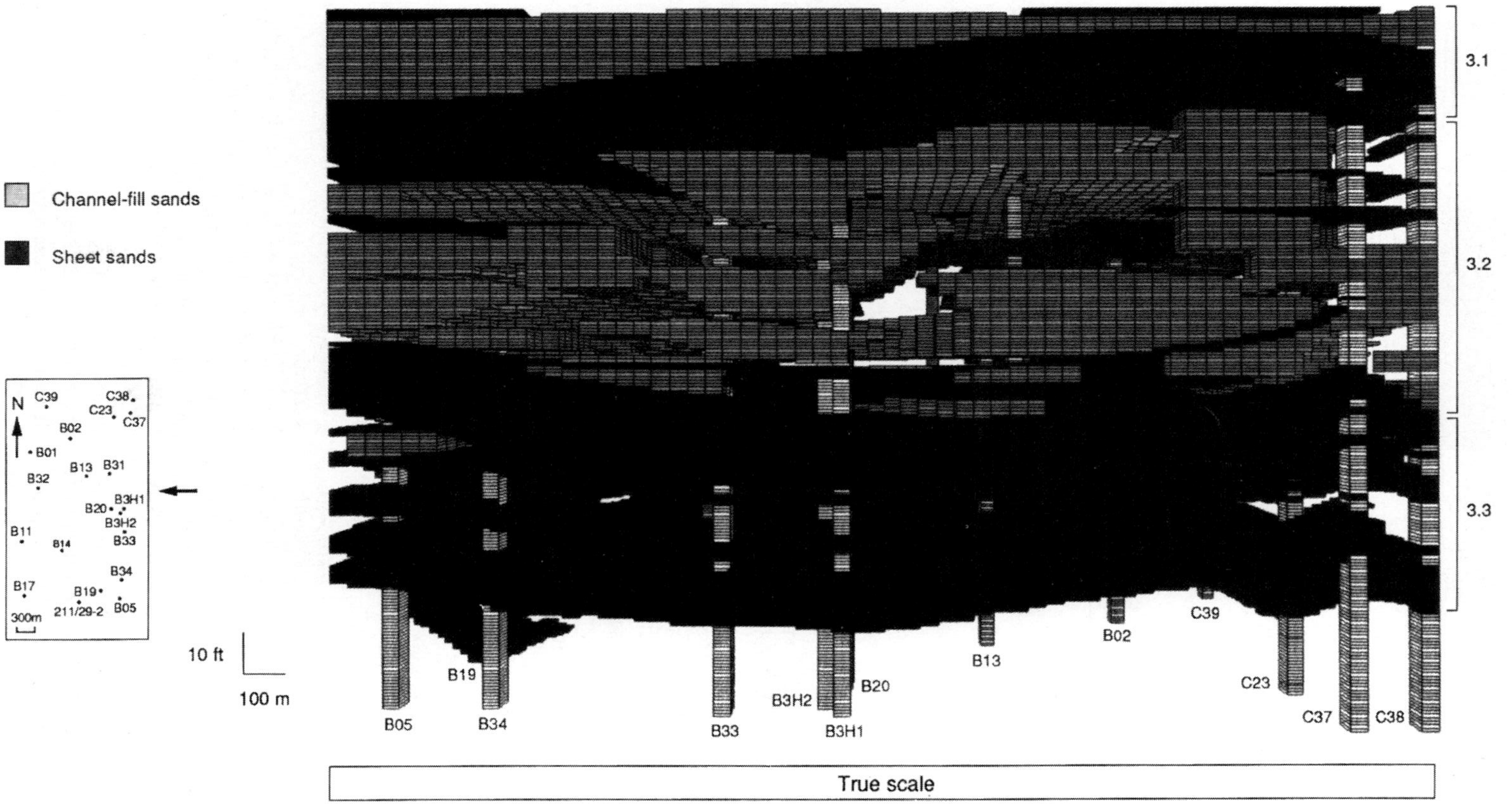

Fig. 3. Object-based geological reservoir model from part of the Ness Formation, Brent field, UK North Sea. The top is defined by a flooding surface overlain by a fieldwide correlatable shale. The base of the model is taken at the top of the coarsening upward sequence of the Rannoch and Etive Formations that is also correlatable throughout the field. The model shows both correlatable reservoir elements (smooth, curvilinear outlines) that have been kriged (conditionally simulated) and non-correlatable elements (rectangular outlines) that have been extrapolated using a Boolean method into the inter-well area parallel to the top of the model. Note that each channel within the model has a defined size and orientation. The transparent parts of the model represent shales. The vertical columns are wells, the locations of which are shown on the inset map. Arrow shows viewing direction. From Bryant *et al.* (1991).

(b) what data are currently available; and
(c) the problems that arise from deficiencies in the current state of knowledge.

DEFINITION OF SPACE

The space occupied by the reservoir is usually defined by a combination of well and seismic data. The top and base of the interval are usually correlatable lithological units, flooding surfaces or unconformities that commonly give rise to mappable seismic events (Vail *et al.*, 1977; Galloway, 1989; Van Wagoner *et al.*, 1990), allowing the reservoir to be set in its basinal context. This step provides a means of integrating regional-scale features (seismic markers) with more local features (log markers) and of tying the low resolution, widespread 3-D seismic data to the high resolution, localized 1-D well data. Furthermore, knowledge of the sequence stratigraphic context of the interval being modelled may help to constrain modelling of the sedimentary architecture within the interval (Van Wagoner *et al.*, 1990; Shanley & McCabe, this volume).

RECOGNITION OF UNITS

The recognition of genetic facies associations within the interval to be modelled is important in order to be able to apply valid qualitative and quantitative models in later stages of the modelling procedure. Commonly recognized objects might be individual shorefaces, channel-fills, overbank mudstones or coals (Fig. 3). In order to optimize the use of quantitative data in modelling geological objects or layers it is necessary to diagnose as precisely as possible the type of object or layer being modelled. At outcrop the sedimentologist is provided with a wide range of possible features to assist in determination of precisely what type of genetic unit is being examined (Fig. 4).

In the subsurface, however, far fewer data are usually available. Cores are usually available from only some of the wells and are used to calibrate interpretations of wireline logs from the remaining wells. The vertical resolution of traditional wireline logging suites (usually 0.5–1.0 m) usually precludes the recognition of fine-scale features, such as bedding styles, that sedimentologists routinely use to distinguish more precisely different types of sedimentary body in outcrop. However, the development of wireline tools with a much greater vertical resolution has enabled detailed observations of sedimentary structures to be obtained even in the absence of cores (Grace & Pirie, 1986; Anxionnaz *et al.*, 1990; Höcker *et al.*, 1990). Furthermore, patterns of sedimentary dips obtained from these tools may be diagnostic for different types of genetic units (Perrin, 1975; Rider, 1978; Goldstein & Hubbard, 1986; Herweijer *et al.*, 1990; Williams & Soek, this volume; Mercadier & Livera, this volume).

It is often more important to characterize shale types than sandstone types. The recognition of laterally extensive shales from discontinuous shales is important in predicting fluid displacement within reservoirs and poses difficult problems for reservoir description (Geehan *et al.*, 1986; Haldorsen *et al.*, 1987). On a large scale (tens to hundreds of kilometres) laterally extensive shales are commonly associated with maximum flooding surfaces (Vail *et al.*, 1977; Galloway, 1989; Van Wagoner *et al.*, 1990). However, at an inter-reservoir scale (hundreds of m to tens of km), it is of critical importance in coastal environments to differentiate in vertical sequence minor flooding surfaces (parasequence boundaries; Van Wagoner, 1985) from local, laterally restricted shales such as channel abandonment fills (see Plate 1, facing p. 16). The flooding surfaces by definition will have much greater lateral extents and are likely to form vertical permeability (kv) barriers and pressure seals (O'Byrne & Flint, 1991).

In order that outcrop-derived data may be used to constrain modelling of subsurface reservoirs, it is necessary that common criteria be used to compare the subsurface reservoirs with potential analogue formations. In this respect the collection of outcrop gamma-ray logs (Myers & Bristow, 1989), outcrop dipmeter profiles (Cameron, 1986; Herweijer *et al.*, 1990; Williams & Soek, this volume) or micropalaeontological data (Carman & Young, 1981; Bryant *et al.*, 1988) may facilitate comparison between parameters measured from reservoir formations and those measured from proposed analogue sequences. It is critical that geometrical outcrop data are synthesized within a sequence stratigraphic, systems tract analysis (e.g. Shanley & McCabe, this volume) as this provides a powerful framework within which to assess the 'fit' of the outcrop analogue to the subsurface data.

From a practical point of view the geological units that are modelled are usually selected to have a genetic geological significance so that sedimentological considerations may be used to infer their properties. It is, however, also important that they should represent the significant elements, or 'flow units' (*sensu* Hearn *et al.*, 1984), that control fluid flow within the reservoir. Commonly there is a close correspondence between genetic geological units and flow units but careful examination of the subsurface data should be undertaken to ensure that this is indeed the case (Hearn *et al.*, 1986; Bryant & Livera, 1991; Hurst, this volume).

ASSIGNMENT OF GEOMETRY AND ORIENTATION

Once genetic geological objects or layers have been identified at the well locations quantitative methods of assigning dimensions to geological units are required. In some cases the density of well data may be so great with respect to the size of the objects being modelled that they closely constrain the sediment body geometries (e.g. Sneider *et al.*, 1977; Kruit, 1987; Miller *et al.*, 1990). In other instances seismic data may be used to delineate precisely genetic geological units (Gelfland *et al.*, 1986; Flint *et al.*, 1988; Heritier *et al.*, 1979; Weber, this volume) but more usually the resolution of the seismic data or spacing of the well data is such that these units require modelling (Weber & van Geuns, 1989). Two common approaches are adopted to model these geometries (Dubrule, 1989; Haldorsen & Damsleth, 1990):

1 Modelling of discrete objects such as shales in sands (e.g. Haldorsen & Lake, 1982) or sands in shales (e.g. Smith & Morgan, 1986; Clemetsen *et al.*, 1990; Gundesø & Egeland, 1990; Wadsley *et al.*, 1990; Hirst *et al.*, this volume).

2 Modelling of continuous variation (e.g. Dubrule & Haldorsen, 1986; Matheron *et al.*, 1987; Fouquet *et al.*, 1989; Bryant & Ben Brahim, 1989; Rudkiewicz *et al.*, 1990).

The former (Boolean) method lends itself to modelling of reservoir/non-reservoir distribution and abrupt changes in attributes such as channel sandbody margins. The latter method is more appropriate to modelling of properties such as porosity and permeability as such data are treated more easily in a conditional simulation. Both methods commonly use cell-based modelling systems. The two approaches have also been combined to model discrete genetic objects using Boolean methods (e.g. channel-fill sandbodies) within which variations of reservoir quality are modelled using cell-based geostatistical methods (Damsleth *et al.*, 1990; Gundesø & Egeland, 1990). Both methods use statistical descriptions of presumed analogous formations to estimate the lateral extent of geological properties; Boolean methods use probability density functions derived from width vs. thickness statistics (Fig. 5) whilst conditional simulation methods use variograms derived from geostatistical analyses (Fig. 6). The application of both methods requires that the orientation of the genetic unit being modelled is also known (except in the rare case where the geological objects being modelled are assumed to be equidimensional), since both the form of the variogram (Fig. 6) and the width of the object vary with respect to the orientation of the object.

Quantitative geometrical data

Geometrical data sets have been collected to constrain reservoir models. Emphasis has been placed on characterizing geological units that are most frequently found to be discontinuous at normal offshore development well spacings (i.e. 700 m–1 km). Foremost in this category are alluvial channel-fill sandbodies (Table 1). Only limited quantitative data regarding the geometries of genetic geological units deposited in other depositional environments are currently available and these data sets are, in many instances, insufficient to obtain reliable results.

Table 1. Quantitative datasets available to guide modelling of fluvial sandbody widths from measured thicknesses

Channel sinuosity	Age	Location	Reference
High	Cretaceous	Colorado, USA	Knutson *et al.*, 1971; Knutson, 1974; Lorenz, 1985; Jones *et al.*, 1984
High + Low	Jurassic	Yorkshire, UK	Nami & Leeder, 1978; Dreyer, 1990; Walderhaug & Mjøs, 1991
High + Low	Various	Global	Fielding & Crane, 1987
High + Low	Various	Global	Martin *et al.*, 1988
High	Cretaceous	Utah, USA	Lowry & Raheim, 1991

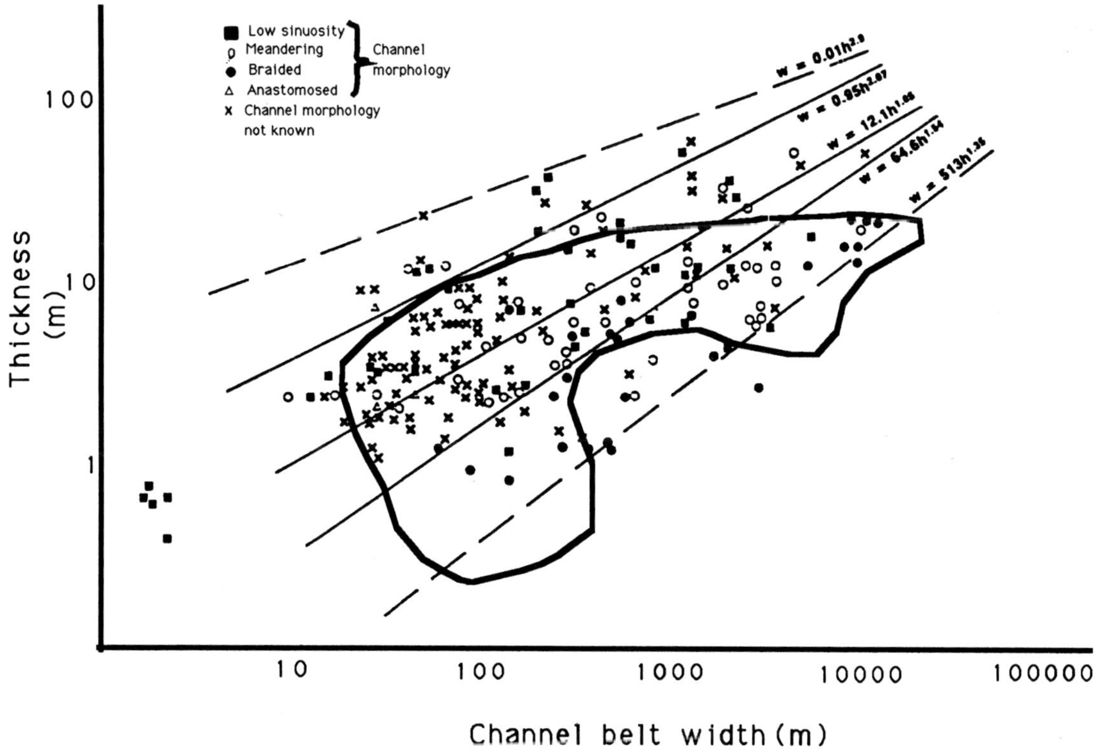

Fig. 5. Cross-plot of fluvial channel-fill sandbody widths as a function of their thicknesses. Note that, although there is a wide scatter in the data, selection of data pertinent to a particular style of channel-fill greatly reduces this scatter. (Main data compiled from Fielding & Crane, 1987.) Thick line envelope is simple meandering stream channel sandbodies (Martin *et al.*, 1988).

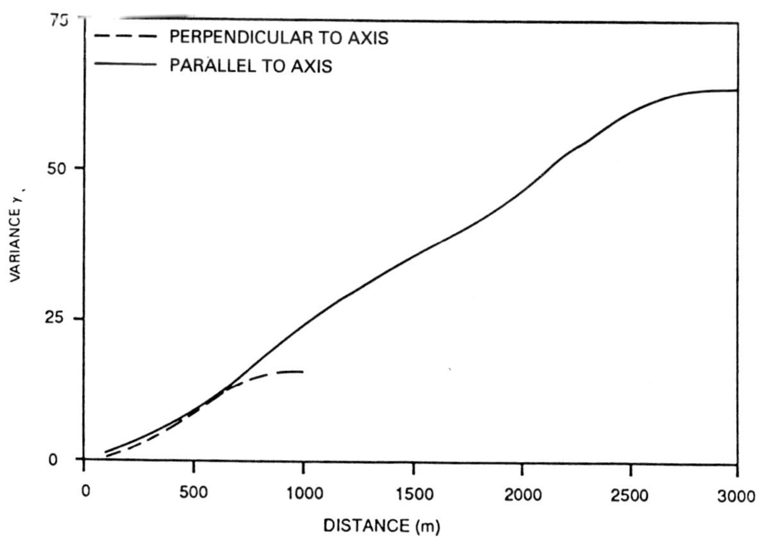

Fig. 6. Semi-variograms of channel-fill sandbody thicknesses measured from outcrop in the Carboniferous Pocahontas basin of Kentucky (taken from descriptions by Ferm & Horne, 1979). (These diagrams represent in a geostatistical form the degree of correlation between thicknesses measured at different separations within the sandbody.) Note that the rate of change of channel thickness (range of the variogram) is dependent on the orientation of the measured section with respect to the sandbody orientation.

Geometrical data may be obtained from a variety of sources:

1 Analysis of data from hydrocarbon reservoirs with a dense well spacing (e.g. Sneider *et al.*, 1977; Kruit, 1987; Miller *et al.*, 1990).

2 Analysis of analogous outcrops (e.g. Montadert, 1963; Lorenz, 1985; Flint *et al.*, 1989; Dreyer *et al.*, this volume).

3 Shallow boreholes in modern environments (e.g. Farrell, 1987).

4 Shallow seismic surveys (e.g. Kolla & Coumes, 1987).

5 Reservoir seismic integrated with well data (e.g. Dromgoole, 1987; Flint *et al.*, 1988).

6 Extrapolation from modern environments.

Integration of data from each of these sources is required in order to provide a better knowledge base for use in constraining 3-D reservoir models. However, care has to be taken in the use of geometries from modern environments owing to the possible modification of such geometries via reworking and compaction. Published data sources are reviewed in the following sections.

Fluvial sandbodies

Given that an empirical relationship appears to exist between the width and depth of alluvial channels (Leopold *et al.*, 1964; Schumm, 1977), there is a possible theoretical basis to expect a relationship between width and thickness for certain classes of alluvial channel-fill sandbodies. This has led to the thesis that channel-belt widths (and by inference sandbody dimensions) might also be related to palaeochannel thicknesses (Leeder, 1973; Fielding & Crane, 1987). The relatively successful application of statistical databases to model the lateral extent of channel-fill sands in the subsurface is in part related to the relative short timespan within which individual channel-fills are generated (10^1–10^2 years). Thus, whilst the stacking of channel-fill sandbodies may be significantly affected by variations in allocyclic mechanisms, such as relative sea-level changes and sediment supply (Allen, 1978; Bridge & Leeder, 1979; Posamentier & Vail, 1988; Bridge & Mackey, this volume), individual channel-fill geometries may be related more closely to autocyclic processes.

Several studies have attempted to obtain empirical relationships between the width and thickness of alluvial sandbodies (Table 1). Quantitative summaries of these width vs. thickness data reveal a wide scatter of data. A promising line of research involves reduction of this scatter by further defining the specific type of channel-fill being modelled (Fig. 5) and, critically, by grouping data into specific systems tracts, via a sequence stratigraphic analysis of the analogue stratigraphy. This approach requires that the type of palaeochannel being modelled may be determined from the available subsurface data, which commonly proves to be problematical (Bridge, 1985). Where it is possible to determine precisely the type of palaeochannels being modelled, then selective use of these quantitative data may give reasonable results (e.g. Lorenz *et al.*, 1985; Dreyer, 1990).

Coastal and shallow marine sandbodies

Whilst fluvial sandbodies accumulate over relatively short time scales, the same is not true of coastal or shallow marine sandbodies, which generally accumulate over much larger areas and are frequently generated over time spans of 10^3–10^5 years. Over these longer time spans allocyclic mechanisms such as tectonic, eustatic and climatic factors may interact to affect the external geometry of the lithosome. Consequently the areal extents of these bodies are unlikely to bear a simple relationship to their thicknesses (Fig. 7). However, deterministic correlation between wells commonly is relatively straightforward, although care must be exercised to correlate marine flooding surfaces first, rather than sand tops (Van Wagoner *et al.*, 1990). Correlation pitfalls also occur due to unawareness of the variable (but predictable) nature of the vertical wireline log profile of individual shoreface sandbodies down depositional dip. Modelling of these sandbodies requires a genetic modelling approach that takes account of changing conditions during their deposition (e.g. Helland-Hansen *et al.*, 1989; Lawrence *et al.*, 1990). In this approach, subsidence rates and patterns deduced from well and seismic data are convolved with sea-level curves to forward model reservoir architectures.

Deep marine sandbodies

Forward modelling has also shown promising results in the modelling of deep-water sandstone reservoirs and carbonate reservoirs (Lawrence *et al.*, 1990). Relatively few empirical geometric data are available (see Schuppers, this volume), although data from shallow seismic surveys of modern deep-sea

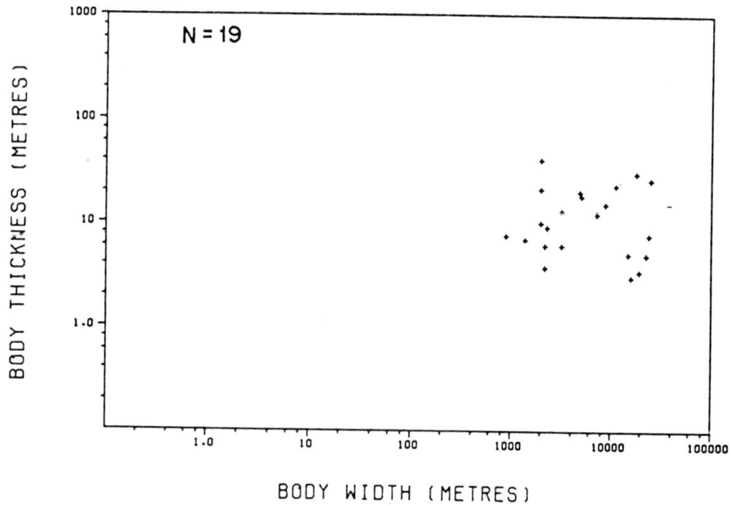

Fig. 7. Cross-plot of shallow marine shelf ridge sandbody widths as function of their thicknesses. Note that there is a wide scatter in the data that is interpreted to result from widely differing thicknesses being generated by geomorphological features of the same width but accumulating under widely varying time scales (Bryant, 1986; unpublished data).

environments may be useful in this respect (e.g. Kolla & Coumes, 1987). Stochastic models have also been employed (Hazeu *et al.*, 1986; Alabert & Massonnat, 1990).

Aeolian sandbodies

Weber (1987) used a width : thickness ratio of 200 : 1 when modelling aeolian cross-set geometries in the Leman gas field based on observations of outcrops in the western USA but comprehensive numerical data based on more extensive outcrop data sets are not currently available. Emphasis in modelling these sand-rich reservoir sequences has focused on modelling of internal heterogeneities rather than external geometries (see below).

Shales

Shales constitute important barriers and baffles to fluid flow within reservoir formations and attempts have also been made to quantify shale dimensions. Zeito (1965) indicated the order of magnitude of the extent of shales deposited in different depositional environments and this has been further elaborated by more recent work (Weber, 1982; Potter *et al.*, 1980; Geehan & Underwood, this volume). The difficulty of obtaining statistics concerning 3-D shale dimensions from 2-D outcrop data (Haldorsen & Chang, 1986) might be overcome by using data from hydrocarbon reservoirs themselves, where well, seismic and pressure data combine to constrain shale dimensions (Grout *et al.*, 1965; Delhomme & Giannesini, 1979; Skaug & Gundesø, 1986; Geehan *et al.*, 1986; Brewster & Jeangeot, 1987). Shales are discussed in more detail below.

Quantitative modelling of faults

Faults are known to be important in compartmentalizing reservoirs and modifying depositional continuity (Bouvier *et al.*, 1989; Smith, 1990). Nevertheless, the published 3-D quantitative reservoir models referenced above generally take only limited account of the role of faults on connectivity. Faults are easy to model if visible on seismic; architectural models can be created in an unfaulted state and then 'laid over' a seismic structure map of an appropriate horizon. However, theoretical and empirical studies have indicated scaling laws to fault systems, which allow prediction of populations of faults with displacements below the resolution of seismic (Childs *et al.*, 1990). Thus, it is critical that statistical modelling of sub-seismic fault systems is developed further to provide realistic modifications on connectivity and simulation behaviour calculated from 3-D architectural models alone (Gauthier *et al.*, 1991). A further important factor is the sealing capacity of faults over the production life of a field (Lindsay *et al.*, this volume).

Quantitative modelling of orientation

Sedimentary bodies are rarely equidimensional, thus the application (and derivation) of width vs. thickness data or forward modelling of reservoir geometries requires a knowledge of the orientation of the genetic units being modelled. Many potentially useful geometric data sets are rendered unusable for modelling purposes because the geometries are not closely related to the orientation of the geological units measured. Determination of the orientation of genetic geological units in the subsurface is a far from trivial problem. Seismic or regional data may often provide an indication of an overall transport direction; however, individual sandbodies may show a wide dispersal of orientation especially in low relief areas (e.g. coastal plains and continental shelves). It is widely assumed that sedimentary dips may be used to determine the orientations of sandbodies, although very few quantitative models exist to allow confident interpretation of sandbody orientation from dips measured by wireline logging tools.

Traditional dipmeter interpretation by log analysts has relied on the interpretation of patterns of upward-decreasing or upward-increasing dip magnitude. Sedimentologists, however, favour interpretation of both dip magnitude and dip azimuth (Perrin, 1975). Furthermore, the processing of the dipmeter data should be optimized with respect to cross-bed set thicknesses (Rider, 1978; Cameron, 1986; Goldstein & Hubbard, 1986; Kerzner, 1986; Höcker *et al.*, 1990). Even when reliable sedimentary dips have been obtained, the relationship of cross-bed dip orientation to overall sandbody orientation is commonly ambiguous. Forward modelling of dip characteristics from the form of sedimentary structures (Rubin, 1987) provides a framework for beginning to interpret complex dip patterns but it still remains to understand the way in which the different sets combine to form a sandbody (Miall, 1988). Quantitative models of dip variation will improve through detailed studies of outcrops (Williams & Soek, this volume) but are currently poorly defined for many types of reservoir sandstones.

ARCHITECTURAL ARRANGEMENT

A basic tenet of sedimentological studies is that reservoir sediments are not randomly ordered in space but are arranged in a systematic manner as a consequence of the interaction of auto- and allocyclic processes affecting the depositional environment. Formalization of this concept has led to the development of (one dimensional) vertical facies sequence analysis using Markov chains and attempts to model (in two dimensions) alluvial architecture as a function of subsidence rate vs. avulsion rate (Allen, 1978; Bridge & Leeder, 1979; Bridge & Mackey, this volume). However, sedimentary architecture is *three dimensional* and the failure of many reservoir models to predict adequately reservoir behaviour under production is directly related to their failure to adequately represent this phenomenon. Quantitative models of three-dimensional architecture are difficult to obtain from outcrops and rely either entirely on subsurface studies or a combination of outcrop and borehole data (e.g. Ravenne *et al.*, 1989; Van Wagoner *et al.*, 1990; Falkner & Fielding, this volume). Quantification of these models (e.g. Mijnssen *et al.*, this volume) may then be used to further our understanding of the processes that control sedimentary architecture and, either by direct application, or through forward modelling programs, act to constrain the way in which three-dimensional models are constructed.

Application of sequence stratigraphic concepts

Sequence stratigraphic analysis stresses the importance of chronostratigraphic correlation rather than lithostratigraphic correlation of reservoir sequences (Vail *et al.*, 1977). This approach is particularly relevant to seismic stratigraphy since seismic reflections tend to reflect chronostratigraphic boundaries rather than lithostratigraphic units. Chronostratigraphic markers may be used to define the boundaries of seismic sequences (Vail *et al.*, 1977; Galloway, 1989) within which smaller scale parasequences may exist (Van Wagoner, 1985; Van Wagoner *et al.*, 1990). A knowledge of the type of seismic sequence within which a given reservoir interval was deposited may be used to infer the types of parasequences and their likely stacking pattern within the larger scale sequence. Several predictive statements can be made via the analysis of architecture in terms of accommodation space. For example, alluvial sequences deposited during low stands of relative sea-level would be expected to show a higher degree of sandbody interconnectedness than those deposited during rising sea-level (e.g. Shanley & McCabe, this volume), given a constant sediment supply.

Furthermore, parasequence boundaries are often associated with laterally extensive shales (Fig. 3, Plate 1) or cemented horizons that significantly affect reservoir drainage. Recognition of these units and their incorporation within the reservoir models should improve the capability of the models to predict dynamic reservoir behaviour. The use of chronostratigraphic markers to define flooding surface- or unconformity-bounded sequences also provides a more reliable framework for modelling intra-reservoir variations but requires either detailed palaeontological control (e.g. Carman & Young, 1981; Osborne & Evans, 1987) or fine-scale log correlation, supported by reservoir pressure measurements (e.g. Johnson & Stewart, 1985; Bryant & Livera, 1991). As an example, Plate 1 shows part of the Cretaceous (Campanian) Blackhawk Formation, Book Cliffs, Utah. The upper Kenilworth Member (base of photograph) is separated from the superjacent parasequence set (Sunnyside Member) by a 25 m thick tongue of marine shale, easily recognizable and correlatable on conventional wireline logs. Within the Sunnyside parasequence set are much thinner shales (arrowed). These apparently minor shales thicken down-dip to tens of metres thickness and would represent fieldwide vertical permeability barriers in an oil field (O'Byrne & Flint, 1991). The uppermost sandbody is younger, being the up-dip section of a separate sequence set (Grassy Member). Sequence stratigraphic analysis thus avoids simple correlation of all the parasequences as a single gross flow unit and allows deterministic modelling of shales with confidence.

These techniques have begun to be applied both within object-based modelling systems (Bryant *et al.*, 1991) and variogram-based models (Rudkiewicz *et al.*, 1990). In both instances this results in a geologically more realistic distribution of reservoir units away from the well control points and reduces the stochastic element of such reservoir models.

MODELLING RESERVOIR PROPERTIES

It is well established that reservoir properties are rarely uniformly distributed within individual sandbodies and generalized models are well known. However, these models tend to be both qualitative and to emphasize vertical trends (e.g. fining-upward channel-fills commonly show upward-decreasing reservoir quality). Quantitative data on reservoir properties have been recorded by several workers and commonly show that reservoir properties scale with sedimentary structures. The importance of these effects is dependent upon the scale of the heterogeneity with respect to the size of the reservoir drainage unit (Haldorsen, 1986; Weber, 1986), formation fluid properties and production mechanism (Szpakiewicz *et al.*, 1987). Only a detailed understanding of the internal sedimentary architecture of reservoir sandbodies can lead to a rigorous modelling of internal variations of reservoir quality.

Flow barriers

Shale beds and laminae vary from millimetres to tens of metres in thickness and from millimetres to kilometres in lateral extent (Potter *et al.*, 1980). They form the most common type of flow barrier in clastic reservoirs. Laterally continuous shales not only constitute barriers to fluid flow but may also form pressure seals, whilst discontinuous shales increase the tortuosity of fluid flow paths (Haldorsen & Chang, 1986). No simple relationship exists between shale thickness and shale length; however, average shale lengths have been shown to differ according to the environments in which they were deposited (Hutchinson *et al.*, 1961; Zeito, 1965; Richardson *et al.*, 1978; Weber, 1986; Geehan *et al.*, 1986). Furthermore, shales which are below the resolution of standard logging tools (typically 0.5 m) may be of considerable lateral extent and have a major effect on reservoir drainage (Grout *et al.*, 1965). Characterization of the effects of shale barriers therefore requires fine-scale analysis of well data and the use of environment-specific data sets.

Reservoir intervals that are tightly cemented by diagenetically precipitated minerals can also form extensive flow barriers (Omre *et al.*, 1990). The lateral extent of carbonate cements is controlled not only by the depositional environment of the sediments but also by their subsequent diagenetic history. Once again fine-scale examination of well data (including detailed petrographic analyses) and reservoir-specific models are required to predict the effect of these tight zones on reservoir drainage.

Shale dimensions

Typically the modelling of flow barrier distributions requires two types of data set. Firstly, the thickness and vertical density of the flow barriers encountered by the wells needs to be established. Secondly, data sets describing the 3-D geometry and orientation of flow barriers from an analogous environment are required in order to make predictions away from the wells (Haldorsen & Lake, 1982; Haldorsen & Chang, 1986).

Data concerning the 3-D geometry of shales are extremely difficult to obtain (Haldorsen & Chang,

Table 2. Quantitative data to guide modelling of reservoir properties within sandbodies

Sandbody type	Heterogeneity data type	Age	Location	Reference
Fluvial	Shale lengths	Cambrian	Algeria	Verrien *et al.*, 1967; Dupuy & LeFebvre du Prey, 1968
Fluvial	Shale lengths/ Poroperm	Cretaceous	Colorado, USA	Jones *et al.*, 1984
Fluvial	Poroperm	Holocene	Netherlands	Weber, 1982
Fluvial	Poroperm	Jurassic	Yorkshire, UK	Dryer *et al.*, 1990; Jacobsen & Rendall, 1991; Ravenne & Beucher, 1988
Fluvioglacial	Poroperm	Pleistocene	Louisiana, USA	Wu *et al.*, 1973
Fluvioglacial	Poroperm	Cambro–Ordovician	Oman	Martin & Cooper, 1984
Aeolian	Poroperm	Various	USA	Lupe & Ahlbrandt, 1979
Aeolian	Poroperm	Permian	UK	Weber, 1982
Aeolian	Poroperm	Jurassic/ Triassic	Utah/Wyoming, USA	Lindquist, 1983, 1988
Aeolian	Perm. barriers	Permian	UK	Martin & Evans, 1988
Aeolian	Poroperm	Jurassic	Arizona, USA	Goggin, 1988; Goggin *et al.*, 1986; Chandler *et al.*, 1989
Coastal	Poroperm	Cretaceous	Wyoming, USA	Jackson *et al.*, 1991
Coastal	Poroperm	Cretaceous	Utah, USA	Stalkup & Ebanks, 1986
Coastal/sh. marine	Poroperm	Cretaceous	W.USA	Hutchinson *et al.*, 1961; Polasek & Hutchinson, 1967
Coastal/sh. marine	Poroperm	Cretaceous	Texas, USA	Dodge *et al.*, 1971
Tidal	Shale distribution	Cretaceous	UK	Allen, 1982
Sh. marine	Cement lengths	Jurassic	UK	Bryant *et al.*, 1988
Sh. marine	Poroperm	Cretaceous	Wyoming, USA	Tomutsa *et al.*, 1986
Deep marine	Poroperm	Tertiary	California, USA	Hsu, 1977
Various	Poroperm	Holocene	USA	Pryor, 1973
Various	Shale lengths	Various	Various	Zeito, 1965; Weber, 1982

1986) and the consequent paucity of quantitative data is a major constraint on the reliability of existing shale modelling procedures (Richardson *et al.*, 1978; Haldorsen & Lake, 1982; Begg *et al.*, 1985; Weber, 1986; Haldorsen *et al.*, 1987; Kossack, 1989). Generalized statements concerning the relative extent of shales in different environments have been published by Zeito (1965) and Weber (1982). However, environmentally specific, statistically valid data sets are very scarce (Table 2). Thus, despite the problems of collecting 3-D shale data the successful application of elaborate, and often expensive, shale modelling routines can only be meaningful if more data are collected (Geehan & Underwood, this volume).

Cemented horizons

Data concerning the 3-D geometry of laterally extensive cemented horizons are even less well known but exist for carbonate-cemented, shallow marine, storm-deposited sands (Bryant *et al.*, 1988; Bjørkum & Walderhaug, 1990). Clearly more data concerning the 3-D geometry of diagenetically-formed flow barriers will be required before such barriers can be statistically modelled with any confidence. However, current work indicates that in clastic shoreline sequences at parasequence boundaries, initial flooding surfaces are characterized by shales which pass up-dip into preferentially cemented zones and eventually coals, as accommodation space progressively decreases to zero (O'Byrne & Flint, 1991). Such data again increase confidence in deterministic modelling procedures within clastic shoreline sequences.

Permeability variations

Permeability variation may be considered at a number of different scales (Haldorsen, 1986; Lasseter *et al.*, 1986). The small-scale (pore-sized) variation of permeability has a fundamental effect on reservoir drainage through its influence on residual hydrocarbon saturations. At a medium-scale (centimetres to tens of metres) permeability is a function of grain size, the amount and type of admixed clay, diagenetic alteration and sedimentary structures. These features are usually combined during core analysis in the description of the 'lithofacies' which constitute the reservoir interval. In an ideal situation core-defined lithofacies have a clearly defined narrow permeability distribution and unique log response. Where this is the case, and rates of change of reservoir quality are low with respect to inter-well distances and log response, then well data may be used to map the spatial distribution of permeability within the reservoir provided that adequate models of lithofacies distributions in 3-D dimensions are available (Sneider *et al.*, 1977; Carman & Young, 1981; Hearn *et al.*, 1986). However, depositional architecture is commonly complex and reservoir-scale permeability is then dependent upon the spatial distribution of reservoir elements.

Areal variation in permeabilty has been investigated by outcrop studies (Goggin *et al.*, 1986; Goggin, 1988; Chandler *et al.*, 1989; Dreyer *et al.*, 1990; Hartkamp-Bakker & Donselaar, this volume). In some reservoirs horizontal permeability may be anisotropic as a result of preferentially oriented sedimentary structures (Montadert, 1963; Verrien *et al.*, 1967; Weber, 1982; Lindquist, 1983; Nurmi, 1985; Kortekaas, 1985). Data concerning the spatial variation of permeability have been collected from closely spaced outcrop measurements and well tests. These studies generally show that permeability variation is random or structured depending upon the scale of the observation. Closely spaced permeability measurements can show an apparently random scatter but define zones of similar permeability. These zones of similar permeability are of finite dimensions, which are of different sizes in different environments. Published data sets describing permeability variation are summarized in Table 2.

Whilst these data sets are useful in establishing the structure of permeability variations in different geological settings, absolute values and rates of change are likely to be reservoir specific and require the use of core-derived data sets.

CONCLUSIONS

New developments in industry practise (particularly the increasing acquisition of 3-D seismic data, drilling of horizontal wells and use of formation imaging devices) have provided more 3-D data concerning reservoir formations than hitherto was available. Concomitantly the increased availability of powerful computing tools has facilitated the numerical description of reservoirs in three dimensions. Rapid developments in these fields have led to a computing capability that, in many cases, is beyond the level of current quantitative geological knowledge. Thus an industry requirement for quantitative geometric data bases and rule bases to guide model construction is frequently not satisfied by conventional geological wisdom. Detailed outcrop studies and forward modelling programs will go someway to meet these needs in the future. Integration of these studies with industry acquired data has the potential to generate 3-D models that provide a link between basin-scale and bedform-scale geological observations by building fully integrated sedimentological models. The pursuit of these goals should therefore result in mutually beneficial partnerships between industry and academic workers.

ACKNOWLEGDEMENTS

Much of this review was carried out during 1985–1989, a period when both authors were employed in the Production Geology Research Department, Koninklijke/Shell Exploratie en Produktie Laboratorium, Rijswijk, The Netherlands. We are grateful to Greg Geehan (BP) and Andrew Hurst (Unocal) for their constructive comments on the manuscript. This paper is published with the permission of Shell Internationale Petroleum Maatschaapij although the views expressed herein remain the sole responsibility of the authors.

Note added in proof

During the preparation of this book there has been a rapid growth in the application of sequence stratigraphy at *high resolution* to outcrops and oil fields. This technique is a critical stage in reservoir modelling as it facilitates recognition of high frequency sequence boundaries and related incised valley fills. These features may control the spatial distribution of extreme permeability values.

REFERENCES

Abib, O., Moretti, F.J., Cen Mei & Yang Yuzhe (1988) Application of geological modelling and reservoir simulation to West Saertu area of the Daqing oil field. *SPE Paper* **17560**.

Alabert, F.G. & Massonnat, G.J. (1990) Heterogeneity in a complex turbititic reservoir: stochastic modelling of facies and petrophysical variability. *SPE Paper* **20604**.

Allen, J.R.L. (1978) Studies in fluviatile sedimentation: an exploratory quantitative model for the architecture of avulsion-controlled alluvial suites. *Sedim. Geol.*, **21**, 129–147.

Allen, J.R.L. (1982) Mud drapes in sand-wave deposits: a physical model with application to the Folkstone Beds (early Cretaceous, southeast England). *Phil. Trans. R. Soc. Lond.*, **A 306**, 291–345.

Anxionnaz, H., Delfiner, P. & Delhomme, J.P. (1990) Computer-generated corelike descritions from open-hole logs. *Bull. Am. Assoc. Petrol. Geol.*, **74**, 375–393.

Begg, S.H., Chang, D.M. & Haldorsen H.H. (1985) A simple statistical model for calculating the effective vertical permeability of a reservoir containing discontinuous shales. *SPE Paper* **14271**.

Bjørkum, P.A. & Walderhaug, O. (1990) Lateral extents of calcite-cemented zones in shallow marine sandstones. In: *North Sea Oil and Gas Reservoirs–II* (Ed. by A.T. Buller, E. Berg, O. Hjelmeland, J. Kleppe, O. Torsæter & J.O. Aasen), pp. 331–336. Graham and Trotman, London.

Bouvier, J.D., Kaars-Sijpesteijn, C.H., Kluesner, D.F., Onyejekwe, C.C. & van der Pal, R.C. (1989) Three-dimensional seismic interpretation and fault sealing investigations, Nun River field, Nigeria. *Bull. Am. Assoc. Petrol. Geol.*, **73**, 1397–1414.

Brewster, J. & Jeangeot, G. (1987) The production geology of the Frigg Field. In: *North Sea Oil and Gas Reservoirs* (Ed. by J. Kleppe, E.W. Berg, A.T. Buller, O. Hjelmeland & O. Torsæter), pp. 75–88. Graham and Trotman, London.

Bridge, J.S. (1985) Paleochannel patterns inferred from alluvial deposits: a critical evaluation. *J. sedim. Petrol.*, **55**, 579–589.

Bridge, J.S. & Leeder, M.R. (1979) A simulation model of alluvial stratigraphy. *Sedimentology,* **26**, 617–644.

Bryant, I.D. & Ben Brahim, A. (1989) Characterisation of facies-related variation of reservoir quality in the Tazerka field, offshore Tunisia. *J. Pet. Sci. Eng.*, **2**, 361–377.

Bryant, I.D., Kantorowicz, J.D. & Love, C.F. (1988) The origin and recognition of laterally continuous carbonate-cemented horizons in the Upper Lias Sands of southern England. *Marine Petrol. Geol.*, **5**, 108–133.

Bryant, I.D. & Livera, S.E. (1991) Identification of unswept oil volumes in a mature field by using integrated data analysis: Ness Formation, Brent Field, UK North Sea. In: *Generation, Accumulation and Production of Europe's Hydrocarbons* (Ed. by A.M. Spencer). Spec. Pub. Europ. Ass. Petrol. Geosci. No. 1, pp. 75–88. Oxford University Press, Oxford.

Bryant, I.D., Paardekam, A.H.M., Davies, P. & Budding, M.C. (1991) Integrated reservoir characterisation of Cycle III, Brent Group, Brent Field, UK North Sea for reservoir management. In: *The Integration of Geology, Geophysics and Reservoir Engineering in Reservoir Delination, Description and Management*. pp. 405–422. Am. Assoc. Petrol. Geol.

Cameron, G.I.F. (1986) Confidence and identification of foresets in stratigraphic dipmeter surveys. *Trans. 10th European Formation Evaluation Symp., Aberdeen*. Paper O.

Carman, G.J. & Young, R. (1981) Reservoir geology of the Forties oilfield. In: *Petroleum Geology of the Continental Shelf of North-West Europe* (Ed. by V.C. Illing & G.D. Hobson), pp. 371–379. Institute of Petroleum, London.

Chandler, M.A., Kocurek, G., Gogan, D.J. & Lake, L.W. (1989) Effects of stratigraphic heterogeneity on permeability in eolian sandstone sequences, Page Sandstone, northern Arizona. *Bull. Am. Assoc. Petrol. Geol.*, **73**, 658–668.

Childs, C.J., Walsh, J.J. & Watterson, J. (1990) A method for estimation of the density of fault displacements below the limits of seismic resolution. In: *North Sea Oil and Gas Reservoirs–II* (Ed. by A.T. Buller, E. Berg, O. Hjelmeland, J. Kleppe, O. Torsæter & J.O. Aasen), pp. 309–318. Graham and Trotman, London.

Clemetsen, R., Hurst, A.R., Knarud, R. & Omre, H. (1990) A computer program for evaluation of fluvial reservoirs. In: *North Sea Oil and Gas Reservoirs–II* (Ed. by A.T. Buller, E. Berg, O. Hjelmeland, J. Kleppe, O. Torsæter & J.O. Aasen), pp. 373–385. Graham and Trotman, London.

Damsleth, E., Tjølsen, C.B., Omre, K.H. & Haldorsen, H.H. (1990) A two-stage stochastic model applied to a North Sea reservoir. *SPE Paper* **20605**.

Delhomme, A.E.L. & Giannesini, J.F. (1979) New reservoir description techniques improve simulation results in Hassi-Messaoud field, Algeria. *SPE Paper* **8435**.

Dodge, C.F., Holler, D.P. & Meyer, R.L. (1971) Reservoir heterogeneities of some Cretaceous sandstones. *Bull. Am. Assoc. Petrol. Geol.*, **55**, 1814–1828.

Dreyer, T. (1990) Sand body dimensions and infill sequences of stable, humid-climate delta plain channels. In: *North Sea Oil and Gas Reservoirs–II* (Ed. by A.T. Buller, E. Berg, O. Hjelmeland, J. Kleppe, O. Torsæter & J.O. Aasen), pp. 335–351. Graham and Trotman, London.

Dreyer, T., Scheie, Å & Walderhaug, O. (1990) Mini-

Plate 1. Shoreface parasequence sets with internal, laterally extensive shales (parasequence boundaries; PB), Blackhawk Formation, Book Cliffs, Utah. Central sandy interval is approx. 40 m in thickness. The parasequence bounding shales (arrowed) are 0.5–2 m thick here but thicken down depositional dip into major marine shale tongues and pass up-dip into preferentially cemented sand-on-sand surfaces and eventually thin coals (see text for further details). The contact between the Sunnyside and Grassy Members is an interfluve sequence boundary, re-used as a flooding surface. Recognition of such surfaces is critical as 3 km to the north (left) a coeval incised valley system truncates the upper Sunnyside Member shoreface deposits (O'Byrne, 1991, pers. comm.).

[*facing page 16*]

permeameter-based study of permeability trends in channel sand bodies. *Am. Assoc. Petrol. Geol. Bull.*, **74**, 359–374.

DROMGOOLE, P.W. (1987) Reservoir geophysics: an overview. In: *North Sea Oil and Gas Reservoirs* (Ed. by J. Kleppe, E.W. Berg, A.T. Buller, O. Hjelmeland & O. Torsæter), pp. 103–108. Graham and Trotman, London.

DUBRULE, O.R.F. (1989) A review of stochastic models for petroleum reservoirs. In: *Geostatistics* (Ed. by M. Armstrong), vol. 2, pp. 493–506. Kluwer, Dordrecht.

DUBRULE, O.R.F. & HALDORSEN, H.H. (1986) Geostatistics for permeability estimation. In: *Reservoir Characterisation* (Ed. by L.W. Lake & H.B. Carroll Jr.), pp. 223–247. Academic Press, London.

DUPUY, M. & POTTIER, J. (1963) Application de méthodes statistiques a l'étude détaillée des réservoirs. *Rev. l'Instit. Français du Pétrol*, **18**, (Dec., Hors série), 258–283.

DUPUY, M. & LEFEBVRE DU PREY, E. (1968) L'anisotropie d'écoulement en milieu poreux presentant des intercalations horizontales discontinues. *IIe Colloque de l'Assoc. Recherche sur les Techn. de Forage et de Prod., Pau, France*, pp. 643–676.

FARRELL, K.M. (1987) Sedimentology and facies architecture of overbank deposits of the Mississippi River, False River region, Louisiana. In: *Recent Developments in Fluvial Sedimentology* (Ed. by F.C. Ethridge, R.M. Flores & M.D. Harvey). Spec. Pub. Soc. Econ. Pal. Miner., **39**, 111–120.

FERM, J.C. & HORNE, J.C. (eds) (1979) Carboniferous Depositional Environments in the Appalachian Region. *Am. Assoc. Petrol. Geol. Field. Guidebook*, 760 pp.

FIELDING, C.R. & CRANE, R.C. (1987) An application of statistical modelling to the prediction of hydrocarbon recovery factors in fluvial reservoir sequences. In: *Recent Developments in Fluvial Sedimentology* (Ed. by F.C. Ethridge, R.M. Flores & M.D. Harvey). Spec. Pub. Soc. Econ. Pal. Miner., **39**, 321–327.

FLINT, S., STEWART, D.J., HYDE, T., GEVERS, E.C.A., DUBRULE, O.R.F. & VAN RIESSEN, E.D. (1988) Aspects of reservoir geology and production behaviour of Sirikit oil field, Thailand: an integrated study using well and 3-D seismic data. *Bull. Am. Assoc. Petrol. Geol.*, **72**, 1254–1269.

FLINT, S., VAN ROSSEM, S.J. & WILLIAMS, H. (1989) Quantitative outcrop analysis of coastal plain sequences from eastern Kentucky: example of high quality data input to three dimensional reservoir modelling. *Bull. Am. Assoc. Petrol. Geol.*, **73**, 356 (abstract).

FOUQUET, C. DE., BEUCHER, H., GALLI, A. & RAVENNE, C. (1989) Conditional simulation of random sets—application to an argillaceous sandstone reservoir. In: *Geostatistics* (Ed. by M. Armstrong), vol. 2, pp. 517–530. Kluwer, Dordrecht.

GALLOWAY, W.E. (1989) Genetic stratigraphic sequences in basin analysis I: architecture and genesis of flood-surface bounded depositional units. *Bull. Am. Assoc. Petrol. Geol.*, **73**, 125–142.

GAUTHIER, B.D.M., LAKE, S. & JENSEN, Y. (1991) Probabilistic modelling of subseismic faults in the Pelican field (Block 211/26a), North Sea, offshore U.K. *Bull. Am. Assoc. Petrol. Geol.*, **75**, 579 (abstract).

GEEHAN, G.W., LAWTON, T.F., SAKURAI, S., KLOB, H., CLIFTON, T.R., INMAN, K.F. & NITZBERG, K.E. (1986) Geologic prediction of shale continuity Prudhoe Bay Field. In: *Reservoir Characterisation* (Ed. by L.W. Lake & H.B. Carroll Jr.), pp. 63–82. Academic Press, London.

GELFLAND, V.A., TAYLOR, G., TESSMAN, J. & LARNER, K.L. (1986) Delineation of river-channel reservoirs by seismic lithological modeling of 3-D seismic data in southern Alberta. *SPE Paper* **14118**.

GOGGIN, D.J. (1988) Patterns of permeability in eolian deposits. *SPE Formation Evaluation*, **3**, 297–306.

GOGGIN, D.J., CHANDLER, M.A., KOCUREK, G.A. & LAKE, L.W. (1986) Patterns of permeability in eolian deposits. *SPE Paper* **14893**.

GOLDSTEIN, B.A. & HUBBARD, J.J. (1986) Comparison and geologic interpretation of alternative dipmeter analyses from a Permo-Carboniferous sequence in the Fitzroy Graben. In: *The Canning Basin, W.A.* (Ed. by P.G. Purcell). Geol. Soc. Australia & Petrol. Explor. Soc. Australia, 505–519.

GRACE, L.M. & PIRIE, R.G. (1986) Stratigraphic interpretation using formation imaging and dipmeter analyses. *SPE Paper* **15611**.

GROUT, J., REISS, L.H. & MONTADERT, L. (1965) Reservoir inhomogeneities deduced from outcrop observations and production logging. *SPE Paper* **1279**.

GUNDESØ, R. & EGELAND, O. (1990) SESIMIRA—A new geological tool for 3-D modelling of heterogenous reservoirs. In: *North Sea Oil and Gas Reservoirs–II* (Ed. by A.T. Buller, E. Berg, O. Hjelmeland, J. Kleppe, O. Torsæter & J.O. Aasen), pp. 363–371. Graham and Trotman, London.

HALDORSEN, H.H. (1986) Simulator parameter assignment and the problem of scale in reservoir engineering. In: *Reservoir Characterisation* (Ed. by L.W. Lake & H.B. Carroll, Jr.), pp. 293–340. Academic Press, London.

HALDORSEN, H.H. & CHANG, D.M. (1986) Notes on stochastic shales; from outcrop to simulation model. In: *Reservoir Characterisation I* (Ed. by L.W. Lake & H.B. Carroll, Jr.), pp. 341–372. Academic Press, London.

HALDORSEN, H.H., CHANG, D.M. & BEGG, S.H. (1987) Discontinuous vertical permeability barriers: a challenge to engineers and geologists. In: *North Sea Oil and Gas Reservoirs I* (Ed. by J. Kleppe, E.W. Berg, A.T. Buller, O. Hjelmeland & O. Torsæter), pp. 127–151. Graham and Trotman, London.

HALDORSEN, H.H. & DAMSLETH, E. (1990) Stochastic modeling. *J. Petrol. Sci.*, **42**, 404–412.

HALDORSEN, H.H. & LAKE, L.W. (1982) A new approach to shale management in field scale simulation models. *SPE Paper* **10976**.

HAZEU, G.J.A., KRAKSTAD, O.S., RIAN, D.T. & SKAUG, M. (1986) The application of new approaches for shale management in a 3-D simulation study of the Frigg Field. *SPE Paper* **15608**.

HEARN, C.L., EBANKS, W.J., TYE, R.S. & RANGANATHAN, V. (1984) Geological factors influencing reservoir performance of the Hartzog Draw Field, Wyoming. *J. Petrol. Technol.*, **36**, 1335–1344.

HEARN, C.L., HOBSON, J.P. & FOWLER, M.L. (1986) Reservoir characterisation for simulation, Harztog Draw Field, Wyoming. In: *Reservoir Characterisation* (Ed. by L.W. Lake & H.B. Carroll Jr.), pp. 341–372. Academic Press, London.

HELLAND-HANSEN, W., STEEL, R.J., NAKAYAMA, K. & KENDALL, C.G. St. C. (1989) Review and computer modelling of the Brent Group stratigraphy. In: *Deltas Sites and Traps for Fossil Fuels* (Ed. by M.K.G. Whateley & K.T. Pickering). Spec. Publ. geol. Soc. London, Blackwell Scientific Publications, Oxford, **41**, 237–252.

Heritier, F.E., Lossell, P. & Wathne, E. (1979) Frigg Field: large submarine fan trap in lower Eocene rocks of North Sea Viking Graben. *Bull. Am. Assoc. Petrol. Geol.*, **63**, 1999–2020.

Herweijer, J.C., Höcker, C.F.W., Williams, H. & Eastwood, K.M. (1990) The relevance of dip profiles from outcrops as reference for the interpretation of SHDT dips. In: *Geological Applications of Wireline Logs* (Ed. by A. Hurst, M.A. Lovell & A.C. Morton), Spec. Publ geol. Soc. London, **48**, 39–43.

Höcker, C.F.W., Eastwood, K.M. & Adams, J.T. (1990) Use of dipmeter data in clastic sedimentological studies. *Bull. Am. Assoc. Petrol. Geol.*, **74**, 105–118.

Hsu, K.J. (1977) Studies of Ventura Field, California II: lithology, compaction, and permeability of sands. *Bull. Am. Assoc. Petrol. Geol.*, **61**, 169–191.

Hutchinson, C.A., Dodge, C.F. & Polasek, T.L. (1961) Identification, classification and prediction of reservoir nonuniformities affecting production operations. *J. Petrol. Technol.*, **13**, 223–230.

Jackson, S.R., Tomutsa, L., Szpakiewicz, M., Chang, M.M., Honarpour, M.M. & Schatzinger, R.A. (1991) Construction of a reservoir model by integrating geological and engineering information—Bell Creek field, a barrier/strandplain reservoir. In: *Reservoir Characterization–II* (Ed. by L.W. Lake, H.B. Carroll Jr. & T.C. Wesson), pp. 524–556. Academic Press, London.

Jacobsen, T. & Rendall, H. (1991) Permeability patterns in some fluvial sandstones. An outcrop study from Yorkshire, northeast England. In: *Reservoir Characterisation–II* (Ed. by L.W. Lake, H.B. Carroll Jr. & T.C. Wesson), pp. 315–339, Academic Press, London.

Johnson, H.D. & Stewart, D.J. (1985) Role of clastic sedimentology in the exploration and production of oil and gas in the North Sea. In: *Sedimentology: Recent Developments and Applied Aspects* (Ed. by P.J. Brenchley & B.P.J. Williams), Spec. Publ. geol. Soc. London, Blackwell Scientific Publications, Oxford, **18**, 249–310.

Jones, J.R., Scott, A.J. & Lake, L.W. (1984) Reservoir characterisation for numerical simulation of Mesaverde meanderbelt sandstone, northwestern Colorado. *SPE Paper* **13052**.

Kerzner, M.G. (1986) Obtaining structural and stratigraphic dip information from dipmeter logs using segmentation trees and hierarchical multilevel optimisation. *SPE Paper* **15539**.

Knutson, C.F. (1974) Modelling of noncontinuous Fort Union and Mesaverde Sandstone reservoirs, Piceance Basin, northwestern Colorado. *SPE Paper* **5024**.

Knutson, C.F., Maxwell, E.L. & Millheim, K. (1971) Sandstone continuity in the Mesaverde Formation, Rulison Field area, Colorado. *J. Petrol. Technol.*, **23**, 911–919.

Kolla, V. & Coumes, F. (1987) Morphology, internal structure, seismic stratigraphy, and sedimentation of Indus Fan. *Am. Assoc. Petrol. Geol. Bull.*, **71**, 650–677.

Kortekaas, Th.F.M. (1985) Water/oil displacement characteristics in cross-bedded reservoir zones. *SPE Paper* **12112**.

Kossack, C.A. (1989) Prediction of layer lengths from layer heights for reservoir simulation: a statistical analysis of outcrop data. *J. Petrol. Technol.*, **41**, 867–871.

Kruit, C. (1987) Sedimentologic reservoir study of a steamdrive project in deltaic reservoir sands, East Tia Juan Field, Venezuela. In: *Reservoir Sedimentology* (Ed. by R.W. Tillman & K.J. Weber), Spec. Publ. Soc. Econ. Pal. Miner., **40**, 293–310.

Lasseter, T.J., Waggoner, J.R. & Lake, L.W. (1986) Reservoir heterogeneities and their influence on ultimate recovery. In: *Reservoir Characterisation* (Ed. by L.W. Lake & H.B. Carroll Jr.), pp. 545–559. Academic Press, London.

Lawrence, D.T., Doyle, M. & Aigner, T. (1990) Stratigraphic simulation of sedimentary basins: concepts and calibration. *Bull. Am. Assoc. Petrol. Geol.*, **74**, 273–295.

Leeder, M.R. (1973) Fluviatile fining-upwards cycles and the magnitude of palaeochannels. *Geol. Mag.*, **110**, 265–276.

Leopold, L.B., Wolman, M.G. & Miller, J.P. (1964) *Fluvial Processes in Geomorphology*. Freeman & Co., London. 522 pp.

Lindquist, S.J. (1983) Nugget Formation reservoir characteristics affecting production in the overthrust belt of southwestern Wyoming. *J. Petrol. Technol*, **35**, 1355–1365.

Lindquist, S.J. (1988) Practical characterization of eolian reservoirs for development: Nugget Sandstone, Utah–Wyoming thrust belt. *Sediment. Geol.*, **56**, 315–339.

Lorenz, J.C. (1985) Predictions of size and orientation of lenticular reservoirs in the Mesaverde Group, northwestern Colorado. *Proceedings of SPE/DOE Joint Symposium on Low Permeability Reservoirs, Denver*, pp. 23–26.

Lorenz, J.C., Heinze, D.M., Clark, J.A. & Searls, G.A. (1985). Determination of widths of meander-belt sandstone reservoirs from vertical downhole data, Mesaverde Group, Piceance Creek basin, Colorado. *Bull. Am. Assoc. Petrol. Geol.*, **69**, 710–721.

Lowry, P. & Raheim, A. (1991) Characterisation of delta front sandstones from a fluvial-dominated delta system. In: *Reservoir Characterisation–II* (Ed. by L.W. Lake, H.B. Carroll Jr. & T.C. Wesson), pp. 665–676. Academic Press, London.

Lupe, R. & Ahlbrandt, T.S. (1979) Sediments of the ancient eolian environment—reservoir inhomogeneity. In: *A Study of Global Sand Seas* (Ed. by E.D. Mckee). *U.S.G.S. Prof. Pap.* **1052**, 241–251.

Luthi, S.M. & Banavar, J.R. (1988) Application of borehole images to three-dimensional geometric modelling of eolian sandstone reservoirs, Permian Rotliegende, North Sea. *Bull. Am. Assoc. Petrol. Geol.*, **72**, 1074–1089.

Martin, J.H. & Cooper, J.A. (1984) An integrated approach to the modelling of permeability barrier distribution in a sedimentologically complex reservoir. *SPE Paper* **13051**.

Martin, J.H. & Evans, P.F. (1988) Reservoir modelling of marginal aeolian/sabkha sequences, southern North Sea (UK Sector). *SPE Paper* **18155**.

Martin, J.H., Evans, A.J. & Raper, J.K. (1988) Reservoir modelling of low-sinuousity channel sands: a network approach. *SPE Paper* **18364**.

Matheron, G., De Fourquet, Ch., Beucher, H., Galli, A., Guerillot, D. & Ravenne C. (1987) Conditional simulation of the geometry of fluvio-deltaic reservoirs. *SPE Paper* **16753**.

Miall, A.D. (1988) Reservoir heterogeneities in fluvial sandstones: lessons from outcrop studies. *Bull. Am. Assoc. Petrol. Geol.*, **72**, 682–694.

Miller, D.D., McPherson, J.G. & Covington, T.E. (1990) Fluvio-deltaic reservoir, South Belridge field, San Joachin valley, California. In: *Sandstone Petroleum Reservoirs* (Ed. by J.H. Barwis, J.G. McPherson & J.C. Studlick), pp. 109–130. Springer-Verlag, New York.

Montadert, L. (1963) La sédimentologie et l'étude détaillée des hétérogénéité's d'un réservoir. *Rev. l'Instit. Français du Pétrol.*, **18** (Dec., Hors série), pp. 241–257.

Myers, K.J. & Bristow, C. (1989) Detailed sedimentology and gamma-ray log characteristics of a Namurian deltaic succession. II: Gamma-ray logging. In: *Deltas Sites and Traps for Fossil Fuels* (Ed. by M.K.G. Whateley & K.T. Pickering). Spec. Publ. geol. Soc. London, Blackwell Scientific Publications, Oxford, **41**, 75–80.

Nami, M. & Leeder, M.R. (1978) Changing channel morphology and magnitude in the Scalby Formation (M. Jurassic) of Yorkshire, England. In: *Fluvial Sedimentology* (Ed. by A.D. Miall). Can. Soc. Petrol. Geol. Mem., **5**, 431–440.

Nestvold, E.O. (1987) The use of 3-D seismic in exploration, appraisal and field development. *Proc. 12th World Petroleum Congress, Houston, 1987*, **2**, 125–132. Wiley, Chichester.

Nestvold, E.O. (1989) The use of 3-D seismic in exploration and production. *Proc. Energy Industries Council—Oil, Gas and Petrochemicals Seminar, Baghdad, Iraq*, Oct. 1989.

Nurmi, R.D. (1985) Eolian sandstone reservoirs: bedding facies and production geology modelling. *SPE Paper* **14172**.

O'Byrne, C.J. & Flint, S. (1991) Anatomy of a parasequence: quantitative heterogeneity analysis for reservoir modelling, Book Cliffs, Utah. *Bull. Am. Assoc. Petrol. Geol.*, **75**, 647–648 (Abstract).

Omre, H., Solna, K. & Tjelmeland, H. (1990) Calcite cementation: description and production consequences. *SPE Paper* **20607**.

Oomkens, E. (1967) Depositional sequences and sand distribution in a deltaic complex. *Geol. Mijnb.*, **46**, 265–278.

Osborne, P. & Evans, S. (1987) The Troll Field: reservoir geology and field development planning. In: *North Sea Oil and Gas Reservoirs* (Ed. by J. Kleppe, E. Berg, A.T. Buller, & O. Hjelmeland O. Torsæter), pp. 39–60. Graham and Trotman, London.

Perrin, G. (1975) Comparaison entre des structures sédimentaires à l'affleurement et les pendageometries de sondage. *Bull. Centre Rés. Pau–SNPA*, **9**, 147–181.

Polasek, T.L. & Hutchinson, C.A. (1967) Characterization of non-uniformities within a sandstone reservoir from a fluid mechanics standpoint. *Proc. 7th World Petrol. Cong., Mexico*, **II**, 397–407.

Posamentier, H.W. & Vail, P.R. (1988) Eustatic controls on clastic deposition II—sequence and systems tract models. In: *Sea Level Changes—an Integrated Approach* (Ed. by C.H. Wilgus, B.S. Hastings, C.G. St. C. Kendall, H.W. Posamentier, C.A. Ross & J.C. Van Wagoner). SEPM Spec. Publ. **42**, 125–154.

Potter, P.E., Maynard, J.B. & Pryor, W.A. (1980) *Sedimentology of Shale*, pp. 61–64. Springer-Verlag, New York.

Pryor, W.A. (1973) Permeability-porosity patterns and variations in some Holocene sand bodies. *Bull. Am. Assoc. Petrol. Geol.*, **57**, 162–189.

Ravenne, C. & Beucher, H. (1988) Recent developments in description of sedimentary bodies in a fluvio-deltaic reservoir and their 3-D conditional simulation. *SPE Paper* **18310**.

Ravenne, C., Eschard, R., Galli, A., Mathieu, Y., Montadert, L. & Rudkiewicz, J.L. (1989) Heterogeneities and geometry of sedimentary bodies in a fluviodeltaic reservoir. *SPE Formation Evaluation*, **4**, 239–246.

Richardson, J.G., Harris, D.G., Rossen, R.H., Van Hee, G. (1978) The effect of small, discontinuous shales on oil recovery. *J. Petrol. Technol.*, **28**, 1531–1537.

Rider, M.H. (1978) Dipmeter log analysis. An essay. *SPWLA 19th Annual Log. Symp. Trans.*, Paper G.

Rubin, D. (1987) Cross-bedding, bedforms and paleocurrents. *Soc. Econ. Pal. Miner. Concepts in Sedimentology and Paleontology*, **1**. 187 pp.

Rudkiewicz, J.L., Guerillot, D., Galli, A. & Group Heresi (1990) An integrated software for stochastic modelling of reservoir lithology and property with an example from the Yorkshire Middle Jurassic. In: *North Sea Oil and Gas Reservoirs–II* (Ed. by A.T. Buller, E. Berg, O. Hjelmeland, J. Kleppe, O. Torsæter & J.O. Aasen), pp. 400–406. Graham and Trotman, London.

Rudkiewicz, J.L., & Eschard, R. (1991) Geological knowledge in the geostatistical modelling of reservoir properties. In: *Reservoir Characterisation–II* (Ed. by L.W. Lake, H.B. Carroll Jr. & T.C. Wesson), pp. 686–688. Academic Press, London.

Ruitenberg, P.A., Buchanan, R. & Marke, P. (1990) Three-dimensional data improve reservoir mapping. *J. Petrol. Technol.*, **42**, 22–61.

Serra, O. (1987) Sedimentological applications of wireline logs to reservoir studies. In: *North Sea Oil and Gas Reservoirs* (Ed. by J. Kleppe, E.W. Berg, A.T. Buller, O. Hjelmeland & O. Torsæter), pp. 277–299. Graham and Trotman, London.

Schumm, S.A. (1977) *The Fluvial System*. Wiley, New York. 338 pp.

Skaug, M. & Gundesø, R. (1986) Geological modelling of the Frigg Field with special emphasis on shale mapping. *SPE Paper* **15859**.

Smith, D.A. (1990) Sealing and non-sealing faults in Louisiana Gulf Coast salt basin. *Bull. Am. Assoc. Petrol. Geol.*, **64**, 145–172.

Smith, P.J. & Morgan, D.T.K. (1986) Modelling faulted and fluvial reservoirs. *SPE Paper* **15853**.

Sneider, R.M., Richardson, F.H., Paynter, D.D., Eddy, R.E. & Wyant, I.A. (1977) Predicting reservoir rock geometry and continuity in Pennsylvanian reservoirs, Elk City Field, Oklahoma. *J. Petrol. Technol.*, **29**, 851–866.

Stalkup, F.I. & Ebanks, W.J. Jr. (1986) Permeability variation in a sandstone barrier island-tidal channel-tidal delta complex, Ferron Sandstone (Lower Cretaceous), Central Utah. *SPE Paper* **15532**.

SZPAKIEWICZ, M., MCGEE, K. & SHARMA, B. (1987) Geologic problems related to characterization of clastic reservoirs for E.O.R. *SPE Formation Evaluation*, **2**, 449–460.

TOMUTSA, L., JACKSON, S.R. & SZPAKIEWICZ, M. (1986) Geostatistical characterisation and comparison of outcrop and subsurface facies: Shannon shelf sand ridges. *SPE Paper* **15127**.

VAIL P.R., MITCHUM, R.M. Jr., TODD, R.G., WIDMIER, J.M. THOMPSON, S. III., SANGREE, J.B., BUBB, J.N. & HATLELID, W.G. (1977) Seismic stratigraphy and global changes of sea-level. In: *Seismic Stratigraphy and Applications to Hydrocarbon Exploration* (Ed. by C.E. Clayton). Mem. Am. Assoc. Pet. Geol., **26**, 49–212.

VAN WAGONER, J.C. (1985) Reservoir facies distribution as controlled by sea-level change. *Proc. SEPM Midyear Mtg., Golden, CO*, 91–92 (Abstract).

VAN WAGONER, J.C., MITCHUM, R.M., CAMPION, K.M. & RAHMANIAN, V.D. (1990) Siliciclastic sequence stratigraphy in well logs, cores and outcrops: concepts for high-resolution correlation of time and facies. *Am. Assoc. Petrol. Geol. Methods in Exploration Series*, **7**, 54 pp.

VERRIEN, J.P., COURAND, G. & MONTADERT, L. (1967) Applications of production geology methods to reservoir characterisitic analysis from outcrop observations. *Proc. 7th World Petrol. Cong., Mexico*, **II**, 425–446.

WADSLEY, A.W., ERLANDSEN, S. & GOEMANS, H.W. (1990) HEX—a tool for integrated fluvial architecture modelling and numerical simulation of recovery processes. In: *North Sea Oil and Gas Reservoirs-II* (Ed. by A.T. Buller, E. Berg, O. Hjelmeland, J. Kleppe, O. Torsæter & J.O. Aasen), pp. 387–397. Graham and Trotman, London.

WALDERHAUG, O. & MJØS, R. (1991) Sandbody geometry in fluvial systems—an example from the Jurassic of Yorkshire, England. In: *Reservoir Characterisation–II* (Ed. by L.W. Lake, H.B. Carroll Jr. & T.C. Wesson), pp. 701–703. Academic Press, London.

WEBER, K.J. (1982) Influence of common sedimentary structures on fluid flow in reservoir models. *J. Petrol. Technol.*, **34**, 665–672.

WEBER, K.J. (1986) How heterogeneity affects oil recovery. In: *Reservoir Characterisation–I* (Ed. by L.W. Lake & H.B. Carroll), pp. 487–544. Academic Press, London.

WEBER, K.J. (1987) Computation of initial well productivities in aeolian sandstone on the basis of a geologic model, Leman gas field, U.K. In: *Reservoir Sedimentology* (Ed. by R.W. Tillman & K.J. Weber), Spec. Publ. Soc. econ. Pal. Miner., **40**, 333–354.

WEBER, K.J., EIJPE, R., LEIJNSE, D. & MOENS, C. (1972) Permeability distribution in a Holocene distributary channel-fill near Leerdam (The Netherlands). *Geologie Mijnb.*, **51**, 53–62.

WEBER, K.J. & VAN GEUNS, L.C. (1989) Framework for constructing clastic reservoir simulation models. *J. Petrol. Technol.*, **42**, 1248–1297.

WU, T.H., VYAS, S.K. & CHANG, N.Y. (1973) Probabilistic analysis of seepage. *J. Soil Mech. and Foundation Eng. Div., Proc. Am. Soc. Civ. Eng.*, **99**, 323–340.

ZEITO, G.A. (1965) Interbedding of shale breaks and reservoir heterogeneities. *J. Petrol. Technol.*, **17**, 1223–1228.

Spec. Publs int. Ass. Sediment. (1993) **15**, 21–56

Alluvial architecture in a sequence stratigraphic framework: a case history from the Upper Cretaceous of southern Utah, USA

K.W. SHANLEY* *and* P.J. MCCABE[†]

**Department of Geology and Geological Engineering, Colorado School of Mines, Golden, Colorado 80401, USA;*
†US Geological Survey, Federal Center, MS 972, PO Box 25046, Denver, Colorado 80225, USA

ABSTRACT

Fluvial architecture has been studied within the context of unconformity-bounded depositional sequences in Turonian through Campanian strata within the Kaiparowits Plateau in southern Utah, USA. Because the rate of base-level rise determines the rate at which accommodation space is created, changes in alluvial architecture may be interpreted in terms of changes in the rate of base-level rise. As a result, criteria are developed that allow time-significant surfaces, such as sequence boundary unconformities, and the landward equivalent of maximum-flooding surfaces to be recognized within terrestrial strata. This results in a better understanding of alluvial architecture, leading to improved models of alluvial sandstone geometry, sandstone interconnectedness, and sand/shale ratios.

Upper Turonian to Coniacian coarse-grained and pebbly sandstones, interpreted as braided stream deposits, reflect deposition following a significant seaward shift in facies tracts. These strata are approximately 10 m thick, laterally amalgamated, and interpreted to reflect low rates of base-level rise. These coarse-grained fluvial deposits grade vertically into tidally-influenced, finer-grained fluvial strata that are encased within an increasing proportion of floodplain mudstones. This change in depositional style is interpreted to reflect an increase in the rate of base-level rise such that laterally persistent, well-connected meanderbelt sandstones do not develop. The influence of tidal processes within fluvial strata deposited many tens of kilometres from a coeval shoreline further supports a rapid base-level rise during the Coniacian. Within the southwestern Kaiparowits Plateau, the tidally-influenced fluvial sandstones are erosionally overlain by 25 m of upper Coniacian–lower Santonian fluvial strata that are laterally amalgamated, finer grained, and coeval with down dip progradational and aggradational shoreface parasequences. These strata are interpreted to represent slow rates of fluvial aggradation following a base-level fall and development of a sequence-boundary unconformity. A subsequent increase in the rate of base-level rise during the late Coniacian and Santonian resulted in approximately 150 m of isolated meanderbelt sandstones and thick, floodplain mudstones. These strata correlate to crevasse splay and thick coal-bearing units that in turn grade abruptly into approximately 200 m of vertically-stacked shoreface deposits. Within these isolated meanderbelt sandstones and overbank deposits, sand–shale ratios decrease markedly and meanderbelt sandstones are poorly connected. Following a significant seaward shift in facies tracts during the early Campanian, coarse-grained and pebbly sandstones, interpreted as braided river deposits, erosionally overlie shoreface, coal-bearing and fine-grained floodplain strata.

The ability to relate changes in depositional architecture to changes in base level and position within a depositional sequence has important implications. At the scale of depositional sequences, these concepts may greatly aid in chronostratigraphic correlation between marine and nonmarine strata. At a smaller scale, such as that employed in reservoir studies, parameters such as sand–shale ratios and sandstone connectedness may be predicted by understanding the position of fluvial strata within a depositional sequence. Finally, this study provides an initial calibration data set (albeit qualitative) for quantitative stratigraphic models that attempt to simulate alluvial architecture.

* Present address: Geology Research, Shell Development Co., PO Box 481, Houston, Texas 77001–0481, USA

INTRODUCTION

On a world-wide basis, alluvial reservoirs account for more than 324 billion barrels of oil-equivalent, proven reserves (Larry Kellison and Frank Ethridge, pers. comm., based on Carmalt & St. John, 1986), are the host rock for major mineral deposits associated with placers and low-temperature secondary deposits (e.g. Miall, 1981; Turner-Peterson, 1986), and are important aquifers for both groundwater and contaminant waste disposal. The ability to predict alluvial architecture on the basis of scattered data sets, therefore, has important economic implications. Furthermore, reconstructions of basin evolution requiring a detailed chronostratigraphic framework between alluvial and shallow-marine strata may be significantly improved by understanding the relationship between alluvial architecture and sequence stratigraphy.

Exposures of Turonian through Campanian strata in the Kaiparowits Plateau of south-central Utah (Figs 1 and 2) afford an opportunity to study coeval shallow-marine, alluvial, and coal-bearing strata across almost 80 km of depositional-dip oriented exposure. Changes in alluvial architecture can be compared directly with parasequence stacking patterns of coeval shoreface strata. As a result, the more thoroughly described sequence stratigraphic models that have evolved in shoreface deposits (e.g. Van Wagoner *et al.*, 1990) can be compared directly with observations regarding alluvial architecture. This comparison allows sequence-stratigraphic

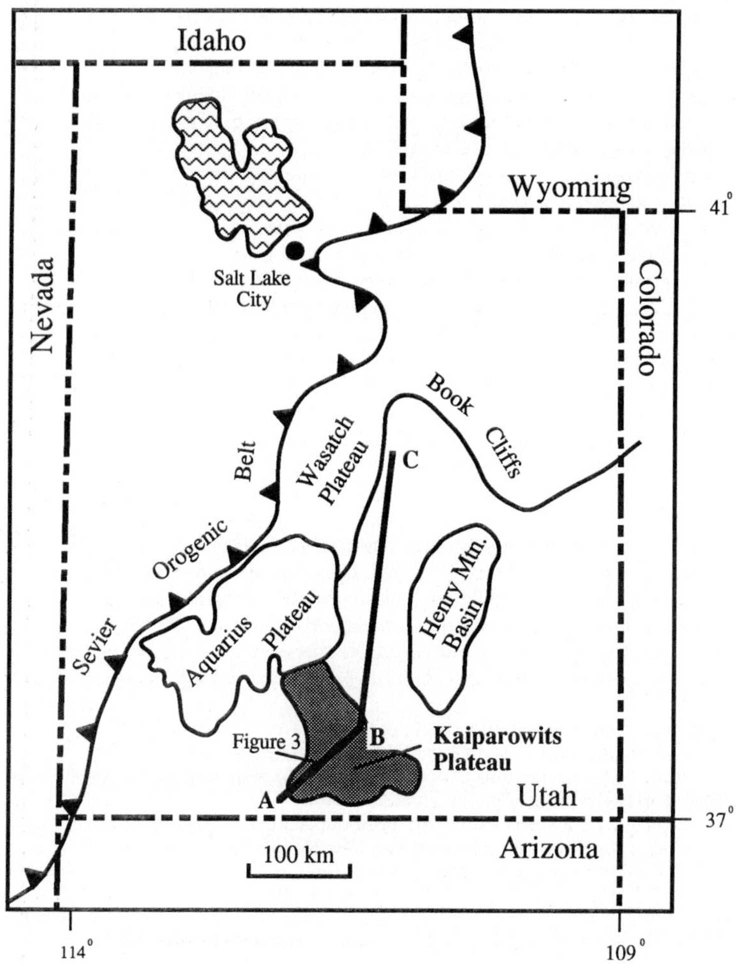

Fig. 1. Regional map illustrating the location of the Kaiparowits Plateau within the state of Utah. Also shown is the trend of the leading edge of the Sevier orogenic belt, and locations of the Wasatch Plateau, Henry Mountain Basin and the Book Cliffs. The line of section A–B–C is shown in Fig. 3.

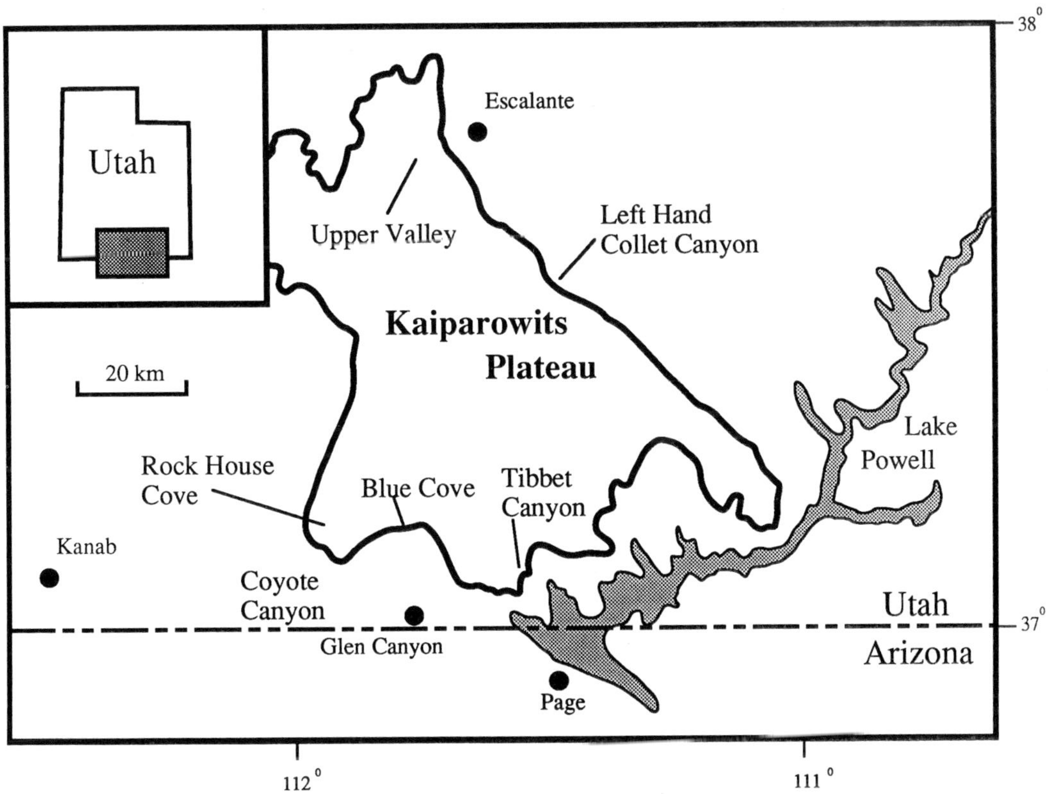

Fig. 2. Map showing the locations of Rock House Cove, Blue Cove, Coyote Canyon, Tibbet Canyon, and Left Hand Collet Canyon. In general, the northeastern part of the plateau consists of shoreface deposits which grade into coal-bearing strata in the central plateau. These in turn grade into alluvial strata towards and west and southwest.

models to be modified and extended into alluvial plain strata.

Recently developed sequence stratigraphic models (e.g. Vail *et al.*, 1977; Haq *et al.*, 1988; Jervey, 1988; Posamentier 1988; Posamentier & Vail, 1988; Posamentier *et al.*, 1988; Ross, 1990; Van Wagoner *et al.*, 1990), as well as quantitative alluvial models (e.g. Allen, 1978, 1979; Bridge & Leeder, 1979; Alexander & Leeder, 1987) suggest that fluvial architecture may be interpreted in terms of base-level change and its effect on accommodation space. Despite the proliferation of both quantitative and qualitative models, there are few outcrop-based studies that describe alluvial architecture in terms of sequence-stratigraphic evolution and fewer still that directly compare alluvial stacking patterns with coeval shoreface and shallow-marine stacking patterns. Although studies of outcrops and subsurface data sets over the last 15 years have encouraged the use of sequence-stratigraphic concepts in shallow-marine strata (e.g. Van Wagoner *et al.*, 1990), similar studies of alluvial strata are still in their infancy. The recognition of sequence boundaries, condensed sections and depositional systems tracts in shallow-marine strata (under conditions of constant sedimentation rates) has produced a methodology suggesting greatly improved stratal predictability and temporal resolution. The presence of abrupt lateral facies changes, typically poor biostratigraphic resolution and limited absolute age-dating, numerous internal erosion surfaces, the absence of through-going marker horizons, and limited outcrops, however, are some of the many difficulties that have frustrated attempts to extend these models to alluvial strata. This frustration is summarized by Walker (1990) who suggests that '. . . sequence stratigraphy, with its emphasis on bounding unconformities and marine flooding surfaces, cannot easily be applied to non-marine rocks (if at all)' (p. 781).

The emphasis of this paper, therefore, is to describe the architecture of alluvial strata within the context of unconformity-bounded depositional sequences as illustrated by the exposures in the Kaiparowits Plateau. Mitchum *et al.* (1977) defined a depositional sequence as '. . . a stratigraphic unit composed of a relatively conformable succession of genetically related strata and bounded at its top and base by unconformities or their correlative conformities' (p. 53). We find that sediment architecture of both shoreface and alluvial strata varies in a systematic and, therefore, predictable manner depending on position within a depositional sequence. This paper focuses on the alluvial strata. Our approach allows chronostratigraphic correlations to be extended inland for at least 60 km, greatly improving temporal resolution for purposes of basin analysis and regional correlation. The recognition of an evolutionary trend in alluvial architecture within depositional sequences provides the necessary framework for further detailed, quantitative analyses of fluvial architecture. This study strongly suggests that detailed studies of fluvial architecture within discrete channel belts, without an understanding of the sequence stratigraphic architecture, are of limited utility beyond the immediate study area or in areas of limited data. The development of analogue models and the application of probabilistic data sets, especially for petroleum reservoirs, is most successful when placed in a sequence-stratigraphic framework.

Physical setting

During the Cretaceous an extensive Cordilleran orogenic fold and thrust belt could be traced from northern Canada and Alaska into Mexico. In the central and southern Rocky Mountains of the United States this feature is known as the Sevier orogenic belt (Armstrong, 1968) (Fig. 1). A foreland basin developed parallel to this orogenic belt in response to flexural loading of the lithosphere by successive thrust sheets and synorogenic sediment (e.g. Armstrong, 1968; Jordan, 1981; Cross, 1986). From the Albian through Maastrichtian the foreland basin was inundated by a vast epicontinental sea. During maximum eustatic highstand the seaway extended from present-day Arctic Canada to the present Gulf of Mexico, a distance of some 5000 km, and from western Utah to Iowa, a distance of some 1500 km (Williams & Stelck, 1975). During the Turonian to Campanian, the region presently known as the Kaiparowits Plateau occupied a position along the western margin of the Cretaceous interior seaway, approximately 120 km to the east of the leading edge of the Sevier orogenic belt (Fig. 1).

Timing of structural deformation within the Sevier orogenic belt continues to be a topic of considerable debate. Although the use of synorogenic strata as a means of dating thrust-fault movement has been questioned (e.g. Blair & Bilodeau, 1988; Heller *et al.*, 1988; Jordan *et al.*, 1988), the onset of thrusting in central Utah is interpreted as mid–late Albian with episodic thrusting throughout the Late Cretaceous (e.g. Wiltschko & Dorr, 1983; Lawton, 1985; Villien & Kligfield, 1986). Similar age-dates have been obtained from geohistory or backstripping analysis (Cross, 1986; Heller *et al.*, 1988).

The Kaiparowits Plateau covers some 3600 km^2 in the southwestern portion of the Colorado Plateau structural and physiographic province (Fig. 2). The present outcrop configuration was established largely by the Miocene (Lucchitta, 1979, 1989) as the Colorado River assumed its present course and began to control headward erosion into the broad plateaus of the Colorado Plateau. The structural configuration of the Kaiparowits Plateau itself is one of gentle, north to northwest plunging asymmetrical anticlines and synclines. Structural dip within the plateau is generally less than 3°. The plateau is locally cut by several high-angle normal faults; however, displacement is minimal and they do not deter detailed stratigraphic and sedimentological investigations.

Stratigraphic setting

Exposures within the study interval belong to the Straight Cliffs Formation, which ranges from Middle Turonian to early Campanian in age and from 300 to 370 m in thickness (Fig. 3). The northeastern escarpment of the plateau, known as the Straight Cliffs, provides approximately 80 km of depositional strike-oriented exposure extending from Lake Powell in the south to the town of Escalante, Utah, to the north (Fig. 2). The southern margin trends east-northeast, is dissected by numerous canyons, and provides almost 80 km of exposure oriented slightly oblique to depositional-dip (Fig. 2). Because of its remote nature, there have been relatively few geological investigations within the Kaiparowits Plateau.

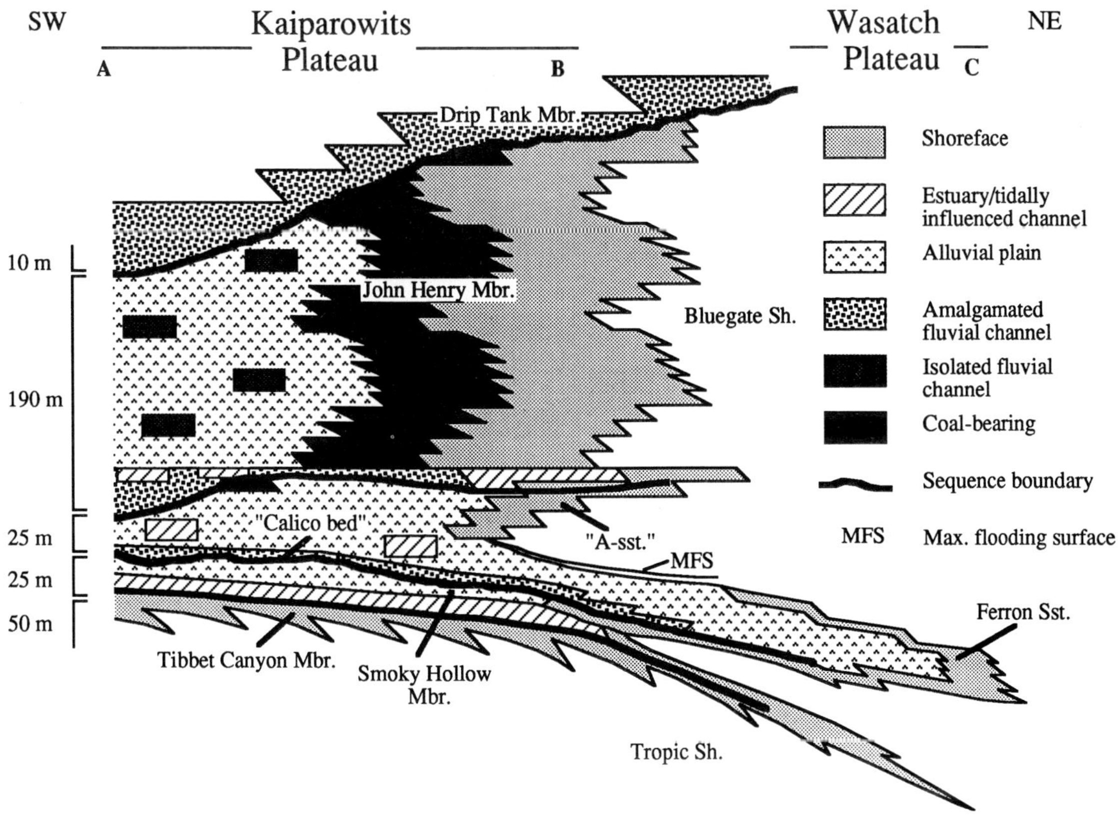

Fig. 3. Sequence-stratigraphic cross-section of Turonian through lower Campanian strata from the Kaiparowits Plateau (southwest) to the Wasatch Plateau (northeast) from Shanley (1991). The position of the line of section, A–B–C, is shown in Fig. 1. Within the Kaiparowits Plateau, the diagram is to scale with a horizontal datum at the top of the A-sandstone flooding event. Stratigraphic thicknesses in the southwestern part of the Kaiparowits Plateau are shown along the left margin of the figure. The portion of the diagram in the Wasatch Plateau is not to scale but is drawn to portray stratigraphic relationships. Sequence boundaries, or their correlative conformities, allow alluvial and shoreface strata to be subdivided into chronostratigraphic units across almost 80 km of depositional dip.

The initial stratigraphic framework for the plateau was established during a series of geologic and geographic expeditions conducted between 1913 and 1923 (Gregory & Moore, 1931). Type sections and facies studies were designated by Peterson (1969a,b), and Vaninetti (1978) suggested a mosaic of coastal plain, lagoonal, deltaic and barrier island sediments for the Straight Cliffs Formation. These earlier interpretations are similar in many respects to the classic interpretations of the Book Cliffs region of central Utah (e.g. Balsley, 1980). More recently, Shanley & McCabe (1989, 1990, 1991) and Shanley (1991), combined with the earlier studies by Peterson (1969a,b) and Vaninetti (1978), have delineated broad, coeval lithofacies belts of alluvial plain, mire, coastal plain, shoreface, and open marine strata within the John Henry Member of the Straight Cliffs Formation (Fig. 3). This overall arrangement of lithofacies types provides an unusual opportunity to compare sedimentary architecture across a broad spectrum of coeval depositional environments. The lithofacies arrangement in the John Henry Member, combined with adjacent strata in the Straight Cliffs Formation, the limited structural complexity and sparse vegetation, provide an ideal field laboratory for the study of alluvial facies architecture and sequence stratigraphy.

Age determinations of the marine strata in the Kaiparowits Plateau are reasonably well constrained by biostratigraphic data based on ammonite and inoceramid collections (summarized in Shanley, 1991). Similar assessments of alluvial and coal-bearing strata, however, are poorly constrained. The lack of magnetostratigraphic variation during the Turonian to Campanian, the paucity of regionally extensive ash beds and the lack of detailed palynological investigations reduce the effectiveness of conventional chronostratigraphy. As a result, we have focused not only on the detailed sedimentology of these strata, but also on basinward shifts in facies tracts and deepening trends that can be traced throughout the plateau and which have regional significance (Shanley & McCabe, 1989; Shanley, 1991). We interpret five unconformity-bounded depositional sequences and their associated highstand and transgressive systems tracts within the Straight Cliffs Formation. These sequences are recognized throughout the plateau and, when combined with available biostratigraphic data, provide a detailed level of chronostratigraphic resolution. A detailed description of the sequence-stratigraphic evolution of the plateau, including a more direct comparison between marine and alluvial strata, and their correlation to adjacent areas is contained in Shanley (1991) and illustrated in Fig. 3.

Field methods

To develop an understanding of sequence-stratigraphic relationship, detailed sections were measured from the Tropic Shale to the Drip Tank Member of the Straight Cliffs Formation (Fig. 3). Significant facies-tract offsets associated with sequence boundaries, transgressive surfaces of erosion (ravinement surfaces), condensed sections and their nonmarine counterparts, and shoreface parasequence stacking patterns were interpreted from these sections. Stratal correlations are based on analysis of the stacking patterns of marine parasequences, visual correlation both in the field and on extensive panorama photographs, and review of low-angle aerial photographs obtained from a helicopter survey of the entire plateau.

Treatment of the facies architecture and its relationship to sequence stratigraphy required a combination of closely spaced detailed sections and panorama photographs within the framework established by the larger scale sequence-stratigraphic study. All sections have been measured to record stratigraphic variation at the centimetre scale. Where talus obscured the fine-grained interval, mattocks were used to trench through the talus to that as complete a section as possible could be measured. All measured sections have more than 95% stratigraphic completeness. Lateral variations have been described and recorded through careful field photography, acquisition of smaller detailed sections, and by walking beds out.

DESCRIPTION AND INTERPRETATION OF SEDIMENTARY FACIES

Introduction

Exposures of both sheet and ribbon alluvial sandbodies (terminology from Friend *et al.*, 1979) dominate the western and west-central portions of the plateau and are emphasized in this study (Fig. 4). The nature of these outcrops is such that the sandstones form prominent ledges whereas the mudrocks are generally obscured by talus material. Observations regarding the mudrocks are, therefore, limited to exposures in trenches, gullies, and locations where overlying sandstones have sheltered the mudrocks from talus. To the east, these alluvial strata interfinger with crevasse splay deposits, thick coals and mudrocks, and thin, ribbon-like channel sandstones. Because thick coals are present in this region, extensive surface burns are common. We have relied on petrologic data collected by Peterson (1969a) and refer to the rocks in terms of Folk's (1974) classification. For the purposes of this paper, we have delineated five facies associations. It is the geometric arrangement of these facies associations that characterizes the architectural style of these strata and allows for comparison with marine strata along the eastern margin of the plateau.

Studies of ancient alluvial sequences are replete with attempts to classify strata on the basis of channel morphology, i.e. meandering, braided, straight, or anastomosed channel patterns, or sediment load, i.e. suspended, mixed, or bedload channels (e.g. Miall, 1978; Rust, 1978; Collinson & Lewin, 1983; Ethridge *et al.*, 1987), despite the fact that a continuum exists between various channel patterns (Schumm, 1977, 1981). This continuum of channel patterns and the autocyclic processes responsible for these patterns, combined with longer term allocyclic processes and their effect on alluvial preservation, suggest that realistic reconstructions of channel morphology are difficult at best. Nevertheless, detailed facies models based on studies of both modern and ancient alluvial strata have been developed

Fig. 4. View to the east of Rock House Cove, an area dominated by alluvial strata (see Fig. 2 for location). The stratigraphic position of this photograph is illustrated on Fig. 3. The prominent sandstone in the foreground (1) is the top of the amalgamated sandstone (A-sandstone) complex which forms the upper part of a transgressive systems tract. The majority of the photograph (2) illustrates the highstand systems tract that comprises isolated sheet and ribbon-channel sandbodies within fine-grained overbank sediments. The skyline of the photograph shows the coarse-grained, slightly darker weathering, fluvial deposits of the Drip Tank Member (3) which unconformably overlie finer-grained, isolated channels of the John Henry Member. Approximately 165 m of section is shown between the top of the amalgamated sandstone and the Drip Tank Member.

which suggest criteria to distinguish alluvial channel patterns (e.g. Ethridge & Flores, 1981; Cant, 1982; Galloway & Hobday, 1983; Walker, 1984; Collinson, 1986). The continuum of alluvial processes and products led Collinson (1978), Friend (1983) and Bridge (1985) to conclude that some end-member channel styles might be recognized with a certain measure of confidence; however, the vast majority of fluvial deposits are difficult to categorize. Friend (1983) and Bridge (1985) considered the following criteria useful in determination of palaeochannel patterns: proportion of channel-fill relative to lateral accretion deposits, the mean grain size of channel-fills relative to lateral accretion deposits, palaeocurrent variance, study of strata interpreted as bedload deposits, and quantitative reconstructions of channel hydraulics. With these criteria in mind, the following interpretations are offered for the facies associations of the Kaiparowits Plateau. Much of the descriptive material for the fluvial sandstones is summarized in Table 1.

Pebbly and coarse-grained to medium-grained sheet sandstone facies

Description

Sandstones comprising this facies belong to both the Calico bed, an informal unit of the Smoky Hollow Member of the Straight Cliffs Formation and the lower portions of the Drip Tank Member of the Straight Cliffs Formation (Fig. 3). Because of the

Table 1. Generalized characteristics and interpretations for the channel sandstones in the Kaiparowits Plateau. Sinuosity estimates are based on Schumm (1963). The sandbody classifications are based on both Friend *et al.* (1979) and Puigdefabregas *et al.* (1989)

Facies	Grain size and sediment structure	Bed and storey thickness	Calculated sinuosity	Width-thickness and classification	Environmental interpretation	Sequence stratigraphic interpretation
Pebbly and coarse-to-medium-grained sheet sandstones	Poorly sorted, pebble and granule material along basal and storey scours. Multiple fining-upwards units from coarse-grain to medium-grain sandstone. Trough cross-beds grade vertically to ripple-lam. and massive sandstone. Scattered sets of planar-tabular cross-bedded sandstones	Less than 1 m thick abbreviated storeys at base. 5 m thick storeys in the upper part Trough sets 0.3–0.5 m near the basal scour of the Calico. 0.25–1.5 m sets near the base of the Drip Tank. Trough sets 0.1–0.15 m thick in the upper parts of both sandstones. Ripple-laminated cosets up to 2.5 m thick	1.2	>50:1 sheet sandstone, or multistorey, lenticularly bedded sheet sandstone	Bar and channel fill deposits of low sinuosity, multi-channel rivers	Fluvial terrace deposits and aggradation of fluvial transgressive systems tract. Basal erosion surfaces are regionally extensive sequence boundaries that can be correlated to adjacent outcrop belts (Shanley, 1991)
Med–fine grain amalgamated sandstones	Med–coarse grain feldspathic litharenite. Overall fining upward trend, disrupted by numerous internal storey scour surfaces. Trough and planar-tabular cross-beds grade vertically to ripple laminated and massive sandstones. Lateral accretion surfaces are preserved in the upper part. Ripple and climbing ripple lamination common in the upper part	Less than 1.5 m abbreviated storeys at the base. Upper storeys 3–8 m thick. Trough sets decrease from 0.5 m near the base to 1.5 m at the top. Convolute beds 0.5–3.0 m thick	1.5–1.8	>50:1 sheet sandstone, or multistorey, lenticularly bedded sheet sandstone	Channel fill deposit of moderate to high sinuosity, perennial rivers	Basal erosion surface has the morphology of a valley suggesting the amalgamated sandstone is an incised valley fill deposit. The increase in preserved fine-grain material is interpreted to reflect increased accommodation space in an alluvial transgressive systems tract

Med–fine grain isolated sandstones	Med–fine grained feldspathic lith arenites. Overall fining-upwards successions within the multistorey sandstones. Granule lags identify the internal storey scours. Trough cross-beds grade vertically to ripple laminated and massive or convolute sandstones	Complete storeys approximately 9 m thick Trough sets decrease from 0.4 m at the base to 0.15 m at the top. Ripple cross-laminated sets 0.25–2.0 m thick occur at the top	1.5	10–60 Multistorey-multilateral sheet and ribbon sandstones. Composite tabular sandbodies	Point bar and channel deposits of moderate to high sinuosity rivers	Fluvial aggradation associated with an early highstand systems tract. Critical to this interpretation is the correlation of these fluvial deposits to aggradational nearshore marine parasequences (Shanley, 1991; Shanley & McCabe, 1991)
Heterolithic Facies	Diverse suite of structures including ripple-laminated sandstones, wavy and lenticular bedded sandstones, sigmoidal bedding geometries, shrinkage cracks, multiple reactivation surfaces, and compound cross-bedding	Inclined heterolithic units up to 12 m thick Mudrocks and sandstone generally on decimetre scale		Composite, tabular sandbodies	Moderate to high sinuosity tidally influenced fluvial deposits	Aggradation of tidally influenced deposits. Temporally equivalent to maximum flooding surfaces in marine strata to the east

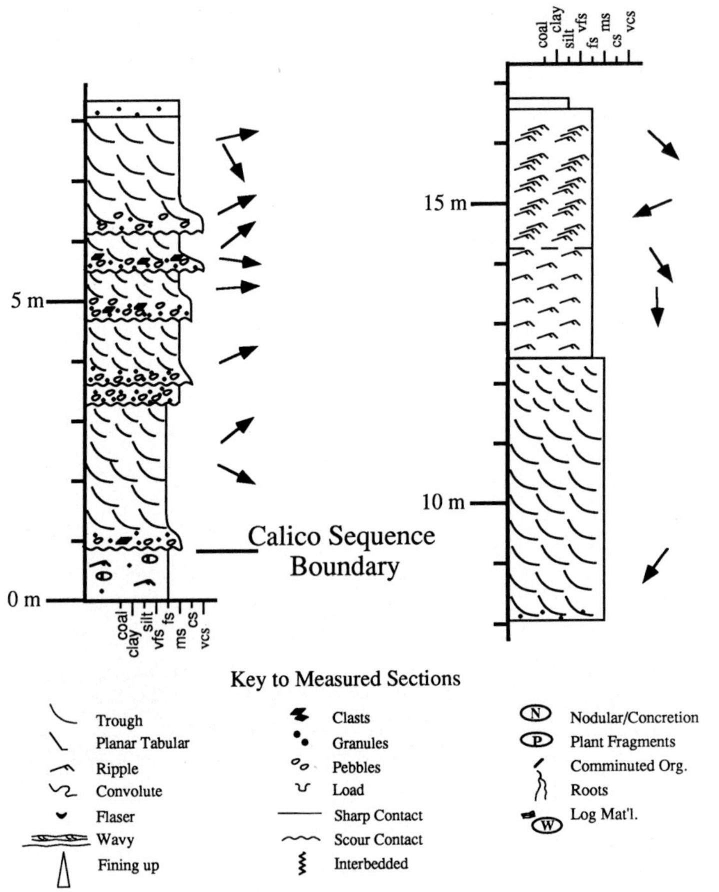

Fig. 5. Measured section of the Calico bed from Tibbet Canyon (see Fig. 2 for location). The sequence boundary is interpreted where coarse-grained and pebbly sandstones overlie fine-grained, rooted alluvial plain strata. This surface marks a major basinward shift in facies tracts. Note the thinly-bedded, multi-storey nature of the lower part of the Calico and the overall fining-up character. These strata comprise a transgressive systems tract and are interpreted as low-sinuosity, possibly braided river deposits. The legend shown in this figure is also used for measured sections in subsequent figures.

emphasis of this paper, we do not attempt to describe the entire Drip Tank Member. Our measured sections include the lowermost 5–15 m of Drip Tank strata and our comments, therefore, are limited to these strata. The Calico bed commonly has a bleached to light grey appearance on outcrop providing a superb visual marker throughout most of the plateau. The Drip Tank Member commonly weathers a brown to brownish-orange colour and is also easily recognized throughout the plateau because it commonly forms the uppermost resistant sandstone ledge along the skyline of the plateau.

The Calico bed, a poorly sorted, friable, granule to medium-grained sandstone with interbedded pebble-conglomerate lenses, erosionally overlies fine-grained channel sandstones, rooted mudrocks, carbonaceous shales and thin, discontinuous coal beds of the Smoky Hollow Member. Erosional relief locally exceeds 2 m (Fig. 5). Thickness of the Calico generally varies across the plateau from 0 to 15 m, although thicknesses as great as 40 m have been reported by Bobb & Ryer (1990) from the extreme northeastern part of the plateau. Conglomerate lenses are common, and generally occur along scour surfaces at the base of the Calico as well as along internal (storey) scour surfaces. These conglomerates are predominantly composed of chert, quartz, and feldspar pebbles whose long axis rarely exceeds 2.5 cm. Conglomerates are both clast and matrix supported, range in thickness from a few centimetres to 0.4 m, and range in length from a few decimetres to 1–2 m (Fig. 6).

The Calico bed comprises several distinct sandstone bodies, or storeys (terminology after Friend *et al.*, 1979), each of which has a scour base and which, in general, fines up from a coarse-grained and granule-bearing sandstone with scattered conglomerate lenses to a medium-grained sandstone with scattered

Fig. 6. Photograph of the Calico bed in Tibbet Canyon illustrating the intermixed pebble conglomerates and medium- to coarse-grained, trough cross-bedded sandstones that are common in the lower part of the Calico bed (3.3 m level in Fig. 5). The lens cap is 70 mm in diameter.

granules and conglomerates (Fig. 5). Mud clasts up to a few centimetres long, finely comminuted plant material, and woody debris are commonly incorporated along scour surfaces, especially the basal scour of the Calico bed. At the base of the Calico, storeys range from decimetres to 1 m in thickness and are abbreviated in the sense that the upper portions of fining-upward units are removed by subsequent scour surfaces. Higher in the Calico, more complete fining-upward units are preserved, ranging up to 5 m in thickness (Fig. 5). Sedimentary structures within fining-upward units of the Calico bed consist primarily of trough cross-stratified and some planar-tabular cross-bedded sandstones that grade up to ripple-laminated sandstones, minor planar-tabular cross beds, and thin massive-bedded sandstones (Fig. 5). Thickness trends of both cross-beds and storeys are summarized in Table 1.

Palaeocurrents in the Calico bed are quite variable when viewed collectively (Fig. 7a); however, this variance decreases markedly when viewed in the context of discrete outcrops. Nevertheless, these data suggest overall transport directions to the east. Similar palaeocurrents were also reported by Peterson (1969a) and Bobb & Ryer (1990).

The Drip Tank Member (Fig. 3) erosionally overlies fine-grained, rooted mudrocks, carbonaceous shales, thick coal-bearing strata, and fine-grained channel deposits of the John Henry Member (Fig. 8). Erosional relief at the base of the Drip Tank Member locally exceeds 5 m. Regional correlations across the Kaiparowits Plateau, however, suggest that the base of the Drip Tank Member has cut down towards the west by almost 60 m (Shanley & McCabe, 1991). The Drip Tank Member is a poorly sorted, granule to medium-grained, feldspathic/litharenite with interbedded pebble-conglomerate lenses. Conglomerates in the Drip Tank are princi-

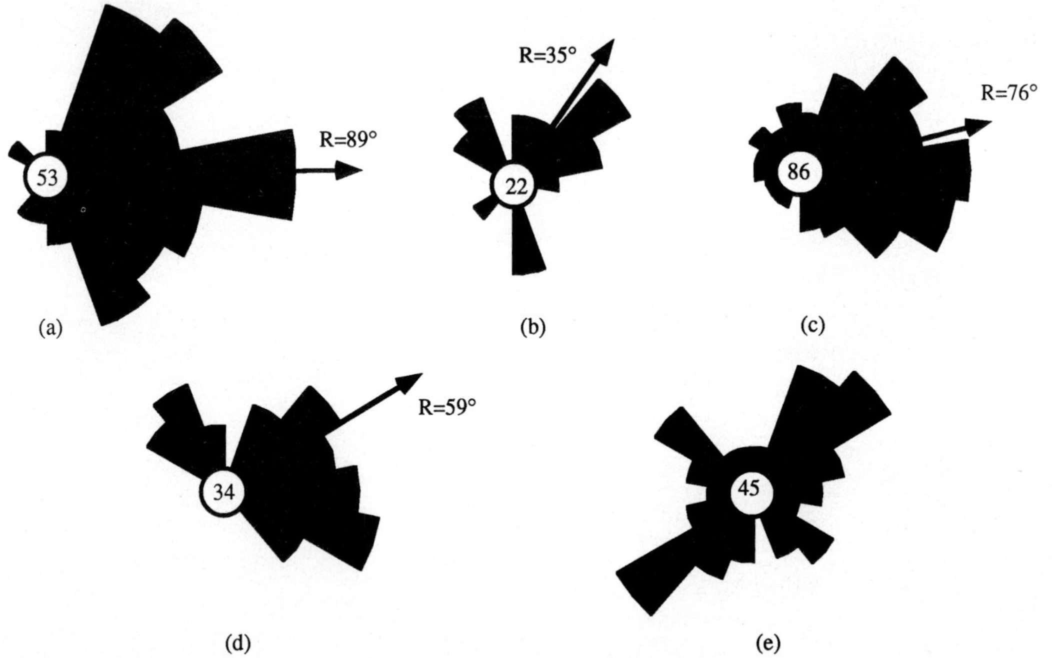

Fig. 7. Palaeocurrent data collected on a regional basis for (a) Calico bed, (b) Drip Tank Member, (c) medium- to fine-grained amalgamated sandstone facies, (d) medium- to fine-grained isolated sandstone facies, and (e) tidally-influenced heterolithic facies. The azimuth of resultant vectors (R) was calculated according to the method outlined in Curray (1956). The number of readings used is shown in the centre of each rose diagram.

pally composed of chert, quartz, and feldspar pebbles whose long axis rarely exceeds 4 cm. Both clast and sand-matrix supported pebble conglomerates are present and range in thickness from a few centimetres to 0.25 m.

Similar to the Calico bed, the Drip Tank Member sandstones are multi-storey in nature with individual storeys ranging from decimetres to a few metres in thickness. Each storey fines upward from coarse-grained sandstones and thin pebble conglomerates to medium-grained sandstones (Fig. 9). Within the lower 15 m the thickness of individual storeys progressively increases upwards from 0.3 m at the base to as much as 1.75 m near the top. Mud-clast conglomerates, woody fragments, and pebble-conglomerate lags outline the basal scour surface as well as internal storey-scour surfaces (Fig. 9). Sedimentary structures within the Drip Tank Member consist primarily of trough cross-stratified and planar-tabular cross-stratified rocks. Palaeocurrents in the Drip Tank Member suggest flow was primarily to the northeast (Fig. 7b), consistent with results reported by Peterson (1969a).

Sandstones within both the Calico and Drip Tank have width/thickness ratios in excess of 50 : 1, which combined with the internal architecture suggest the Calico bed and the lower portions of the Drip Tank Member are sheet sandstones (Friend *et al.*, 1979) or multi-storey, lenticularly-bedded sheet sandstones (Puigdefabregas *et al.*, 1989).

Interpretation

The sandstones comprising the Calico bed and the Drip Tank Member are interpreted as bar and channel-fill deposits of low-sinuosity (terminology from Rust, 1978), perennial rivers. The multi-storey, sheet-like geometry of these sandstones make estimates of channel width and depth difficult. Nevertheless, calculated sinuosity values (Schumm, 1963) are less than 1.2. The very low proportion of lateral accretion to channel-fill deposits, combined with the relatively coarse-grain size of channel-fill deposits, the unidirectional palaeocurrents with low variance, and sheet-sand geometry support the interpretation of low-sinuosity, probably multi-channel rivers.

Fig. 8. Photograph of coarse-grained, amalgamated fluvial sandstones of the Drip Tank Member unconformably overlying fine-grained fluvial sandstones and carbonaceous mudstones of the John Henry Member from Rock House Cove in the southwestern part of the plateau (see Fig. 2 for location). This contact is interpreted as a widespread sequence boundary that can be traced into adjacent outcrop areas.

Peterson (1969a) and Bobb & Ryer (1990) arrived at similar conclusions for the Calico bed. The predominance of trough cross-stratification suggests that during flood stage the river bed was largely covered with three-dimensional dunes although the presence of planar-tabular cross-stratification suggests that straight-crested, two-dimensional dunes were also present. The lack of thick planar-tabular sets, with palaeocurrent orientations divergent from adjacent trough cross-beds, however, may suggest that there were few braid bars migrating transverse to flow and may argue against an extensive multi-channel system. Ripple laminated, finer-grained sandstones were confined to the upper portions of bars in both the Calico and Drip Tank. The presence of both clast and matrix-supported pebble conglomerates along the basal scours and the internal storey scours may suggest discharge variations leading to changes in shear stress and sorting. The abundant mud intraclasts, mud-clast conglomerates and woody debris, especially at the base of both the Calico bed and the Drip Tank Member indicate that these river systems were eroding fine-grained, mudrock-dominated, vegetated floodplains. The multi-storey, sheet-like geometry of both sandstones further suggests that rates of floodplain and valley aggradation were relatively low, consistent with quantitative-model experiments (e.g. Bridge & Leeder, 1979) that suggest how fluvial architecture might vary due to changes in accommodation space.

It is worth speculating on the origin of the bleached appearance of the Calico bed not only for its importance as a visual marker but also for its role in sequence stratigraphic interpretations. Outcrop examination of the Calico bed suggests that both the friable nature and bleached appearance are in part due to preferential leaching of feldspar grains and the precipitation of kaolinite clays. Although it is

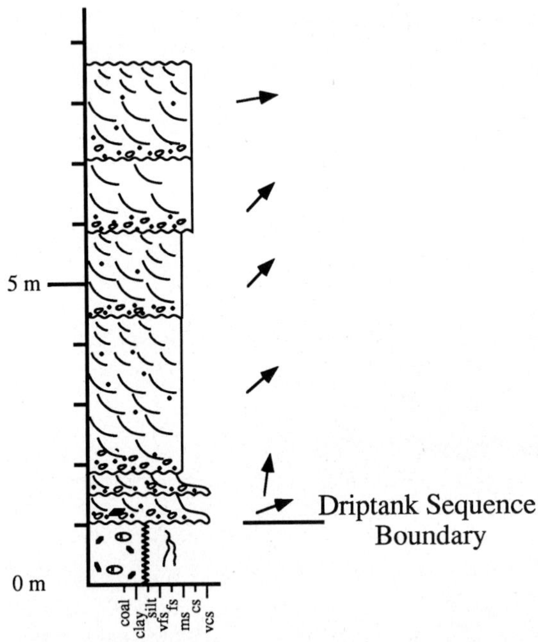

Fig. 9. Measured section of the lower part of the Drip Tank Member from Tibbet Canyon (Fig. 2). The sequence boundary interpreted at the base is placed where pebbly and coarse-grained sandstones erosionally overlie fine-grained and rooted alluvial plain strata. This contact marks a significant seaward shift in facies tracts similar to the base of the Calico bed. The thin, multi-storey nature of the lower part of the Drip Tank is interpreted to reflect low rates of base-level rise resulting in the slow creation of accommodation space. These strata comprise the basal part of a transgressive systems tract and are interpreted as low-sinuosity, possibly braided river deposits. (See Fig. 5 for the legend and Fig. 2 for the location of this section.)

difficult to constrain the timing of this alteration without conducting a paragenetic investigation (not part of this study), field relationships suggest this alteration may have occurred shortly after deposition during the late Turonian to early Coniacian; channel deposits of similar composition occur within the upper part of the Calico bed and are not bleached. Regional relationships as well as sedimentological observations suggest that the bleached part of the Calico bed was likely preserved as a series of terrace deposits during a period of base-level lowering before significant onlap (Shanley & McCabe, 1990; Shanley, 1991). The alteration of the Calico bed, therefore, occurred during periods of prolonged subaerial exposure prior to significant fluvial aggradation of the John Henry Member.

Medium- to fine-grained amalgamated sandstone facies

Description

Sandstones comprising this facies occur near the base of the John Henry Member of the Straight Cliffs Formation (Fig. 3). These sandstones form a prominent cliff-face that can be correlated along the southwestern margin of the plateau from Rock House Cove to Blue Cove, a distance of approximately 12 km (Fig. 2). Thickness of the amalgamated sandstone facies varies from 0 to 30 m. The abrupt change in grain size that commonly marks the basal scour surface of the amalgamated sandstone facies, combined with a marked change in alluvial stacking pattern, facilitates the recognition of the lower part of this sandstone complex throughout the study area. The gradual but irregular increase in proportion of fine-grained strata in the upper part of this facies, however, renders the upper contact of this facies difficult to place (Fig. 10).

These sandstones cut into fine-grained, rooted, carbonaceous mudrocks, and heterolithic facies. Erosional relief at the base of the amalgamated sandstone locally exceeds 2 m. The basal scour surface is covered with clay-clast conglomerates, with individual clay clasts up to 3 cm in length, scattered chert pebbles and granules, and woody debris. Coarse-grained sandstone lags also occur along the basal scour surface (Fig. 11).

This sandstone complex is composed of several distinct storeys that can be traced laterally for distances of 10 to 100 m. At the base, storeys range from less than 1 m to 1.5 m thick with individual storey-scours marked by an increase in grain size and a thin granule or pebble lag (Fig. 11). Internal sedimentary structures of these lower storeys consist primarily of trough and planar-tabular cross-bedding. In contrast to the lower part of this sandstone complex, upper storeys are thicker (3–8 m), have more complete fining upward units, and have greater variation in sedimentary structures (Fig. 11). The basal portions of upper storeys commonly contain clay clasts, scattered chert granules and, rarely, bone material. Several sandstones within the upper portion of the amalgamated complex contain laterally inclined surfaces. Where these surfaces are more completely preserved, they are low-angle in nature near the top of the sandstones, have maximum dips between 5 and 15° within the middle of a sandstone, and become asymptotic at the base

Fig. 10. View to the west of the amalgamated channel sandstone complex in Blue Cove (see Fig. 2 for location). The thick sandstone cliff in the centre of the photograph is approximately 10 m thick. The sequence boundary (SB) at the base of the amalgamated sandstone complex is placed at the base of the thick sandstone cliff in the centre of the photograph. This sequence boundary can be traced throughout the southwestern part of the plateau and is referred to as the A-sequence boundary. The upper part of the amalgamated sandstone complex is gradational with isolated channel sandstones and fine-grained overbank strata. The increase in fine-grained strata towards the top of the amalgamated sandstone complex is seen clearly in this photograph. We interpret the upward increase in fine-grained strata to reflect increased accommodation space within a gradually aggrading valley complex. This particular valley sandstone can be traced from Rock House Cove to Blue Cove. (TST = transgressive systems tract; HST = highstand systems tract.)

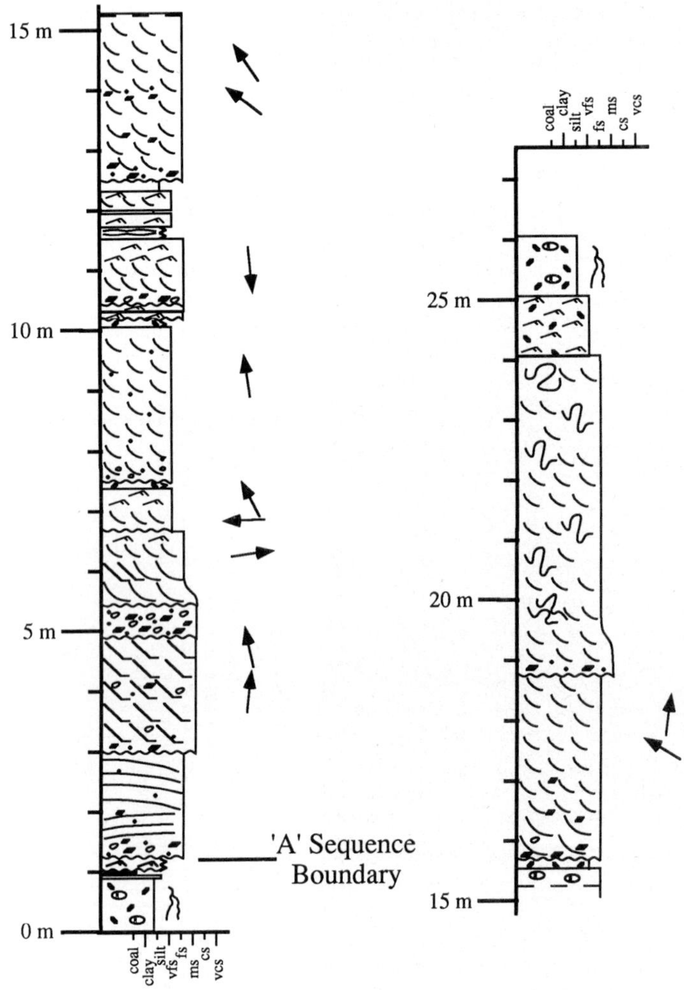

Fig. 11. Measured section of the medium- to fine-grained, amalgamated sandstone facies from Blue Cove in the vicinity of the photograph shown in Fig. 10. The sequence boundary is interpreted where coarser-grained, multi-storey, amalgamated sandstones erosionally overlie fine-grained and rooted alluvial plain strata. Note the numerous thin, fining-up sandstone complexes at the base of the measured section which become thicker towards the top of the section. This trend in preservation is interpreted to reflect the gradual filling of a broad incised valley complex. These strata comprise a transgressive systems tract and are interpreted as moderate- to high-sinuosity river deposits. (See Fig. 5 for the legend and Fig. 2 for the location of this section.)

of the sandstone. Internal sedimentary structures consist of trough cross-beds (Fig. 11). Where more complete fining upward units are preserved, trough cross-bedded sandstones commonly grade upward into ripple-laminated or massive sandstones and finally into ripple laminated siltstones or very-fine-grained sandstones. Some trough and planar-tabular cross-bedded sandstones have been convoluted into beds 0.5–3 m thick. Thickness data for both storeys and cross-bed sets are summarized in Table 1.

Within the upper part of this facies association are thinly-bedded, ripple and climbing ripple-laminated sandstones that grade into ripple-laminated siltstones and very-fine-grained sandstones. These sandstones contain finely comminuted organic material that is concentrated along laminae; they are interbedded with rooted mudrocks. Inclined beds of ripple-laminated sandstones and thin mudrocks also occur within the upper part of this facies.

Palaeocurrents in the amalgamated facies (Fig. 7c) suggest transport directions to the northeast through southeast. Although palaeocurrents are clearly unidirectional, there is a spread of almost 140° in the data set. Palaeocurrent orientations taken from planar-tabular sets that exceed 1 m in thickness diverge by almost 50–70° from the orientations taken from adjacent trough cross-beds. Palaeocurrents from laterally inclined strata are approximately normal to the direction of inclination. Based on thickness and areal extent, width/thickness ratios exceed

50 : 1, suggesting the amalgamated sandstone complex may be considered a sheet sandstone (Friend *et al.*, 1979) or a multi-storey, lenticularly-bedded sheet sandstone (Puigdefabregas *et al.*, 1989).

Interpretation

The amalgamated sandstone facies association of the John Henry Member is interpreted as channel-fill deposits of moderate- to high-sinuosity (Rust, 1978), perennial rivers. Although lateral-accretion deposits occur near the base of this facies, estimates of sinuosity (Table 1) are principally derived from exposures in the upper part, where both channel preservation and lateral accretion are better developed. The relatively high proportion of lateral accretion to channel-fill deposits combined with the finer grain sizes of the lateral accretion deposits, and the 140° variation in palaeocurrents all support the interpretation of moderate- to high-sinuosity rivers. The presence of planar-tabular sets and palaeocurrent orientations that significantly diverge from adjacent trough cross-beds may reflect the local preservation of straight-crested bars. Planar-tabular sets within the upper part of these fining-upward sandstones may have formed as scroll bars. The preservation of discrete fining-upward units that grade from trough cross-beds to ripple and massive bedded sandstones and siltstones are interpreted as the preservation of the downstream portion of point-bar complexes. During flood stage, the channel base was probably covered with three-dimensional dunes that graded into current ripples on the margins of the point bar. Convolute lamination within trough cross-bedded sandstones is interpreted as partial liquifaction and deformation of water-saturated sediment under the influence of lateral shear. Such conditions may be due to the migration of large bedforms over an unconsolidated substrate during conditions of high discharge (Plint, 1983). Channel sandstones that are completely dominated by convolute lamination most probably reflect catastrophic sediment liquefaction, perhaps via earthquakes. The abundant mud-clasts and woody debris incorporated along scour surfaces within this sandstone complex indicate that these river systems eroded fine-grained, mudrock-dominated, vegetated flood plains.

Based on vertical and lateral relationships, the amalgamated sandstone facies is interpreted to occur within an incised valley. Within this valley, fine-grained strata, interpreted as overbank deposits and lateral accretion deposits, become increasingly common towards the top of the sandstone complex. As a result, the multi-storey, multi-lateral morphology grades vertically into smaller multi-lateral sand-bodies more typical of sandstone bodies found in the isolated sandstone facies. The increased preservation of discrete channel forms and higher proportion of fine-grained strata within the upper part of the valley is thought to reflect increased accommodation space combined with changes in rates of fluvial aggradation similar to models proposed by Bridge & Leeder (1979). Alternatively, this vertical change in preservation may reflect an evolutionary change in the fluvial system itself; however, the local presence of lateral accretion deposits and minor fine-grained intervals within the lower part of this facies association suggests that similar rivers are present throughout the amalgamated facies.

Medium- to fine-grained isolated sandstone facies

Description

Sandstones comprising this facies association occur throughout the middle and upper portions of the John Henry Member and locally within the Smoky Hollow Member (Fig. 3). Unlike the previous two facies that form prominent cliffs, these sandstones are noted for their isolated exposures. Individual sandstones range from 5 to 12 m in thickness and from 65 m to approximately 300 m in width (measured normal to palaeoflow) and are encased in interbedded siltstones and very-fine-grained sandstone facies (Fig. 12). To the east, in the vicinity of Tibbet Canyon, the isolated sandstone facies association interfingers with thick coal-bearing strata, crevasse-splay deposits, and thin, ribbon-like channel sandstone strata.

These sandstones erosionally overlie fine-grained, rooted, carbonaceous mudrocks of the floodbasin facies with erosional relief locally exceeding 3 m. The basal scour surface of these sandstones have tool and flute marks, and are commonly covered with woody debris, small clay clasts less than 3 cm long, scattered granules and thin coarse-grained sandstone lags between 0.1 and 0.2 m thick. Clay-clast conglomerates between 0.2 and 0.65 m thick are present at the base of some sandstones (Fig. 13).

These sandstones have the geometric arrangement of both multi-storey, multi-lateral ribbon sands, as well as sheet sandstones (Friend *et al.*, 1979) and are classified as composite, tabular sand-bodies (Puigdefabregas *et al.*, 1989). In the case

Fig. 12. View to the east at Blue Cove of medium- to fine-grained isolated channel-belt sandstones encased in overbank siltstones and sandstones (see Fig. 2 for the location of this photograph). Note the laterally discontinuous nature of the sheet and ribbon sandbodies (4). This photograph also illustrates the upward change in sandbody connectedness between channel sandstones of the amalgamated facies (high overall net/gross) (3) versus those of the isolated facies (low overall net/gross) (4). The amalgamated sandstones are part of a transgressive systems tract whereas the isolated channels comprise a highstand systems tract. In the foreground are shoreface sandstones of the Tibbet Canyon Member (1) and alluvial deposits of the Smoky Hollow Member (2). The skyline of the photograph illustrates the sharp contact at the base of the coarse-grained and pebbly sandstones of the Drip Tank Member.

of multi-storey sandbody exposures, two to three storeys are normally present. Complete single storeys are approximately 9 m thick and grade upward from medium- to fine-grained sandstone and interbedded very-fine-grained sandstone and siltstone. Internal sedimentary structures within these sandstones consist of trough cross-beds, ripple-lamination, massive bedding, and convolute lamination (Fig. 13). The dimensions of cross-bed sets are reported in Table 1. In a single case an entire storey, approximately 9 m thick, consists of convolute lamination. Towards the top of sandstone storeys, trough cross-beds decrease in set thickness and are replaced by current ripple-lamination. Near the margins of these sandbodies where they can be observed to scour into fine-grained floodbasin strata, the sandstones are dominated by ripple- and climbing-ripple stratification. The uppermost part of these sandbodies is often characterized by massive and mottled sandstones approximately 0.3 m thick.

Palaeocurrents in this facies association are consistent within a single storey, but may show as much as 90° variation between storeys, indicating a range in transport directions from northwest to east with a strong northeast component (Fig. 7d). The primary distinction between sandstones belonging to this facies versus the amalgamated-sandstone facies is the degree of sandstone amalgamation. Consequently, the upper contact of the amalgamated sandstone facies in some locations is difficult to place because of the gradational change to the more isolated sandstone facies.

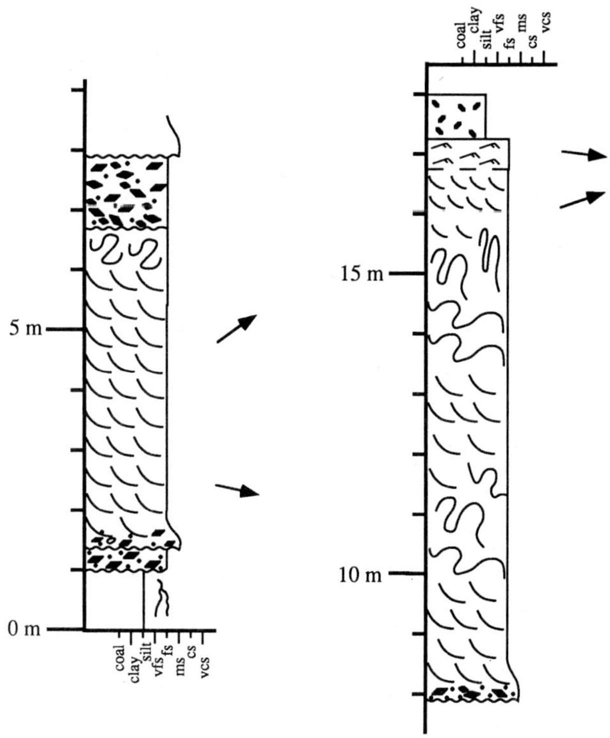

Fig. 13. Measured section of the medium- to fine-grained, isolated sandstone facies from Rock House Cove (see Fig. 2 for the location of this section). These sandstones typically are comprised of two storeys and form sheet-like deposits. Individual storeys are more complete in this facies than in the amalgamated sandstone facies illustrated in Fig. 11. These strata comprise a highstand systems tract and are interpreted as sinuous river deposits.

Interpretation

The isolated sandstones of the John Henry Member are interpreted as channel deposits of rapidly aggrading moderate- to high-sinuosity, perennial rivers. Lateral accretion deposits are not well developed in many of these sandstones and estimates of sinuosity are largely based on storey thicknesses. Because of the nature of the outcrop, fine-grained, abandoned-channel plugs are generally poorly exposed. The high proportion of fine-grained sediment adjacent to these sandstones, combined with the internal architecture of these sandstones, support the interpretation of a single-channel, sinuous fluvial system. Where complete abandoned-channel strata can be observed, estimates of channel depth and width are possible. Such exposures, although rare, indicate channel depths of 5–8 m and channel widths of 75–125 m. The erosional bases, abundant clay intraclasts, clay-clast conglomerates, and woody debris suggest that these channels cut into vegetated, fine-grained floodplain sediments. Minor variations in discharge are suggested by the presence of thin, coarse-grained sandstone lags and scattered granules that are interpreted to represent the traction load under flood discharge. The channel sandstones are dominated by fining-upward units that grade from trough cross-stratified sandstones to ripple-laminated sandstones and finally to thin-rooted mudstones and massive-bedded sandstones. These units are interpreted as the deposits of the downstream side of point bars. The rooted mudstones and massive-bedded sandstones that cap these units suggest the upper point-bar surfaces had established vegetation.

We interpret the isolated channel architecture to reflect the high rates at which accommodation space was created, which approximately equalled the rate of sedimentation, similar to models developed by Bridge & Leeder (1979).

Heterolithic facies

Description

Strata comprising this facies association occur in the lower one-third of the John Henry Member (Fig. 3). As the name implies, this facies association contains

Fig. 14. Photograph of inclined heterolithic strata (1) in the lower part of the John Henry Member at Rock House Cove in the southwestern part of the plateau (see Fig. 2 for the location of this photograph). The unit of inclined strata is approximately 10 m thick. The section illustrated in Fig. 15 was measured at this locality. These strata are interpreted as lateral accretion deposits within a tidally-influenced river. Note the persistent mud drapes that can be traced into the lower part of the channel deposits. The top of the photograph (2) records amalgamated fluvial deposits associated with the lower transgressive portion of the A-sequence.

a wide variety of sedimentary rocks such that any outcrop may contain some or all of the characteristics described here.

In general, there is an overall fining trend in grain size from the eastern side of the plateau at Left Hand Collet Canyon to the western side of the plateau at Rock House Cove (Fig. 2). These strata erosionally overlie a variety of facies ranging from fine-grained, rooted carbonaceous mudrocks to medium-grained trough cross-bedded sandstones. The basal contact of the heterolithic facies is erosional with maximum depth of scour approaching 10 m. Despite the depth of scour, the overlying strata are dominated by interbedded ripple-laminated sandstones and thin mudrocks. The internal geometry of these heterolithic strata is complex. Inclined strata with palaeocurrents normal to the direction of stratal inclination are common as are large internal scour surfaces (Fig. 14). Although the internal sedimentary structures are dominated by ripple-laminated sandstones and thin mudrocks, there is a variety of other sedimentary features (Fig. 15). These include wavy and lenticular-bedded sandstones, in sets approximately 0.2–0.5 m thick, pseudoplanar laminated sandstones (Smith, 1971; or transcurrent lamination of Allen, 1984), sigmoidal beds that are outlined by thin mud drapes, ripple-laminated sandstones with shrinkage-crack casts on the base, trough and planar-tabular-bedded sandstones containing multiple reactivation surfaces, and compound cross-stratification (Fig. 15). Of particular significance are the thin mudrocks that commonly drape ripple-laminated sandstones and which can be traced from the upper part of a sandstone complex along inclined surfaces into the basal part of the sandstone (Fig. 14). Based on geometric criteria, sandstones within the hetero-

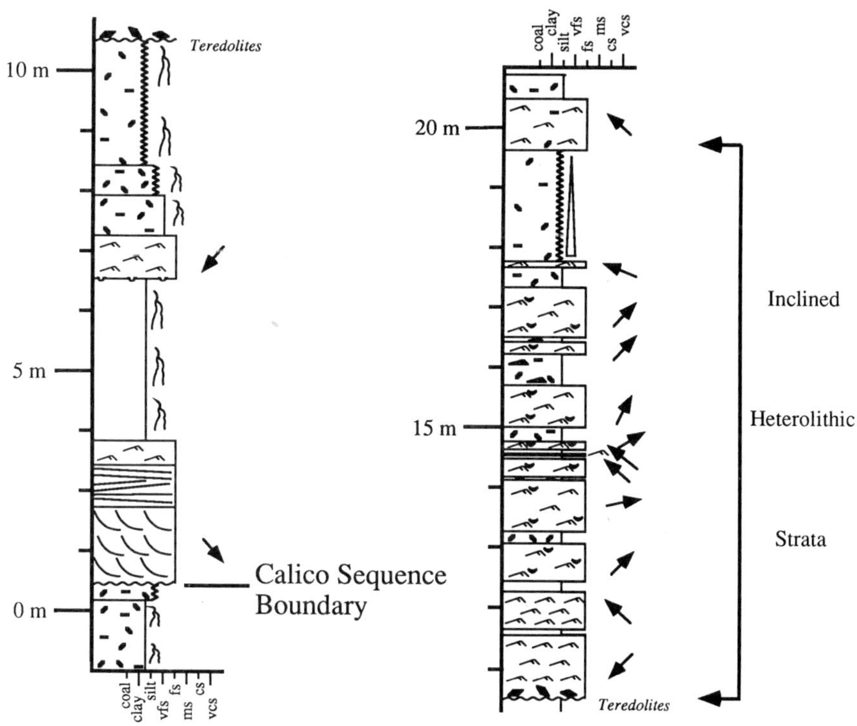

Fig. 15. Measured section of the inclined heterolithic strata shown in Fig. 14 from Rock House Cove. Note the alternations of thin, ripple-laminated and flaser-bedded sandstones with thin, yet persistent, mud drapes that can be traced into the base of the channel deposits. These strata are interpreted as lateral accretion deposits of a tidally-influenced meandering river located approximately 60 km from a coeval shoreline. These heterolithic strata are temporally equivalent to marine condensed section deposits located on the eastern margin of the plateau.

lithic facies are classified as composite, tabular sandbodies (Puigdefabregas *et al.*, 1989).

Trace fossils within this facies association are the most diverse of all the strata described in this paper and include *Teredolites*, *Arenicola*, and *Diplocriterion*. *Teredolites* is commonly found in the basal portions of this facies association, whereas *Arenicola* and *Diplocriterion* are more commonly found in the upper portions of sandstones. Body fossils may include whole or broken inoceramid shells. Palaeocurrents in these sandstones are taken primarily from ripple-laminated sandstones and show a pronounced bidirectional palaeocurrent orientation (Fig. 7e).

Interpretation

The heterolithic facies assemblage occurs at discrete horizons within the John Henry Member. Along the northeastern margin of the plateau where these strata are overlain by a transgressive lag deposit, they are interpreted to reflect sedimentation in open to restricted estuaries. Across the vast majority of the plateau, however, these strata are entirely contained within nonmarine strata and were previously interpreted as fluvial deposits (Peterson, 1969a,b; Vaninetti, 1978). We interpret these strata as tidally-influenced river deposits; they represent maximum flooding into the alluvial plain (Shanley & McCabe, 1990, 1991; Shanley, 1991) and are temporally equivalent to maximum flooding surfaces in nearshore marine strata. An examination of lateral accretion surfaces and the associated mud drapes suggest that many of the tidally-influenced rivers were of moderate to high sinuosity. The overall westward fining trend in grain sizes and reduction in hydraulic energy, as indicated by the sedimentary structures, is interpreted to reflect increasing distance inland from an open estuary or coeval shoreline. The presence of sigmoidal bedding geometries, double-mud drapes, wavy and lenticular bedding, multiple reactivation surfaces, and shrinkage cracks, interpreted as synaeresis cracks, all support the interpretation of fluvial deposits with a pronounced tidal influence. The palaeocurrent data suggest fluvial processes modified by ebb and flood oriented currents that are normal to the regional trend of the palaeoshoreline. The trace fossil assemblage suggests at least brackish water conditions (e.g. Ekdale *et al.*, 1984).

While these complex channel deposits are interpreted as the product of tidally-influenced rivers, they almost certainly do not reflect the result of daily tidal processes or even spring–neap cycles. Rather, we interpret these channels to reflect the interaction of tidal processes, channel and valley morphology, variations in seasonal runoff conditions and development of turbidity maxima within the fluvial system (Shanley, 1991). Rahmani (1988) came to similar conclusions for heterolithic channel-fill deposits in Upper Cretaceous strata in Alberta, and Allen (George Allen, pers. comm., 1990), and Gelfenbaum (1983) reported similar observations and interpretations from Holocene studies along the Gironde River in France and the Columbia River in Washington, USA, respectively.

Interbedded siltstone–sandstone facies

Description

Strata comprising this facies association are the principal component of the Smoky Hollow Member and the John Henry Member of the Straight Cliffs Formation throughout the western and west-central portions of the Kaiparowits Plateau (Fig. 3). Cumulative thickness is approximately 150 m. Because of the fine-grained nature of this facies association, exposures are generally poor and the geometric relationships between the various fine-grained lithologies that comprise this facies association are difficult to evaluate. Interbedded siltstones and sandstones have gradational contacts with the upper part of the amalgamated sandstone facies association, and an erosional relationship with sandstones of the isolated sandstone facies association.

This facies association is primarily composed of siltstones, interbedded very-fine-grained sandstones and siltstones, and thin, laterally discontinuous very-fine-grained sandstones; claystones and thin coals, although present, are rare. The siltstones and interbedded very-fine-grained sandstones contain scattered carbonaceous material, have well-developed root traces, contain slickensides, and scattered plant material. Several siltstone beds have a fractured texture and contain purple-coloured siderite concretions. Rooted siltstones and sandstones varying from 0.5 m to 4 m in thickness are interbedded throughout this facies association. The thin, very-fine-grained sandstones that occur throughout this interval vary in thickness from 0.2 m to 1.2 m, have sharp bases, are ripple-laminated, and grade into mottled and massive sandstone in the uppermost 0.1–0.6 m. Sandstones less than 0.2 m thick contain massive and mottled bedding textures. Claystones in this facies association are up to 3.7 m thick and contain no discernible sedimentary structures. The coals in this facies are less than 0.1 m thick and are laterally discontinuous.

Interpretation

These strata are interpreted as fine-grained flood-basin deposits that accumulated on broad alluvial plains predominantly during periods of overbank sedimentation. Discontinuous thin, fining-upward sandstones capped with mottled and massive sandstones are interpreted as the distal portions of crevasse splays that have subsequently become vegetated and root penetrated. Thin claystone intervals with little or no comminuted plant material are interpreted as local lacustrine deposits that developed on the alluvial floodplain. The large volume of clastic overbank sedimentation from adjacent rivers probably prevented organic-rich peat deposits from forming (McCabe, 1984). Soil horizons were locally well developed as indicated by the abundant root traces, sideritic nodules, and concretions within some of the siltstone intervals. The quality of the exposures, however, has prevented a detailed correlation of these soil horizons.

FLUVIAL ARCHITECTURE AND SEQUENCE STRATIGRAPHY

Introduction

The geometric arrangement of the facies associations provides the basis for developing a sequence-stratigraphic model for alluvial strata. In the Kaiparowits Plateau the geometry of alluvial strata can be directly compared with coeval coal-bearing strata in the central and eastern part of the plateau, as well as coeval shallow-marine strata exposed along the eastern margin of the plateau (Figs 2 and 3). Such comparisons place needed constraints on our models of alluvial sedimentation and offer insights into how significant stratal surfaces in shallow-marine strata are expressed in alluvial strata. Prominent sequence boundaries related to the Calico bed and the Drip Tank Member can be visually traced across the entire plateau. Within this framework, Shanley (1991) and Shanley & McCabe (1991) recognized additional sequences and described how sequence boundaries, maximum flooding surfaces in marine strata, parasequence-stacking patterns of nearshore strata, and relative thicknesses of strata along the Straight Cliffs escarpment are correlated to the southwestern part of the Kaiparowits Plateau. In addition, upward-deepening successions that record a transition from alluvial to offshore and condensed section deposits were described in the Smoky Hollow and lower part of the John Henry Member along the Straight Cliffs escarpment (Shanley, 1991). Application of these concepts in alluvial strata in the central and southwestern plateau, within the framework provided by visually-correlatable surfaces, is the subject of the remainder of this paper.

Sequence boundaries in alluvial strata

Recognition of sequence boundaries in alluvial strata is difficult because basinward shifts in facies tracts and related incision can be difficult to distinguish from more localized channel scour. Both the Calico and Drip Tank sequence boundaries are associated with major basinward shifts in facies tracts, can be traced throughout the plateau, and are recognized in adjacent outcrop belts (Fig. 3) (Shanley, 1991). These basinward shifts are interpreted on the basis of an abrupt contrast in facies tracts above and below the sequence boundary as well as a change in the degree of fluvial sandstone amalgamation. It is emphasized that these changes are observed not only in vertical section, but can also be traced along the outcrop belt. Coarse-grained and pebbly, laterally amalgamated fluvial deposits (Figs 5 and 9) directly overlie finer-grained fluvial and alluvial-plain strata, carbonaceous shales and, to the east, shoreface strata. The regional extent of these erosional surfaces, coupled with the lack of coarse-grained sandstone and pebble-conglomerates in underlying channel and overbank strata, suggest these coarse-grained, bedload rivers are not coeval with underlying deposits. In addition to the abrupt change in facies development across the sequence boundary unconformities, there is a concomitant change in fluvial architecture. Alluvial deposits that underlie both the Calico and Drip Tank sequence boundary unconformities are multi-storey sandbodies that contain complete fining-upward units in the uppermost storey (i.e. Fig. 13). These strata belong to the isolated channel-sandstone facies and the fine-grained floodbasin facies associations. The abrupt change to amalgamated, coarse-grained and pebbly deposits of the Calico bed and Drip Tank Member in which preservation of fining-upward storeys is incomplete signifies a change in the rate of fluvial aggradation and/or a change in accommodation potential (after Bridge & Leeder, 1979). Abrupt changes in composition and degree of amalgamation (or fluvial stacking pattern) across erosional surfaces that have regional extent also suggest a pronounced change in hydraulic character as well as rates of alluvial aggradation and together support the interpretation of sequence-boundary unconformities.

A sequence boundary has also been identified at the base of the medium- to fine-grained amalgamated-sandstone facies (Shanley, 1991) near the base of the John Henry Member (Fig. 3). Criteria that allow identification of this sequence boundary, however, are more subtle than those used for the Calico and Drip Tank sequence boundaries. To the east, along the Straight Cliffs, fluvial and tidally-influenced fluvial strata overlie a sequence boundary (A-sequence boundary) that truncates shoreface parasequences and which has been identified as an incised valley (Hettinger *et al.*, 1990) (Fig. 3). This incised valley fill is overlain by a widespread marine-flooding surface (Hettinger *et al.*, 1990; Shanley, 1991). The A-sequence boundary itself is slightly younger than a maximum flooding surface observed in somewhat older marine strata (Shanley, 1991). Within the alluvial-dominated portion of the western and southwestern plateau, amalgamated fluvial sandstones erosionally overlie tidally-influenced fluvial strata that are coeval with this maximum marine flooding surface (Shanley, 1991), and fine-grained alluvial-plain strata containing well-developed palaeosols. Lateral relationships indicate these amalgamated sandstones comprise a broad, incised valley, portions of which can be traced from Rock House Cove to Blue Cove (Fig. 2). Although there is a significant change in alluvial stacking patterns from isolated meanderbelt and tidally-influenced fluvial deposits below the A-sequence boundary to coarser-grained, amalgamated sandstones above, the degree of facies tract dislocation is not as severe as with the Calico bed and Drip Tank Member. It is the change in fluvial amalgamation over a broad area that signifies an abrupt change in aggradation rate and accommodation space and, when combined with observations in near-shore marine strata, serve to identify the sequence boundary.

Figure 3 also illustrates a sequence boundary within the Tibbet Canyon Member of the Straight Cliffs Formation (Shanley, 1991). Because this unconformity erodes into marine strata and juxtaposes predominantly tidal channel deposits on offshore and shoreface strata, we do not emphasize this surface in a discussion of alluvial deposits.

Valley fill and alluvial-transgressive deposits

Laterally amalgamated, fluvial sheet-sand deposits with a high relative proportion of interconnected, coarser-grained channel-fill sandstone (referred to as high net/gross reservoir sandstone) are characteristic of three distinct stratigraphic levels in the Kaiparowits Plateau. These are the Calico bed, the amalgamated sandstone complex (A-sandstone) in the lower part of the John Henry Member, and the basal portion of the Drip Tank Member (Fig. 3). Each of these three sandstone intervals overlie sequence boundaries that have been traced throughout the plateau.

Throughout the western and southern margins of the plateau, amalgamated sandstones of the Calico bed and A-sandstone grade vertically into isolated channel deposits that are interbedded with fine-grained alluvial plain strata, and heterolithic, tidally-influenced channel-fill deposits. The transition from coarse-grained, pebbly deposits of the Calico bed to tidally-influenced heterolithic river deposits reflects a vertical progression from low-sinuosity, coarser-grained river deposits to more sinuous, fine-grained channel-fill sandstones deposited within tidally-influenced rivers. This 'upward-deepening' facies succession reflects coastal transgression, and maximum marine incursion into alluvial environments (Shanley & McCabe, 1990). These finer-grained alluvial strata are correlated with retrogradational shoreface parasequences, such as those at the top of the Ferron Sandstone Member of the Mancos Shale, that pass upward to a widespread condensed section deposit within the Mancos Shale (Shanley, 1991) (Fig. 3). Tidally-influenced fluvial deposits, therefore, are temporally equivalent with the maximum flooding surface in marine strata that is represented by a condensed section and that can be traced throughout the eastern margin of the plateau (Shanley & McCabe, 1990; Shanley, 1991).

A similar set of upward-deepening fluvial deposits occur in association with the A-sandstone. Sinuous river deposits that have a high degree of lateral amalgamation grade upward into more isolated channel sandstones and finally into tidally-influenced heterolithic strata. This succession is interpreted to reflect filling of the incised valley associated with the A-sequence boundary, coastal transgression and maximum marine flooding into alluvial depositional environments (Fig. 3). Tidally-influenced fluvial deposits that cap the A-sandstone in the western and southwestern plateau are coeval with development of a maximum marine flooding surface to the east (Shanley & McCabe, 1990; Shanley, 1991).

Readers will note that the fluvial deposits immediately overlying sequence boundaries in the Kaiparowits Plateau comprise an alluvial transgressive systems tract and that no lowstand deposits are described. Although the reasons for this are more a matter of nomenclature than of depositional processes, some discussion is warranted. Posamentier & Vail (1988), Posamentier *et al.* (1988) and Van Wagoner *et al.* (1990) suggest that the term transgressive systems tract be restricted to strata bounded by the maximum marine-flooding surface above, and the first major marine-flooding surface (or transgressive surface) below. Lowstand deposits, on the other hand, are located below the first widespread transgressive surface. Based on our regional stratigraphic correlations (Shanley, 1991; Shanley & McCabe, 1991) we suggest that the sequence boundaries can be traced to the east of the Kaiparowits Plateau for a considerable distance. Initial marine flooding of these sequence boundaries is thought to have onlapped well to the east of the plateau. The maximum flooding surface in marine strata and its alluvial equivalents, however, are recognized and provide a reliable, laterally persistent, time-significant horizon. Recognition of true lowstand deposits requires an accurate, three-dimensional understanding of the palaeogeography associated with sequence boundary unconformities. We suggest that the lowstand deposits associated with the sequence boundaries described in this paper are located to the east of the Kaiparowits Plateau and that the fluvial deposits in the plateau belong to transgressive systems tracts.

Highstand deposits

Tidally-influenced fluvial strata associated with transgression over the Tibbet Canyon, Calico, and A-sequence boundaries are overlain by fine-grained floodbasin strata, isolated fluvial sandstones, and thin, discontinuous coals and carbonaceous shales. In the case of the strata overlying the A-sandstone complex, these deposits account for the vast majority of strata exposed in the John Henry Member (Fig. 3). This thick succession of isolated channels and overbank siltstones and sandstones have been physically correlated into the central portion of the plateau where they interfinger with thick crevasse-splay deposits and thick, laterally continuous coal seams. The coal measures, in turn, have been physically traced into thick, aggradational nearshore deposits along the Straight Cliffs escarpment (Shanley & McCabe, 1991). This architectural style of aggradational shoreface parasequences, isolated fluvial channel belts, and thick fine-grained overbank deposits is interpreted to reflect increased rates of accommodation that are balanced by rates of deposition. Our work in shoreface strata along the eastern margin of the plateau suggests that additional high frequency sequence boundaries may be present within these strata; however, we have yet to recognize these stratal surfaces within the alluvial deposits (Shanley & McCabe, 1991).

CONTROLS ON FLUVIAL STACKING PATTERNS

Background

The emphasis on geometric arrangement through geologic time has advanced alluvial-architecture studies beyond facies studies (Walker, 1990). Efforts to describe alluvial architecture have proceeded along two somewhat parallel fronts—those based on conceptual or numerical modelling of fluvial systems, and those based on outcrop observations. Some of the earliest models for fluvial architecture were those of Allen (1965) who portrayed the three-dimensional architecture of alluvial strata through time in a number of different geologic settings through the use of block diagrams. Although these models did not attempt to portray the effects of any particular allocyclic or autocyclic variable, they clearly illustrate how changes in fluvial stacking might occur. More rigorously constrained numerical models (e.g. Allen, 1978, 1979; Leeder, 1978; Bridge & Leeder, 1979; Alexander & Leeder, 1987) allow the changes in alluvial architecture to be studied as specific variables are adjusted. Numerical models such as these share a common theme by emphasizing the role of tectonic subsidence in creating accommodation space for alluvial sedimentation. By changing rates of subsidence, avulsion, and aggradation, varying degrees of sandstone amalgamation and sandbody interconnectedness can mathematically be created. Although these early models were not tested against outcrop-based data sets, they are of paramount importance in our understanding of alluvial architecture because they demonstrate the sensitivity of alluvial sequences to changes in accommodation space. These models suggest that the degree of channel-sandstone amalgamation and interconnectedness increase as the rate at which accommodation space is created decreases. This rather simple relationship is critical in relating fluvial architecture to sequence stratigraphy.

Lawrence *et al.* (1990) and Ross (1990) incorporated the relationship between alluvial architecture and subsidence rates predicted by Bridge & Leeder (1979) into their basin-scale stratigraphic models. As a result, these models predict specific architectural relationships between alluvial and shoreface strata. These larger scale, stratigraphic models reflect the delicate interplay between sediment supply and changes in accommodation space due to base-level changes that underpin sequence stratigraphic concepts (e.g. Jervey, 1988; Posamentier *et al.*, 1988; Posamentier & Vail, 1988). In the context of sequence-stratigraphic studies, base level can be thought of as an '. . . equilibrium surface . . . above which a particle can not come to rest and below which deposition and burial are possible' (Sloss, 1962, p. 1051). In more familiar terms, base level can be approximated by the concept of a graded fluvial profile in alluvial settings, sea level in nearshore regions, and the concept of a graded shelf in shallow-marine environments (Mackin, 1948; Swift, 1970). Understanding stratigraphic models in terms of base-level change and accommodation space allows models to be applied in a variety of geologic settings regardless of whether the available space was created by tectonic subsidence, eustatic change, or some combination of the two (the most likely case!). Field verification of such models, however, requires outcrops that contain a sufficiently broad spectrum of facies tracts so that significant stratal surfaces can be traced between facies tracts and so that stacking patterns can be analysed in these different facies tracts; the Kaiparowits Plateau presents this opportunity (Fig. 3).

Previous outcrop studies that sought to understand changes in sandstone amalgamation and alluvial architecture have invoked changes in base level, climate, subsidence, and sea level as well as changes

Table 2. Outcrop studies of alluvial strata have invoked changes in base level, climate, lake level, and subsidence to account for changes in sandstone amalgamation and interconnectedness. Of these mechanisms, most authors favour changes in subsidence to account for the observed changes in architecture

Age and lithologic unit	Location	Reference
Subsidence controlled architecture		
Holocene—alluvial fans	Death Valley, California, USA	Butler, 1984
Pliocene—Glenns Ferry Fm.	Idaho, USA	Kraus & Middleton, 1987
Oligocene—fluvial strata	Jaca Basin, Spanish Pyrenees	Turner, 1989
Late Cretaceous to Eocene—Chuckanut Fm.	Washington, USA	Johnson, 1984
Eocene—Capella Fm.	Tremp/Grauss Basin, Spanish Pyrenees	Atkinson, 1986
Palaeocene and Eocene—Willwood Fm.	Wyoming, USA	Kraus & Middleton, 1987
Eocene—Castissent Sandstone	Tremp/Grauss Basin, Spanish Pyrenees	Marzo *et al.*, 1988
Latest Cretaceous/Tertiary fluvial strata	Tremp/Ager Basin, Spanish Pyrenees	Puigdefabregas *et al.*, 1989
Campanian—lenticular sst. and shale	Green River Basin, Wyoming, USA	Schuster & Steidtmann, 1987
Aptian/Albian—Eumerala Fm.	Otway Basin, Australia	Felton, 1989
Aptian/Albian—Cloverly Fm.	Wyoming, USA	Meyers *et al.*, 1989
Triassic—Otter Sandstone	Devon, England	Clarey & Tunbridge, 1989
Triassic—Chinle Fm.	Colorado Plateau, USA	Blakey & Gubitosa, 1984
Frasnian—Catskill delta complex	New York, USA	Gordon & Bridge, 1987
Late Devonian—fluvial strata	Munster Basin, Ireland	Graham *et al.*, 1989
Sea-level, lake-level and base-level controlled architecture		
Miocene—fluvial strata	Ridge Basin, California, USA	Link, 1984
Jurassic—Scalby Fm.	Yorkshire, England	Nami & Leeder, 1978
Triassic—Ivishak Fm.	Prudhoe Bay, north slope Alaska, USA	Atkinson *et al.*, 1988
Devonian—Catskill delta complex	New York, USA	McCave, 1969
Climate controlled architecture		
Holocene alluvial strata	Southern USA	Hall, 1990
Quaternary alluvial strata	Colorado River, Texas, USA	Blum, 1990
Cenozoic—fluvial strata	South Africa	de Wit, 1989
Famennian—Kap Graah Group	eastern Greenland	Olsen, 1990

in avulsion rates and other autocyclic processes (Table 2). Although the sedimentary rock record incorporates the effect of all these variables, there is a preponderance of papers in which authors suggest changes in tectonic subsidence as the driving mechanism for changes in alluvial architecture. Despite the fact that many of these tectonic models require abrupt changes in rates of tectonic subsidence, they still reflect variations in fluvial architecture due to changes in accommodation space. The inability of many of these studies, however, to trace regionally significant stratal surfaces between alluvial and marine strata limit their use in evaluating changes in fluvial architecture due to changes in base level within the context of sequence stratigraphy.

In a series of conceptual models, Posamentier (1988), Posamentier *et al.* (1988) and Posamentier & Vail (1988), proposed a model outlining the timing of fluvial aggradation within the context of depositional sequences. These models incorporated concepts such as the equilibrium point (position on a depositional profile where the rate of eustatic change is equal to the rate of tectonic subsidence), the bayline (line containing points to which stream profiles are adjusted—under certain circumstances may be equivalent to the shoreline), and stream equilibrium profiles. Although Posamentier *et al.* (1988) and Posamentier & Vail (1988) express their results in terms of eustasy, careful consideration of their variables allows the reader to view these same models in terms of base-level change (relative sea level). These models indicate that significant alluvial aggradation requires seaward migration of the bayline during a slow base-level rise. Furthermore, Posamentier *et al.* (1988) and Posamentier & Vail (1988) suggest that the rate at which alluvial accommodation space is created controls the degree of sandstone amalgamation and fluvial sandstone interconnectedness. Within late-highstand systems tract deposits, fluvial strata are thought to be laterally continuous whereas early-highstand systems tract deposits are more likely to be characterized by isolated channel deposits.

Insights into the timing of alluvial aggradation and the response of fluvial systems to base-level change can also be acquired by examining dam-engineering and reservoir siltation projects (e.g. Eakin, 1936; Stevens, 1936, 1946; Strand, 1968; Lara & Sanders, 1970; Lara 1972, 1983; Condit *et al.*, 1978; Ferrari, 1988; Lyons & Randle, 1988; Posamentier & Vail, 1988; Orvis, 1989). Similar to the results proposed by Posamentier *et al.*, (1988) and Posamentier & Vail (1988), these studies indicate that during rapid base-level rise, significant fluvial aggradation does not occur until the rate of rise decreases allowing fluvial systems to readjust to grade as the stream profile is translated in a basinward direction. Under conditions of more slowly rising base level, however, fluvial aggradation may occur depending on rates of sedimentation versus rates of base-level rise (Shanley & McCabe, 1990).

More recently, there has been a resurgence in studying the response of fluvial systems to climatic changes (e.g. Schumm & Brakenridge, 1987; Blum, 1990; Hall, 1990). These studies suggest that, at significant distances inland, fluvial architecture is greatly affected by climatic change. The climatic effect may be sufficient to overwhelm any change in the fluvial system due to base-level change in the marine basin. We suggest, however, that fluvial systems near the coast (within perhaps 100–150 km of the coast) have the potential to be greatly affected by base-level/relative sea level change. Where incision into shoreface and offshore deposits can be demonstrated, we further suggest that base-level changes are required to change the elevation to which streams grade at the coast.

A model from the Kaiparowits Plateau

Alluvial strata in the Kaiparowits Plateau are characterized by regionally extensive sequence boundaries overlain by fluvial deposits that grade from laterally-amalgamated to more isolated channel deposits and heterolithic or tidally-influenced strata. When incorporated into the sequence-stratigraphic model developed for the Kaiparowits Plateau (Shanley, 1991; Shanley & McCabe, 1991) a synergistic model emerges within which variations in fluvial architecture may be better understood (Fig. 16).

During slow rates of base-level rise, and the initial part of a base-level fall, progradational shoreface parasequences are produced. As these slow rates of rise cease, and the rate of base-level fall accelerates, sediment is increasingly bypassed in a basinward direction with associated incision. As a result, shoreface parasequences are truncated and overlain by a regionally significant sequence boundary (Fig. 16A). Within fluvial systems the base-level fall has the effect of increasing fluvial gradients and causing headward erosion that is in part controlled by nickpoints (Schumm *et al.*, 1987; Butcher, 1990). Recent flume studies further suggest that incision resulting from base-level lowering results in increased sediment supply as drainage basins are progressively expanded (Schumm, 1977; Schumm *et al.*, 1987; Koss *et al.*, 1990). As a result of these increases in both fluvial gradients and sediment supply, sediment may bypass the alluvial realm and be redistributed in a more basinward position in the form of lowstand deposits. Under certain circumstances, these lowstand deposits appear spatially detached from the previous highstand shoreface. In a chronostratigraphic sense, there is little fluvial sediment preservation during base-level lowering and the sequence boundary represents a time of erosion and sediment bypass. Where fluvial deposits are preserved during the base-level fall, they are often in the form of river-terrace deposits that exhibit a complex relationship with the sequence boundary (Leopold *et al.*, 1964). The result of sediment bypass is a basinward shift and often an abrupt facies tract dislocation across the sequence boundary. The severity of the facies dislocation may be used as a proxy for estimating the magnitude of base-level change in a qualitative fashion (i.e. braided-river deposits incised into marine shales might suggest a greater base-level change than if they were incised into meandering-river deposits). In the Kaiparowits Plateau, for example, we have interpreted three

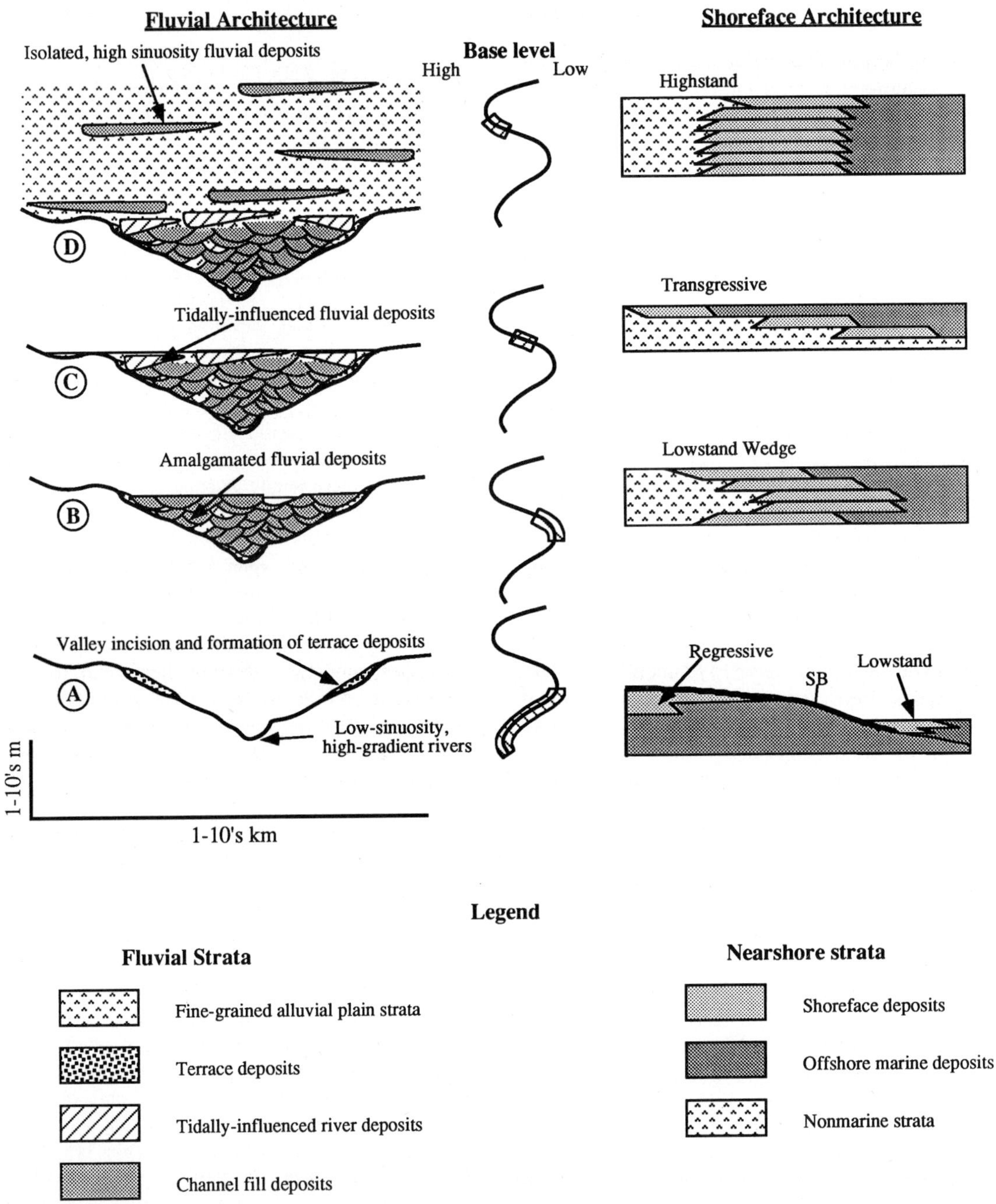

Fig. 16. Summary diagram illustrating the relationships between shoreface and fluvial architecture as a function of base-level change. Shoreface-architecture is shown in cross-sections oriented normal to the coastline whereas fluvial architecture is depicted for transverse sections of incised valleys. (A) Slow rates of base-level rise leading to base-level fall. Progradational shoreface geometries, sequence boundaries, and lowstand deposits are illustrated for nearshore marine strata whereas fluvial incision and the development of alluvial terraces are illustrated in nonmarine strata. (B) During reduced rates of base-level fall and a change to slowly rising base level, incised valleys initially aggrade with amalgamated fluvial deposits. These are temporally equivalent with shoreface deposits that initially prograde and subsequently begin to aggrade and/or retrograde. (C) Increased rates of base-level rise result in retrogradational shoreface parasequences, overlain by a maximum flooding surface, that is temporally equivalent to tidally-influenced fluvial deposits. (D) Reduced rates of base-level rise that are approximately balanced by rates of sedimentation result in aggradational shoreface parasequences and the development of isolated meanderbelt sandstones that 'float' in a matrix of alluvial plain deposits.

sequence boundaries within the alluvial succession of strata. The most dramatic facies dislocations are those associated with the Calico bed and the Drip Tank Member of the Straight Cliffs Formation. We have correlated these to major eustatic changes suggested by Haq *et al.* (1988) (Shanley, 1991). The sequence boundary associated with the A-sandstone has a less dramatic facies tract dislocation, which we have also correlated to eustatic changes suggested by Haq *et al.* (1988).

As the rate of base-level fall decreases, and base-level begins to rise, significant fluvial aggradation initially takes place near the river mouth and progressively onlaps in a landward direction (Fig. 16B). These amalgamated fluvial deposits are temporally equivalent with shoreface deposits that initially prograde and subsequently begin to aggrade and/or retrograde. The shoreface deposits of the Ferron Sandstone Member are an example of lowstand and transgressive shoreface deposits. Fluvial sedimentation during the base-level rise propagates in an upstream direction, dependant on the rate of base-level rise and sediment supply. Our research suggests that significant fluvial sedimentation requires equilibrium profiles to translate in a basinward direction similar to the models developed by Posamentier *et al.* (1988) and Posamentier & Vail (1988) and that somewhat reduced rates of base-level rise are required. We interpret alluvial strata in the Kaiparowits Plateau to have been deposited during a rise in base level; the degree of sandstone amalgamation and interconnectedness is interpreted to reflect changes in the overall rate of base-level rise. Sequence boundaries in the Kaiparowits Plateau represent periods of widespread erosion and the development of broad incised valleys with terrace deposits. During the initial phases of alluvial aggradation, while the rate of base-level rise was relatively low, rivers were confined to the lower portions of valleys. The rapid switching of fluvial channels combined with low rates of alluvial aggradation allowed channels to cannibalize much of their coeval floodplain strata resulting in amalgamated channel sandstones such as the Calico bed, A-sandstone, and lower part of the Drip Tank Member (Fig. 16B). Progressive infilling of incised valleys, combined with increased rates of base-level rise (increased rate of base-level rise also enhanced by reduced topographic gradients), result in increased accommodation space. These effects, combined with a decrease in sediment supply as stream gradients were reduced, result in tidally-influenced fluvial deposits, an increasing proportion of preserved fine-grained floodplain strata and more isolated channel sandstone typical of the upper part of the John Henry Member (Figs 3 and 16B).

Increased rates of base-level rise commonly result in retrogradational shoreface parasequences that are overlain by a maximum flooding surface represented by marine condensed section deposits. The point of maximum marine transgression at the coastline is marked within alluvial strata by the most landward extend of tidally-influenced rivers (Fig. 16C). Within drowned-river valleys tidal processes combine with fluvial processes to produce a complex, heterolithic facies assemblage (Shanley & McCabe, 1990). Tidally-influenced river deposits are, therefore, temporally equivalent with maximum-flooding surfaces in marine strata providing a time-significant surface (Shanley & McCabe, 1990) (Fig. 16C). Within the Kaiparowits Plateau, alluvial strata bounded by an underlying sequence boundary and the top of the tidally-influenced deposits are placed in an alluvial transgressive systems tract. The tidally-influenced river deposits at the top of the Calico and the top of the A-sandstone, in the western side of the plateau, illustrate the landward extent of the marine condensed interval, while the retrogradational shoreface parasequences at the top of the Ferron Sandstone illustrate the geometry due to a rapid base-level rise (also see Fig. 3).

Following maximum flooding, decreased rates of base-level rise are matched or slightly exceed sediment supply (Schumm *et al.*, 1987). This results in aggradational stratal geometries. Alluvial channels become increasingly isolated with an increasing proportion of preserved fine-grained sediment such that meanderbelt sandstones appear to 'float' in a matrix of alluvial plain deposits (Fig. 16D). These strata comprise the alluvial portion of the highstand systems tract and correspond with a period of shoreface aggradation. The isolated meanderbelts and aggradational shoreface parasequences of the John Henry Member illustrate both of these fluvial and nearshore marine relationships (Fig. 3).

DISCUSSION

The ability to predict fluvial architecture has important economic implications. Exploration and development of petroleum reservoirs, groundwater aquifers, and sediment-hosted mineral deposits is greatly aided by an understanding of the controls on alluvial architecture. Previous models, invoked to explain changes in alluvial architecture, commonly resorted to facies changes that were interpreted to reflect abrupt changes in factors such as tectonic subsidence and sediment supply. As a result of this approach, lithofacies within alluvial successions were interpreted broadly to follow Walther's Law, and regionally extensive, time-significant surfaces were not recognized (Warme & May, 1990). Our work in the Kaiparowits Plateau, however, suggests that regionally extensive, time-significant sequence-boundary unconformities can be recognized in alluvial strata and that sediment architecture may be predicted. Furthermore, we suggest that these models can be successfully applied to subsurface studies where data consist of widespread cores and well logs.

Studies of many petroleum reservoirs, hosted in fluvial deposits, reveal that they are frequently composed of alternating amalgamated, persistent channel sandstones with a relatively high net/gross reservoir sandstone value, and more isolated, poorly connected, laterally discontinuous sandstones with a low net/gross sandstone value (e.g. Johnson & Krol, 1984; Ravenne *et al.*, 1987; Atkinson *et al.*, 1988, 1990; Ravenne & Beucher, 1988; Brown & Richards, 1989; Helland-Hansen *et al.*, 1989; Livera, 1989; Miller *et al.*, 1990; McPherson & Miller, 1990). Furthermore, these fluvial deposits are generally thought to occur within a few tens of kilometres of their coeval marine basin. We suggest that sequence boundary unconformities may underlie many of these laterally amalgamated sandstones and that these sequence boundaries correspond to a basinward shift in depositional environments that might be better observed in a coastal position. The amalgamated sandstones themselves probably comprise the lower and middle portion of a transgressive systems tract. The more isolated channel sandstones probably correspond to late transgressive and early highstand systems tracts that are time equivalent to retrogradational and aggradational shoreface parasequences. Recognition of the sequence stratigraphic framework within an alluvial succession allows three-dimensional alluvial sandstone architecture to be predicted from limited one-dimensional data sets. These concepts are most readily applied to petroleum reservoir appraisal and development where an appreciation of alluvial architecture allows for more efficient planning of drilling programmes (e.g. Budding, 1988). In the later stages of reservoir life, for example, an improved understanding of alluvial architecture can greatly improve the construction of detailed reservoir models allowing for more successful secondary and tertiary recovery methods (Weber, 1986; Weber & van Geuns, 1990). Three-dimensional reservoir models can be produced with greater confidence from well data, and geostatistical and probabalistic models can be applied more intelligently by recognizing the chronostratigraphic architecture within fluvial successions.

CONCLUSIONS

Extensive, laterally amalgamated fluvial deposits overlie regionally significant sequence boundary unconformities. These sequence boundaries can be traced into shoreface strata and correspond to a basinward shift in depositional environments.

Amalgamated fluvial sandstones reflect aggradation during slowly rising base-level and correspond to early transgressive systems tract deposits. The transition from amalgamated deposits to more fine-grained isolated ones with an increasing proportion of preserved fine-grained strata and finally tidally-influenced strata correspond to coastal transgression and the development of a maximum flooding surface. Together, these deposits form the transgressive systems tract.

Thick sequences of isolated channel-belt sandstones and abundant fine-grained floodbasin strata are equivalent to aggradational shoreface strata and comprise the early highstand systems tract.

Correlation of individual nearshore marine parasequences into thick alluvial deposits that are several kilometres inland of a coeval shoreline *is not* presently possible. Correlation of parasequence-sets (terminology from Van Wagoner *et al.*, 1990), or systems tracts, however, *is* possible from nearshore strata to a position as much as 80 km inland of a coeval shoreline within an entirely alluvial succession.

The architecture of alluvial strata deposited within 100–150 km of a marine or lacustrine basin within a depositional sequence is largely controlled by changes in base-level and accommodation space. This strongly suggests that detailed, quantitative studies of alluvial architecture should be placed within a sequence-stratigraphic framework if they are to be of maximum utility. Failure to do so may reduce the effectiveness with which these studies can be applied to areas outside of where the study was conducted.

Consideration of fluvial architecture in terms of base-level and changes in accommodation space allows these models to be applied to a variety of tectonic settings throughout geologic time.

ACKNOWLEDGEMENTS

This research represents part of the senior author's PhD dissertation at the Colorado School of Mines, Golden, Colorado, USA. This research has been carried out in cooperation with the US Geological Survey, Branch of Coal Geology. Shanley gratefully acknowledges financial support as well as stimulating discussions from the following organizations: Shell Development Company, in particular Norm McIver and John Barwis; Koninklijke/Shell Exploratie en Productie Laboratorium, in particular Huw Williams, Marc Budding and Steve Flint (presently at the University of Liverpool, Department of Geology); Exxon Production Research Company, in particular John Van Wagoner and Gordon Moir; Canadian Hunter Exploration Ltd., in particular Mac Jervey and Jack Wendte; Marathon Oil Company, in particular Bill Ross, Jeff May, and Rich Spaw; Amoco Production Company, in particular Art Berman and Gary Taylor; and British Petroleum, in particular Mick Casey, Mike Mayall, and Gus Gustason. Financial support is also gratefully acknowledged from the Colorado School of Mines, Department of Geology, the Geological Society of America, the SIPES Foundation, and the Amoco Foundation. This paper benefitted significantly from reviews provided by Doug Cant, Bill Ross, and Paul Weimer. We are extremely grateful for their time and efforts.

REFERENCES

Alexander, J. & Leeder, M.R. (1987) Active tectonic control on alluvial architecture. In: *Recent Developments in Fluvial Sedimentology* (Ed. by F.G. Ethridge, R.M. Flores & M.D. Harvey), Spec. Publ. Soc. econ. Paleont. Miner., Tulsa, **39**, 243–252.

Allen, J.R.L. (1965) A review of the origin and characteristics of recent alluvial sediments. *Sedimentology*, **5**, 89–191.

Allen, J.R.L. (1978) Studies in fluviatile sedimentation: an exploratory quantitative model for the architecture of avulsion-controlled alluvial suites. *Sedim. Geol.*, **21**, 129–147.

Allen, J.R.L. (1979) Studies in fluviatile sedimentation: an elementary geometrical model for the connectedness of avulsion-related channel sand bodies. *Sedim. Geol.*, **24**, 253–267.

Armstrong, R.L. (1968) Sevier orogenic belt in Nevada and Utah. *Bull. geol. Soc. Am.*, **79**, 429–458.

Atkinson, C.D. (1986) Tectonic control on alluvial sedimentation as revealed by an ancient catena in the Capella Formation (Eocene) of northern Spain. In: *Paleosols, Their Recognition and Interpretation* (Ed. by V.P. Wright), pp. 139–179. Princeton University Press, Princeton.

Atkinson, C.D., Trumbley, P.N. & Kremer, M.C. (1988) Sedimentology and depositional environments of the Ivishak sandstone, Prudhoe Bay Field, North Slope, Alaska. In: *Giant Oil and Gas Fields, a Core Workshop* (Ed. by A.J. Lomando & P.M. Harris), Soc. econ. Paleont. Miner., Tulsa, Core Workshop, **12**, 561–613.

Atkinson, C.D., McGowen, J.H., Bloch, S., Lundell, L.L. & Trumbley, P.N. (1990) Braidplain and deltaic reservoir, Prudhoe Bay Field, Alaska. In: *Sandstone Petroleum Reservoirs—Casebooks in Earth Science* (Ed. by J.H. Barwis, J.G. McPherson, & J.R.J. Studlick), pp. 7–29. Springer-Verlag, New York.

Balsley, J.K. (1980) *Cretaceous Wave-Dominated Delta Systems, Book Cliffs, East-Central Utah*. Continuing Education course, field guide, Am. Ass. Petrol. Geol., Tulsa, 163 pp.

Blair, T.C. & Bilodeau, W.L. (1988) Development of tectonic cyclothems in rift, pull-apart, and foreland basins: sedimentary response to episodic tectonism. *Geology*, **16**, 517–520.

Blakey, R.C. & Gubitosa, R. (1984) Controls of sandstone body geometry and architecture in the Chinle Formation (Upper Triassic), Colorado Plateau. *Sedlm. Geol.*, **38**, 51–86.

Blum, M.D. (1990) Climatic and eustatic controls on Gulf Coastal Plain fluvial sedimentation: an example from the Late Quaternary of the Colorado River, Texas. Extended abstract—*Sequence Stratigraphy as an Exploration Tool, Concepts and Practices in the Gulf Coast*, Gulf Coast Section, Soc. econ. Paleont. Miner., Eleventh Annual Research Conference, pp. 71–83.

Bobb, M.C. & Ryer, T.A. (1990) Upper Cretaceous Calico bed, southwestern Utah: sedimentology, sea level and paleotectonics. *Bull. Am. Assoc. Petrol. Geol.* **74**, 1316 (Abstract).

Bridge, J.S. (1985) Paleochannel patterns inferred from alluvial deposits: a critical evaluation. *J. Sedim. Petrol.*, **5**, 579–589.

Bridge, J.S. & Leeder, M.R. (1979) A simulation model of alluvial stratigraphy. *Sedimentology*, **26**, 617–644.

Brown, S. & Richards, P.C. (1989) Facies and development of the Middle Jurassic Brent delta near the northern limit of its progradation, UK North Sea. In: *Deltas: Sites and Traps for Fossil Fuels* (Ed. by M.K.G. Whateley & K.T. Pickering), Spec. Publ. geol. Soc. London, Blackwell Scientific Publications, Oxford, **41**, 253–267.

Budding, M.C. (1988) Prediction of sediment geometries in sandstone reservoirs. *Quantification of Sediment Body Geometries and their Internal Heterogeneities*. (British Sedimentological Research Group meeting, convened by M. Mayall and T. Astin), London, March 1–2, 1988 (Abstract).

Butcher, S.W. (1990) The nickpoint concept and its implications regarding onlap to the stratigraphic record. In: *Quantitative Dynamic Stratigraphy* (Ed. by T.A. Cross), pp. 375–385. Prentice Hall, Englewood Cliffs.

BUTLER, P.R. (1984) Fluvial response to on-going tectonism and base level changes, lower Amargosa River, southern Death Valley, California. *Sedim. Geol.*, **38**, 107–125.

CANT, D.J. (1982) Fluvial facies models. In: *Sandstone Depositional Environments* (Ed. by P.A. Scholle & D. Spearing), Mem. Am. Ass. Petrol. Geol., Tulsa, **31**, 115–137.

CARMALT, S.W. & ST. JOHN, B. (1986) Giant oil and gas fields. In: *Future Petroleum Provinces of the World* (Ed. by M.T. Halbouty), Mem. Am. Ass. Petrol. Geol., Tulsa, **40**, 11–53.

CLAREY, T. & TUNBRIDGE, I. (1989) Geometry and architecture of a Triassic sheet sandstone. *Fourth International Conference on Fluvial Sedimentology*, Sitges, Spain, Abstract Volume, 99 pp.

COLLINSON, J.D. (1978) Vertical sequence and sand body shape in alluvial sequences. In: *Fluvial Sedimentology* (Ed. by A.D. Miall), Mem. Can. Soc. Petrol. Geol., Calgary, **5**, 577–586.

COLLINSON, J.D. (1986) Alluvial sediments. In: *Sedimentary Environments and Facies*, 2nd edn (Ed. by H.G. Reading), pp. 20–62. Blackwell Scientific Publications, Oxford.

COLLINSON, J.D. & LEWIN, J. (eds) (1983) *Modern and Ancient Fluvial Systems*. Spec. Publs int. Ass. Sediment. Blackwell Scientific Publications, Oxford, **6**, 575 pp.

CONDIT, W., DRAKE, C.L., MAYER, L., & SPYDELL, R. (1978) Sedimentation in Lake Powell. *Lake Powell Research Project Bulletin*, **64**, National Science Foundation.

CROSS, T.A. (1986) Tectonic controls of foreland basin subsidence and Laramide style deformation, western United States. In: *Foreland Basins* (Ed. by P.A. Allen and P. Homewood), Spec. Publs int. Ass. Sediment., Blackwell Scientific Publications, Oxford, **8**, 15–39.

CURRAY, J.R. (1956) The analysis of two-dimensional orientation data. *J. Geol.*, **64**, 117–131.

DE WIT, M.C.J. (1989) Coarse-grained Plio-Pleistocene fluvial sediments and vertical facie changes as indicators of climatic change at Vanwyksvlei, Boesmanland, South Africa. *Fourth International Conference on Fluvial Sedimentology*, Sitges, Spain, Abstract Volume, 120 pp.

EAKIN, H.M. (1936) *Silting of Reservoirs*, United States Department of Agriculture, Technical Bulletin **524**.

EKDALE, A.A., BROMLEY, R.G., & PEMBERTON, S.G. (1984) *Ichnology—The Use of Trace Fossils in Sedimentology and Stratigraphy*. Soc. econ. Paleont. Miner., Short Course **15**, Tulsa, 317 pp.

ETHRIDGE, F.G. & FLORES, R.M. (eds) (1981) *Recent and Ancient Nonmarine Depositional Environments—Models for Exploration*. Spec. Publ. Soc. econ. Paleont. Miner., Tulsa, **31**, 349 pp.

ETHRIDGE, F.G., FLORES, R.M. & HARVEY, M.D. (eds) (1987) *Recent Developments in Fluvial Sedimentology—Contributions from the Third International Fluvial Sedimentology Conference*. Spec. Publ. Soc. econ. Paleont. Miner., Tulsa, **39**, 389 pp.

FELTON, E.A. (1989) Evolution of Early Cretaceous braided fluvial depositional systems, eastern Otway Basin, Southern Australia. *Fourth International Conference on Fluvial Sedimentology*, Sitges, Spain, Abstract Volume, 128 pp.

FERRARI, R.L. (1988) *1986 Lake Powell Survey*. United States Department of Interior, Bureau of Reclamation, REC-ERC-88-6.

FOLK R.L. (1974) *Petrology of Sedimentary Rocks*. Hemphill Publishing Co., Austin, Texas, 182 pp.

FRIEND, P.F. (1983) Towards the field classification of alluvial architecture or sequence. In: *Modern and Ancient Fluvial Systems* (Ed. by J.D. Collinson & J. Lewin), Spec. Publs int. Ass. Sediment., Blackwell Scientific Publications, Oxford, **6**, 345–354.

FRIEND, P.F., SLATER, M.J. & WILLIAMS, R.C. (1979) Vertical and lateral building of river sandstone bodies, Ebro Basin, Spain, *J. geol. Soc. London*, **136**, 39–46.

GALLOWAY, W.E. & HOBDAY, D.K. (1983) *Terrigenous Clastic Depositional Systems—Applications to Petroleum, Coal, and Uranium Exploration*. Springer-Verlag, New York, 423 pp.

GELFENBAUM, G. (1983) Suspended-sediment response to semidiurnal and fortnightly tidal variations in a mesotidal estuary: Columbia River, USA. *Mar. Geol.*, **52**, 39–57.

GORDON, E.A. & BRIDGE, J.S. (1987) Evolution of Catskill (Upper Devonian) river systems: intra- and extrabasinal controls. *J. sedim. Petrol.*, **57**, 234–249.

GRAHAM, J.R., JAMES, A. & RUSSEL, K.J. (1989) Basin history deduced from subtle changes in fluvial style: an example from the Upper Devonian of SW. Ireland. *Fourth International Conference on Fluvial Sedimentology*, Sitges, Spain, Abstract Volume, 141 pp.

GREGORY, H.E. & MOORE, R.C. (1931) *The Kaiparowits Region, a Geographic and Geologic Reconnaissance of Parts of Utah and Arizona*. Prof. Pap. U.S. geol. Surv., **164**, 161 pp.

HALL, S.A. (1990) Channel trenching and climatic change in the southern U.S. Great Plains. *Geology*, **18**, 342–345.

HAQ, B.U., HARDENBOL, J. & VAIL, P.R. (1988) Mesozoic and Cenozoic chronostratigraphy and cycles of sea-level change. In: *Sea Level Changes: an Integrated Approach* (Ed. by C.K. Wilgus, B.S. Hastings, C.G. St. C. Kendall, H.W. Posamentier, C.A. Ross & J.C. Van Wagoner), Spec. Publs Soc. econ. Paleont. Miner., Tulsa, **42**, 71–108.

HELLAND-HANSON, W., STEEL, R., NAKAYAMA, K. & KENDALL, C.G. St. C. (1989) Review and computer modelling of the Brent Group stratigraphy. In: *Deltas: Sites and Traps for Fossil Fuels* (Ed. by M.K.G. Whateley & K.T. Pickering), Spec. Publ. geol. Soc. London, Blackwell Scientific Publications, Oxford, **41**, 237–252.

HELLER, P.L., ANGEVINE, C.L., WINSLOW, N.S. & PAOLA, C. (1988) Two-phase stratigraphic model of foreland basin sequences. *Geology*, **16**, 501–504.

HETTINGER, R.D., MCCABE, P.J. & SHANLEY, K.W. (1990) Clastic facies architecture associated with a major marine transgression—the mid-Cretaceous of the Kaiparowits Plateau, Utah. *Bull. Am. Assoc. Petrol. Geol.*, **74**, 1327 (Abstract).

JERVEY, M.T. (1988) Quantitative geological modeling of siliciclastic rock sequences and their seismic expressions. In: *Sea Level changes: an Integrated Approach* (Ed. by C.K. Wilgus, B.S. Hastings, C.G. St. C. Kendall, H.W. Posamentier, C.A. Ross & J.C. Van Wagoner), Spec. Publ. Soc. econ. Paleont, Miner., Tulsa, **42**, 47–69.

JOHNSON, H.D. & KROL, D.E. (1984) Geological modeling

of a heterogeneous sandstone reservoir: Lower Jurassic Statfjord Formation, Brent Field. *SPE Paper* **13050**.

JOHNSON, S.Y. (1984) Cyclic fluvial sedimentation in a rapidly subsiding basin, northwestern Washington, *Sedim. Geol.*, **38**, 361–391.

JORDAN, T.E. (1981) Thrust loads and foreland basin evolution, Cretaceous, western United States. *Bull. Am. Assoc. Petrol. Geol.*, **65**, 2506–2520.

JORDAN, T.E., FLEMINGS, P.B. & BEER, J.A. (1988) Dating thrust fault activity by use of foreland basin strata. In: *New Perspectives in Basin Analysis* (Ed. by K.L. Kleinspehn & C. Paola), pp. 307–330. Springer-Verlag, New York.

KOSS, J.E., ETHRIDGE, F.G. & SCHUMM, S.A. (1990) Effects of base-level change on coastal plain and shelf systems—an experimental approach. *Bull. Am. Assoc. Petrol. Geol.*, **74**, 697 (Abstract).

KRAUS, M.J. & MIDDLETON, L.T. (1987) Contrasting architecture of two alluvial suites in different structural settings. In: *Recent Developments in Fluvial Sedimentology, Contributions from the Third International Fluvial Sedimentology Conference* (Ed. by F.G. Ethridge, R.M. Flores & M.D. Harvey), Spec. Publ. Soc. econ. Paleont. Miner., Tulsa, **39**, 253–262.

LARA, J.M. (1972) *The 1969 Elephant Butte Reservoir Sediment Survey*. United States Department of Interior, Bureau of Reclamation, REC-ERC-72-13, 56 pp.

LARA, J.M. (1983) *Elephant Butte reservoir, 1980 Sedimentation Survey*. United States Department of Interior, Bureau of Reclamation, Sedimentation and River Hydraulics Section, Denver, Colorado.

LARA, J.M. & SANDERS, J.I. (1970) *The 1963–64 Lake Mead Survey*. United States Department of Interior, Bureau of Reclamation, REC-OCE-70-21, 155 pp.

LAWRENCE, D.T., DOYLE, M. & AIGNER, T. (1990) Stratigraphic simulation of sedimentary basins: concepts and calibrations. *Bull. Am. Assoc. Petrol. Geol.*, **74**, 273–295.

LAWTON, T.F. (1985) Style and timing of frontal structures, thrust belt, central Utah. *Bull. Am. Assoc. Petrol. Geol.*, **69**, 1145–1159.

LEEDER, M.R. (1978) A quantitative stratigraphic model for alluvium with special reference to channel deposit density and interconnectedness. In: *Fluvial Sedimentology* (Ed. by A.D. Miall), Mem. Can. Soc. Petrol. Geol., **5**, 587–596.

LEOPOLD, L.B., WOLMAN, M.G. & MILLER, J.P. (1964) *Fluvial Processes in Geomorphology*. WH. Freeman and Co., San Francisco, 522 pp.

LINK, M.H. (1984) Fluvial facies of the Miocene Ridge Route Formation, Ridge Basin, California. *Sedim. Geol.*, **38**, 263–285.

LIVERA, S.A. (1989) Facies associations and sand-body geometries in the Ness Formation of the Brent Group, Brent Field. In: *Deltas: Sites and Traps for Fossil Fuels* (Ed. by M.K.G. Whateley & K.T. Pickering), Spec. Publ. geol. Soc. London, Blackwell Scientific Publications, Oxford, **41**, 269–286.

LUCCHITTA, I. (1979) Late Cenozoic uplift of the southwestern Colorado Plateau and adjacent lower Colorado River region. *Tectonophys.*, **61**, 63–95.

LUCCHITTA, I. (1989) Overflight of the tectonic boundary between the Colorado Plateau and the Basin Range provinces—IGC Field Trip T116/389. *Sedimentation and Tectonics of Western North America*, vol. 3. Publications for the 28th International Geological Congress, Washington, D.C., July, 1989, American Geophysical Union, Washington, D.C., pp. T116/389: 1–21.

LYONS, J. & RANDLE, T. (1988) *Platte River Channel Characteristics in the Big Bend Reach*. United States Department of the Interior, Bureau of Reclamation, Sedimentation and River Hydraulics Section, Denver, Colorado.

MACKIN, J.H. (1948) Concept of the graded river. *Bull. geol. Soc. Am.*, **59**, 463–512.

MARZO, M., NIJMAN, W. & PUIGDEFABREGAS, C. (1988) Architecture of the Castissent fluvial sheet sandstones, Eocene, South Pyrenees, Spain. *Sedimentology*, **35**, 719–738.

MCCABE, P.J. (1984) Depositional environments of coal and coal-bearing strata. In: *Sedimentology of Coal and Coal-bearing Sequences* (Ed. by R.A. Rahmani & R.M. Flores), Spec. Publs int. Ass. Sediment. Blackwell Scientific Publications, Oxford, **7**, 13–42.

MCCAVE, I.N. (1969) Correlation of marine and nonmarine strata with examples from Devonian of New York State. *Bull. Am. Assoc. Petrol. Geol.*, **53**, 155–162.

MCPHERSON, J.G. & MILLER, D.D. (1990) A subsurface study of fluvial sand-body geometries and connectedness. *Sediments 1990, 13th International Sedimentological Congress, Nottingham, England*, Abstracts volume (papers), pp. 349–350.

MEYERS, J.H., SUTTNER, L.J., SOREGHAN, M. & MIDDLEMAN, B. (1989) Architecture of fluvial sandstone complexes and partitioning of the Early Cretaceous Rocky Mountain foreland basin, central Wyoming. *Fourth International Conference on Fluvial Sedimentology*, Sitges, Spain, Abstract Volume, 182 pp.

MIALL, A.D. (ed.) (1978) *Fluvial Sedimentology*, Mem. Can. Soc. Petrol. Geol., **5**, 859 pp.

MIALL, A.D. (1981) *Analysis of Fluvial Depositional Systems*. Am. Ass. Petrol. Geol., Education Course Note Series **20**, 75 pp.

MILLER, D.D., MCPHERSON, J.G. & COVINGTON, T.E. (1990) Fluviodeltaic reservoir, South Belridge Field, San Joaquin Valley, California. In: *Sandstone Petroleum Reservoirs—Casebooks in Earth Science* (Ed. by J.H. Barwis, J.G. McPherson & J.R.J. Studlick), pp. 109–130. Springer-Verlag, New York.

MITCHUM, R.M. JR., VAIL, P.R. & THOMPSON III, S. (1977) Seismic stratigraphy and global changes of sea level, part 2: the depositional sequence as a basic unit for stratigraphic analysis. In: *Seismic Stratigraphy—Applications to Hydrocarbon Exploration* (Ed. by C.E. Payton), Mem. Am. Assoc. Petrol. Geol., Tulsa, **26**, 53–62.

NAMI, M. & LEEDER, M.R. (1978) Changing channel morphology and magnitude in the Scalby Formation (M. Jurassic) of Yorkshire, England. In: *Fluvial Sedimentology* (Ed. by A.D. Miall), Mem. Can. Soc. Petrol. Geol., **5**, 431–440.

OLSEN, H. (1990) Astronomical forcing of meandering river behavior: Milankovitch cycles in Devonian of east Greenland. *Palaeogeogr. Palaeoclim. Palaeoecol.*, **79**, 99–115.

ORVIS, C.J. (1989) *Elephant Butte Reservoir 1988 Sedimentation Survey*. United States Department of Interior,

Bureau of Reclamation, Sedimentation Section, Denver, Colorado.

PETERSON, F.E. (1969a) *Cretaceous Sedimentation and Tectonism in the Southeastern Kaiparowits Region, Utah.* United States Geological Survey, Open-File Report 1314, 259 pp.

PETERSON, F.E. (1969b) *Four New Members of the Upper Cretaceous Straight Cliffs Formation in the Southeastern Kaiparowits Region, Kane County, Utah.* Bull. U.S. geol. Surv., **1274–J**, pp. J1–J28.

PLINT, A.G. (1983) Sandy fluvial point-bar sediments from the Middle Eocene of Dorset, England. In: *Modern and Ancient Fluvial Systems* (Ed. by J.D. Collinson & J. Lewin), Spec. Publs int. Ass. Sediment., **6**, 355–368.

POSAMENTIER, H.W. (1988) Fluvial deposition in a sequence stratigraphic framework. In: *Sequences, Stratigraphy, Sedimentology: Surface and Subsurface* (Ed. by D.P. James & D.A. Leckie), Mem. Can. Soc. Petrol. Geol., **15**, 582–583.

POSAMENTIER, H.W., JERVEY, M.T. & VAIL, P.R. (1988) Eustatic controls on clastic deposition I—Conceptual framework. In: *Sea Level Changes: an Integrated Approach* (Ed. by C.K. Wilgus, B.S. Hastings, C.G. St. C. Kendall, H.W. Posamentier, C.A. Ross & J.C. Van Wagoner), Spec. Publ. Soc. econ. Paleont. Miner., Tulsa, **42**, 109–124.

POSAMENTIER, H.W. & VAIL, P.R. (1988) Eustatic controls on clastic deposition II—Sequence and systems tract models. In: *Sea Level Changes: an Integrated Approach* (Ed. by C.K. Wilgus, B.S. Hastings, C.G. St. C. Kendall, H.W. Posamentier, C.A. Ross & J.C. Van Wagoner), Spec. Publ. Soc. econ. Paleont. Miner., Tulsa, **42**, 125–154.

PUIGDEFABREGAS, C., COLLINSON, J., CUEVAS, J.L., DREYER, T., MARZO, M., MELLERE, D., MERCADE, L., MUNOZ, J.A., NIJMAN, W. & VERGES, J. (1989) Alluvial deposits of the successive foreland basin stages and their relation to the Pyrenean thrust sequences. In: *Guidebook Series of the Fourth International Conference on Fluvial Sedimentology, Convened in Sitges, Spain* (Ed. by M. Marzo & C. Puigdefabregas). Publication of the Geological Survey of Catalunya, Spain, 175 pp.

RAHMANI, R.A. (1988) Estuarine tidal channel and nearshore sedimentation of a Late Cretaceous epicontinental sea, Drumheller, Alberta, Canada. In: *Tide-Influenced Sedimentary Environments and Facies* (Ed. by P.L. de Boer, A. van Gelder & S.D. Nio), pp. 433–474, D. Reidel, Dordrecht.

RAVENNE, C. & BEUCHER, H. (1988) Recent development in description of sedimentary bodies in a fluvio-deltaic reservoir and their 3D conditional simulations. *SPE Paper* **18310**.

RAVENNE, C., ESCHARD, R., GALLI, A., MATHIER, Y., MONTADERT, L. & RUDKIEWICZ, J.L. (1987) Heterogeneities and geometry of sedimentary bodies in a fluvio-deltaic reservoir. *SPE Paper* **16752**.

ROSS, W.C. (1990) Modeling base-level dynamics as a control on basin-fill geometries and facies distribution: a conceptual framework. In: *Quantitative Dynamic Stratigraphy* (Ed. by T.A. Cross), pp. 387–399. Prentice Hall, Englewood Cliffs.

RUST, B.R. (1978) A classification of alluvial channel systems. In: *Fluvial Sedimentology* (Ed. by A.D. Miall), Mem. Can. Soc. Petrol. Geol., **5**, 187–198.

SCHUMM, S.A. (1963) *A Tentative Classification of Alluvial River Channels*. Circular U.S. geol. Surv., **477**.

SCHUMM, S.A. (1977) *The Fluvial System*. John Wiley and Sons, New York, 338 pp.

SCHUMM, S.A. (1981) Evolution and response of the fluvial system, sedimentologic implications. In: *Recent and Ancient Nonmarine Depositional Environments: Models for Exploration* (Ed. by F.G. Ethridge & R.M. Flores), Spec. Publ. Soc. econ. Paleont. Miner., Tulsa, **31**, 19–29.

SCHUMM, S.A. & BRAKENRIDGE, G.R. (1987) River responses. In: *The Geology of North America, vol. K-3: North America and the Adjacent Oceans during the Last Deglaciation* (Ed. by W.F. Ruddiman & H.E. Wright, Jr.), pp. 221–240. Geological Society of America, Boulder, Colorado.

SCHUMM, S.A., MOSLEY, M.P. & WEAVER, W.E. (1987) *Experimental Fluvial Geomorphology*. John Wiley and Sons, New York, 413 pp.

SHANLEY, K.W. (1991) Sequence stratigraphic relationships and facies architecture of Turonian–Campanian strata, Kaiparowits Plateau, south-central Utah. PhD dissertation, Department of Geology and Geological Engineering, Colorado School of Mines, Golden, Colorado, USA, 390 pp.

SHANLEY, K.W. & MCCABE, P.J. (1989) Sequence stratigraphic relationships and facies architecture of Turonian–Campanian strata, Kaiparowits Plateau, south-central Utah. *Bull. Am. Assoc. Petrol. Geol.*, **73**, 410–411.

SHANLEY, K.W. & MCCABE, P.J. (1990) Tidal influence in fluvial strata—a key element in high-resolution sequence stratigraphic correlation. *Bull. Am. Assoc. Petrol. Geol.*, **74**, 762.

SHANLEY, K.W. & MCCABE, P.J. (1991) Predicting facies architecture through sequence stratigraphy—an example from the Kaiparowits Plateau, Utah. *Geology*, **19**, 742–745.

SHUSTER, M.W. & STEIDTMANN, J.R. (1987) Fluvial-sandstone architecture and thrust-induced subsidence, northern Green River Basin, Wyoming. In: *Recent Developments in Fluvial Sedimentology, Contributions from the Third International Fluvial Sedimentology Conference* (Ed. by F.G. Ethridge, R.M. Flores & M.D. Harvey), Spec. Publ. Soc. econ. Paleont. Miner., Tulsa, **39**, 279–285.

SLOSS, L.L. (1962) Stratigraphic models in exploration. *Bull. Am. Assoc. Petrol. Geol.*, **46**, 1050–1057.

SMITH, N.D. (1971) Pseudo-planar stratification produced by very low amplitude sand waves. *J. sedim. Petrol.*, **41**, 69–73.

STEVENS, J.C. (1936) The silt problem. *Transactions of the American Society of Civil Engineers*, **101**, 207–288.

STEVENS, J.C. (1946) Future of Lake Mead and Elephant Butte reservoir. *Transactions of the American Society of Civil Engineers*, **111**, 1231–1342.

STRAND, R.I. (1968) *Behaviour of the Middle Loup River Channel as Influenced by the Milburn Diversion Dam*. United States Department of the Interior, Bureau of Reclamation, Sedimentation Branch, Denver, Colorado.

SWIFT, D.J.P. (1970) Quaternary shelves and the return to

grade. *Mar. Geol.*, **8**, 5–30.

TURNER, J.P. (1989) Evolving alluvial stratigraphy and thrust front development in the west Jaca piggyback basin, Spanish Pyrenees. *Fourth International Conference on Fluvial Sedimentology*, Sitges, Spain, Abstract Volume, p. 235.

TURNER-PETERSON, C.E. (1986) Fluvial sedimentology of a major uranium-bearing sandstone—A study of the Westwater Canyon Member of the Morrison Formation, San Juan Basin, New Mexico. In: *A Basin Analysis Case Study: the Morrison Formation Grants Uranium Region, New Mexico* (Ed. by C.E. Turner-Peterson, E.S. Santos & N.S. Fishman), Am. Ass. Petrol. Geol., Tulsa, Studies in Geology **22**, 47–75.

VAIL, P.R., MITCHUM JR., R.M., TODD, R.G., WIDMIER, J.M., THOMPSON III, S., SANGREE, J.B., BUBB, J.N. & HATLELID, W.G. (1977) Seismic stratigraphy and global changes in sea level, parts 1–11. In: *Seismic Stratigraphy—Applications to Hydrocarbon Exploration* (Ed. by C.E. Payton), Mem. Am. Ass. Petrol. Geol., Tulsa, **26**, 49–212.

VAN WAGONER, J.C. (1985) *Reservoir Facies Distribution as Controlled by Sea-level Change*. Soc. econ. Paleont. Miner., Annual Midyear Meeting, Golden, Colorado, Abstracts **2**, 91.

VAN WAGONER, J.C., MITCHUM JR., R.M., CAMPION, K.M. & RAHMANIAN, V.D. (1990) *Siliciclastic Sequence Stratigraphy in Well Logs, Cores, and Outcrops: Concepts for High-resolution Correlation of Time and Facies*. Am. Ass. Petrol. Geol., Tulsa, Methods in Exploration Series **7**, 55 pp.

VANINETTI, G.E. (1978) Coal stratigraphy of the John Henry Member of the Straight Cliffs Formation, Kaiparowits Plateau, Utah. Unpublished M.Sc. Thesis, University of Utah, Salt Lake City, Utah.

VILLIEN, A. & KLIGFIELD, R.M. (1986) Thrusting and synorogenic sedimentation in central Utah. In: *Paleotectonics and Sedimentation in the Rocky Mountain Region, United States* (Ed. by J.A. Peterson), Mem. Am. Ass. Petrol. Geol., Tulsa, **41**, 281–307.

WALKER, R.G. (ed.) (1984) *Facies Models*, 2nd edn. Geosciences Canada Reprint Series 1. Geological Association of Canada, Toronto, 317 pp.

WALKER, R.G. (1990) Facies modeling and sequence stratigraphy. *J. sedim. Petrol.*, **60**, 777–786.

WARME, J.E. & MAY, J.A. (1990) Sequence boundaries versus facies changes: correlation alternatives for the same rocks. *Sediments 1990, 13th International Sedimentological Congress, Nottingham, England*, Abstracts volume (papers), pp. 589–590.

WEBER, K.J. (1986) How heterogeneity affects oil recovery. In: *Reservoir Characterization* (Ed. by L.W. Lake & H.B. Carroll Jr.), pp. 487–544, Academic Press, New York.

WEBER, K.J. & VAN GEUNS, L.C. (1990) Framework for constructing clastic reservoir simulation models. *J. Petrol. Technol.*, **42**, 1248–1297.

WILLIAMS, G.D. & STELCK, C.R. (1975) Speculations on the Cretaceous paleogeography of North America. In: *The Cretaceous System in the Western Interior of North America* (Ed. by W.G.E. Caldwell), Spec. Pap. geol. Ass. Can., **13**, 1–20.

WILTSCHKO, D.V. & DORR, J.A. (1983) Timing of deformation in overthrust belt and foreland of Idaho, Wyoming, and Utah. *Bull. Am. Assoc. Petrol. Geol.*, **67**, 1304–1322.

Spec. Publs int. Ass. Sediment. (1993) **15**, 57–80

Sedimentary architecture of field analogues for reservoir information (SAFARI): a case study of the fluvial Escanilla Formation, Spanish Pyrenees

T. DREYER*, L.-M. FÄLT†, T. HØY**, R. KNARUD‡, R. STEEL§ *and* J.-L. CUEVAS¶

**Norsk Hydro Research Centre, Sandsliveien 90, N-5020, Bergen, Norway;*
† *Den norske stats oljeselskap a.s. (Statoil), Geolab, PO Box 300, N-4001 Stavanger, Norway;*
***Norwegian Petroleum Directorate, PO Box 600, N-4001 Stavanger, Norway;*
‡*Saga Petroleum a.s., Kjørboveien 16, PO Box 490, N-1301 Sandvika, Norway;*
§*University of Bergen, Geological Institute, Dep. A, Allegaten 41, N-5007 Bergen, Norway;*
¶*Servei Geologic de Catalunya, Diputacio 92, 08015 Barcelona, Spain*

ABSTRACT

Quantified information on the geometries of sandbodies and their internal heterogeneities is a necessary prerequisite for the realistic modelling of reservoirs and their behaviour during production. A database for sedimentary architectures of field analogues for reservoir information (SAFARI) is being set up, based on the analysis of critical problems within reservoirs of the Norwegian continental shelf.

The Upper Eocene Escanilla Formation of the Ainsa Basin in the Spanish Pyrenees was selected as one of several analogue fluvial study objects. Excellent outcrop conditions allowed continuous tracing of sedimentary units over large areas, and sandbody geometries could be studied in three dimensions. Data documenting the large-scale sedimentary architecture and smaller scale sandbody heterogeneity patterns were collected, along with various point-related measurements and general characteristics of the formation. Principles of bounding surface hierarchies and classification of architectural elements were adopted to ensure systematic data handling.

The Escanilla Formation has a maximum preserved thickness of 750 m in the study area, and contains sediments formed in medial/distal alluvial fan to axial fluvial system settings. The coarse member deposits, which constitute 30–55% of the formation, consist of a number of different channel/channel-belt types, the most common ones being channel- and sheet-braided bodies, coarse-grained point bar bodies and deeply incised trunk stream sandbodies. The fine member deposits are characterized by crevasse splays and immature palaeosols. Deposition occurred in response to thrust activity and was influenced by differential subsidence in a foreland basin.

The outcrop data are stored in the SAFARI database system, where they may be analysed in order to quantify critical parameters such as sandbody geometries and interconnectedness, channel-body orientation and flow-unit distribution.

INTRODUCTION

The SAFARI project

The 'sedimentary architectures of field analogues for reservoir information' (SAFARI) project was initiated in 1988 as part of the Norwegian efforts in improved oil recovery, a task now given high priority on the Norwegian continental shelf. The project focuses on improved geological reservoir description, part of the basis for improved reservoir simulation and management.

The general effort towards improved reservoir description includes the development of mathematical programmes for geological modelling, for example GEOPROBE, SESIMIRA, (Gundesø &

Egeland, 1990; Damsleth *et al.*, 1990), FLUREMO and DESIRE (Damsleth *et al.*, 1990). The geomathematical models, which take account of reservoir heterogeneity during simulated production, are now commonly used in connection with various improved recovery methods. Application of such models have, however, revealed an acute lack of quantitative data concerning heterogeneities of varying types and at various scales.

Reservoir heterogeneities may be of depositional, diagenetic or tectonic origin. SAFARI focuses on macroscopic depositional inhomogeneities, such as those that generate high permeability zones and fluid-flow barriers in reservoirs.

The SAFARI work is organized as a cooperative effort between the Norwegian Petroleum Directorate, Saga, Statoil and Norsk Hydro. This cooperation both reduces costs and improves the quality of the resulting product. Figure 1 illustrates the role of SAFARI within the reservoir evaluation process.

SAFARI aims

The main aim of the SAFARI project is to build a database for quantitative information concerning sedimentary heterogeneities in reservoirs, largely through the documentation of sedimentary bodies and boundaries in sequences from various depositional environments. Study objects are well-exposed, recent and ancient reservoir analogue-sequences, chosen with reference to Norwegian Shelf analogues. The main reasons for collecting such data are:

1 to reduce uncertainty in well-to-well correlation at a stage when well data are limited;
2 to create terms of reference for conceptual geological modelling;
3 to generate basic data for geomathematical modelling of reservoirs.

SAFARI activities

The main activities of the SAFARI project are *reviews of critical problems in Norwegian reservoirs*, the *collection of data sets* from selected outcrop analogues and satellite images of modern fluvial systems and the *construction of a computer database system*.

Reservoir data reviews and outcrop data collection have so far concerned Norwegian fluvial and shallow-marine reservoir sequences. Selected outcrop analogues in Utah and in the Spanish Pyrenees have have been studied, and field work standards and procedures were established for each environment prior to data collection.

Table 1. The different data types included in the standard SAFARI field analogue database. Each data type can be assigned a number of different attributes

General information
1 Basin type
2 Formation
3 Lithofacies
4 Architectural elements
5 Correlation markers
6 Stratigraphy
7 Maps

Geographically localized information
8 Cross-sections
9 Bodies
10 Bounding surfaces
11 Samples
12 Palaeocurrents
13 Photographs
14 Vertical profiles

Table 2. Fluvial reservoirs on the Norwegian continental shelf (completely or partly fluvial)

Field	Fluvial formation
Brage	Statfjord
Frøy	Sleipner
Gullfaks	Ness, Statfjord
Gullfaks Sør	Ness, Statfjord, Lunde
Heidrun	Åre
Hild	Ness
Huldra	Ness
Murchison	Ness, Statfjord
Oseberg	Ness
Sleipner Satelitter	Sleipner
Sleipner Vest	Sleipner
Snorre	Lunde, Statfjord
Snorre Vest	Ness
Statfjord	Ness, Statfjord
Statfjord Nord	Ness
Statfjord Øst	Ness
Veslefrikk	Ness, Statfjord
25/2–4	Sleipner
30/6 Beta	Ness
30/6 Kappa	Statfjord
30/9 Omega	Ness
30/9–6	Ness
34/7–B	Ness
34/8–1	Ness
34/10 Beta	Ness
34/10 Gamma	Ness, Statfjord
35/11–2	Ness
35/8–1	Ness
35/8–2	Ness
7122/6–1	Snadd

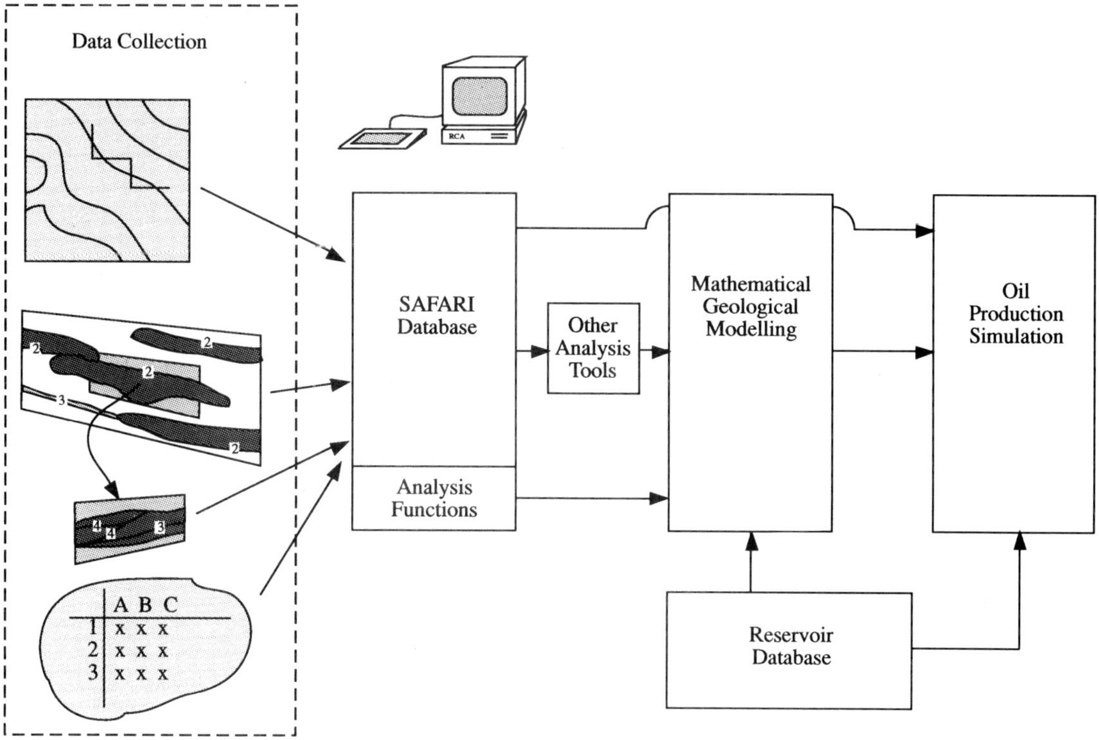

Fig. 1. The role of the SAFARI database within the reservoir evaluation process.

A computer-based system for registration, storage, retrieval and interactive analysis of field data has been developed. The system's ability to handle two-dimensional outcrop data is an important feature. The system is developed for a workstation, and is based on an ARC/INFO geographical information system linked to a database (Giffen *et al.*, 1990).

Table 1 illustrates the data types of the standard database used by SAFARI. Aspects of the database are examplified below in relation to a field study of the Escanilla Formation.

Norwegian fluvial reservoirs

The data collection from fluvial field analogues was preceded by an inventory of fluvial reservoir sequences on the Norwegian continental shelf and a review of critical problems associated with these. This was done to provide a basis for the selection of suitable field analogues and for the definition of parameters and data categories to be collected in the field.

The discovery of the gigantic Statfjord Field in 1974 established Triassic and Jurassic deltaic and fluvial reservoirs as an important play type on the Norwegian shelf. In the northern part of the North Sea, a considerable amount of oil and gas has been discovered in the fluvial Ness, Statfjord and Lunde Formations (Table 2). These formations include a wide range of different fluvial environments; from lower delta plain in parts of the Ness Formation to braided and coarse-grained meandering systems in the Lunde and Statfjord Formations. In the southern part of the North Sea, smaller amounts of hydrocarbons are proven in the Sleipner Formation, which contains coastal plain sediments partly similar to the Ness Formation. On the Haltenbanken, offshore mid-Norway, most of the hydrocarbon resources have been discovered in shallow marine strata. Only a small amount of oil and gas is proven in fluvial reservoirs in the Lower Jurassic Åre Formation. In block 7121/06 in the Barents Sea, a small gas discovery was made in the upper, fluvial part of the Snadd Formation.

The uncertainties associated with resource estimates and recovery factors are often more pronounced in fluvial reservoir sequences than in other, more homogeneous reservoir types. Through careful reservoir description and modelling, there is great potential for improving reliability of resource estimates and recovery factors. This has been an important incentive for the selection of fluvial deposits as pilot study objects within the SAFARI project.

The evaluation of critical factors associated with fluvial reservoirs on the Norwegian continental shelf revealed that collection of analogue data from braided stream systems was in most demand. The analogue data are primarily intended to support the geological models for the Upper Lunde and Statfjord Formations in the Statfjord, Gullfaks, Snorre and other fields (Table 2). A comparison of the sedimentary characteristics of these formations and numerous outcropping alluvial sequences resulted in the selection of the Escanilla Formation in the Spanish Pyrenees (described below) as one of several field study objects.

ESCANILLA FORMATION: GEOLOGICAL SETTING AND MAIN CHARACTERISTICS

Stratigraphy and depositional setting

The Escanilla Formation (Garrido-Mejias, 1968) forms the upper part of a major shallowing-upward sequence that was deposited in the southern Pyrenean foreland basin during the Eocene (Fig. 2; Mutti *et al.*, 1988). It belongs to the Campodarbe Group, which is a thick and laterally extensive fluvial-dominated unit formed in response to the main phase of thrust-sheet emplacement in the Pyrenees (Puigdefabregas, 1975; Puigdefabregas *et al.*, 1989). The formation was studied in the extreme western part of the Tremp-Graus Basin complex, in a synclinal structure referred to as the Ainsa Basin (Fig. 3). The axis of the syncline is oriented approximately north–south, and the basin is bounded by the Mediano anticline to the east and the Boltana anticline to the west. Previously, the basin has also been referred to as the Santa Maria de Buil syncline (Nijman & Nio, 1975; Mutti *et al.*, 1988). The two anticlines are thought to be the surface expressions of major lateral thrust-ramps (e.g. Atkinson, 1986; Hirst & Nicholls, 1986), and were active before and possibly during sedimentation of the Escanilla Formation (Mutti *et al.*, 1988; Puigdefabregas *et al.*, 1989). Below, it is suggested that the anticlines influenced both the palaeo-drainage pattern and the coarse member/fine member ratio of the formation.

In the central part of the Ainsa Basin, the fluvial sediments of the Escanilla Formation conformably overlie the grade laterally into shallow marine deposits of the Belsue (or Sobrarbe) Formation (Fig. 4). However, as one approaches the anticlines, a progressive unconformity is developed between the Escanilla Formation and the strata below, eventually resulting in the superposition of the Escanilla Formation upon marine limestone of mainly Lutetian age (Puy de Cinca and Guara Formations; Puigdefabregas, 1975; Mutti *et al.*, 1988). The Escanilla Formation has a maximum preserved thickness of approximately 750 m in the

GROUP*	FORMATION	ENVIRONMENT	STAGE	SERIES
	Collegats	Alluv. fan		Oligocene
Campodarbe	?38 M.a.*			
Campodarbe	Escanilla	Fluvial	Priabonian	Eocene
Campodarbe	39.5 M.a.*			
Campodarbe	Belsue	Shallow marine	Bartonian	
Campodarbe	San Vincente	Deeper marine, turbidite-complexes		
Campodarbe	44 M.a.*		Lutetian	
Santa Liestra				

Fig. 2. Middle to Upper Eocene stratigraphy in the Ainsa Basin (*after Mutti *et al.*, 1988).

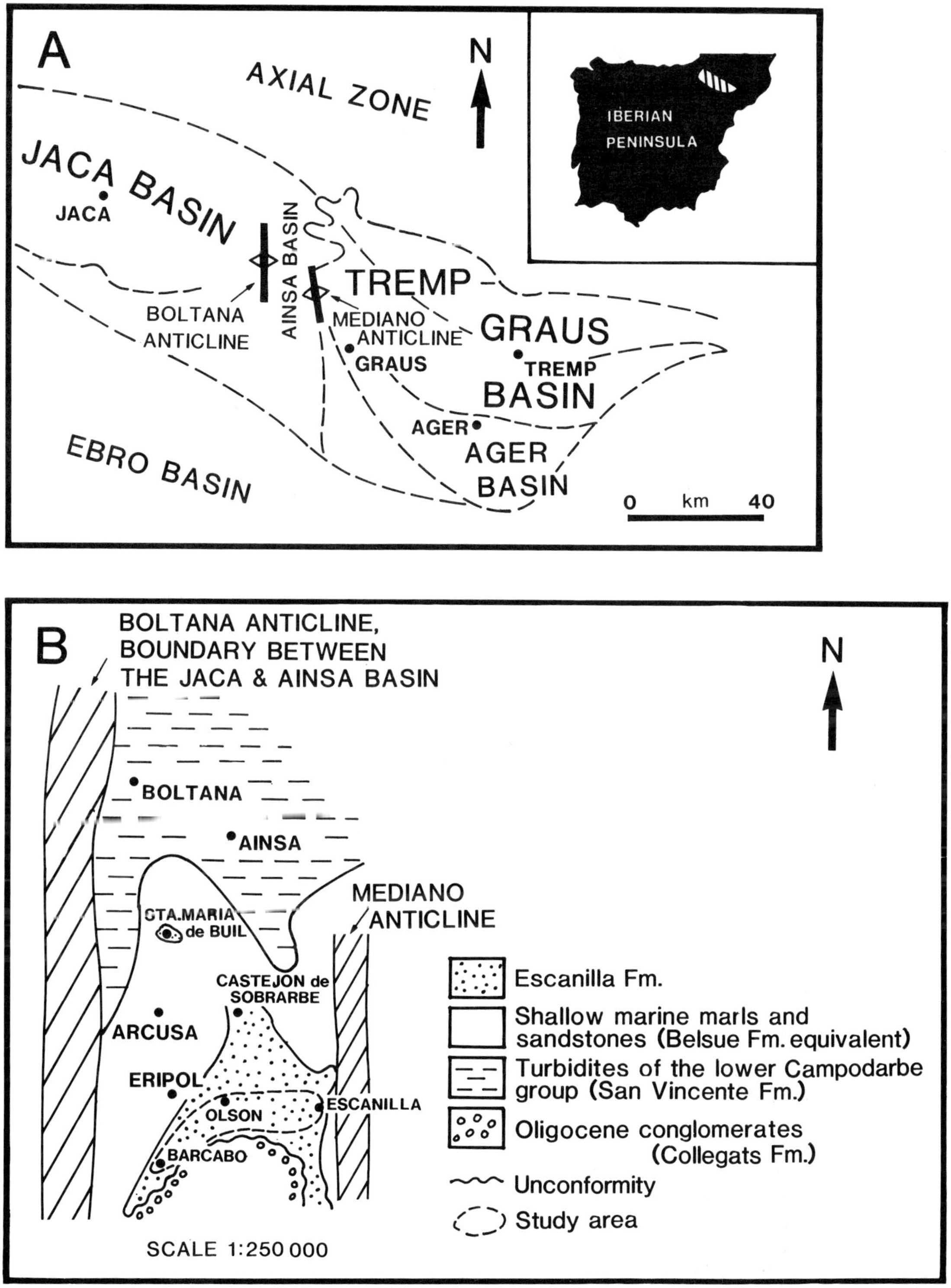

Fig. 3. (A) Location of the Ainsa Basin in the south Pyrenean foreland region. Inset: Map of the Iberian Peninsula with the south Pyrenean foreland basin cross-hatched. (B) Simplified geological map of the Ainsa Basin, with the study area shown.

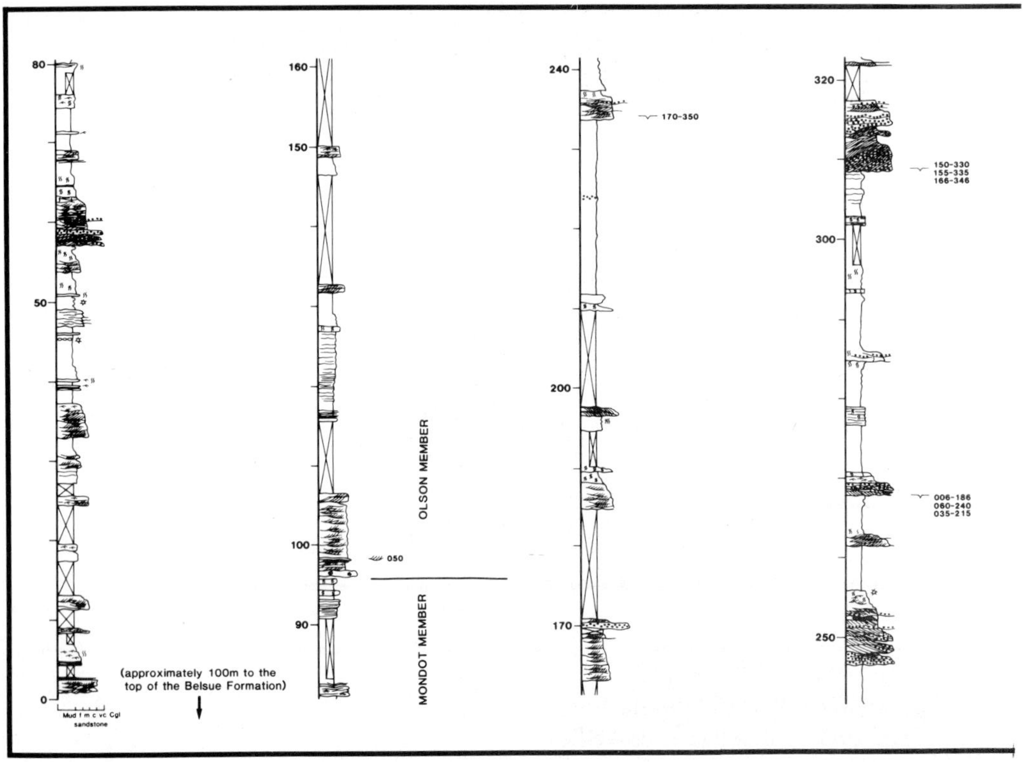

Fig. 4. Vertical log through most of the Escanilla Formation in the Olson area. Note the alternation between channel sandbodies and interchannel deposits, and the subdivision of the formation into informal members.

central part of the Ainsa Basin (Fig. 4) but thins onto the flanks of the anticlines, and is not present in the cores of the anticlines. Approximate timelines, constructed from a combination of lateral tracing in the field and investigation of aerial photographs, converge towards the anticlines. These factors indicate that the sediment accommodation potential decreased towards the anticline axes, and that this area at times may have been characterized by non-deposition and erosion. The Escanilla Formation is unconformably overlain by Oligocene alluvial fan conglomerates of the Collegats Formation (Atkinson, 1983).

Mutti *et al.* (1988) tentatively suggested that the Escanilla Formation was deposited between 38 and 39.5 Ma (i.e. over a period of 1.5 million years) and, with a preserved thickness of 800 m in the core of the Ainsa Basin, the calculated sediment accumulation rate for the formation is 0.53 m/1000 years. This is above the average for fluvial environments (Sadler, 1981). The high sedimentation rate is reflected by the absence of mature palaeosols (see below). As discussed above, the Escanilla Formation thins towards the anticlines, implying that the rate of sediment accumulation was lower in these areas. This may have been due to deflection of rivers away from these positive elements, and/or to overall lower rates of subsidence in the anticlinal areas. The deflection of rivers is clearly indicated by the palaeocurrent pattern (see below). However, the strong westward components of palaeocurrent, even in areas close to the anticlines, suggest that rivers did cross these areas of low subsidence. This is also suggested by the lack of mature palaeosol profiles in these regions, related to clastic contamination from overbank flooding in nearby channels.

Fine member characteristics

The Escanilla Formation in the Ainsa Basin comprises alternations between fine member and coarse member deposits, formed in a Mediterranean to

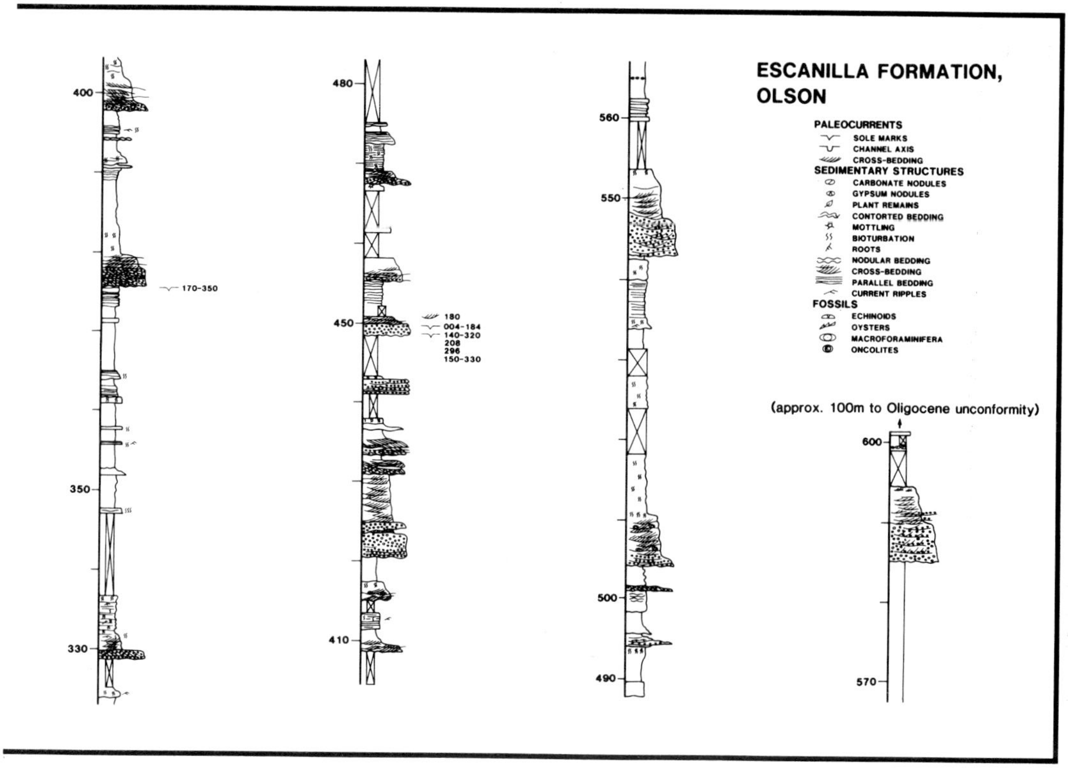

semi-arid climate (Atkinson, 1983). The fine member is dominated by monotonous sequences of pale brownish-red to orange coloured mudstones. Colour mottling, caliche nodules and beds, and alabastrine nodular gypsum are infrequently developed. Thin and bioturbated micritic limestones of variable lateral extent occur at a few levels (Fig. 4). The mudstones are mostly destratified with a weakly developed horizontation (Wright, 1989), and occur in intervals 0.5–100 m thick. They are interpreted as overbank fines which were subjected to soil-forming processes of limited duration and intensity (entisols) (Atkinson, 1986; Retallack, 1988). The micritic limestones appear to be similar to those described by Nickel (1982) from the Escanilla Formation further east in the Tremp-Graus Basin; by analogy they are interpreted to have formed in abandoned channels and floodplain lakes. Thin (0.1–2.5 m), sharp-based beds of very-fine-grained sandstone to conglomerate constitute approximately 25% of the fine member-dominated intervals (Fig. 4). These sandstone beds are tabular to broadly lenticular, single-storey to multi-storey, and can normally be traced laterally into coarse member channel margins. Normal grading is common, but ungraded beds also occur. They show a distinct fining away from the channels, accompanied by a gradual change from primary sedimentary structures (parallel lamination, current ripples and small-scale cross-bedding) into secondary sedimentary structures (pedo- and bioturbation). At their distal ends, these sandbodies merge into fine member mudstones. These fine-member sandstone beds are interpreted as crevasse splay and sheetflood deposits.

Coarse member characteristics

The coarse member deposits consist of lenticular to tabular bodies of sandstones and conglomerate, with very minor amounts of mudstone as drapes and thin beds. The sediments are poorly to moderately sorted, angular grains are common, and petrographically the coarse member deposits can be classified as calclithites and lithic arenites. Individual lenticular bodies have variable, but in general rather low, width/thickness (W/T) ratios measured in a direction normal to the palaeotransport direction, with a mean value of around 40. However, the lenticular units are commonly vertically stacked or laterally linked, and these composite coarse member bodies have W/T ratios frequently exceeding 100. The thickness applied in the calculation of W/T ratios is the maximum thickness. The most common sedimentary structures in the coarse member sediments are trough and tabular cross-beds, parallel lamination, normal grading, massive beds, gutter casts and lateral accretion bedding. Directional features (see below) tend to be unidirectional within individual coarse member bodies. Fining upward units are ubiquitous both on bed and bedset scale, and the bases of the coarse member bodies are erosive. The above features suggest subaerial deposition under high-energy, unidirectional flow conditions in a position relatively close to the source area, and consequently the coarse member bodies are interpreted as fluvial channel deposits. According to Puigdefabregas (1975), Nickel (1982), and Atkinson (1983, 1986), the coarse and fine member sediments of the Escanilla Formation probably represent medial to distal alluvial fan deposits, fed by major distributary channels/palaeovalleys, sourced in the Pyrenees to the north (Fig. 5).

Vertical trends

At first glance, the spatial distribution of sandbodies in the Escanilla Formation seems to be random, suggesting merely autocyclic control on sedimentation. However, upon closer scrutiny, some distinct vertical trends can be seen (Fig. 4). Centrally in the Ainsa Basin, the most notable trends are the overall coarsening upwards in the coarse member bodies from base to top, the disappearance of marginal marine facies-interfingerings some 200 m above the base of the formation, the low density of coarse member bodies in the lower 200 m of the formation, and the appearance of more laterally persistent (up to several kilometres) multi-storey/multi-lateral coarse member bodies in the middle to upper parts of the formation. These vertical trends suggest that the Escanilla Formation is regressive overall, containing coastal plain deposits near the base and more proximal alluvial sediments near the top. Further, the distinct change occurring at about 200 m above the base makes it convenient to subdivide the Escanilla Formation in the Ainsa Basin into two informal members. The lower 200 m, characterized by marginal marine intercalations, pale fine member colours, scarcity of conglomerate size clasts, and a low sandbody density, is referred to as the Mondot Member. The upper 550 m, characterized by more brownish-red fine member colours, a higher sandbody density and lateral continuity, coarser grain sizes and an absence of marine-influenced intervals, is referred to as the Olson Member. Compositionally, there is also a difference between the two members. The Mondot Member contains a number of loosely cemented sandbodies with white weathering colours. These are especially abundant near the top of the member, and the white coloration seems to be caused by a significant percentage of diagenetically altered (kaolinitized) feldspars and granitic rock fragments. In the Olson Member, 'white' sandbodies are rare, and the coarse member bodies are dominated by Mesozoic and lower Paleogene limestone and sandstone clasts set in an extensively carbonate-cemented groundmass.

CONDITIONS AND TECHNIQUES FOR DATA ACQUISITION IN THE FIELD

Data collection from outcrops

The aim of the field work was to collect all available and relevant data as defined by the tabulation of data types to be included in the database (Table 1).

A set of general data was established, describing the geological setting and main characteristics of the Escanilla Formation. These data are primarily considered for easy comparison with reservoir sections, and are reported above. The main part of the data set collected from outcrops are the continuous descriptions of the geometries of bounding surfaces and sedimentary bodies which compose the alluvial deposits. These were collected at a range of scales, and provide data pertaining to reservoir communication (large scale) and local 'near-well' distribution of flow units (small scale).

In the Escanilla Formation at Olson, the excellent exposure and the ravined morphology of the outcrops provide favourable conditions for measurement and documentation of this kind of data set (Fig. 6). The sedimentary geometries were primarily described in 2-D cross-sections along outcrops, and

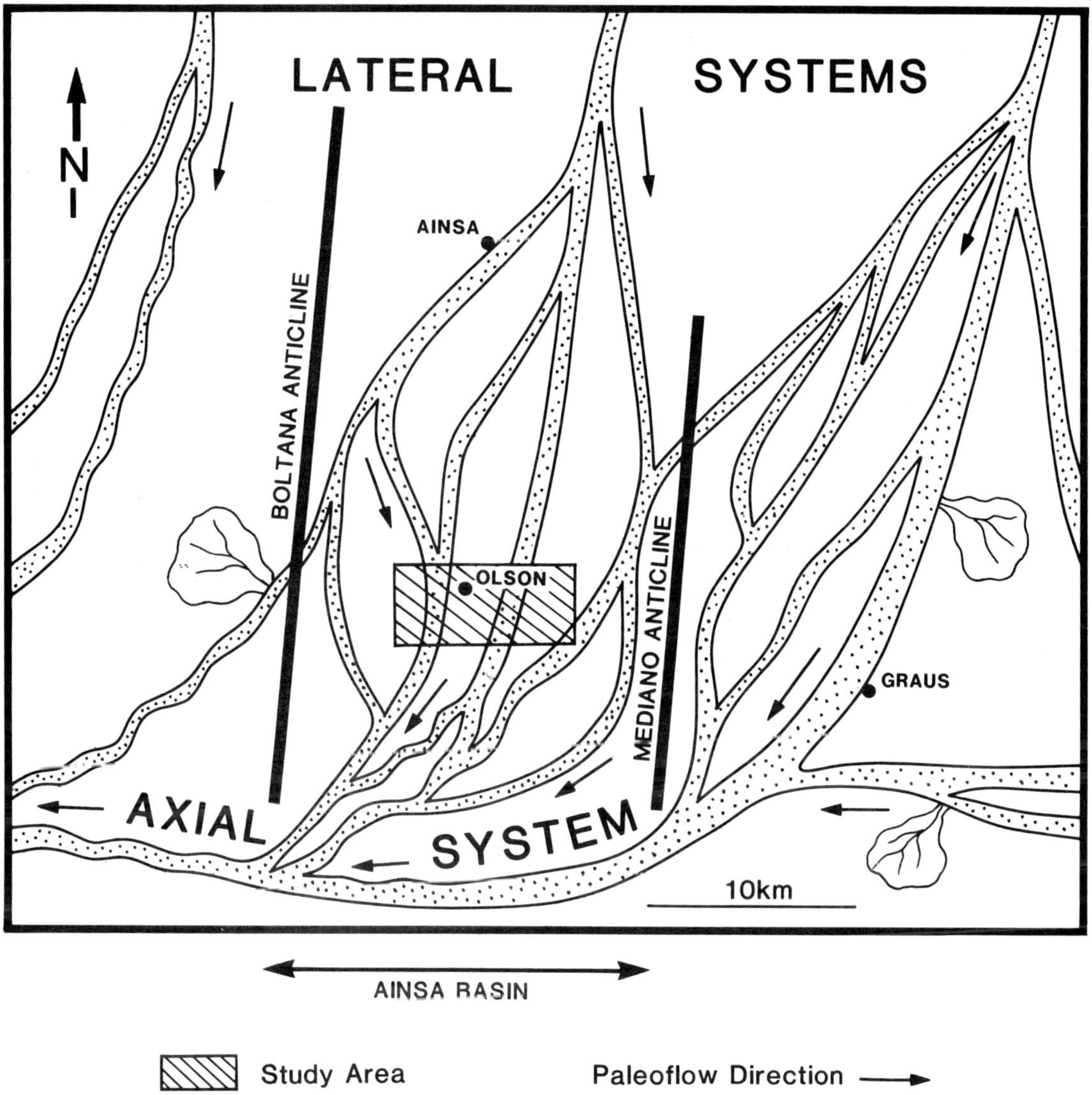

Fig. 5. Palaeogeographic reconstruction of the fluvial system in which the Escanilla Formation (Olson Member) was deposited, showing only the general arrangement of channels (stippled), major crevasse lobes and interchannel fines. Note the interfingering between the lateral and axial drainage systems in the foreland basin and the effect of the anticlines on palaeoflow directions.

when combined, they provide a 3-D image of the selected interval.

For large-scale descriptions, approximately 100 m thick intervals were selected and traced laterally for several kilometres, in order to give documentation over distances comparable with typical well spacings in offshore reservoirs. In the Escanilla Formation, more than 20 km of measured large-scale cross-sections form the basis for the spatial model (see below). These cross-sections focus on the extent and degree of contact between the individual channel sandstone bodies.

Documentation of geometries and properties at a more detailed scale (see Fig. 9) was made on selected portions of the large-scale cross-sections, or in some cases outside these. These detailed cross-sections

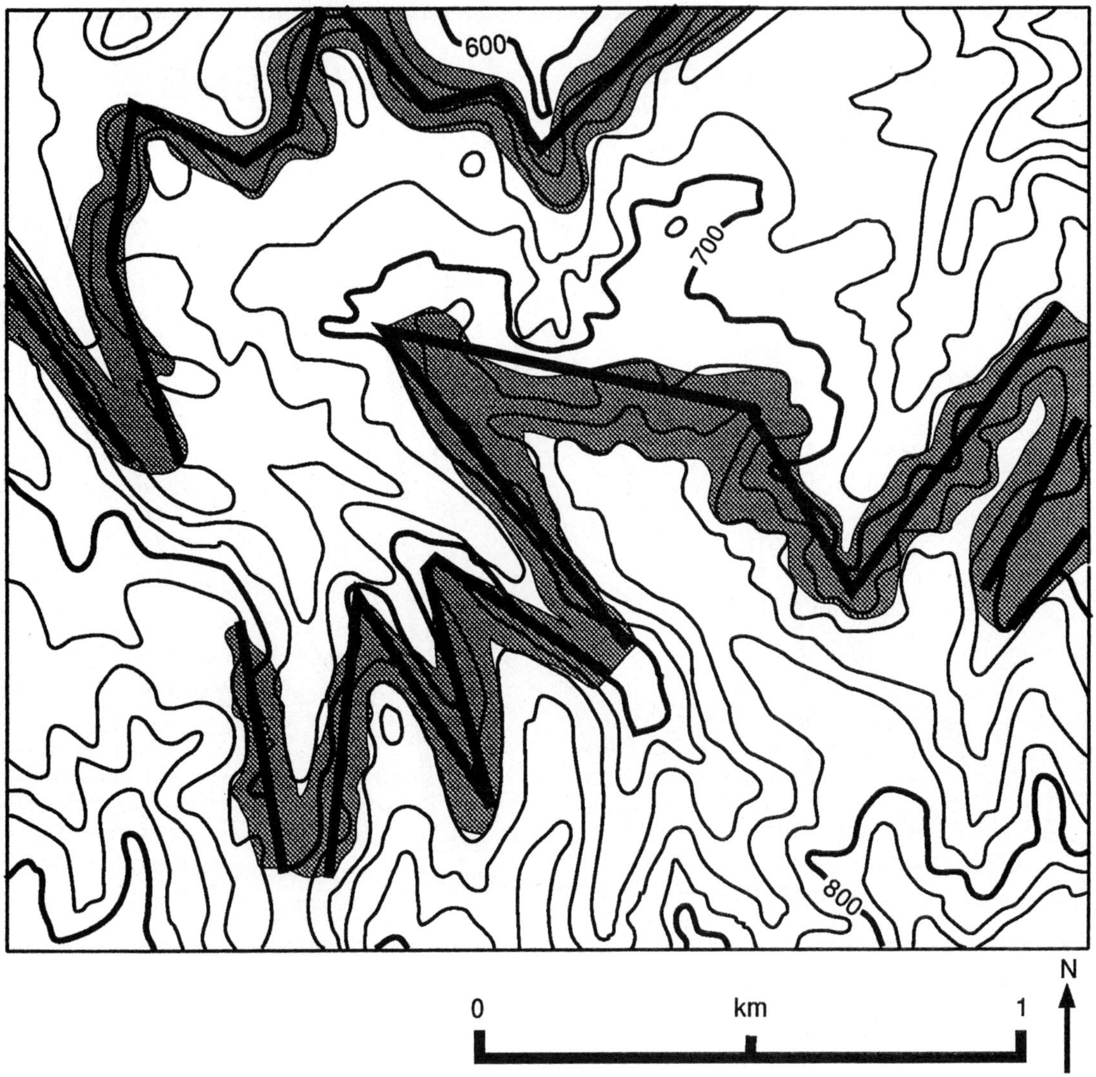

Fig. 6. Example of the topography of Escanilla Formation outcrops at Olson. The extent of some of the outcrops described in cross-sections is stippled, and the positions of cross-section projection lines are indicated.

typically represent single channelbelt sandstone bodies, covering distances adequately large to describe the geometries and organization of the architectural elements (*sensu* Miall, 1988) (100–700 m). These data sets provide small-scale spatial models of the distribution of architectural elements associated with the different types of coarse member bodies.

A variety of techniques were adopted for the documentation in the field of outcrop cross-sections, the most important being simple measurement by means of measuring tapes and rods. Where outcrops are straight and regular, measurement from photographs (mosaics and panoramas) was possible, when combined with proper on-site inspection. Field notes on polaroid photographs proved to be an efficient means of temporary documentation. It was an important aim of the field work to minimize the need for reorganization of the data before it could be registered in the computer database.

In order to simplify the collection and further

handling of the 2-D cross-section data, outcrop measurements were documented as if projected into a plane perpendicular to palaeohorizontal strata (Fig. 6). This simplification introduces some errors in the cross-section documentation, which adds to those arising from the measurement methods. For the purpose of quantification of sedimentary reservoir architecture and heterogeneities, however, these errors are considered small, and they do not affect the validity of the data set. The database system does not constrain the accuracy at which measurements can be registered. It is possible to combine utilization of advanced measuring techniques and detailed representation of the outcrop configuration to provide a higher degree of accuracy if this is considered important.

Apart from the 2-D cross-sections, point-related data of various types were recorded. One important category of this data type is data on orientation of channel sandbodies. The most reliable indicators of channel orientation were considered to be measurements of gutter casts and channel margin orientations.

Due to pervasive carbonate cementation in the Escanilla Formation, permeability mapping by means of minipermeametry had to be abandoned. Otherwise, this is an important aspect of data collection from outcrop analogues for the SAFARI database.

In addition to the data types described above, vertical sedimentary logs were collected for characterization of the whole formation and to provide a basis for comparison with well data. Furthermore, interpretation of aerial photographs was performed in order to investigate regional variability in the formation, and to establish stratigraphic correlation between different parts of the area.

Bounding surfaces and architectural elements

A standardized system of bounding surface hierarchy has been adopted, in order to provide a systematic tool for the field description of sedimentary geometries. All bounding surfaces and the sediment bodies delineated by these have also been assigned properties related to a set of classification schemes. These classifications include parameters such as lithology, grain-size composition, architectural element and heterogeneity. Principles for the classification of bounding surfaces and of sedimentary architectural elements are in general agreement with those of Miall (1988), with some alterations and additions (Fig. 7). For the bounding surface hierarchy, the order of numbering has been reversed and one extra level (3rd order, bounding single channel deposit) has been introduced as compared to Miall's (1988) classification. Some extra architectural elements have also been added.

It is of critical importance that the data collected using the principles of bounding surface hierarchy and architectural element classification are closely related to the subdivision of similar reservoir rocks into flow units. Although impossible to demonstrate by permeability measurements on the Escanilla Formation outcrops (due to diagenetic destruction of porosity) we consider that the elements and properties described are important components for description of fluid flow in any analogous reservoir. This is based on the examination of properties such as grain size, sorting, stratification and internal heterogeneity, combined with experience from analogous Norwegian offshore reservoirs. It has been demonstrated by several workers that sedimentary architectural elements form flow units in reservoirs of various sedimentary origins (Hearn *et al.*, 1984; Kortekaas, 1985; Ebanks, 1987; van de Graaff & Ealy, 1989; Høimyr *et al.*, 1990; Dreyer *et al.*, 1990).

For the Escanilla Formation outcrops, the method outlined has proved to be a very efficient approach to attain a consistency in the data set, and the classification system also forms a good basis for the systematic handling of the data set in the database system.

DATA COLLECTED IN THE ESCANILLA FORMATION

Overview

In the design of the SAFARI database, the types of data to be collected in fluvial formations were specified. In general, the 14 data types in Table 1 can be grouped into three main categories: general data, data pertaining to reservoir communication, and data on small-scale heterogeneity distribution in coarse member bodies. The three data categories were introduced in the section on data acquisition. Moreover, the general data were discussed in some detail in the introductory section on the Escanilla Formation. In the remainder of this section, the two other data categories will be discussed.

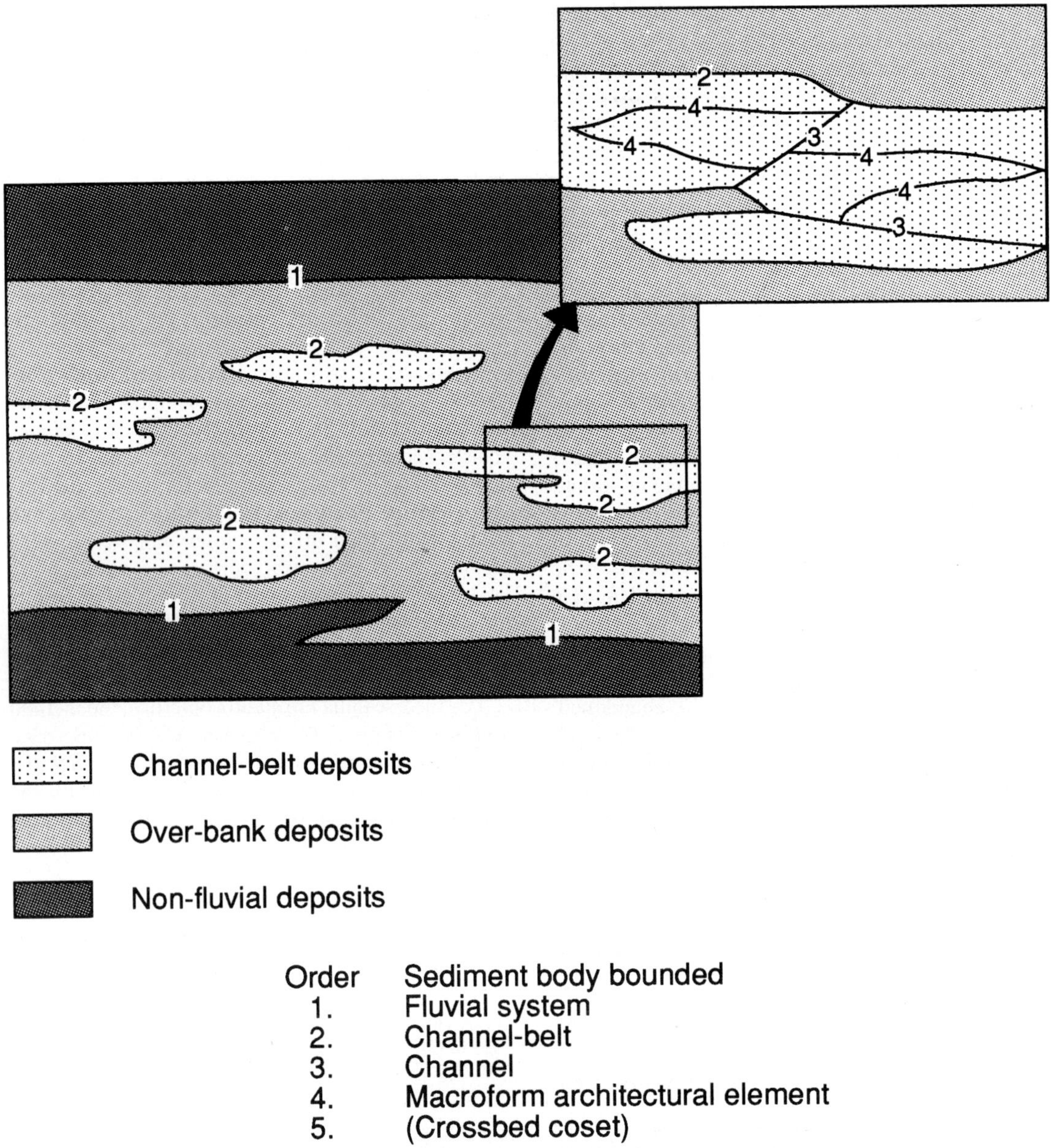

Fig. 7. Bounding surface hierarchy adopted for the description of cross-sections.

Large-scale cross-sections

The main objective of the large-scale cross-sectional data is to obtain information on reservoir communication in fluvial sequences. This objective was primarily addressed by collecting data concerning the dimensions of, and spatial relationships between, various fluvial lithofacies. The large-scale (1:500) cross-sections were constructed in the field according to the techniques outlined above. The endpoints of cross-sections represent positions where the apparent strike direction of the beds change (Fig. 6).

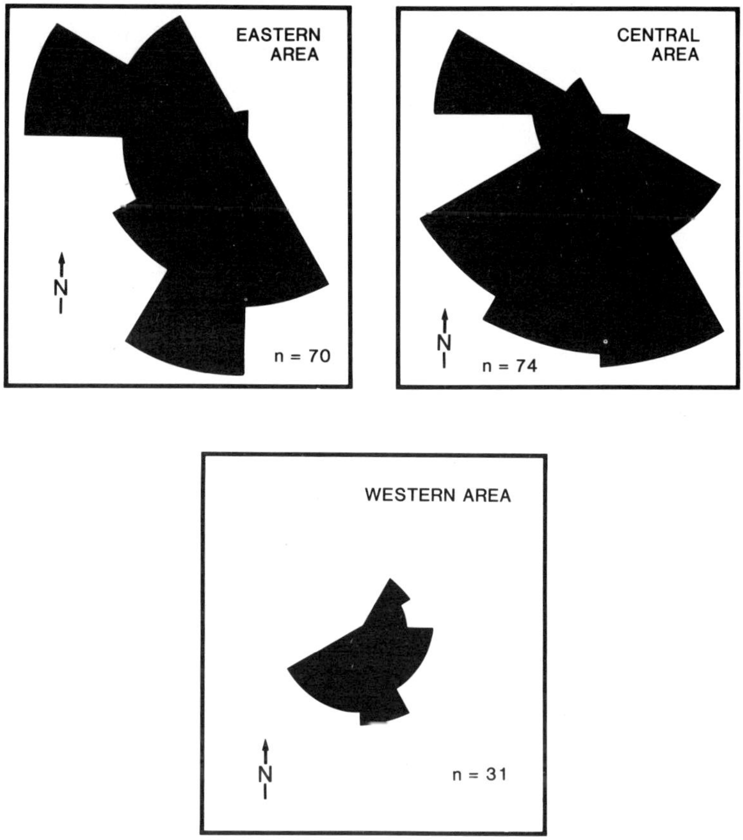

Fig. 8. Diagram summarizing the palaeocurrent information from the Escanilla Formation. Note the large overall spread in palaeoflow directions, and the different palaeocurrent patterns in the three regions of the study area.

Large-scale cross-sections were made in stratigraphic levels from the lower, middle and upper parts of the Olson Member, in areas where the dissected topography created exposures oriented oblique to, normal to and parallel with the local palaeotransport direction (Fig. 8). In total, approximately 20 km of large-scale cross-sections were collected. An example of these is given in Plate 1 (facing p. 76).

The main elements of the large-scale cross-sections are sedimentary bodies (lithofacies) separated by a hierarchy of bounding surfaces. In Plate 1, each lithofacies is given a specific colour, and the bounding surfaces are marked by their respective hierarchical numbers. Only bounding surfaces of rank 2 and 3 (Fig. 7) are mapped on these cross-sections, delineating the extent and spatial arrangement of channel-belts, floodplain fines, individual channels, and lithofacies types. A simple lithofacies subdivision has been used in the cross-sections, consisting of only six different types which, nevertheless, highlight the main lithological heterogeneity in the formation. The six types are:

A Pebbly sandstones and conglomerates in the coarse member.

B Fine to coarse sandstones in the coarse member.

C Mudstones in the coarse member.

D Fine member mudstones.

E Fine member sandstones.

F Arkosic, white-weathered coarse member sandstones.

Types A, B and D are common, and represent the bulk of the channel and floodplain deposits in the Escanilla Formation. Type E is moderately common, and for the most part represent crevasse-like sandstones, normally attached to channel bodies. Type C is uncommon, whereas type F is common only in the Mondot Member.

The large number of 1:500-scale cross-sections

were digitized into a GIS (Geographical Information System) format so that analyses of the data set could be made (see below).

Small-scale cross-sections

Heterogeneities occurring within coarse member bodies were described by detailed 1 : 100-scale cross-sections (Fig. 9). This scale of inhomogeneity affects reservoir performance by influencing fluid flow patterns, and is particularly important for the modelling of near-well flow-unit distributions. In the small-scale cross-sections, the bodies and bounding surfaces have been recorded in such a manner that the potential flow-units of the various coarse member types are highlighted (Fig. 9). A fundamental element of this method is the notion that the architectural elements of the channel bodies (see Table 4) in a broad sense correspond to flow units. Although this should only be regarded as a first approach, the differences in grain-size distribution and structural assemblage between architectural elements appear to justify this assumption. Architectural elements are separated by fourth-order bounding surfaces within individual channels. In the example in Fig. 9, all surfaces are of fourth-order, except for the second-order surfaces separating the channel body from the surrounding floodplain fines. Third-order surfaces are absent due to the single-storey nature of the channel body (Fig. 7). Small-scale cross-sections like the one in Fig. 9 were made for all the coarse member types in the Olson Member (see below).

The channel body in Fig. 9 belongs to coarse member type 3 (Fig. 10), coarse-grained meandering channel deposits. It was found to be a few hundred metres in width, the right-hand part of Fig. 11 being close to one of the channel margins. Palaeocurrent readings imply transport to the east, in a direction normal to the sense of migration indicated by the prominent lateral accretion surfaces (Fig. 11). The channel bodies that can be seen below the study object are of the same type, but have transport directions towards the south. Detailed mapping of the coarse-grained meandering channel body revealed that the main internal heterogeneity is related to the spatial arrangement of architectural elements. The lower part of the body is dominated by elements GB and LA(g) (Fig. 9 and Table 4), forming a basal 'gravel flow-unit' of high initial permeability. In the upper part, elements LA(s) and SB dominate, forming a 'sandy flow-unit' of lower initial permeability. Additionally, there is a pronounced fining upwards in the upper sandy part, indicating an upwards deterioration in reservoir quality (Fig. 9). The same pattern of heterogeneity is also present in sections cutting obliquely through the channel body, along the dipping lateral accretion surfaces (Figs 9 and 11). The gravel-to-sand transition probably reflects deposition on the lower and upper part of a coarse-grained point bar (McGowen & Garner, 1970). Note the pronounced decrease in the dip of the lateral accretion sets towards the top of the upper point bar (Fig. 11). It should also be noted that the boundaries between lateral accretion sets are free of fine-grained drapes, making fluid communication between individual sets relatively straightforward.

INTERPRETATION AND ANALYSIS OF ESCANILLA FORMATION DATA

The SAFARI database is primarily designed to provide quantitative information to support geomathematical modelling tools and/or simulation tools. These modelling tools require either primary outcrop data sets or statistical information based on analysis of these. Examples of such data from the Escanilla Formation are illustrated in Plate 2 (facing p. 77) and Fig. 12. Important types of data derived from analysis of the Escanilla data set are relative facies distributions, thickness and width distributions for architectural elements (and relations between these), characterization of spatial relationships between architectural elements, and quantification of sandbody interconnectedness and communication. The categories of interpreted and analysed data, derived from the study of the Escanilla Formation, are:

1 the general characteristics of the Escanilla Formation;

2 the geometrical description of coarse member bodies and their internal organization of architectural elements and heterogeneities;

3 data on orientations and palaeodrainage patterns;

4 data on sandbody interconnectedness and reservoir communication.

General characteristics

The general characteristics of the Escanilla Formation are summarized in Table 3. Data derived from wells in the reservoir may be compared with similar types of data from the Escanilla Formation to determine the degree of analogy between the sequences.

Realistically, the analogues studied in this project will in some aspect differ from the North Sea reservoirs with which they are compared. Given a reasonable degree of similarity, however, parts of the comparative data set may be applicable for reservoir modelling purposes. With adequate knowledge about factors controlling sandbody geometry, heterogeneity distribution, palaeodrainage/sandbody orientation, etc., the quantitative part of the analogue data set may be converted to fit better the conditions in the reservoir(s).

Coarse member types

An essential part of the information derived from the Escanilla Formation is the knowledge about the geometries and arrangement of architectural elements in the coarse member bodies. These parameters have considerable influence on sandbody volumes, interconnectedness and flow unit distribution in an analogous reservoir.

Based on geometrical characteristics, grain-size trends, stacking patterns, and distribution of architectural elements (Table 4), four types of coarse member bodies have been recognized in the Escanilla Formation (Fig. 10). Arranged in order of descending frequency of occurrence, these are:

Type 1: single-storey and multi-storey/multi-lateral sand dominated bodies with moderately high width/thickness (W/T) ratios

Around 50% of the coarse member bodies in the Escanilla Formation belong to this type. Irregular to well-defined fining-upward trends typically occur, and conglomeratic intervals are commonly present only in the lower parts. The sandbodies are 2–16 m thick, the mean thickness being 4 m. Stories are stacked both vertically and laterally, but whereas the number of vertically stacked stories at any point does not exceed three, the number of laterally stacked bodies varies from one to eight. W/T-ratios range from 24 to 150, with 60 as the mean. CH, SB, GB and DA are the most common architectural elements (Table 4). Type 1 coarse member bodies are interpreted as the deposits of braided streams. In the cases where these bodies consist of only one or a few storeys with a pronounced lenticular form, they may be termed channel-braided (Cotter, 1978). In most cases, though, the term sheet-braided is better suited.

Table 3. General information on the Escanilla Formation in the Ainsa Basin

Age	Upper Eocene (Priabonian)
Basin type	Foreland Basin
Source area	Thrust sheets in the Pyrenees to the north
Depositional setting	Medial to distal part of alluvial fan, and axial fluvial system
Transport directions	Mainly to the S, SW, SE and W (influenced by anticlines)
Formation below	Belsue Fm.—shallow marine
Lower contact	Gradational (central parts) to unconformable (marginal parts) over marine beds
Formation above	Collegats Fm.—proximal alluvial fan
Upper contact	Unconformably overlain by alluvial fan conglomerate
Thickness	400–800 m
Sand/shale ratio	30–55
Mean sediment accumulation rate	0.53 m/1000 yrs
Lithologies	Common: mudstones, fine- and coarse-grained sandstones, conglomerates. Rare: caliche, gypsum and lacustrine limestones
Climate of deposition	Semi-arid to temperate with moderate amounts of rainfall
Vertical trends	Overall progradational. Divisible into two terms: Mondot Member (coastal plain/marginal marine) and Olson Member (fluvial)

Type 2: multi-storey and single-storey sand-dominated bodies with low W/T-ratios

These sandbodies are 3–42 m thick (mean: 7 m) and rarely more than a few hundred metres wide. W/T-ratios vary from 5 to 35, with a mean about 20. The grain size ranges from fine-grained sand to conglomerate, and the bodies do not exhibit well-developed fining-upward trends. Multi-storey bodies are most common and the number of stories ranges from 2 to 10. In the Olson Member, up to 30 stories may stack in a slightly offset pattern within local areas of approximately one square km. In these cases, there

Fig. 10. Panoramic photo of a stratigraphic interval corresponding to one of the large-scale cross-sections described and analysed in this study. The photo also gives an impression of the well-exposed nature of the Escanilla Formation. In addition, the four different coarse member types are indicated.

Fig. 11. Photographic overview of the stratigraphic interval containing the coarse-grained point bar sandbody of Fig. 9 (arrow). Person for scale (boxed) at left end of sandbody. Note the prominent lateral accretion surfaces with upward decreasing dips and the tabular shape of the sandbody. The multi-storey sandbody below the mapped one was also formed in a meandering stream, but the lateral accretion surfaces are difficult to discern in this section because they dip in a direction approximately normal to the strike of the exposure.

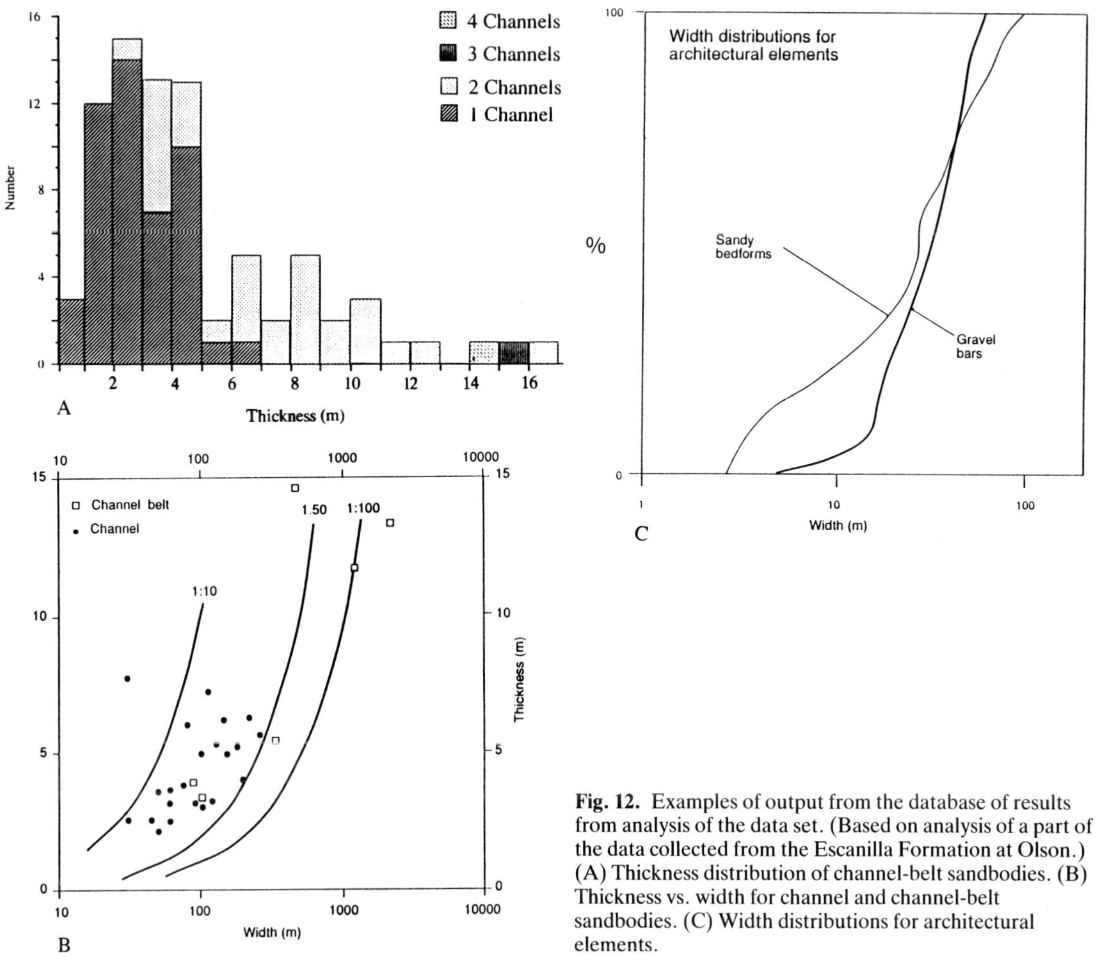

Fig. 12. Examples of output from the database of results from analysis of the data set. (Based on analysis of a part of the data collected from the Escanilla Formation at Olson.) (A) Thickness distribution of channel-belt sandbodies. (B) Thickness vs. width for channel and channel-belt sandbodies. (C) Width distributions for architectural elements.

is a tendency for an upwards increase in the W/T-ratio of individual storeys. The multi-storey sandbodies always have palaeocurrents directed towards the south, southwest or southeast. Type 2 coarse member bodies are associated with the deepest basal erosional scours in the Escanilla Formation, and incisions up to 15 m in depth may occur in the lower parts of thick, multi-storey bodies. SB, DA, PI, GB and CH are the most abundant architectural elements (Table 4). Type 2 coarse member bodies are thought to reflect deposition in stabilized, vertically-aggrading channels. In the cases where these bodies are thick, strongly multi-storey and deeply incised, they may represent major distributaries of the alluvial fans which were present along the northern margin of the foreland basin.

Type 3: sand- and gravel bodies with lateral accretion bedding

The most striking feature of these bodies is large-scale cross-strata that have migrated in a direction normal to the main channel orientation. These cross-strata may penetrate the entire body, or they may only be present in the upper sandy part of these distinctly upward-fining sand- and gravel bodies. The lateral accretion strata terminate upwards by laterally fingering into flat-lying floodplain fines. The upward-fining in these bodies is characterized by a lower gravelly part (1–6 m thick) which is overlain by an upper sandy part of equal or slightly greater thickness. The transition between the two zones may be sharp, or it may occur as a 0.2–1.5 m

Table 4. Architectural elements distinguished in the fluvial deposits of the Escanilla Formation

Code	Name	Grain size	Heterogeneity*	Sedimentary structures/features	Depositional processes
LA (s)	Laterally accreted sandstone	Fine to very coarse sand, mean: coarse sand	PH	Epsilon XB, massive, current ripples Present in upper/middle parts of channel, 1.5–4.8 m thick bedsets, fining in grain size updip along epsilon XB, XB dip increases towards lower part of point bar	Sand-deposition on upper to middle parts of laterally migrating coarse-grained point bar
LA (g)	Laterally accreted gravel	Very coarse sand to very large pebbles; mean: medium pebbles	MH	Epsilon XB, massive, graded bedding Present in lower/middle parts of channel, 1.2–4.5 m thick bedsets, fining in grain size upwards in beds (normal grading) and updip along epsilon XB, usually low XB dip angles	Gravel deposition on lower to middle parts of laterally migrating coarse-grained point bar. Fluctuating flow energy
PI	Passive infill	Medium silt to fine sand; mean: very fine sand	SH	Planar strat., current ripples, mud drapes, bioturbated Occupies upper part of individual channel bodies, 0.2–2.7 m thick bedsets	Settling of fine-grained material in partly abandoned channel reaches, lower flow regime deposition
SB	Sandy bedforms	Medium sand to granules; mean: coarse sand	PH	Tabular XB, trough XB, planar strat. Random occurrence in coarse member bodies, 0.6–2.9 m thick bedsets, may amalgamate into thicker, superimposed bedsets. Very common in braided stream deposits	Migration of sandy bedforms along channel floors, in crevasse splays, on bartops, etc.
GB	Gravel bars and bedforms	Very coarse sand to cobbles; mean: medium pebbles	He	Massive, planar strat., tabular XB, trough XB, graded bedding, solemarks Commonly occupies basal parts of individual channels, 0.3–5.1 m thick bedsets, amalgamation common, strong basal scours	Migration of gravelly bars and bedforms within channels, also rapid deposition of coarsest bedload in scours, channel lags

CH	Channels	Medium sand to small pebbles; mean: very coarse sand	He	Massive, planar strat., tabular XB, trough XB, graded bedding (FU), solemarks Lens-shaped, coarse-grained sedimentary bodies present *within* larger coarse member channel bodies, contains a variety of rapidly alternating sedimentary structures, 1–10 m thick (largest thicknesses repr. amalgamated CHs), strongly erosive bases, may be genetically linked with other architectural elements (e.g. substantial gravel lags at the base of channelforms are classified as GB)	Formed as a result of high-energy cut-and-fill episodes that took place in channel-belts, without the establishment of significant amounts of bars and bedforms
DA	Downstream accreting macroform	Fine sand to coarse sand; mean: medium sand	He	Large-scale XB, bio/pedoturbation, current ripples, mud drapes, graded bedding. Trough and tabular XB, planar strat. Most common in upper and marginal parts of coarse member bodies, 2–8 m thick bedsets, communication between bedsets may at times be tortuous due to mud drapes, the normal-graded nature of many beds cause heterogeneous flow properties, common in braided stream deposits, low to moderate dip angles of large-scale XB	Migration of large-scale sandy bars and sandflats in a downstream/oblique direction, avalanche processes and migration of smaller bedforms along masterbedding, mud drapes forming at low-flow stage, the channel islands were sometimes vegetated
LS	Laminated sand sheets	Very fine sand to coarse sand; mean: fine sand	MH	Planar strat., tabular XB, current ripples, bio/pedoturbated, solemarks Normally occurs at the margins of coarse member bodies, extending into overbank fines, 0.2–3.3 m thick bedsets, often shows fining upwards	Deposition of crevasse splays at channel margins, in-channel accumulation of sediments formed under upper flow-regime plane bed conditions
OF	Overbank fines	Clay to very fine sand; mean: silt	PH	Similar to lithofacies D (Table 2)	Similar to lithofacies D (Table 2).

* The abbreviations in this column refer to the various classes of heterogeneity within the different architectural elements. In order of decreasing heterogeneity, these are: VH (very heterogeneous), He (heterogeneous), MH (moderately heterogeneous), PH (partly heterogeneous), SH (slightly heterogeneous), Ho (homogeneous).

thick transition zone where sandy and gravelly beds interfinger. Type 3 sand- and gravel bodies are usually single-storey in a vertical sense, but may consist of 2–3 laterally interlocking bodies (multi-lateral). Oblique stacking trends have also been oberved. Geometrically, the laterally accreted bodies have moderate W/T-ratios, ranging from 20 to 85 with a mean of 43. Both tops and bases of type 3 bodies tend to be flat. The most common architectural elements are LA(s), GB, LA(g) and PI (Table 4). Type 3 coarse member bodies probably represent coarse-grained meandering stream deposits, typified by lower (gravelly) and upper (sandy) point bars (e.g. McGowen & Garner, 1970).

Type 4: sheetlike conglomeratic bodies

Conglomerate is the dominant grain size in all but the uppermost metre or so of these bodies, and they have W/T ratios in excess of 100. The conglomeratic sheets are transitional to the coarsest and most laterally persistent bodies of type 1, but have been differentiated because they tend to form thick (20–30 m) levels which stand out as marker horizons in the Ainsa Basin. One of these levels, occurring about 200 m above the base of the Olson Member, appears to have a basinwide extent. Type 4 conglomeratic bodies are not present in the Mondot Member, but become increasingly dominant towards the top of the Olson Member. The bodies are highly multi-lateral and multi-storey, and tend to have a massive to slightly upwards fining character, except near the top, where the clast size decreases markedly. GB and CH are the dominant architectural elements. Type 4 coarse member bodies are thought to have been formed in a gravelly braided stream environment, probably associated with an alluvial fan (Atkinson, 1983). The thickness and lateral continuity of these bodies reflect deposition in periods when the ratio between coarse-grained sediment supply and accommodation space was high. It is possible that reduced subsidence caused by uplift in the anticlines or the source area may have influenced the development of this coarse member type.

Orientation data and palaeodrainage

Variations in the orientation of elongate sandbodies is a factor of major concern for reservoir communication, as it influences the degree of interconnectedness between sandbodies. Important factors such as the total pore volume connected to a well and the fluid-flow pathways between wells depend upon the degree of interconnectedness.

By reconstructing palaeodrainage patterns from outcropping fluvial sequences, data sets pertaining to the orientation of elongate channel sandbodies may be obtained. Palaeocurrent measurements are essential in these reconstructions. Moreover, studies of palaeodrainage patterns may reveal how factors such as differential subsidence, base-level changes and climate exert control on palaeodrainage in a particular depositional system.

Palaeocurrent information from the Escanilla Formation in the studied area is summarized in Fig. 8. Measurements have been taken from channel margins, gutter casts and large- to medium-scale cross-beds. In general, transport is to the south and southwest, with deviations to the west and southeast. The general southward sense of fluvial flow suggests that the Pyrenean thrust-belt to the north acted as the main source area. This is also supported by palaeocurrent and compositional data reported in this study and by Atkinson (1983). In Fig. 8, it is evident that the palaeodrainage pattern varies geographically. To the east, in the vicinity of the Mediano anticline, westerly to southwesterly components dominate. In the central part of the Ainsa Basin, palaeocurrents are mostly to the south/southwest, whereas near the Boltana anticline to the west, southerly to southeasterly components are prominent. An example from this area can be seen in the coarse-grained meandering channel body illustrated in Figs 9 and 11, which has a palaeocurrent direction towards 096 degrees.

The palaeocurrent variability can be related to the influence of the anticlines on the drainage, and perhaps to the interaction between axial and lateral fluvial systems in the foreland basin (Fig. 5). Palaeocurrents to the south, southeast and southwest can be explained by radial-pattern fluvial flow along the alluvial fans that entered the foreland basin from the north (Atkinson, 1986). The more pronounced eastwards components near the Boltana anticline suggest that this positive element to some degree deflected southward-flowing river systems to the east (and to the west on the western side of the anticline; Puigdefabregas, 1975). A similar westward deflection probably took place on the western side of the Mediano anticline. However, it is possible that both the westerly and southwesterly components also reflect the influence of a west/northwest-flowing axial system in the foreland basin (Puigdefabregas 1975;

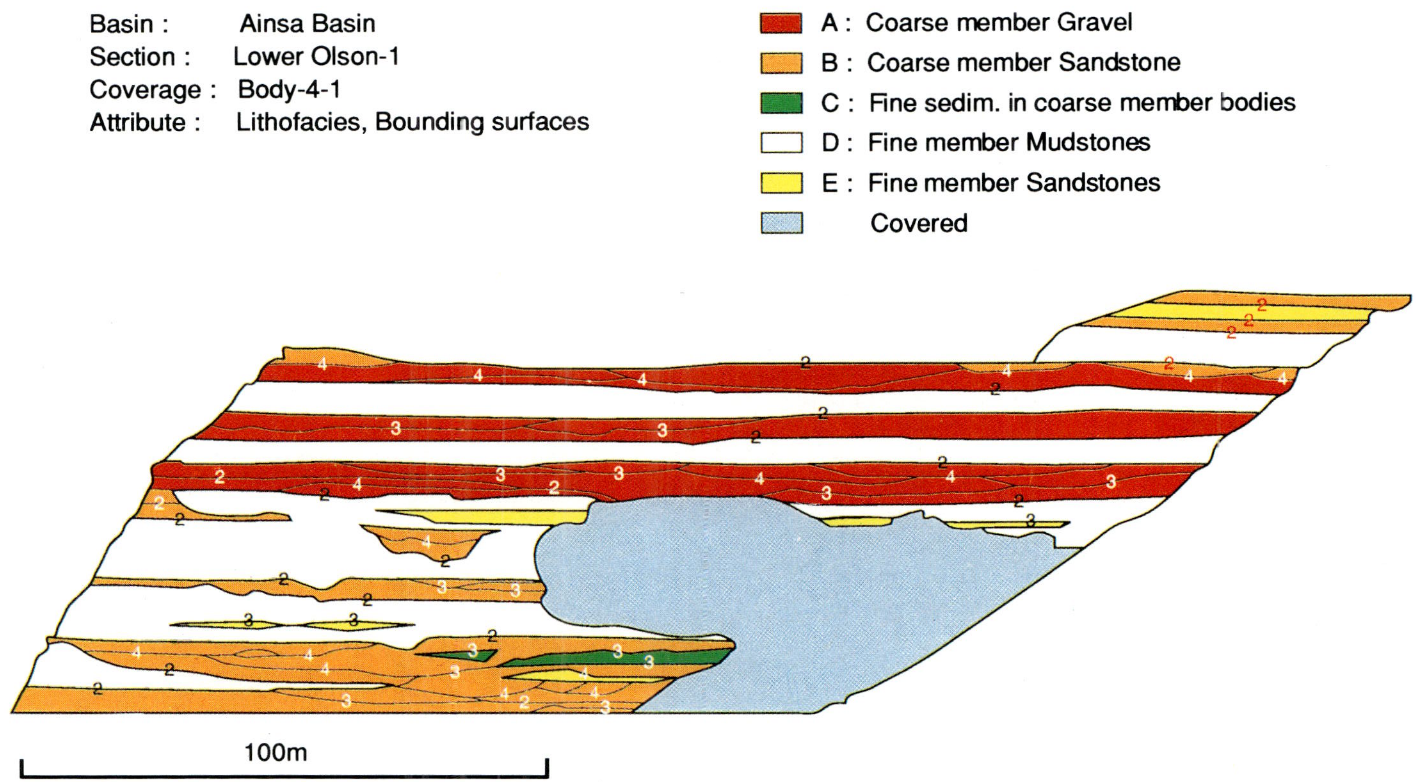

Plate 1. Large-scale cross-section documenting the sandbody architecture, at the level of channel and channel-belt deposits. Lithofacies units are indicated. The cross-section, from the lower part of the Olson Member is part of approximately 6 km of continuous lateral outcrop descriptions from the same stratigraphic level. Diagrams of this type are digitized into the SAFARI database where they are subjected to analysis of, for example, sandbody density, interconnectedness and frequency of the different lithofacies.

[*facing page 76*]

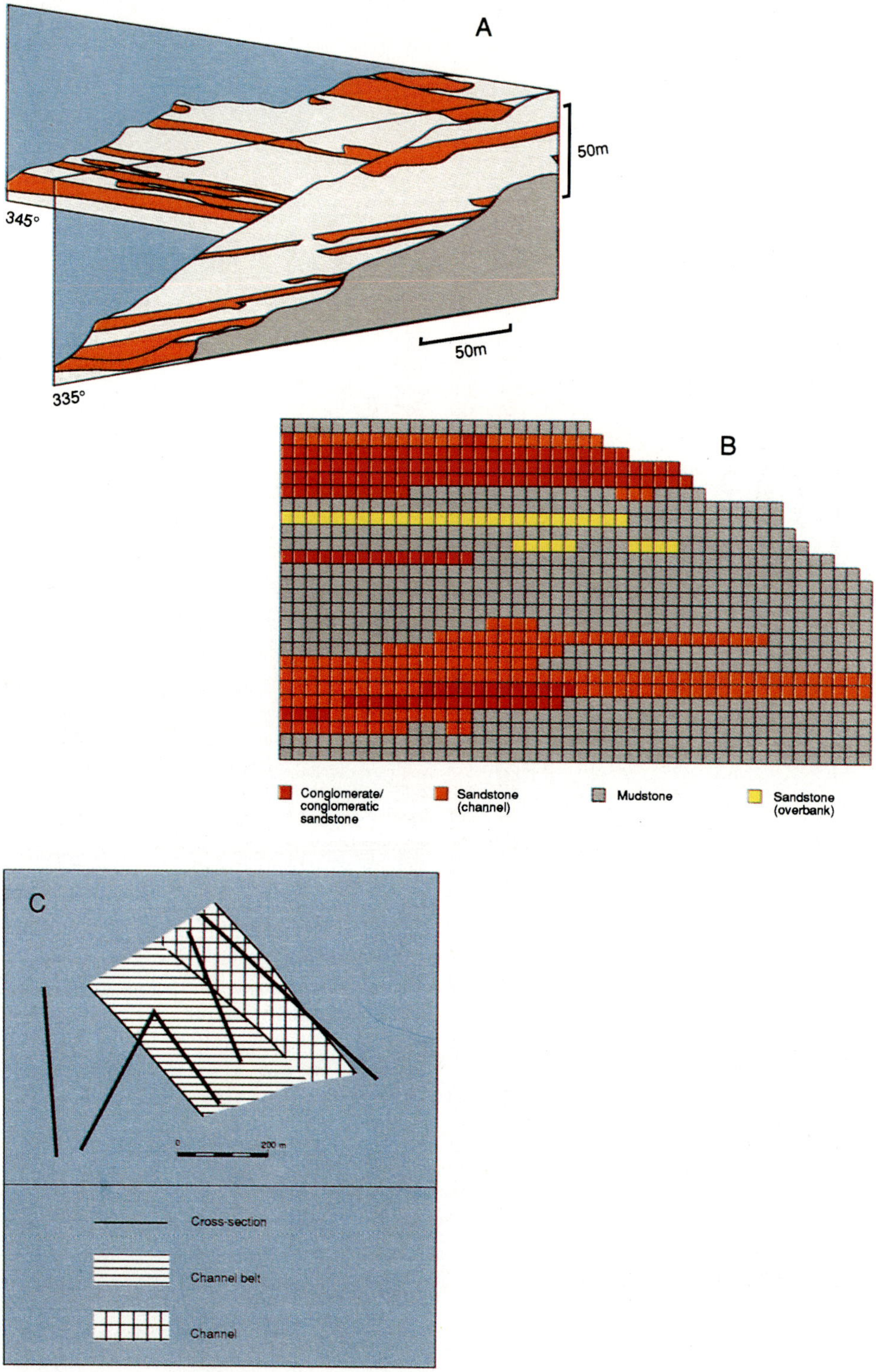

Plate 2. Examples of output from the database system of various types of primary data. (A) Fence diagram illustrating channel-belt morphologies. (B) Gridded generalization of cross-section with lithofacies data. (C) Sediment body map showing channel-belt and channel in plan view.

Puigdefabregas *et al.*, 1989). This system may have captured the southward-directed lateral drainage in a transition zone where the southerly-flowing rivers gradually (?) were deflected westwards (Fig. 5).

Sandbody interconnectness—reservoir communication

A factor of critical importance for hydrocarbon production is internal communication, or the reservoir 'plumbing system'. This factor may be studied and quantified, based on the data from outcrop analogues, by means of specific analysis functions in the SAFARI database system.

For instance, the sandbody volume which may be accessed by imaginary wells penetrating the Escanilla fluvial sequence can be calculated (cf. Haldorsen & McDonald, 1987). One or a combination of several cross-sections may be used, complimented by poroperm values derived from minipermeametry and architectural element analysis. It is possible to quantify the effects on reservoir depletion of variations in sandbody density, well-spacing and number, and location of wells (Fig. 13). Moreover, the fluid-flow pathways between imagined pairs of wells (injector–producer) may be mapped and quantified in terms of, for example, distance and tortuosity.

The communication in a reservoir is in reality a three-dimensional problem, and the 2-D structure of the cross-sectional data does not represent an ideal basis for communication analyses. Therefore, an interactive analysis function is integrated in the SAFARI database system which allows for correlation and delineation of channel bodies between cross-sections (Fig. 12). In this manner, map views of any channel body may be generated. By combining the cross-sectional and horizontal data, a fully three-dimensional geometrical model can be produced and analysed in terms of sandbody interconnectedness, angular dispersion of channels, and variations in channel width.

By performing the analyses outlined above, valuable guidelines as to the choice of well spacing, well number and positioning, and enhanced oil recovery (EOR) methods may be attained for fluvial reservoir sequences.

CONCLUSIONS

Quantitiative geological data for reservoir modelling purposes may be derived from studies of analogue outcrop sequences. Collection of large data sets is necessary, which can be analysed in order to provide data to support geomathematical modelling of heterogeneous reservoir sequences.

The 400–750 m thick Escanilla Formation, Ainsa Basin, Spanish Pyrenees, has many features in common with the Upper Triassic and Lower Jurassic fluvial sequences of offshore Norway. In an overall sense, the Escanilla Formation is progradational, and it can be subdivided into two informal members: the Mondot Member (coastal plain/marginal marine) and the Olson Member (fluvial). This study focuses on the Olson Member, which was deposited both on the medial to distal parts of an alluvial fan, and in the axial drainage system of the South Pyrenean foreland basin.

Braided and coarse-grained meandering stream deposits are most common in the Olson Member, with subordinate amounts of palaeovalley infills. The lack of mature palaeosols in the fine member testifies to rapid rates of deposition. The drainage pattern of the Olson fluvial system was significantly influenced by the presence of positive structural elements (anticlines) in the basin. A considerable variation in the lateral extent of and degree of interconnectedness between channel bodies characterizes the Escanilla Formation.

In the field, various mapping techniques were used to generate a large number of two-dimensional cross-sections along selected stratigraphic levels. The cross-sections can be joined to form lateral profiles several kilometres in length. Principles of bounding surface hierarchy and architectural element classifications have been applied to ensure systematic data handling.

Fourteen different types of data were collected in the Escanilla Formation (Table 1). Main emphasis was put on data pertaining to reservoir communication (i.e. large-scale cross-sections) and flow-unit distribution (i.e. small-scale cross-sections), as well as general data for comparative purposes. Palaeocurrent data have also been emphasized in order to understand the palaeodrainage pattern and its influence on sandbody orientation.

A GIS-type database system has been designed for storage and analysis of the outcrop data. Its primary purpose is to provide quantitative information to support geomathematical modelling tools and/or reservoir simulation tools. Output include data on the angular dispersion of channel bodies, estimates of sandbody interconnectedness and well contact volumes, W/T diagrams, gridded versions of

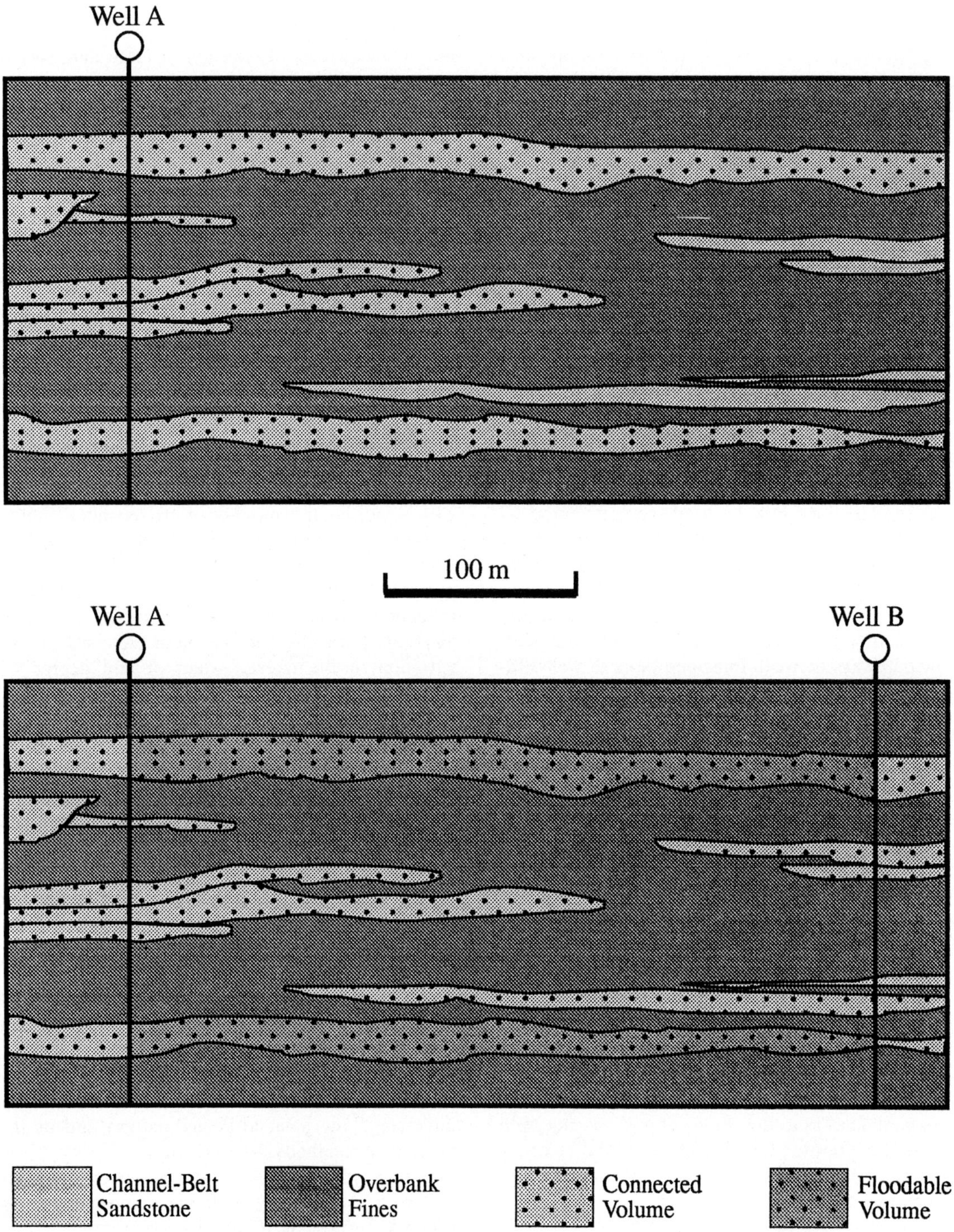

Fig. 13. Illustration of the analysis of sandbody interconnectedness and reservoir communication. The volumes connected to one (A) and two (B) imaginary wells are illustrated, and the floodable volume between an injector and a producer well is indicated (B). From such cross-sectional documentation of sandbody architecture in the SAFARI database the connected and floodable volumes may be analysed and quantified in relation to variable well spacings, configurations and perforation intervals.

cross-sections, and maps showing the distribution of channel bodies and architectural elements.

ACKNOWLEDGEMENTS

We thank Norsk Hydro, Saga Petroleum a.s., Den norske stats oljeselskap a.s. (Statoil) and the Norwegian Petroleum Directorate for the permission to publish information on the SAFARI project and the data collected from the Escanilla Formation.

A large number of people from the participating companies have contributed to the database development and the data collection work in the field.

We also gratefully acknowledge the assistance from C. Puigdefabregas in connection with the field work in the Spanish Pyrenees.

The constructive criticism of the manuscript provided by S.S. Flint, N. Tyler and an anonymous reviewer is greatly appreciated.

REFERENCES

Atkinson, C.D. (1983) Comparative sequences of ancient fluviatile deposition in the Tertiary South Pyrenean Basin, Northern Spain. Unpublished PhD Thesis, University of Wales, UK, 350 pp.

Atkinson, C.D. (1986) Tectonic control on alluvial sedimentation as revealed by an ancient catena in the Capella Formation (Eocene) of Northern Spain. In: *Palaeosols—Their Recognition and Interpretation* (Ed. by V.P. Wright), Blackwell Scientific Publications, Oxford, pp. 139–179.

Cotter, E. (1978) The evolution of fluvial style with special reference to the Central Appalachian Paleozoic. In: *Fluvial Sedimentology* (Ed. by A.D. Miall), Can. Soc. Petrol. Geol. Mem., **5**, 361–383.

Damsleth, E., Tjølsen, C.B., Omre, H. & Haldorsen, H.H. (1990) A two-stage stochastic model applied to a North Sea reservoir. *SPE* Paper **2065**.

Dreyer, T., Scheie, Å. & Walderhaug, O. (1990) Minipermeameter-based study of permeability trends in channel sand bodies. *Am. Assoc. Petrol. Geol. Bull.*, **74**, 359–374.

Ebanks, W.J. (1987) Geology in enhanced recovery. In: *Reservoir Sedimentology* (Ed. by R.W. Tillman & K.J. Weber). Spec. Publ. Soc. econ. Paleont. Miner., Tulsa, **40**, 1–14.

Giffen, M.C., Holbæk-Hansen, E., Steinlein, O. & Westad, F. (1990) Geological application of GIS with X, Z-dimensions, *EGIS 90, 1st European Conference on Geographical Information Systems*, Amsterdam, 10–13 April 1990.

Garrido-Mejias, A. (1968) Sobre la estratigrafia de los conglomerates Campanue (Santa Liestra) y formaciónes superiores del Eocene (extremo occidental de la cuenca de Tremp-Graus) Pirineo Central, Provincia de Huesca. *Acta Geológica Hispanica*, **3**, 359–374.

van de Graaff, W.J.E. & Ealy, P.J. (1989) Geological modelling for simulation studies. *Am. Assoc. Petrol. Geol. Bull.*, **73**, 1436–1444.

Gundesø, R. & Egeland, O. (1990) SESIMIRA—A new geological tool for 3-D modelling of heterogeneous reservoirs. In: *North Sea Oil and Gas Reservoirs—II* (Ed. by A.T. Buller, E. Berg, O. Hjelmeland, J. Kleppe, O. Torsæter & J.O. Aasen), pp. 363–371. Graham and Trotman, London.

Haldorsen, H.H. & McDonald, C.J. (1987) Stochastic modelling of underground reservoir facies (SMURF). *SPE* Paper **16751**, 575–589.

Hearn, C.L., Ebanks, W.J., Tye, R.S. & Rangathan, V. (1984) Geological factors influencing reservoir performance of the Hartzog Draw Field, Wyoming. *J. Petrol. Technol.*, **36**, 1335–1344.

Hirst, J.H. & Nichols, G.J. (1986) Thrust tectonic controls on Miocene alluvial distribution patterns, southern Pyrenees. In: *Foreland Basins* (Ed. by P.A. Allen & P. Homewood), Spec. Publs int. Ass. Sediment., **8**, 247–258.

Høimyr, Ø., Kleppe, A. & Nystuen, J.P. (1992) The effect of heterogeneities in braided stream channel sandstones on the simulation of oil revovery: A case study from the Snorre Field, North Sea. In: *Advances in Reservoir Geology* (Ed. by M. Ashton), Spec. Publ. geol. Soc. London. (In Press).

Kortekaas, Th.F.M. (1985) Water/oil displacement in crossbedded reservoir zones. *SPE Paper* **12112**.

McGowen, J.H. & Garner, L.E. (1970) Physiographic features and stratification types of coarse-grained point bars: modern and ancient examples. *Sedimentology*, **14**, 77–111.

Miall, A.D. (1988) Reservoir heterogeneities in fluvial sandstones: lessons learned from outcrop studies. *Am. Assoc. Petrol. Geol. Bull.*, **72**, 682–697.

Mutti, E., Seguret, M. & Sgaveti, M. (1988) *Sedimentation and Deformation in the Tertiary Sequences of the Southern Pyrenees*. Guidebook to Fieldtrip 7. AAPG Mediterranean Basins Conference, Nice, 199 pp.

Nickel, E. (1982) Alluvial-fan-carbonate facies with evaporites, Eocene Guarga Formation, southern Pyrenees, Spain. *Sedimentology*, **29**, 761–796.

Nijman, W.J. & Nio, S.D. (1975) The Eocene Montanana delta. In: *Guidebook to Excursion 19, part B. IXth Int. Sed. Congr. Nice, France* (Ed. by J. Rosell & C. Puigdefabregas), 56 pp.

Puigdefabregas, C. (1975) La Sedimentacion Molasica en la Cuenca de Jaca. *Monografias del Instituto de Estudios Pirenaicos*, No 104, 188 pp.

Puigdefabregas, C., Collinson, J.D., Cuevas, J.L., Dreyer, T., Marzo, M., Mercade, L., Nijman, W., Verges, J., Mellere, D. & Munoz, J.A. (1989) Alluvial deposits of the successive foreland basin stages and their relation to the Pyrenean thrust sequences. *Guidebook to the 4th Int. Conf. on Fluvial Sed.* (Ed. by M. Marzo & C. Puigdefabregas), 175 pp.

Retallack, G.J. (1988) Field recognition of paleosols. In: *Paleosols and Weathering Through Geologic Time: Principles and Applications* (Ed. by J. Reinhardt & W.R. Sigleo), pp. 1–20.

Sadler, P.M. (1981) Sediment accumulation rates and the completeness of stratigraphic sections. *J. Geol.*, **89**, 569–584.

Wright, V.P. (1989) Paleosol recognition. In: *Paleosols in Siliciclastic Sequences*, *PRIS short course notes*, 001, Reading University, 1–25.

Spec. Publs int. Ass. Sediment. (1993) **15**, 81–98

Quantitative facies analysis of coal-bearing sequences in the Bowen Basin, Australia: applications to reservoir description

A. FALKNER *and* C. FIELDING

Department of Earth Sciences, University of Queensland, Queensland 4072, Australia

ABSTRACT

The Late Permian sequence of the foreland Bowen Basin in east-central Queensland, Australia, is extensively mined for coal in large opencut mines. These highwall exposures combined with road and rail cuttings and close-spaced borehole networks present a superb opportunity to collect and interpret high quality, quantitative three-dimensional data on the external geometry and internal heterogeneity of sediment bodies.

Highwall photomosaics combined with information from boreholes were used to produce accurate reconstructions of sediment bodies in three dimensions. Data collected from surface exposures and boreholes have been combined to produce detailed sediment body descriptions and interpretations. Such techniques have been applied to the three major coal-bearing units in the Bowen Basin—German Creek Formation, Moranbah Coal Measures and Rangal Coal Measures—and equivalents at 12 mines. This paper details results from the former two units.

The German Creek Formation was formed by deltaic progradation eastward from the western, cratonic margin of the Bowen Basin. The unit was deposited on the lower delta plain and delta front of a mixed-influence delta moulded by fluvial outflow, waves and tides. The Moranbah Coal Measures are partly laterally equivalent to the German Creek Formation, but were derived from the eastern, orogenic margin of the basin. They accumulated in upper delta plain environments that contained large southerly flowing distributary channels.

Potential reservoir facies within the German Creek Formation comprise: (i) distributary channel belts; (ii) amalgamated, distributary mouth bars; and (iii) foreshore sands. Mouth bars form regionally extensive sand sheets with low internal heterogeneity. Channel belt deposits are less laterally extensive and have higher internal heterogeneity. Foreshore sands have very low internal heterogeneity but form only small rock volumes in comparison to the previously mentioned facies.

Channel belt sands within the Moranbah Coal Measures form the only potential reservoir facies with both high and low sinuosity channels forming belts up to 6 km wide. Low sinuosity channel fills show lower internal heterogeneity and often have marginal overbank wings that enhance interconnectedness of channel sandstones.

A strong linear relationship is displayed between the thickness and width of channel belt sandstone bodies in the two units.

INTRODUCTION

In recent years it has become desirable to improve our understanding of the external geometry and internal characteristics of sediment bodies for the purpose of hydrocarbon reservoir modelling. This trend has coincided with advances in the understanding of natural depositional systems, and in the capacity of computers to handle large and complex data sets.

Following the Geological Society of London meeting 'Quantification of Sediment Body Geometries and their Internal Heterogeneities' in 1988, it was apparent that two approaches to the problem could be adopted: (i) computer modelling of sediment body architecture, based on deterministic and actualistic data sets (e.g. Willis, 1989; Collier, 1989); and (ii) empirical data collection, ideally based on high quality outcrops and closely-spaced subsurface data sets (e.g. Lorenz *et al.*, 1985).

This paper documents a project aimed towards quantification of sediment body parameters in coal-bearing deltaic and alluvial plain sequences within the Permo-Triassic Bowen Basin of east-central Queensland, Australia. It illustrates the variety and quality of data which may be gathered from large exposures, in this case opencut coal mines and associated subsurface operations. This case study also emphasizes the relevance of such empirical data gathering to reservoir modelling, and underlines the value of large-scale, three-dimensional data to the development of facies models.

The Late Permian sequence of the Bowen Basin in central Queensland, Australia, contains measured and indicated coal reserves of 27 000 million tonnes worked in both opencut and underground operations (Rawlings, 1990). Two stratigraphic intervals in particular are exploited by numerous mines that collectively cover several hundred kilometres of strike (Fig. 1).

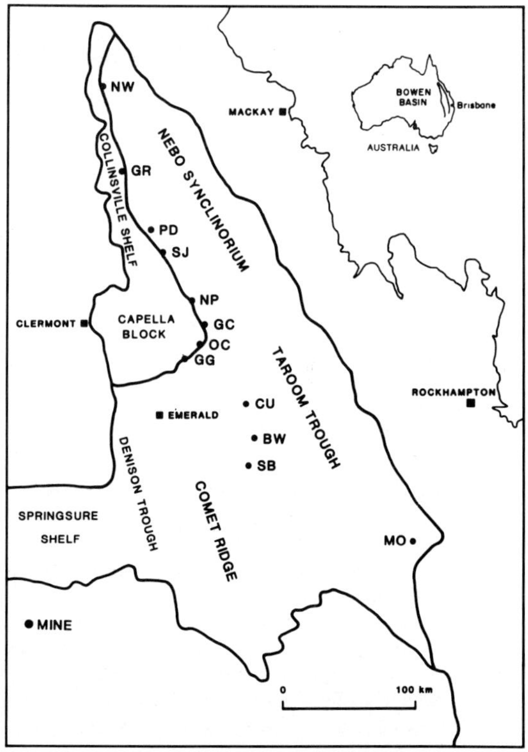

Fig. 1. Map illustrating the location of the Bowen Basin and the distribution of mining operations in the German Creek Formation, Moranbah Coal Measures and Rangal Coal Measures. NW = Newlands; GR = Goonyella/Riverside; PD = Peak Downs; SJ = Saraji; NP = Norwich Park; GC = German Creek; OC = Oaky Creek; GG = Gregory; CU = Curragh; BW = Blackwater; SB = South Blackwater; MO = Moura. Cross-sections illustrated in Figs 4 and 5 are oriented north–south along the western limb of the basin.

Highwall exposures average 40 m in height (Fig. 2) and collectively extend for 10–30 km at each mine. Multiseam operations reveal several interseam intervals that can be confidently correlated between mines. The extensive highwalls are supplemented by other man-made exposures in ramps, road and rail cuttings and creek diversion channels. Close-spaced borehole networks provide another source of information that allows sediment bodies exposed in highwalls to be confidently traced in the subsurface (Fig. 3). This is especially the case when underground mine exploration projects have been undertaken in reserves several kilometres downdip from highwall exposures. Data for this study were collected from 12 opencut mine sites (Fig. 1).

METHODOLOGY

In order to obtain detailed data on sediment bodies, controlled orthogonal photomosaics of Bowen Basin highwalls and other large exposures were produced using a 35 mm camera and 100 ASA colour print film. This allowed photomosaics to be prepared during fieldwork, thus enabling highwalls to be mapped on site shortly after photography. A 2 m scale stick was carried by an assistant and placed at 40 m intervals along the highwall during photography. Highwall mapping was then carried out with the aid of binoculars and involved annotation of transparent overlays. Where access to substantial vertical sections was possible, detailed sedimentological information such as sedimentary structures and palaeocurrent directions was recorded. Such access was typically at the ends of highwalls, at highwall failures and in mine ramps, creek diversions, railway and road cuttings.

A three-dimensional perspective of highwall exposures has been obtained from company data in the form of borehole records and data from previously-excavated strips such as photomosaics, highwall

Fig. 2. A typical Bowen Basin highwall exposure, at Moura mine. The face is approximately 40 m high and shows a levee sequence dipping gently towards the photographer, attached to a channel the fill of which is visible in the far distance (arrowed). The highwall is 1.75 km long and total strike exposure in the Moura mine is 25 km.

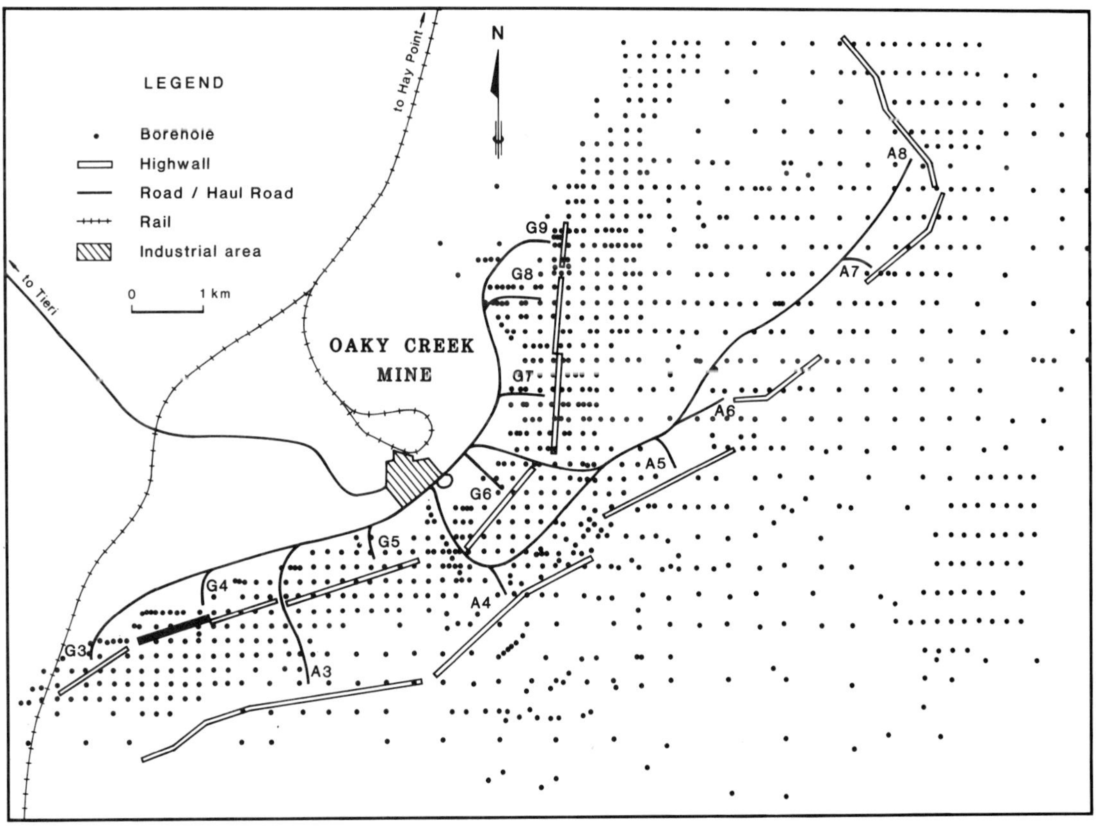

Fig. 3. Map of the Oaky Creek mine illustrating coverage of highwall exposures and borehole networks at a typical Bowen Basin coal mine.

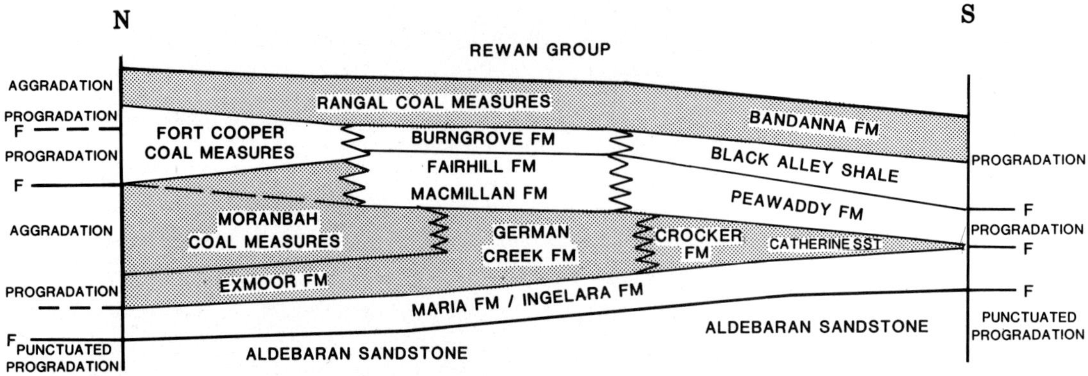

Fig. 4. Schematic, north–south cross-section along the western limb of the Bowen Basin, illustrating the geometry of Late Permian stratigraphic units (based on Draper & Balfe, 1985). Major flooding surfaces are marked F. The Moranbah Coal Measures/German Creek Formation interval can be recognized as a southward-thinning wedge, whereas the later Rangal Coal Measures and equivalents form a basinwide sheet.

maps and survey data. Boreholes typically are logged open hole with wireline logs (gamma ray, sonic, long- and short-spaced density, neutron), allowing calibration against less abundant cored holes or adjacent highwalls. Because of the abundance of boreholes, selected section lines were used to trace sediment bodies laterally.

Highwall maps were drafted utilizing photomosaics, field overlays and lithological logs. Sediment body boundaries and internal bedding surfaces were recorded using a simple hierarchical scheme. Sediment body dimensions were measured directly from photomosaics and selected highwall maps were digitized. Thus, a database of quantitative geometrical data with associated sedimentological information was constructed for a range of different facies.

REGIONAL GEOLOGY

The Bowen Basin is interpreted as a retroarc foreland basin that formed during the final stages of the Tasman Orogenic Phase (Murray, 1985). The fill is of Early Permian to Middle Triassic age with dominantly marine Early Permian strata overlain by a coal-bearing Late Permian section of mainly coastal to alluvial affinity, and a dominantly alluvial Triassic sequence.

The basin initially developed as a series of grabens and half-grabens arising from a phase of back-arc crustal extension, which coincided with a period of arc volcanism (Fielding *et al.*, 1990a). These early sub-basins were filled by thick sequences of continental clastic sediments, which were ultimately transgressed by the sea as a phase of more even, widespread thermal subsidence began. For much of Late Permian times, the Bowen Basin remained largely marine, with modest clastic sediment input from basin margins. In the latest Permian, compressive deformation of the basin began, coinciding with a resurgence in arc volcanism. The marine basin was rapidly infilled, and as thrust sheets propagated from the rising orogen to the east, the basin once again became continental. Sediment accumulation in the Bowen Basin probably ceased during Mid Triassic times, and was followed by a Mid-to-Late Triassic period of intense uplift and erosion (Beeston, 1986). The basin is preserved as a tectonically modified, north–south trending synclinal structure.

The German Creek Formation, Moranbah Coal Measures and Rangal Coal Measures (and equivalents) comprise the three most economically important units in the Bowen Basin, and form the basis for this study (Fig. 4). Recent studies of sediment dispersal patterns in the Bowen Basin by the authors and co-workers have provided a sound understanding of basinal palaeogeography (Fielding *et al.*, 1990b).

The German Creek Formation and Moranbah Coal Measures are partly laterally equivalent and were formed during the transition between the thermal subsidence phase and the onset of thrust loading. The former represents a deltaic and coastal complex attached to the western (cratonic) margin of the basin, whereas the latter forms a more laterally extensive clastic wedge supplied from the north-

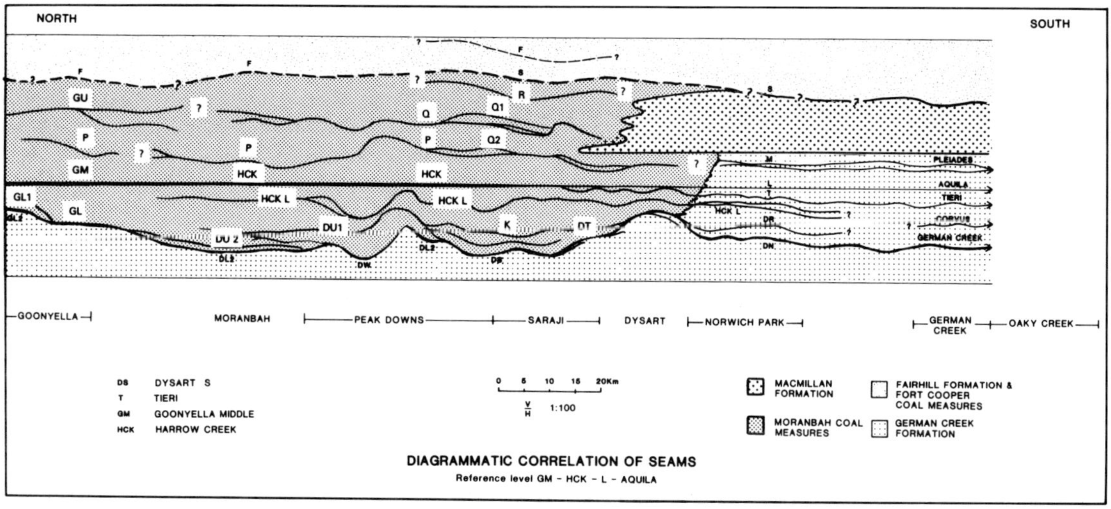

Fig. 5. North–south cross-section along the western limb of the Bowen Basin, illustrating coal seam stratigraphy within the German Creek Formation and Moranbah Coal Measures. Major coals in the latter show a much greater incidence of seam splitting than those in the former. Modified after BHP-Utah Coal Ltd.

ern and volcanically resurgent eastern basin margins (Fielding *et al.*, 1990b). The later Rangal Coal Measures and equivalents are interpreted as areally extensive, largely axially (southward) draining alluvial plain deposits, formed during the early stages of thrust loading and under the influence of active volcanism (Falkner & Fielding, 1990; Fielding *et al.*, 1990b).

The work reported here has involved data gathering from all three units mentioned above. In order to illustrate the results of the project and their usefulness, however, this paper will concentrate on the German Creek Formation and Moranbah Coal Measures.

The German Creek Formation and Moranbah Coal Measures form a regionally extensive, southward thinning stratigraphic package, bounded at base and top by flooding surfaces of regional extent (Fig. 4). The German Creek Formation overlies the marine Maria Formation, grades laterally into the marine Crocker Formation to the south and east and into the terrestrial Moranbah Coal Measures to the north, and is overlain by the marine Macmillan Formation (Figs 4, 5).

The lower German Creek Formation is typically 110 m thick and contains only minor coal. The upper, coal-rich interval is typically 160 m thick and has at its base the laterally extensive Lillyvale-German Creek-Dysart-Goonyella Lower seam (Fig. 5). This constitutes the major economic seam of the formation and is on average 3 m thick. It has a sheet-like geometry with only minor splits (Fig. 5), is low in ash and sulphur and is exported as coking coal. The other seams in the German Creek Formation also show modest splitting, but are thinner and in many cases are higher in ash and sulphur. All coals thin, split and ultimately pinch out as the German Creek Formation is traced eastward into the subsurface.

The German Creek Formation is sandstone dominated with subordinate thin, fine-grained sequences of interbedded siltstone and sandstone. Sandstones display a quartzose or mixed quartz-volcanic lithic mineralogy. The notion of a coastal environment of deposition is supported by the interfingering of coal-bearing and marine fossil-bearing facies, and the unit's stratigraphic context. Coal seams in the upper part of the formation can be traced continuously over hundreds of square kilometres indicating that the coastal plain was areally extensive. Bioturbation is extensive and varied, and rare marine fossils are also present. Falkner & Fielding (in press) interpret the German Creek Formation at outcrop as lower delta plain and delta front deposits of a mixed-influence (wave–tide–fluvial interactive) deltaic complex.

The Moranbah Coal Measures are 200–400 m thick on the western side of the basin, and up to

760 m thick, steeply dipping and structurally complicated to the east. The unit is partly laterally equivalent to the German Creek Formation (Figs 4, 5), though evidently derived from a different source area.

The Moranbah Coal Measures are predominantly fine grained in nature with interbedded siltstone and sandstone the most abundant rock type. Sandstone bodies within the unit are characteristically thick (>10 m) and laterally extensive. Coals are low in ash and sulphur, and up to 9 m thick. Seam geometry is complex with numerous component coal bodies (plies) and abrupt splitting and coalescence over short distances, in contrast with the German Creek Formation where seams are mostly laterally homogeneous over large areas (Fig. 5).

The Moranbah Coal Measures have been interpreted as upper delta plain deposits by Falkner & Fielding (1989, 1990). The fresh volcanic lithic mineralogy of the unit suggests derivation from the resurgent arc to the east of the basin, although dominantly southward directed palaeocurrent data indicate the establishment of axial drainage networks at this time. The volcanic lithic Moranbah Coal Measures prograded far enough southward to interdigitate with the craton-derived German Creek system (Figs 4, 5), producing the petrographic mixing observed in the upper German Creek Formation.

FACIES ASSEMBLAGES

Characteristics of lithofacies or facies associations recognized in the upper German Creek Formation and Moranbah Coal Measures are summarized in Table 1. These facies have been recognized on the basis of lithological and geometrical criteria, including lateral facies relationships observed on the mine lease scale (Fig. 6). Each formation contains a delta plain lithofacies assemblage (Facies 1–5; Table 1); the upper German Creek Formation additionally contains three facies of shoreline and delta front aspect (Facies 6–8; Table 1).

German Creek Formation

Distributary channel fills (Facies 1) comprise channelized, fining-upward sandstone sequences up to 15 m thick with mainly unimodal palaeocurrent directions (Fig. 6). The dominance of this facies at the Gregory and Oaky Creek mines, the occurrence of channel belts up to 2800 m wide, and instantaneous channel depths up to 15 m attest to considerable fluvial outflow. The presence of bimodal palaeocurrent directions and rhythmic siltstone laminae in some distributary facies indicate reversing flows, suggesting tidal flux within some channels. Three subfacies have been recognized on the basis of geometry and internal characteristics (Falkner & Fielding, in press):

1 laterally-accreted sandstone fills with rhythmic siltstone partings and unimodal palaeocurrents;

2 similar bodies with few siltstone partings and bimodal palaeocurrents; and

3 complex, vertically-accreted sandstone fills with unimodal palaeocurrents.

Interdistributary bay deposits (Facies 2–4) display a variety of small-scale sedimentary structures, with bi- or polymodal palaeocurrent distributions. Wave-formed ripples are abundant, together with a restricted suite of ichnofossils. These facies occur laterally adjacent to channel-belt deposits (Fig. 6). Coal seams (Facies 5) accumulated in areally extensive, long-lived coastal peat mires, the blanket geometry of seams suggesting accumulation on abandoned portions of the delta plain.

Sharp-bounded or fining upward, quartz-rich sandstones 1–6 m thick comprise Facies 6 (Fig. 7). These bodies are associated with channel mouths and cover several square kilometres, in at least one case with an ovoid planform, elongate parallel to the palaeocoastline (Fig. 6). They are interpreted as foreshore deposits, arising from wave reworking of river-borne sediments at the mouths of distributaries.

Thick, sharply-based, tabular sandstone sheets covering hundreds of square kilometres (Facies 7) occur immediately eastward of documented distributary mouths (Fig. 6). These bodies contain abundant deep, vertical burrows with rare marine body fossils. They are interpreted as remnants of distributary mouth bars formed by wave reworking of river-borne sands in shoreface water depths. The considerable areal extent of individual Facies 7 bodies implies that they represent amalgamation of several river mouth bars by wave processes (Fig. 8). Towards the lateral margins of mouth bar complexes, the base of Facies D is seen to be coarsening-upward rather than sharp (Fig. 8).

Laterally equivalent to Facies 7, and particularly abundant 20–30 km east (downdip) of distributary mouth sites, are sequences of extensively bioturbated, interbedded siltstone and sandstone (Facies 8) (Fig. 8). This facies becomes more abundant as the German Creek Formation is traced further east

Table 1. Characteristics of upper German Creek Formation and Moranbah Coal Measures lithofacies

Facies	Interpretation	Lithology	Sedimentary structures	Palaeocurrents	Geometry
1	Distributary channel	Erosively based, fine- to coarse-grained sandstone, variable siltstone partings	Epsilon cross-strata, trough cross-bedding, ripple cross-lamination	Unimodal or bimodal/bipolar	Elongate sheet German Creek Formation: width 300–2800 m (N–S) length > 10 km (E–W) thickness 2–15 m Moranbah Coal Measures: width ⩽ 6 km length > 10 km thickness 10–50 m
2	Interdistributary bay/lake	Thinly interbedded fine-grained sandstone and siltstone	Ripple cross-lamination, interlamination structures, wave ripples	Variable—bimodal or polymodal	Irregular sheet width ⩽ 10 km length ⩽ 10 km thickness ⩽ 30 m
3	Crevasse channel	Sharply-based medium- to coarse-grained sandstone	Trough cross-bedding, ripple cross-lamination	Variable—unimodal	Ribbon width ⩽ 50 m length > 0.5 km thickness ⩽ 8 m
4	Crevasse splay	Sharply-based medium- to coarse-grained sandstone	Trough cross-bedding, ripple cross-lamination	Variable—unimodal	Lobate sheet width typically 200 m length? thickness 0.5–3.5 m
5	Peat mire	Coal	Flat bedding	No data	Sheet width ⩽ 150 km (N–S) length ⩽ 50 km (E–W) thickness 1–10 m
6	Foreshore	Fine- to medium-grained quartzose sandstone	Flat and low-angle lamination inclined bedding surfaces ripple cross-lamination	Variable—typically eastward	Sheet (ovoid in some cases) width ⩽ 5 km length ⩽ 5 km thickness 1–6 m
7	Amalgamated mouth bars	Amalgamated fine- to coarse-grained quartzose sandstone beds, rare siltstone partings	Hummocky cross-stratification, flat and low-angle lamination, minor cross-bedding	No apparent trends	Sheet (grossly kidney-shaped) width ⩽ 100 km (N–S) length ⩽ 30 km (E–W) thickness 9–20 m
8	Distal delta front	Interbedded, intensely bioturbated sandstone and siltstone	Graded beds, flat lamination, hummocky cross-stratification	No data	Unknown probably extensive sheet width ? length ? thickness < 30 m

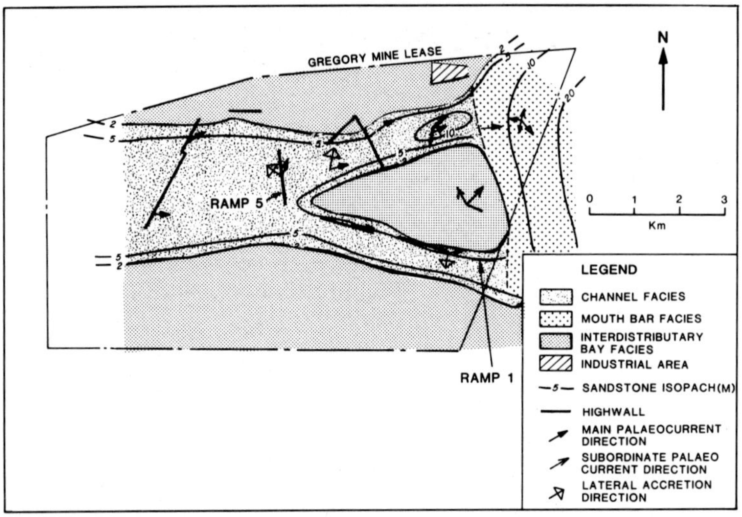

(a)

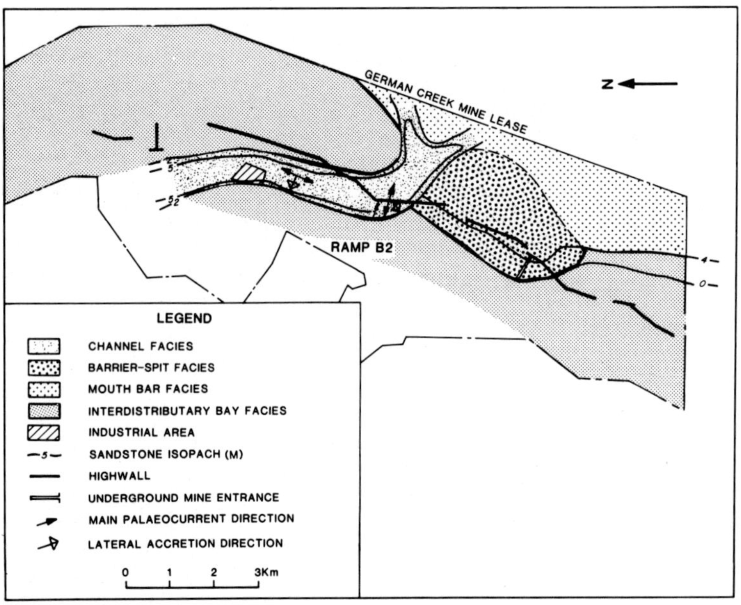

(b)

Fig. 6. Interpreted facies maps for the German Creek to Corvus interseam interval at (a) Gregory and (b) German Creek mines. Both show the lower reaches and mouth of distributary channels (Facies 1), flanked by interdistributary bay areas (Facies 2–4) and passing eastward into proximal mouth bar sands (Facies 7). The German Creek map additionally shows the development of a quartz sandstone body on the southern side of a distributary mouth, interpreted as a barrier-spit (Facies 6) formed by wave reworking and southward longshore drift of river-borne sands. Locations of highwalls illustrated in Fig. 11 are also shown. Gregory map based on highwall exposures and several hundred borehole records, German Creek map on extensive highwall and underground exposure and *c*. 40 boreholes.

into the subsurface. From its lithology and lateral relationships, Facies 8 is interpreted as representing distal portions of the delta front, formed in lower shoreface to offshore water depths.

The overall environment of deposition of the upper German Creek Formation as preserved is interpreted to be the lower delta plain and delta front of a mixed-influence (fluvial–wave–tide inter-active) delta system (Falkner & Fielding, in press). An arcuate shoreline trend and palaeobathymetric contours for the German Creek deltaic system are implied by the sites of distributary mouths and the distribution of Facies 7, respectively (Fig. 8), in turn suggesting that fluvially-supplied sediment was effectively reworked by wave processes in the marine basin. Conversely, the effect of tides is restricted to

Fig. 7. View of foreshore facies (6) overlying the German Creek seam at German Creek mine. Note low-angle dipping bedding surfaces and a thin siltstone parting within the 4 m thick foreshore interval. The overlying section comprises interdistributary bay facies (Facies 2–4) with thin coal seams (Facies 5). Geologist is 2 m high.

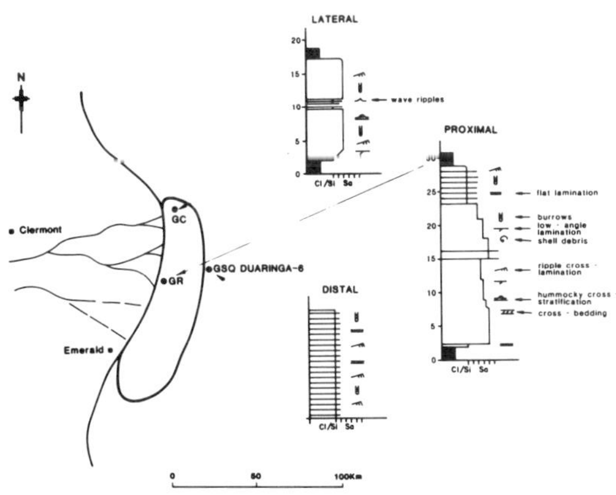

Fig. 8. Map showing the interpreted extent of amalgamated mouth bar sand sheets (Facies 7), variation in their vertical sequence character laterally and the eastward (offshore) transition to distal delta front facies (Facies 8) within one interseam interval of the German Creek Formation. Position of facies boundary is fixed by high density of boreholes in the northern half of the map (cf. Fig. 3), and more widely-spaced boreholes in the southern half. GC = German Creek mine; GR = Gregory mine.

sedimentary structures and bimodal palaeocurrent distributions in some channel and interdistributary sequences, suggesting only a minor role for tides in shaping the deltaic coastline.

Moranbah Coal Measures

Major channel deposits in the Moranbah Coal Measures (Facies 1) can be divided into three subfacies, two of which are sandstone dominated and the third comprises interbedded siltstones and sandstones. These are interpreted as laterally accreted, sandstone-dominated channel fills (Fig. 9), complex vertically-accreted channels, and inclined heterolithic stratified (IHS), laterally-accreted deposits, respectively. Palaeocurrent data from these structures indicate interaction between easterly-flowing and southerly streams as far north as Peak Downs mine, and a dominance of southerly-directed streams further north and east (Fig. 10). Limited evidence for tidal flows in channels is present in north or westward directed palaeocurrent distributions (Fig. 10)

Fig. 9. Highwall exposure of a laterally accreted channel fill (Facies 1) showing a cross-section through a meander bend in the Moranbah Coal Measures at Goonyella mine. The channel fill is 18 m thick at its thickest part and is overlain by interdistributary lake deposits (Facies 2–4) and the 1.5 m thick 'P' Tuff. Palaeocurrent direction trends out of the face and then back into it as the highwall is traversed.

and in the occurrence of rhythmic siltstone laminae in some channel sandstones.

Floodbasin facies (Facies 2–4; Table 1) are similar to those of the German Creek Formation, but show better-developed crevasse channel and crevasse splay deposits (Facies 3, 4). A further difference is a comparative paucity of wave-generated structures with respect to the upper German Creek Formation. Coal seams up to 9 m thick (Facies 5) can be traced for up to 150 km north–south and show complex splitting characteristics (Fig. 5). Consequently, interseam intervals within the Moranbah Coal Measures are highly variable in thickness, in contrast to those of the German Creek Formation (Fig. 5). The more complex seam geometry is suggestive of greater levels of clastic sediment introduction and compactional control on sediment body architecture than in the German Creek Formation.

The Moranbah Coal Measures are interpreted as the deposits of an extensive upper delta plain that was occasionally subjected to tidal inundation, perhaps during marine transgressions recorded in the German Creek Formation. Other than the limited evidence for tidal flux, no indicators of lower delta plain conditions (such as widespread wave-generated sedimentary structures in interdistributary facies) have been noted from the Moranbah Coal Measures. The delta plain was crossed by a variety of large distributary channels, with straight to sinuous morphologies, which flowed mainly towards the south. Along the western margin of the basin, axially (southward) flowing streams interacted with eastward-flowing streams sourced in the adjacent cratonic landmass (Fig. 10).

Most of the delta plain at any given time was the site of fine-grained sediment deposition in floodbasin environments lateral to river channels. These interdistributary lakes were infilled by a combination of overbank and crevasse-derived sediments. Areally extensive, long-lived peat mires formed on aban-

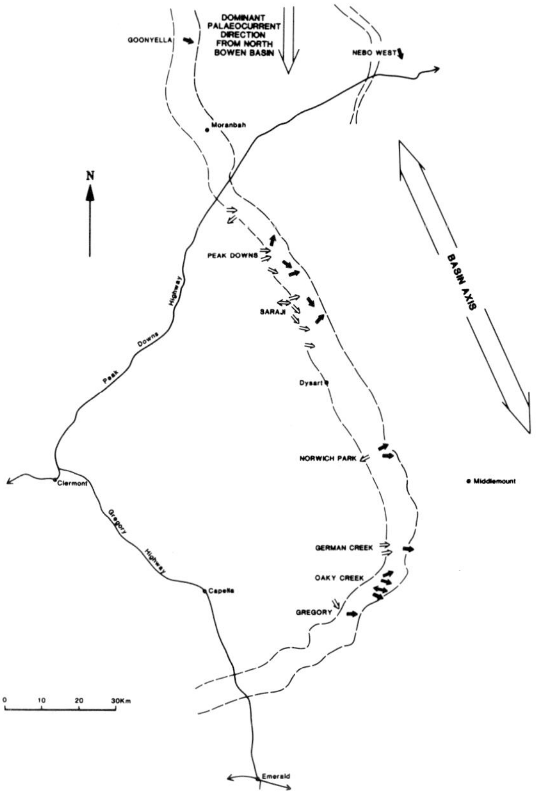

Fig. 10. Map showing palaeocurrent mean directions for channel belt sandstone bodies mapped in the German Creek Formation and Moranbah Coal Measures. Exposures of strata below the German Creek seam and equivalents are denoted by unfilled arrows, those above by filled arrows. Double-headed arrows indicate bipolar distributions. Note the gradual northward change from exclusively eastward-directed streams to a mixture of east and more southward-directed drainage.

doned portions of the plain and were occasionally invaded by major channels, creating seam splits.

RESERVOIR CHARACTERISTICS

The reservoir properties of facies from the units discussed in this paper are here considered with respect to (i) geometry (thickness, width, length); (ii) orientation (within the basin); (iii) internal architecture; and (iv) horizontal and vertical heterogeneity. Reservoir quality cannot be covered adequately, owing to a general lack of poroperm data.

German Creek Formation

Three sandstone facies within the upper German Creek Formation have reservoir potential: (1) distributary channel; (6) foreshore; and (7) amalgamated mouth bar. It is equally important, however, to consider non-reservoir facies in order to assist evaluation of possible multi-pay reservoirs. Each facies has quite distinct geometrical and internal characteristics, which are detailed below.

Distributary channel (Facies 1)

This facies occurs as elongate sheet sandstone bodies (following the scheme of Friend *et al.*, 1979), 2–15 m thick and 300–2800 m wide, elongate perpendicular to the palaeoshoreline (i.e. elongate east–west). Highwall exposures (Fig. 11) and isopach maps (Fig. 6) suggest that channel bodies of subfacies 1 and 2 (laterally-accreted types) are tabular in cross-section with abrupt thickness decreases towards margins. Subfacies 3 bodies, conversely, show a more regular 'double wedge' cross-section with systematic variation in thickness (Fig. 11). Channel bodies thicken and diverge laterally at river mouths and are continuous basinwards with proximal mouth bar sands (Facies 7) (Fig. 6). Because of this association, subfacies 1 and 2 are interpreted as representing the lower tidal reaches of distributaries, with subfacies 2 recording the greater tidal influence. Subfacies 3, which has not been observed in river mouth settings, is interpreted to represent upper delta plain reaches of distributaries, above the limit of tidal inundation.

All channel bodies show signs of internal inhomogeneity (Fig. 11). The laterally-accreted subfacies 1 and 2 display regular internal dipping surfaces (epsilon cross-strata) with variable development of siltstone partings along these surfaces (Fig. 11). The orientation of these lateral accretion surfaces is highly variable. In the case of subfacies 1, the upper part of the channel fill is typically composed of siltstone, its thickness increasing towards the final channel plug (Fig. 11). The internal architecture of subfacies 3 is less predictable, with nesting of broadly channelized erosion surfaces and numerous discontinuous siltstone partings throughout the body (Fig. 11). The overall infilling geometry is close to concentric.

The internal heterogeneity noted above would doubtless make Facies 1 an anisotropic reservoir with respect to fluid flow. While the thickness and

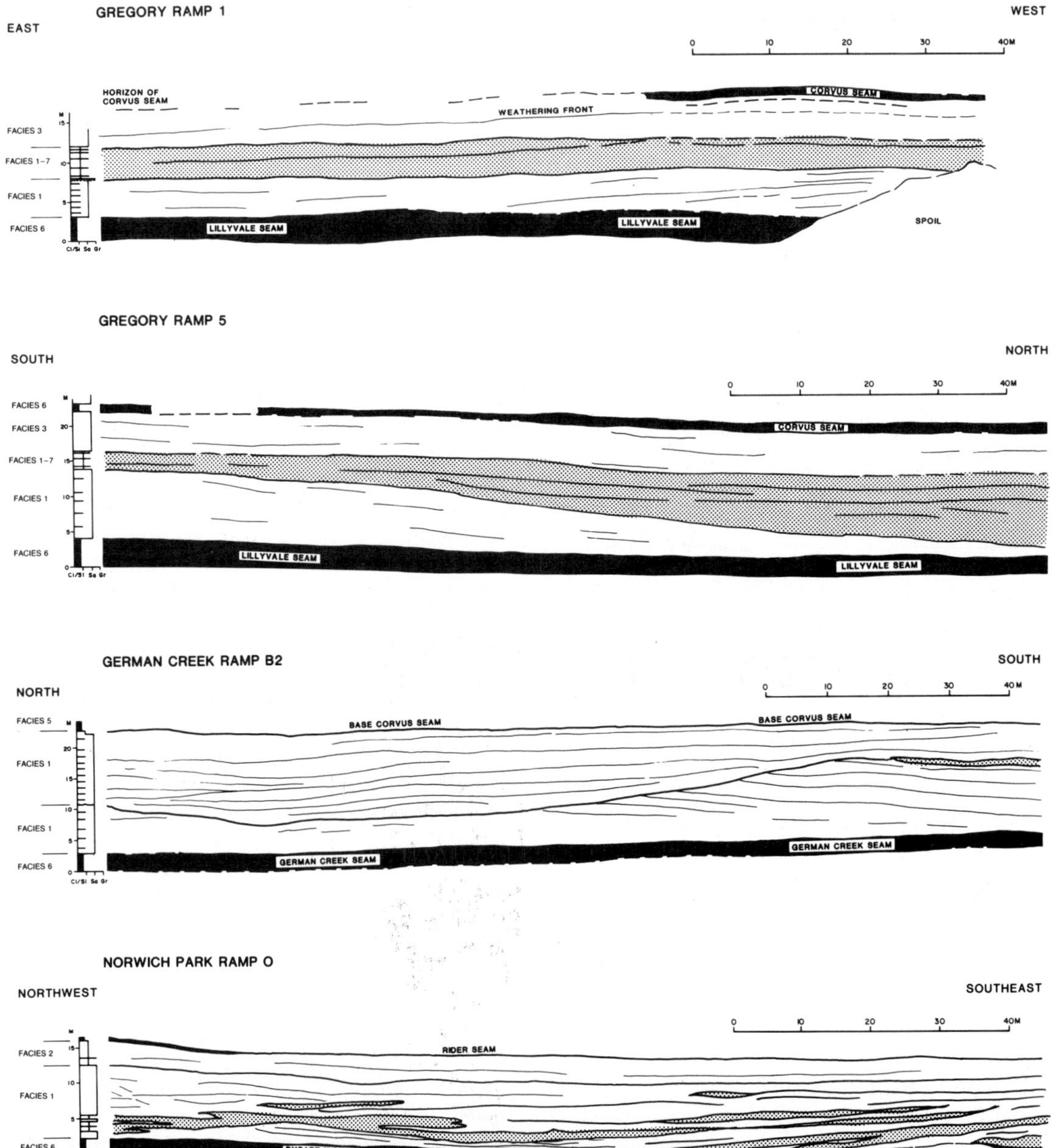

Fig. 11. Examples of sandstone body architecture within distributary channel fills (Facies 1) of the German Creek Formation, traced from controlled photomosaics of highwalls. Coals are shaded with a dense tint, interdistributary bay deposits (Facies 2–4) with a lighter tint and sandstones remain untinted. Representative bedding planes within the various sediment bodies are drawn as thin lines, whereas major bounding surfaces are shown as thick lines. Gregory Ramp 1 and Ramp 5 show longitudinal and transverse sections respectively through a subfacies 1 fill, while German Creek Ramp B2 and Norwich Park Ramp 0 illustrate transverse sections through subfacies 2 and 3 fills, respectively. See Fig. 6 for location of Gregory and German Creek sections.

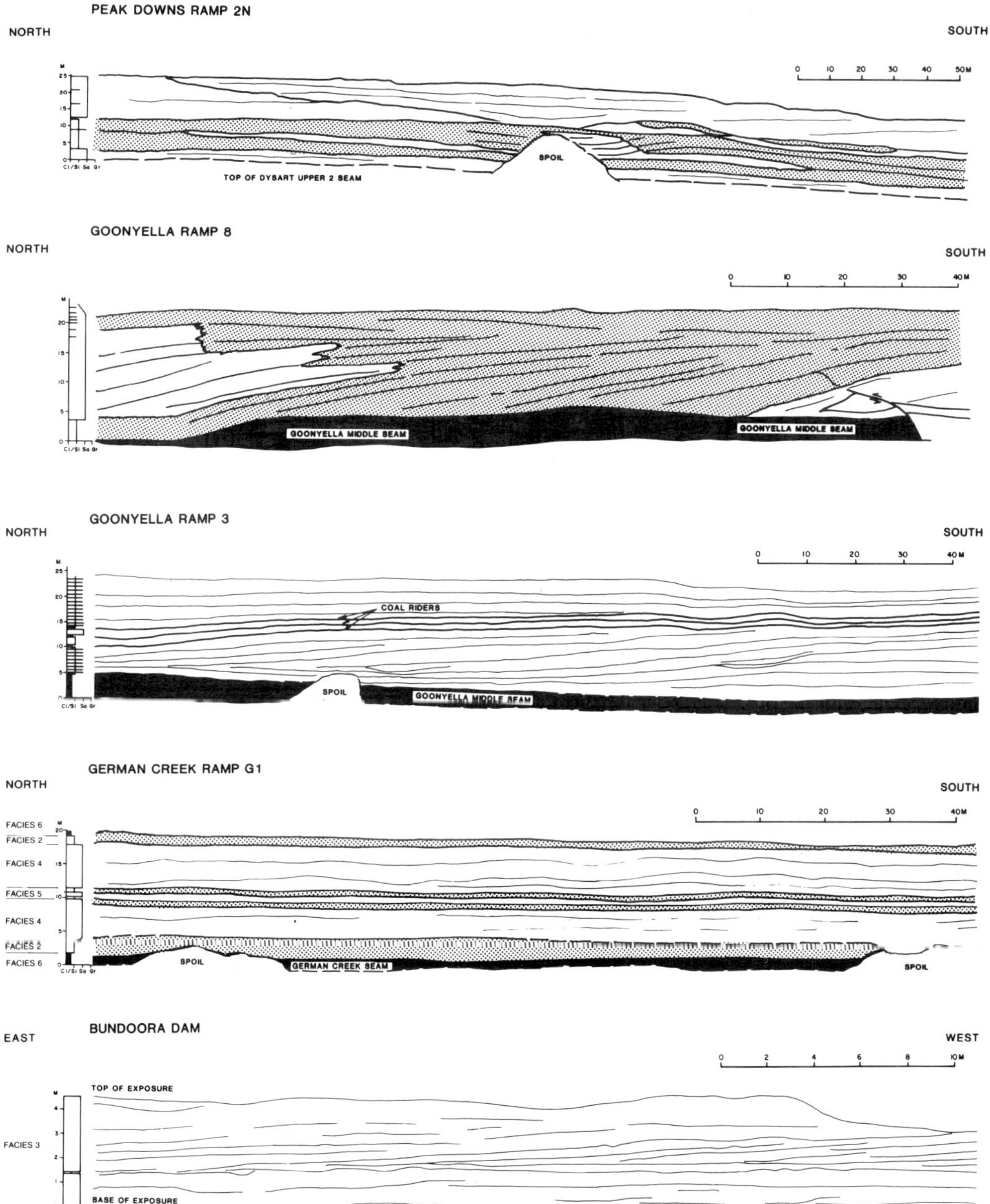

Fig. 12. Examples of sandstone body architecture within distributary channel fills (Facies 1) of the Moranbah Coal Measures, and mouth bar (Facies 7) and foreshore (Facies 6) deposits of the German Creek Formation. Shading and line details are the same as in Fig. 11. Peak Downs Ramp 2N, Goonyella Ramp 8 and Goonyella Ramp 3 show transverse sections through vertically accreted, laterally accreted (sandstone-dominated) and laterally accreted (IHS-dominated) Moranbah Coal Measure channel fill types, respectively. Note that the light tint in the Goonyella Ramp 8 section has been used to denote sandstone-dominated interbedded strata, whereas no tint has been used in the Goonyella Ramp 3 section to aid clarity. German Creek Ramp G1 illustrates two mouth bar (Facies 7) sandbodies within the German Creek Formation and enclosing strata, and Bundoora Dam a foreshore sequence (Facies 6, German Creek Formation) showing seaward-dipping bedding surfaces.

lateral continuity of internal siltstone partings would not be likely to impart any significant permeability barrier, the siltstone plugs noted in some subfacies 1 bodies could act as local seals (Fig. 11).

Foreshore (Facies 6)

These in some cases show ovoid plan geometry and limited lateral extent, located along-strike from a distributary mouth (Fig. 6), and in others a more sheet-like geometry. In the former case, a barrier-spit origin can be inferred (Falkner & Fielding, in press), involving sand accumulation by longshore drift on the south side of a channel mouth. This example passes eastward into proximal mouth bar deposits (Facies 7) (Fig. 6). Internally, foreshore sands appear homogeneous, in some cases with seaward (east)-dipping bedding planes representing prograding foreshore surfaces (Fig. 12). Very few fine-grained partings were noted from this facies.

Amalgamated mouth bar (Facies 7)

Mouth bar sandstone bodies represent the most extensive potential reservoir facies in the German Creek system. Their great lateral extent (up to 30 km east–west by 100 km north–south) probably reflects homogenization of all mouth bars on the delta front at any given time by wave reworking. These broadly kidney-shaped sheet sands are elongate north–south, mirroring the palaeoshoreline trend (Fig. 8), and are located immediately offshore from mapped distributary mouths (Fig. 6). Internally, sands show a layer-cake architecture (Fig. 12), with laterally extensive component sand sheets separated by similarly extensive siltstone partings. These bodies would probably behave as strongly vertically stratified reservoirs, with little anisotropy within component layers (Fig. 12).

Other facies

Facies 7 passes offshore (eastward) into the more interbedded distal delta front deposits of Facies 8. The lithology of the latter would minimize reservoir potential, despite the possibility of their being laterally extensive. Similarly, the mainly thinly interbedded nature of interdistributary bay deposits (Facies 2–4) would limit reservoir potential. Such intervals would be expected to act as low permeability layers within thicker reservoir sections (Fig. 11).

Coal seams (Facies 5) show sheet geometry, and are in places quite thick (up to 10 m). Consequently, it is possible that some German Creek coals could act as partial seals within a thicker reservoir section.

Moranbah Coal Measures

Distributary channel fills (Facies 1) are the main potential reservoir facies within the Moranbah Coal Measures. Both high and low sinuosity types are

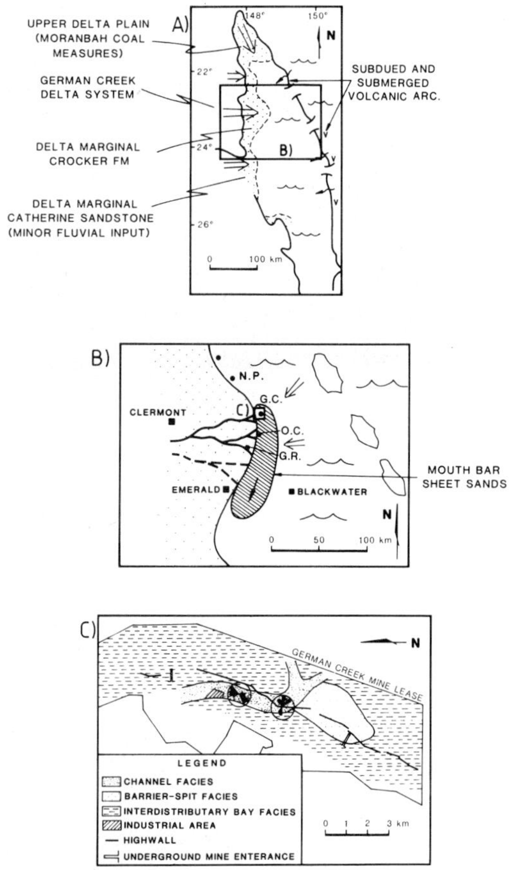

Fig. 13. Palaeogeographic maps for the German Creek Formation, illustrating the different scales of investigation in the study: (A) Basinal, showing sediment dispersal patterns within the Bowen Basin during accumulation of the German Creek Formation and Moranbah Coal Measures. (B) Regional, showing the German Creek delta system prograding eastward into a shallow sea with delta front mouth bar sands. ⇦ = wave direction. ↓ = longshore drift direction. (C) Local, a facies map for the German Creek mine illustrating facies relationships within the German Creek Formation.

Fig. 14. View of a concentrically filled, sandstone-dominated channel in a laterally accreted, mainly IHS-filled channel belt deposit within the Rangal Coal Measures at Blackwater mine. The sandstone ranges up to 9 m thick.

represented as channel-belt deposits up to 50 m thick and 6 km wide. Channel body orientation is more variable than in the German Creek Formation, as noted above.

Laterally accreted sandstone channel fills (Fig. 12) are characterized by thick, continuous sandstone bodies with some interbedded siltstone, and epsilon cross-strata. A moderately anisotropic reservoir behaviour with respect to fluid flow may be expected. Vertically accreted sandstone channel fills are entirely similar to those described from the German Creek Formation (subfacies 3; compare Figs 11 and 12), and may consequently be expected to form more internally heterogeneous reservoirs. Connectivity between bodies of this variety may be improved by overbank 'wings' that are similar lithologically to channel fills and extend for hundreds of metres beyond channel belt margins. Inclined heterolithic stratified (IHS) channel fills comprise epsilon cross-stratified, thinly interbedded sandstone/siltstone (Fig. 12). Such internally complex bodies would be likely to form anisotropic reservoirs, particularly given the variability in dip azimuth of epsilon cross-strata.

As with the German Creek Formation, interdistributary sequences (Facies 2–4) are considered to have only limited reservoir potential, although crevasse splay sandstones (Facies 4) and associated channels (Facies 3) within floodbasin sequences could increase connectivity between neighbouring channel sandstones (Fig. 12). The same observations apply to the thick coals (Facies 5) of the Moranbah Coal Measures as to those of the German Creek Formation.

DISCUSSION AND CONCLUSIONS

The lithological and geometrical characteristics of the German Creek and Moranbah Coal Measures depositional systems are detailed above. Within the German Creek Coal Measures, sediment bodies

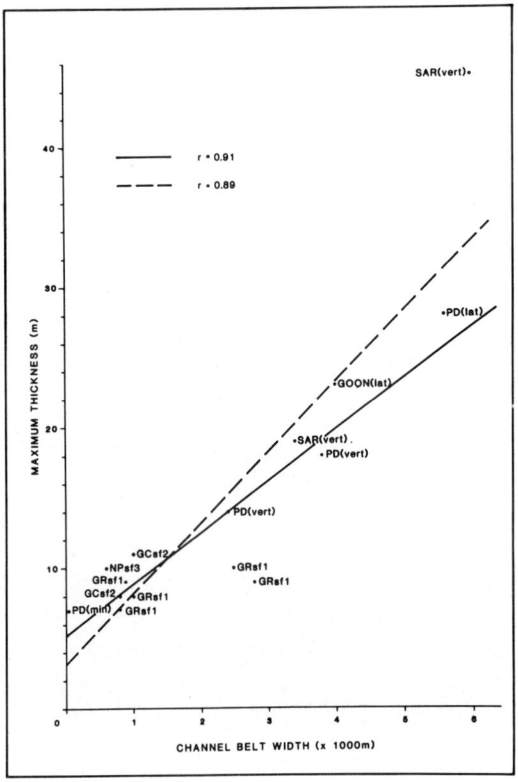

Fig. 15. Crossplot of sandstone body maximum thickness against channel belt width for distributary channel deposits of the German Creek Formation and Moranbah Coal Measures (all data mapped directly from highwalls). Linear regression lines describing the data set are plotted, the solid line excluding the point at the top right of the plot, and the dashed line including it. Pearson's regression coefficient for the two relationships is given at the top left. GR = Gregory; GC = German Creek; NP = Norwich Park; PD = Peak Downs; SAR = Saraji; GOON = Goonyella; (min) = minor crevasse channel; (vert) = vertically-accreted channel fill; (lat) = laterally-accreted channel fill; (sf) = subfacies.

associated with deposition in a mixed-influence (fluvial–wave–tide interactive) deltaic system are recognized. The major potential reservoir facies here are distributary channel, foreshore and mouth bar deposits, each of which have quite distinct characteristics. Of particular interest is the extensive nature of amalgamated mouth bar facies in this system. The upper delta plain Moranbah Coal Measures (and the alluvial Rangal Coal Measures) display a relatively simple lithofacies assemblage, the main variety being in differing styles of channel fills. In these systems channel deposits are the only potential reservoir facies, and some styles of channel fill are considered to have only limited potential.

Considerable quantitative data have been gathered, with respect not only to individual facies but also to geometrical relationships between facies. Indeed, it has been possible in this study to examine the formation of interest at every scale from small-scale outcrop to that of the entire basin (Fig. 13). This introduces the opportunity, for example, to investigate possible relationships between the dimensions of individual facies and those of the entire system.

This paper has focused on two interfingering, coeval depositional systems within the Late Permian of the Bowen Basin. The differences between the two can be attributed to their differing environment of deposition and setting within the basin (Fig. 13A). Some similarity is noted, however, particularly between the distributary channel fills of the two units, which might intuitively be expected. In particular, the vertically-accreted channel fill styles in each unit are similar (Figs 11, 12).

The large-scale, laterally-accreted channel fill types within the Moranbah Coal Measures are not found within the German Creek Formation. These sandstone-filled and IHS-filled channels are peculiar to the volcanic lithic, latest Permian formations in the Bowen Basin, being also abundant in the Rangal and Baralaba Coal Measures. Furthermore, they are restricted to upper delta plain and alluvial plain sequences, possibly implying a base-level control on fluvial style. In the Rangal Coal Measures and equivalents, channel fills are exceptionally heterogeneous involving random interbedding of sandstone and IHS strata, which makes the recognition of channel fill as opposed to floodbasin facies difficult even in highwall exposure. Such heterogeneous channel fills would constitute a most difficult target to characterize or predict ahead of drilling (Fig. 14).

In an attempt to evaluate any possible common links between channel fill dimensions within the German Creek and Moranbah Coal Measures, crossplots were prepared. A strong relationship was found to exist between sandstone body maximum thickness and channel belt width (Fig. 15). Such a relationship has considerable predictive value, as it suggests that the width of a channel belt deposit within these formations may be predicted from its maximum thickness. Indeed, it is plausible that such relationships may describe other deltaic plain and alluvial sequences, a hypothesis that cannot presently be tested owing to the lack of published data sets of this kind.

Characteristics of facies within the formations discussed herein are considered to reflect a combination of autocyclic and allocyclic controls. Possible base-level influences on features such as coal seam geometry and fluvial style are currently under investigation as part of a sequence stratigraphic analysis of the Bowen Basin.

ACKNOWLEDGEMENTS

This project was funded by a grant from BP Petroleum Development Ltd, whose support and encouragement are gratefully acknowledged. BHP-Utah Coal Ltd, Capcoal, Central Queensland Coal Associates and Oaky Creek Coal are thanked for allowing access to exposures and other data. Traceson Drafting produced the illustrations at short notice, and S. Flint, C. O'Byrne and H. Williams reviewed the manuscript.

REFERENCES

Beeston, J.W. (1986) Coal rank variation in the Bowen Basin, Queensland. *Int. J. coal Geol.*, **6**, 163–179.

Collier, R.E.Ll. (1989) Modelling the role of differential compaction and tectonics upon Westphalian architecture in the Northumberland Basin. In: *The Role of Tectonics in Devonian and Carboniferous Sedimentation in the British Isles* (Ed. by R.S. Arthurton, P. Gutteridge & S.C. Nolan) Yorks. Geol. Soc. Occas. Publ. **6**, 189–199.

Draper, J.J. & Balfe, P.E. (1985) Late Permian stratigraphy—western Bowen Basin, Queensland. In: *Advances in the Study of the Sydney Basin, Proceedings of the 19th Symposium*, University of Newcastle, Newcastle NSW, pp. 106–109.

Falkner, A.J. & Fielding, C.R. (1989) German Creek Formation and Moranbah Coal Measures: a transition from marine shelf to upper delta plain. In: *Advances in the Study of the Sydney Basin, Proceedings of the 24th Symposium*, University of Newcastle, Newcastle NSW, pp. 41–48.

Falkner, A.J. & Fielding, C.R. (1990) Late Permian coal-bearing depositional systems of the Bowen Basin. In: *Bowen Basin Symposium 1990 Proceedings*, Mackay Qld., Geol. Soc. Austr. (Qld Div.), 36–41.

Falkner, A.J. & Fielding, C.R. (1993) Geometrical facies analysis of a mixed-influence deltaic system—the Late Permian Creek Formation, Bowen Basin, Australia. In: *Alluvial Sedimentation* (Ed. by M. Marzo & C. Puigdefabregas). Spec. Publs int. Ass. Sediment, Blackwell Scientific Publications, Oxford, **17** (in press).

Fielding, C.R., Gray, A.R.G., Harris, G.I. & Salomon, J.A. (1990a) The Bowen Basin and overlying Surat Basin. In: *The Eromanga to Brisbane Geoscience Transect: a Guide to Crustal Development across Phanerozoic Australia in southern Queensland* (Ed. by D.M. Finlayson), Bur. Min. Resour. Austr. Geol. Geophys. Bull. **232**, 105–116.

Fielding, C.R., Falkner, A.J., Kassan, J. & Draper, J.J. (1990b) Permian and Triassic depositional systems in the Bowen Basin. In: *Bowen Basin Symposium 1990 Proceedings*, Mackay Qld, Geol. Soc. Austr. (Qld Div.), 21–25.

Friend, P.F., Slater, M.J. & Williams, R.C. (1979) Vertical and lateral building of river sandstone bodies, Ebro Basin, Spain. *J. geol. Soc. London*, **136**, 39–46.

Lorenz, J.C., Heinze, D.M., Clark, J.A. & Searls, C.A. (1985) Determination of widths of meander-belt sandstone reservoirs from vertical downhole data, Mesaverde Group, Piceance Creek Basin, Colorado. *Bull. Am. Assoc. Petrol. Geol.*, **69**, 710–721.

Murray, C.G. (1985) Tectonic setting of the Bowen Basin. In: *Bowen Basin Coal Symposium*, Geol. Soc. Austr. Abstr., **17**, 5–16.

Rawlings, C.D. (1990) Geological maintenance: a reactive or preventative management method in Bowen Basin coal mines. In: *Bowen Basin Symposium 1990 Proceedings*, Mackay Qld, Geol. Soc. Austr. (Qld Div.), 101–105.

Willis, B.J. (1989) Palaeochannel reconstructions from point bar deposits: a three-dimensional perspective. *Sedimentology*, **36**, 757–766.

Spec. Publs int. Ass. Sediment. (1993) **15**, 99–112

Quantification of turbidite facies in a reservoir-analogous submarine-fan channel sandbody, south-central Pyrenees, Spain*

J.D. SCHUPPERS

Faculty of Mining and Petroleum Engineering, Delft University of Technology, PO Box 5028, 2600 GA Delft, The Netherlands

ABSTRACT

This study deals with the description and quantification of turbidite deposits from the Lower Ainsa Channel in the south-central Pyrenees, Spain. The channel is part of a turbidite succession which filled an elongated foreland basin in Eocene time. The outcrop indicates a minimum original channel width of 500 m and a minimum depth of approximately 35 m. The channel-fill has been subdivided into five lithofacies. The sediments consist predominantly of an alternation of thick-bedded sandstone and very thin-bedded, bioturbated mudstone and silt. Mean sandstone permeability is 7 mD. Locally, pervasive burrowing (up to 18% of rock volume) and relatively high burrow-fill permeability (up to 36 mD) greatly improved the potential reservoir quality of the otherwise impermeable mudstone/siltstone intervals. The channel-fill is asymmetrical, with a sandier, coarser-grained and more strongly amalgamated left part of the channel compared with the right part of the channel (looking in a downcurrent direction). Data are presented on the nature, geometry and distribution of the distinct lithofacies, mudstone/siltstone lengths and sandstone permeabilities.

INTRODUCTION

Sand-lobes and sand-filled channels on submarine fans and related turbidite systems possess important hydrocarbon reservoir potential. Turbidite characteristics of interest for exploration purposes include the following.

1 The large areal extent of lobe sheet-sands makes lobes attractive exploration targets.

2 Channel-sands are less continuous than lobe-sands, but generally show a better sorting, larger grain size, thicker beds, and fewer interbedded shaly permeability barriers, which lends the channel-fills excellent reservoir quality (Walker, 1980).

3 Turbidite environments with a high sand/shale ratio (channels and proximal lobes) exhibit generally higher depositional porosity than facies with low sand/shale ratio (Bjølykke, 1983), and are likely to retain higher porosity and permeability throughout their diagenetic history (McLean, 1981).

4 Stratigraphic traps are provided by the lateral facies change of channel-sands into fines, the vertical sealing of sandbodies by shale-blankets, and the onlap of sands to slope muds or the flanks of isolated basins (Walker, 1978; Wilde *et al.*, 1985).

In contrast with the large volumes of oil-in-place, turbidite reservoirs generally show a very low recovery efficiency due to their complex reservoir architecture (Tyler *et al.*, 1984). Therefore, hydrocarbon production from turbidite reservoirs is likely to benefit from more detailed knowledge of bed-continuity, interconnectedness and permeability distribution. Submarine-fan deposits are prime candidates for geological characterization, both in producing fields as well as in outcrop analogues (Schuppers *et al.*, 1990).

The influence of depositional architecture on the production behaviour of reservoirs has been widely

* Research Centre for Geological Reservoir Characterization (GeoRes) publ. no. 003.

recognized (Gillund, 1969; Krause *et al.*, 1987). Spatial variation of sedimentary facies creates distinct compartments in reservoirs. The arrangement of these compartments builds the conduit system for fluid flow and strongly dictates the amount of ultimate recovery.

The aim of any geological reservoir description is to unravel the sandbody geometry, lithological continuity, and if possible even bed-to-bed correlation between wells. For reservoir characterization purposes, correlation must identify lithohydraulic units, i.e. distinct lithological/petrophysical entities which can be regarded as reservoir (sub)zonations or flow-units. As lithohydraulic units are defined by their depositional, diagenetic and petrophysical properties, it is these units that, by their spatial distribution, determine the hydraulic behaviour and hence fluid production and injection trends within reservoirs.

Standard sedimentological outcrop description is concerned with lithology, grain size and sorting, sedimentary structures and trends in both lateral and vertical textural and bed-thickness variation. To translate any observed sedimentary facies into lithohydraulic units, data are required on the porosity and permeability of the rock to supplement the above mentioned descriptors.

In simulating the production behaviour of a reservoir it is essential to incorporate the effect of contrasting lithologies, which in a sandstone–shale sequence includes making reliable estimates of the dimensions and spatial distribution of shale beds. It is these shale beds that act as baffles or even barriers to fluid flow, increase the tortuosity of the hydrocarbon flow-path and hence decrease the effective vertical permeability of the reservoir.

Over the last decade, a great effort has been made to model statistically these discontinuous shales in order to generate models of the interwell area (Haldorsen & Lake, 1984; Begg & King, 1985). The relation between vertical permeability and shale occurrence may be expressed as:

$$K_{\mathrm{ve}} = \frac{K_{\mathrm{ss}}\,\alpha}{\{1 + n_{\mathrm{sh}}(L/4)\}^2}, \qquad (1)$$

in which K_{ve} = effective vertical permeability, K_{ss} = sand permeability, α = net/gross, n_{sh} = number of shales per metre and L = average shale length (m) (Hazeu *et al.*, 1988). All shales are assumed to be impermeable.

The calculation of L is then based on data from a cumulative distribution curve of shale lengths representative of the geological unit. Figure 1 compares published shale lengths in different sedimentary environments.

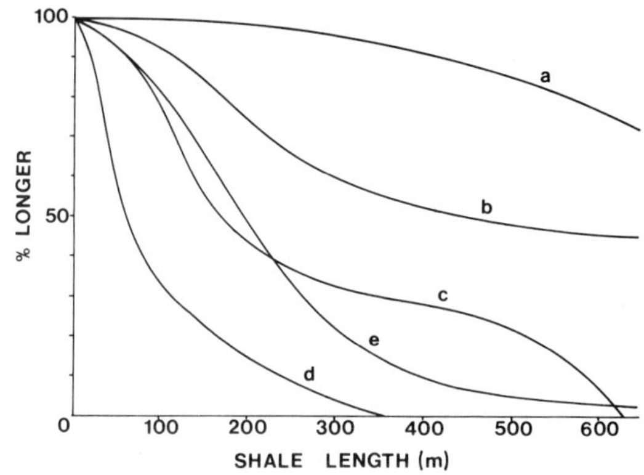

Fig. 1. Continuity of shale intercalations as a function of depositional environment. After Weber (1982) and Hazeu *et al.* (1988). Curves represent the following environments: marine (a), deltaic barrier (b), delta fringe and plain (c), distributary channel (d) and submarine-fan lobe (e).

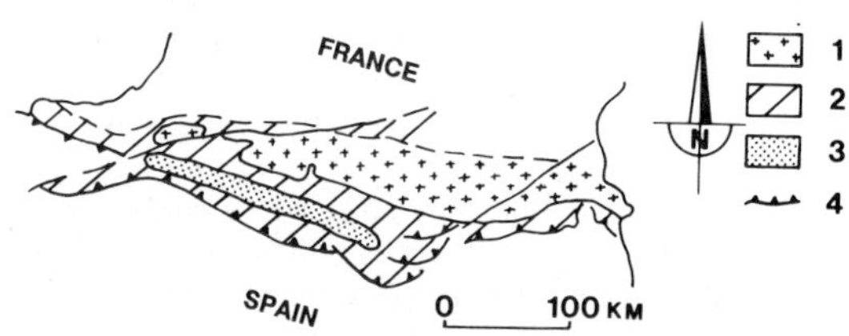

Fig. 2. Simplified geological map of the Southern Pyrenean Foreland Basins (modified after Souquet *et al.*, 1977). 1 = Hercynian basement; 2 = Post-Hercynian sedimentary cover; 3 = Tertiary South-Central Pyrenean Foreland Basin; 4 = thrust.

Fairly little is known about shale lengths in submarine-fan environments, yet it is in turbidite deposits that major hydrocarbon accumulations occur. This highlights the need for a more detailed database. A reservoir study of the North Sea Frigg Field (Hazeu *et al.*, 1988), composed of turbidite lobe deposits, yielded the (estimated) shale length distribution curve shown in Fig. 1.

This paper presents the results of an outcrop study of a submarine-fan channel in the south-central Pyrenees, Spain. Data have been collected on the nature, geometry and distribution of distinct lithofacies, shale lengths and sand permeabilities. In order to define and correlate the different lithofacies

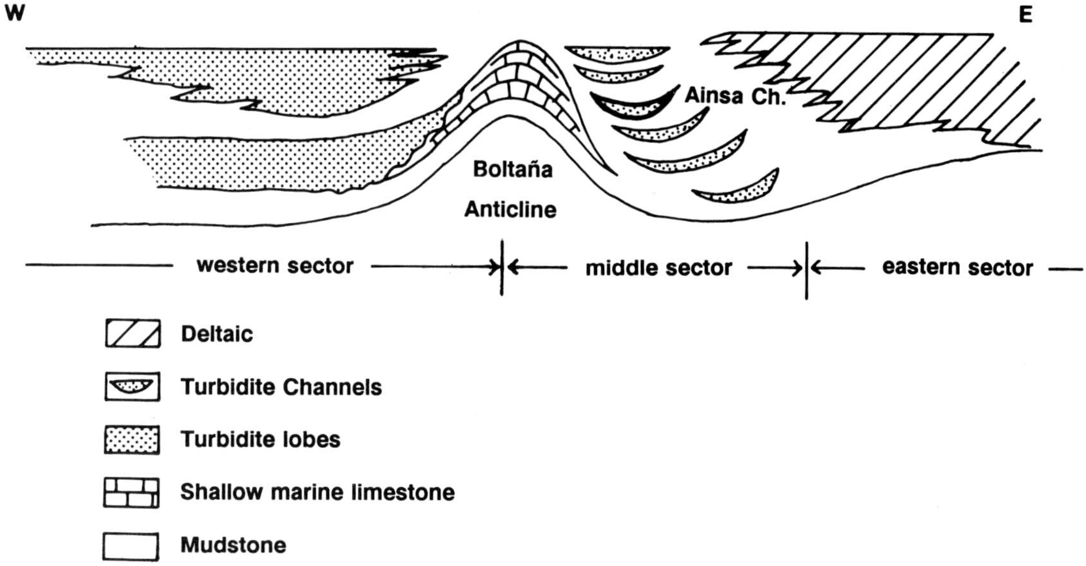

Fig. 3. Diagrammatical stratigraphical cross-section of the South-Central Pyrenean Foreland Basin (modified after Mutti, 1985).

units, 16 vertical sedimentary log-sections with an average log spacing of 35 m were recorded across the outcrop width. The data collected from these logs have been interpolated visually where possible, both on the outcrop and from photo-panels. Permeabilities were measured using a minipermeameter (Eijpe & Weber, 1971). To minimize the effect of weathering on porosity and permeability, measurements were taken either on samples from a recently re-excavated part of the outcrop, or on 5 cm core plugs.

GEOLOGICAL SETTING

A well-known exposure of a turbidite channel-fill is located near the village of Ainsa in the southern Pyrenees, Spain. These turbidite deposits are part of the sediment-fill of the Tertiary South-Central Pyrenean Foreland Basin, which trended parallel to the axial zone of the Pyrenean uplift (Fig. 2). It is bounded to the east by the Segre fault, whereas the western part was in connection with the Atlantic Ocean (Mutti *et al.*, 1972). The basin has a length of about 200 km, a width of 40–50 km, a maximum sediment-fill thickness of about 3500 m, and a total volume of some 25 000 km^3.

The basin can be subdivided into three adjacent palaeogeographical sectors (Fig. 3), bounded by north–south trending structural ramps (Cuevas *et al.*, 1985).

1 In the eastern sector, the sedimentary fill consists of fluvial, deltaic and shallow marine deposits. This part of the basin is commonly referred to as the Tremp-Graus Basin (Nijman & Nio, 1975).

2 The middle sector contains a succession of deep-marine mudstone and sandy turbidite channels (San Vicente Formation of Van Lunsen, 1970; Hecho Group of Mutti *et al.*, 1972). The total thickness of this succession is approximately 2500 m, 10% of which are turbidite sandstone deposits.

3 The western sector contains a stack of turbidite lobes with a total thickness of approximately 3500 m. To the west the lobes grade into mudstone and basin plain turbidites. The sediments in this sector are known as the Hecho Group (Mutti *et al.*, 1972).

The present study focuses on one particular sandy turbidite channel in the middle sector, the Lower Ainsa Channel (Mutti *et al.*, 1981).

The sediments of the channel-fill are derived from fluvial and shallow-marine depositional systems situated to the southeast of the studied area (Mutti *et al.*, 1985; Mutti *et al.*, 1988). The channel is thought to be of late-middle Eocene (Bartonian) age

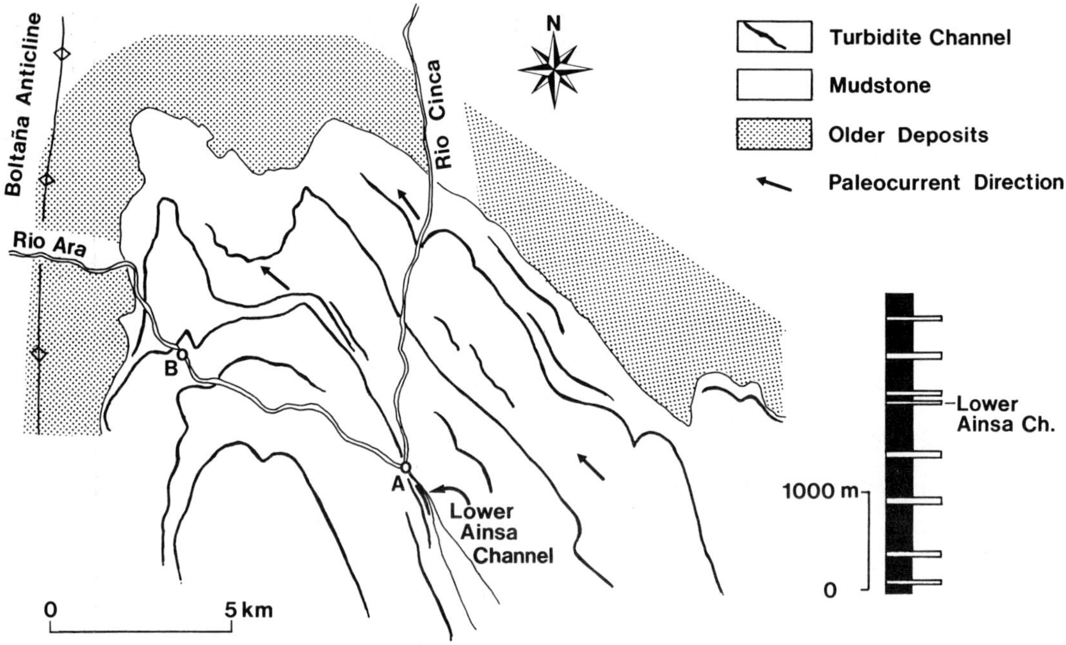

Fig. 4. Geological sketch map of sandy turbidite channels in the middle sector of the South-Central Pyrenean Foreland Basin. A = Ainsa; B = Boltaña. (Modified after Mutti *et al.*, 1981).

(Mutti, 1985; Mutti *et al.*, 1985). Deposition of the channel-fill and related lateral time-equivalent successions took place under structural control, during the main phase of growth of the Boltaña anticline (Fig. 3). As a result, turbidite deposition during Bartonian time became restricted to channel and inter-channel areas in the sector east of the Boltaña anticline (no lobes corresponding to the Lower Ainsa Channel have been reported), and palaeocurrents became roughly parallel to the anticlinal axis (Mutti *et al.*, 1988).

OUTCROP CONDITIONS

The exposure of the Lower Ainsa Channel, south of Ainsa (Figs 4 and 5), is approximately 750 m wide and 30 m high. The southern channel-wing as well as the axial part are superbly exposed in a quarry. The extreme southern margin and the northern wing of the channel are less well-exposed due to scree and vegetation. The channel morphology is clearly demonstrated by the angular unconformity between thick-bedded channel-axis sandstones cutting into thinly-bedded, bioturbated channel-margin deposits.

Diagenesis has led to pervasive calcite cementation. Consequently, the measured permeabilities are low compared with other turbidite deposits (see compilation in Shanmugam & Moiola, 1988).

The average outcrop trend is 160–340° and palaeocurrent directions range from 260° to 320° with a mean of 300° (Fig. 6). Reconstruction of the channel dimension then yields a width of 500 m for the exposed section. Clearly, this is a minimum figure as the extreme northern channel-margin is not exposed. The maximum measured sandbody thickness is 35 m. Assuming a channel-width of 600 m, the width/thickness ratio of the Lower Ainsa Channel fill is approximately 17.

Modern, well-described analogues of comparable size are reported from the Navy Fan, Pacific Ocean (Normark & Piper, 1985), and the Crati Fan, Mediterranean (Ricci Lucchi *et al.*, 1985).

NATURE OF LOWER AINSA CHANNEL-FILL

Figures 5 and 7 show the multi-storey nature of the channel-fill. Two major thinning upwards sequences

Fig. 5. Axial part of the Lower Ainsa Channel. Orientation of outcrop is approximately north–south. A and B refer to sedimentary log sections shown in Fig. 7.

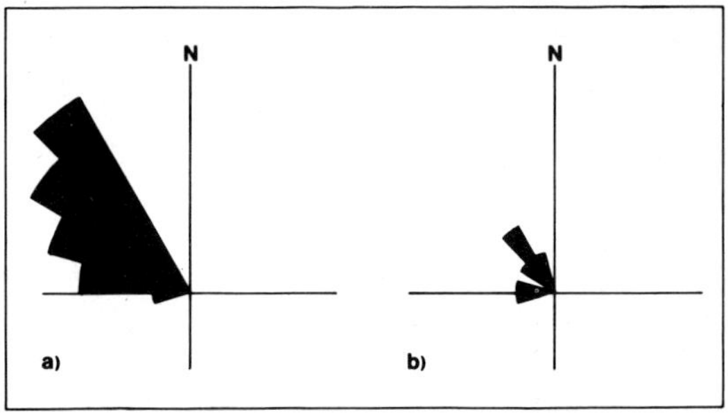

Fig. 6. Rose diagrams of directional data, with 24 segments at 15° interval. (a) Groove marks and flute casts, 19 measurements. Vector mean direction (VMD) 298.9, vector mean length (VML) 0.950. (b) Slump axes, seven measurements. VMD 302.3, VML 0.893.

can be recognized, separated by a channel-wide interval of lithofacies 3. The lower of the two major sequences is more massive and sandier than the upper. The upper part of section B shows the lowermost of a few erratic thick beds separated by thick (pebbly) mudstone intervals, without any apparent sequential arrangements.

The real base of the channel (below lowermost beds of section A, Fig. 5, not visible on photograph) is represented by a relatively thin lag of residual conglomerates. The channel history thus comprises an initial erosive phase followed by a multiphase infilling.

For reservoir characterization purposes, the sediments of the Lower Ainsa Channel-fill are subdivided into five distinct lithofacies. This subdivision is based on lithology, grain size, texture, biogenic structures and gas-permeability (Table 1). These lithofacies units, in descending order of potential reservoir quality, are:

1 massive to graded, medium- to thick-bedded, fine- to very coarse-grained sandstone;
2 sandy conglomerate and pebbly sandstone;
3 bioturbated mudstone, siltstone and very fine-grained sandstone;
4 pebbly mudstone and muddy sandstone;
5 dark mudstone.

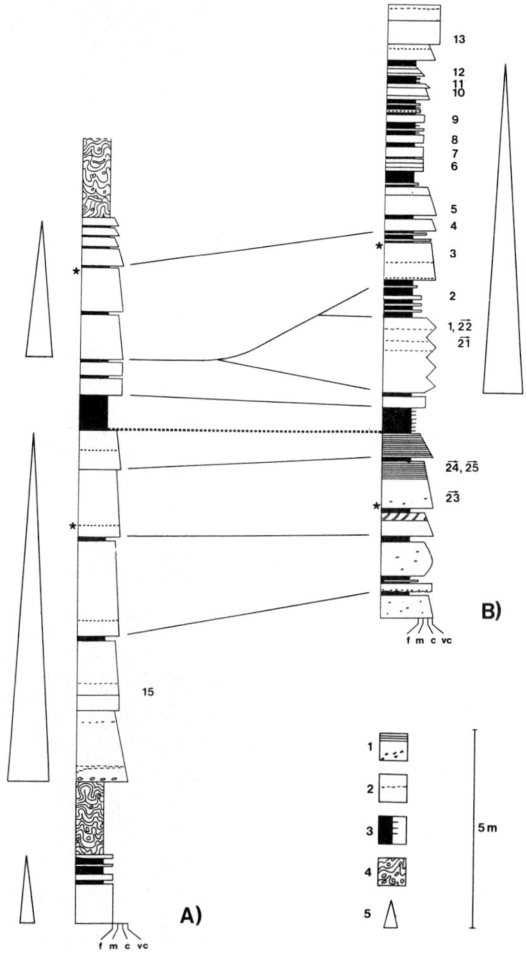

Fig. 7. Sedimentary log sections from the axial part of the Lower Ainsa Channel (for location see Fig. 5). See text for discussion. Numbers refer to samples in Table 2. Asterisks denote interval used for calculation of net sand percentages (see Fig. 9). North is to the right; sections are 70 m apart. 1 = sandstone with dispersed mudstone clasts in lower part and parallel laminae in upper part; 2 = amalgamated sandstone with parting of mudstone chips; 3 = mudstone with very thin siltstone beds; 4 = pebbly mudstone; 5 = thinning upwards sequence.

Lithofacies 1

Lithofacies 1 contains single as well as amalgamated beds. The base of the bed generally contains abundant mudclasts. Most of the beds are internally structureless or display a faint parallel lamination; only a few beds show small-scale ripple cross-lamination at their top. Complete Bouma sequences (Bouma, 1962) are very scarce; beds are normally graded, although massive (i.e. ungraded) beds occasionally occur. Abundant plant remains are concentrated at the top of many sandstone beds. The mudstone of the Bouma E-interval (When preserved) is normally pervasively bioturbated. Many sharp-based beds are not visibly erosive or show

Fig. 8. Horizontal (A) and vertical (B) *Thallassinoides* burrows in silts and shales of lithofacies 3.

low-angle erosion over tens of metres laterally. Irregular cut-and-fill structures are rare.

Trace fossils include *Thallassinoides, Scolicia, Granularia* and *Fucosopsis* (all adapted to a wide grain-size spectrum). These forms are normally lacking in continuously accumulated fine-grained deep-sea sediments, and are explained by Wetzel (1984) to be created by organisms brought in from the shelf by turbidity currents. This has recently been advocated for *Thallassinoides* in particular by Föllmi & Grimm (1990). The abundance of *Thallassinoides* in the Lower Ainsa Channel-fill (Fig. 8) and the scarcity of this burrow in non-channel turbidite environments might be explained by the high concentration of nutrients in the form of preserved organic matter in the channel-fill. Evidence for organisms feeding on the organic debris comes from the deviation of vertical escape burrows into organic-rich horizontal laminae within one turbidite, before continuing their upward path.

The sharp base of the sandstone beds, the eroded mudclasts, the grading and the paucity of sedimentary structures indicate rapid deposition from waning currents. In the Mutti & Ricci Lucchi (1975) facies scheme, this lithofacies can be classified as C_1 when beds are graded, and transitional between B_1 and C_1 when ungraded (Lowe, 1976).

Permeabilities range from 1 to 20 mD with a mean value of 7.2 mD (Table 2); there is no apparent relation between grain size and permeability.

Table 1. Characterization of the sediments of the Lower Ainsa Channel-fill

Lithofacies:	Sandstone	Sandy congl/ pebbly sst	Bioturbated mudst/silt/sst	Pebbly mudst/ muddy sst	Mudstone
Lithofacies unit no:	1	2a/b	3	4a/b	5
Bed thickness (cm)	30–130	5–200	2–95	50–100/200	not appl.
Max. lateral extent (m)[1]	750	100–150	750	100/200	basin wide
Base of bed	sharp	irregular	sharp	irregular	not appl.
Grain size (microns)	125–2000	125–20 000	3–250[2]	3–20 000	3–60
Grading	common	rare	rare	absent	absent
Permeability	high	intermediate	low, variable	zero	zero
Amalgamation	common	common	rare	absent	absent
Mudclasts	common	common	absent	rare	absent
Internal structure	massive, parallel or ripple lamin.	rare clast alignment	parallel, wavy or ripple lamin.	absent	absent
Slumping	rare	absent	rare	rare	common
Bioturbation	common	absent	very common	common	rare

[1] Measured along outcrop trend.
[2] Max. 500 in sand-filled burrows.

A remarkable feature is the observed sudden increase in permeability of a sandstone bed due to post-depositional fluidization in the axial part of the channel. Permeability changes from 7 mD in a graded, structureless sandstone bed (samples 23 and 24, Table 2) to 20 mD where the same bed shows dish structures, fluid escape pipes and upturned laminae (sample 25, Table 2). The presumed filtering-out and subsequent concentration of clay in discontinuous laminae (dishes) by upward flow of pore fluid (Lowe, 1975) may thus turn a homogeneous sandstone with a low permeability into a heterogeneous sandstone with discontinuous low permeability streaks (dishes) and high bulk sandstone permeability. This effect has also been described by Hurst & Buller (1984) from submarine-fan channels in the Norwegian offshore.

Lithofacies 2

Sandy conglomerates (lithofacies subunit 2a) are clast-supported with a medium- to coarse-grained sandy matrix. The clasts are spherical and well-rounded, and are presumedly derived from alluvial fan and fan-delta deposits at the southern basin margin (Mutti *et al.*, 1988). Clasts are mainly limestone, with some sandstone and metamorphics. All clasts are unfractured.

The term 'pebbly sandstone' (lithofacies subunit 2b) is used for sandstone beds in which pebble-sized material forms a minor constituent. Pebbles are generally concentrated at the base of a bed, although occasionally pebbles may be aligned in distinct levels.

The lateral as well as downcurrent transition from conglomerates to pebbly sandstones and sandstones suggests that the conglomerates are residual deposits from highly charged, pebble- and sand-laden density currents.

Whole-rock permeabilities could not be measured in this lithofacies, and permeability for the sandy matrix is assumed to be comparable to that of lithofacies 1. As the clasts are unfractured, and permeability values for clasts will be lower on average (e.g. limestone clasts) than for the matrix, whole-rock permeability is assumed to be lower than that of lithofacies 1, and inversely proportional to clast content.

Lithofacies 3

This lithofacies is characterized by an intercalation of very fine-grained, very thin-bedded (2–6 cm) sandstones and mudstones or siltstones. Bioturbation in this lithofacies is locally pervasive and may constitute up to 18% of the bulk rock volume. Burrows are mainly vertical, 1–2 cm in diameter and

Table 2. Permeabilities and maximum grain size of samples from lithofacies 1. Samples 1–14 are indicated in Fig. 7, section B. Samples 15–20 represent one bed, from channel axis (Fig. 7, section A) to margin (see Fig. 10). Samples 21–25 are located approximately 100 m to the right (north) of section B; 25 is fluidized counterpart of 23–24 (see text)

Sample	K (mD)	Max. grain size (μm)
1	9	600
2	1	200
3	7	850
4	5	250
5	7	700
6	1	250
7	11	250
8	5	500
9	4	150
10	11	200
11	4	100
12	4	250
13	20	350
14	6	150
15	6	2000
16	4	1500
17	11	1000
18	12	1000
19	4	1000
20	10	500
21	7	700
22	10	500
23	7	1000
24	7	400
25	20	1400
Samples 1–24	Mean K 7.2	Standard dev. 4.0

filled with fine- to coarse-grained sand. Sandstone beds are non-erosive and may show parallel and ripple cross-lamination and occasional wavy bedding. Trace fossils are mainly *Granularia*, and pervasive, horizontal and vertical sand-filled *Thallassinoides* burrows (Fig. 8).

This lithofacies occurs interbedded with and as channel-margin equivalents of the sandstones of lithofacies 1. Deposition of channel-wide lithofacies 3 occurred from small-volume, dilute turbidity currents during periods of relatively low accumulation rates.

Permeability of the shale and silt is very low. A value of 1 mD has been measured for a cored silty bed (one sample, five measurements). However, the large vertical and horizontal sand-filled burrows greatly increase the whole-rock permeability, which can be estimated using volumetric averaging:

$$K_{av} = K_{sh}(1 - \rho_{bur}) + K_{bur}\rho_{bur} \quad (2)$$

in which K_{av} is the whole-rock permeability, K_{sh} the silt or shale permeability, K_{bur} the burrow-fill permeability and ρ_{bur} the burrow density (volume of burrows per rock volume). Permeability of the burrows may be as high as 36 mD due to their loose texture, as measured on coarse-sand-filled *Thallassinoides* burrows (Fig. 8). Given a burrow density of 0.18 and a mean silt/shale permeability of 1 mD, equation (2) yields a whole-rock permeability of 7.3 mD, which equals the mean sandstone permeability of lithofacies 1. Note that this is a maximum value, as bioturbation is generally much less than 18%. The degree to which the shaly interbeds in the Lower Ainsa Channel-fill take on the appearance of permeability barriers therefore depends upon the degree of bioturbation; pervasively (18% of bulk rock volume) burrowed intervals should in this case not be regarded as permeability barriers.

Lithofacies 4

Pebbly mudstones (lithofacies subunit 4a) are composed of poorly sorted, well-rounded clasts embedded within a muddy matrix. The framework is matrix-supported. Pebbly mudstones do not show any vertical trends in grain size nor a preferred grain orientation. They are interpreted to represent debris-flow deposits.

Muddy sandstones (lithofacies subunit 4b) consist of a mixture of coarse- to very coarse-grained sand, mudclasts and mud. Grains are matrix-supported. These mixtures are gradational to the sandstone beds below and above. Muddy sandstones closely resemble the intra-sandstone slurries described by Ricci Lucchi & Valmori (1980) and Ricci Lucchi (1981) from lobe and fan fringe environments. The mudclasts, the concomitant deposition of coarse sand and mud and the occasional co-occurrence with minor slumps suggests that muddy sandstones in the Ainsa channel originate from sand-laden, highly erosive and turbulent flows, which were able to take up the hemipelagic and turbiditic mud as well as part of an underlying sand bed, and re-deposited this load almost instantaneously (as inferred from the poor sorting and chaotic nature of the mixture), thus smoothing out a local topography.

Due to the loose and friable nature of this lithofacies, measurement of the permeability by means of a minipermeameter is not possible, and therefore no values can be given. Lithofacies 4a (pebbly mud-

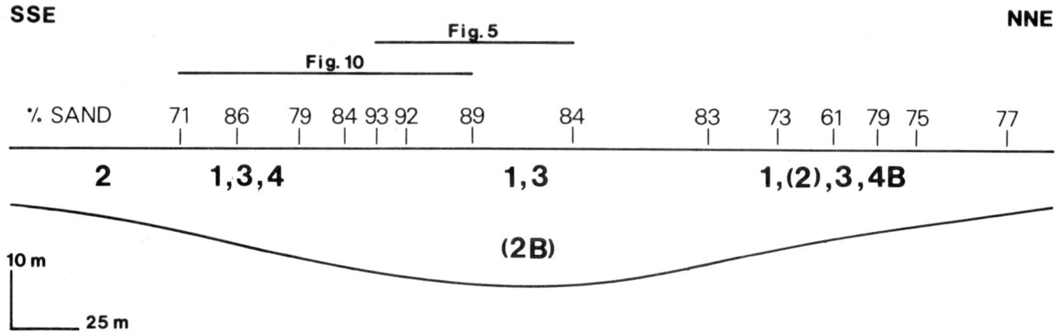

Fig. 9. Lithofacies distribution in the Lower Ainsa Channel. Brackets indicate the relative scarcity of a lithofacies. Net sand percentage is given for 14 log-sections; percentage is calculated for a horizontal slice with good lateral correlation (see Fig. 7). Location of Figs 5 and 10 is indicated.

stone) is regarded as impermeable. Permeability of the muddy sandstone (4b) might be somewhat higher due to loose grain packing and a higher grain/matrix ratio, but can still be regarded as negligible with respect to other lithofacies.

Lithofacies 5

Dark coloured mudstone forms the lithofacies in which the channel-fill lithofacies are embedded. The mudstone does not contain distinct silty laminations, and bioturbation is scarce. Slumping is a common feature, especially below the base of the channel-fill. The muds represent the deposits from continuous hemipelagic background sedimentation. Permeability is assumed to be zero.

Distribution of lithofacies

Figure 9 depicts the gross lithofacies distribution and net sand percentage in the Lower Ainsa Channel. From this figure it is apparent that:

1 lithofacies 1 and 3 occur channel-wide, whereas units 2a and 4 as well as occasional slumps are more or less concentrated in the southern part of the outcrop;

2 the net sand percentage is highest immediately left (south) of the channel axis.

If indeed the major part of the channel is exposed in the outcrop, the lithofacies distribution suggests an asymmetric channel-fill. The coarser-grained, more chaotic and more intensively amalgamated southern margin of the channel (Fig. 10) might then correspond to the outer bend of a curved or possibly meandering channel. Alternatively, the southern margin facies-association may be the result of lateral infilling and/or increased turbulence and erosion as gravity flows derived from east or south-east were deflected along the NNW-trending Boltaña anticline. The asymmetric palaeocurrent distribution (Fig. 6) lends support to the latter interpretation.

Shale lengths

As discussed earlier, the variable extent to which the sediments of lithofacies 3 are bioturbated exerts a strong control on the overall permeability of this lithofacies. A lesser degree of bioturbation will increase the effectiveness of this (and comparable) lithofacies as internal permeability barriers in turbidite sandbodies.

It is therefore interesting to compare the horizontal dimensions of this lithofacies in the outcrop with data from studies on shale lengths in a variety of depositional environments (Fig. 1). The Ainsa outcrop is particularly well suited for this purpose, because the major part of the channel is exposed. This allows for a realistic estimation of the dimensions of what is generally regarded as 'non-reservoir' rock (lithofacies 3) within a 'reservoir' sandbody.

Data on lengths of 30 measured mudstone/siltstone levels (lithofacies 3 and mudstone caps of lithofacies 1 sandstone beds) are compiled in Fig. 11. The lengths were measured along the strike of the outcrop and therefore neither represent dimensions along nor transverse to palaeo-flow. From Fig. 11 it appears that the distribution of the lengths is log-normal, and average length is approximately

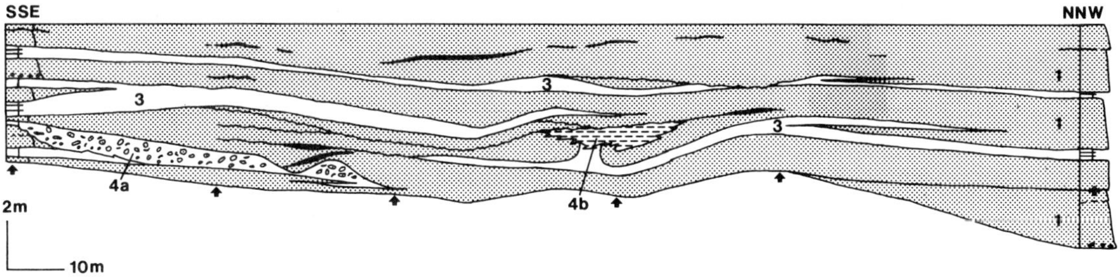

Fig. 10. Part of the southern channel-margin to illustrate interconnectedness and architecture of sandstone units and lithofacies distribution. Numbers indicate lithofacies (Table 1). For location of section see Fig. 9. Arrows indicate location of samples 15 to 20, from right to left respectively (Table 2).

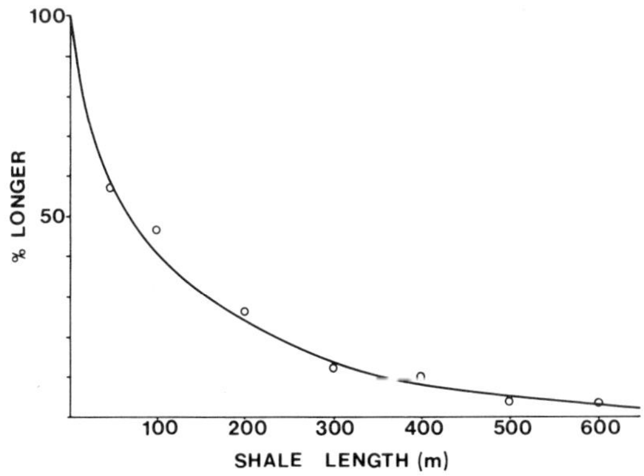

Fig. 11. Shale length distribution curve for the Lower Ainsa Channel.

50 m. The shape of the Lower Ainsa curve shows a marked similarity to the Frigg Field curve for small turbidite lobes (Fig. 1e) compiled by Hazeu *et al.* (1988).

COMPARISON WITH HYDROCARBON PRODUCING FIELDS

Well-described, relatively mud-rich, oil or gas producing channel-fills comparable to the Lower Ainsa Channel in terms of lithofacies, bed thickness, grain size, channel dimensions and width/thickness ratio (Table 3) include the Arbuckle Field, Upper Cretaceous Forbes Formation, Sacramento Basin, California (Imperato & Nilsen, 1990), the Yoluwne Field, Miocene Stevens Sandstone, San Joaquin Basin, California (Berg & Royo, 1990), Sansinena Field, Miocene Puente Formation, Los Angelos Basin, California (Cutler *et al.*, 1990), and, to some extent, the Permian Spraberry Trend, Midland Basin, West Texas (Tyler & Gholston, 1988). Larger producing channel sandbodies are often multi-storey, in which individual channels may be difficult to distinguish (e.g. Forties Field, Kulpecz & van Geuns, 1990; Spraberry Trend meander loops, Tyler & Gholston, 1988).

CONCLUSIONS

The Lower Ainsa Channel-fill can be satisfactorily described by subdividing the sediments into five distinct lithofacies units. These units, in descending order of potential reservoir quality, are:

1 massive to graded, medium- to thick-bedded, fine- to very coarse-grained sandstone;
2 sandy conglomerate and pebbly sandstone;
3 bioturbated mudstone, siltstone and very fine-grained sandstone;
4 pebbly mudstone and muddy sandstone;
5 dark mudstone.

Sandstone permeability of lithofacies 1 may vary within the same bed due to post-depositional fluidization of the soft sediment.

The fine-grained sediments of lithofacies 3 are locally pervasively bioturbated. Burrows are filled with fine to coarse sand and may have relatively high permeabilities. Under these circumstances, therefore, neither this lithofacies nor the thin Bouma E-intervals (when preserved) of the sandstone beds of lithofacies 1 should be considered impermeable barriers to fluid flow.

The channel-fill shows an asymmetry in lithofacies distribution and sand/shale ratio; looking downstream, the left channel-margin and the axial part

Table 3. Comparison of Lower Ainsa Channel with selected hydrocarbon-producing fields. See text for references.

Channels	Width (km)	Thickness (m)	W/Th	Sand (%)	Grain size
Ainsa	0.6	35	17	60–90	Fine–coarse
Yowlumne	0.3–1.5	24–90	12–30	75	Fine–coarse
Arbuckle	1.6–6.0	15–40	40–240	50–100	Fine–coarse
Sansinena	0.90	110	8		Fine–coarse
Spraberry	0.8–1.6	8–15	100	60–90	Very fine

are sandier, coarser-grained and more intensively amalgamated than the right channel-margin. This asymmetry is presumably related to deflection of turbidity currents against the rising Boltaña anticline.

A length distribution curve has been constructed for the mudstone/siltstone levels. The length-distribution is log-normal, and the average length is approximately 150 m.

The Lower Ainsa Channel compares well to a number of hydrocarbon producing fields in terms of size, shape, and nature of the fill.

ACKNOWLEDGEMENTS

The fieldwork for this research was carried out with the continuous support of C.R Geel. I. Blok is kindly acknowledged for his sampling services. The manuscript benefited from critical reviews by H. Dronkert, M.E. Donselaar and C.R. Geel from the Research Centre for Reservoir Characterization (GeoRes) at TU Delft. The comments of I.D. Bryant and S. E. Livera improved the final manuscript and are greatly appreciated.

REFERENCES

Begg, S.H. & King, P.R. (1985) Modeling the effect of shales on reservoir performance: calculation of effective vertical permeability. *SPE Paper* **13529**.

Berg, R.R. & Royo, G.R. (1990) Channel-fill turbidite reservoir, Yowlumne Field, California. In: *Sandstone Petroleum Reservoirs* (Ed. by J.H. Barwis, J.G. McPherson & R.J. Studlick), pp. 467–487. Springer-Verlag, New York, 583 pp.

Bjørlykke, K. (1983) Diagenetic reactions in sandstones. In: *Sediment Diagenesis* (Ed. by A. Parker & B.W. Sellwood), pp. 169–213. Reidel Publishing Company, Dordrecht, 427 pp.

Bouma, A. (1962) *Sedimentology of Some Flysch Deposits*. Elsevier, Amsterdam, 168 pp.

Cuevas, M., Donselaar, M.E. & Nio, S-.D. (1985) Eocene clastic tidal deposits in the Tremp-Graus Basin (provs. of Lerida and Huesca). In: *Excursion Guidebook, 6th Eur. Reg. Mtng Int. ass. Sedimentol., Lerida, Spain* (Ed. by M.D. Mila & J. Rosell), pp. 217–266.

Cutler, W.G., Montoya, G.L. & Ucok, H. (1990) Submarine channel facies control on reservoir performance. *J. Petrol. Technol.* **42**, 909–915.

Eijpe, R. & Weber, K.J. (1971) Mini-permeameters for consolidated rock and unconsolidated sand. *Bull. Am. Assoc. Petrol. Geol.*, **55**, 307–309.

Föllmi, K.B. & Grimm, K.A. (1990) Doomed pioneers: gravity-flow deposition and bioturbation in marine oxygen-deficient environments. *Geology*, **18**, 1069–1072.

Gillund, G.N. (1969) Review of the Lobstick Cardium miscible flood. *J. Can. Petrol. Techn.*, **8**, 66–74.

Haldorsen, H.H & Lake, L.W. (1984) A new approach to shale management in field scale models. *SPE Journal*, **24**, 447–457.

Hazeu, G.J.A., Krakstad, O.S., Rian, D.T. & Skaug, M. (1988) The application of new approaches for shale management in a three-dimensional simulation study of the Frigg Field. *SPE Formation Evalution*, **3**, 493–501.

Hurst, A. & Buller, A.T. (1984) Dish structures in some Paleocene deep-sea sandstones (Norwegian sector, North Sea): origin of the dish-forming clays and their effect on reservoir quality. *J. sedim. Petrol.*, **54**, 1206–1211.

Imperato, D.P. & Nilson, T.H. (1990) Arbuckle Field, Sacramento Basin, California. In: *Sandstone Petroleum Reservoirs* (Ed. by J.H. Barwis, J.G. McPherson & R.J. Studlick), pp. 535–555. Springer-Verlag, New York.

Krause, F.F., Collins, H.N., Nelson, D.A., Machemer, S.D. & French, P.R. (1987) Multiscale anatomy of a reservoir: geological characterization of Pembina-Cardium pool, west-central Alberta, Canada. *Bull. Am. Assoc. Petrol. Geol.*, **71**, 1233–1260.

Kulpecz, A.A. & van Geuns, L.C. (1990) Geological modelling of a turbidite reservoir, Forties Field, North Sea. In: *Sandstone Petroleum Reservoirs* (Ed. by J.H. Barwis, J.G. McPherson & R.J. Studlick), pp. 489–507. Springer-Verlag, New York.

Lowe, D.R. (1975) Water escape structures in coarse-grained sediments. *Sedimentology*, **22**, 157–204.

Lowe, D.R. (1976) Subaqueous liquefied and fluidized sediment flows and their deposits. *Sedimentology*, **23**, 285–308.

McLean, H. (1981) Reservoir properties of submarine-fan facies: Great Valley Sequence, California. *J. Sedim. Petrol.*, **51**, 865–872.

Mutti, E. (1985) Hecho Turbidite System, Spain. In: *Submarine Fans and Related Turbidite Systems* (Ed. by

A.H. Bouma, W.R. Normark & N.E. Barnes), pp. 205–208. Springer-Verlag, New York.

MUTTI, E. & RICCI LUCCHI, F. (1972) Le torbiditi dell'Appennino settentrionale: Introduzione all'analisi di facies. *Mem. Soc. Géol. Ital.*, **11**, 161–199. (English translation by Nilsen, T.H., 1978, *Int. Geol. Review*, **20**, 125–166.

MUTTI, E. & RICCI LUCCHI, F. (1975) Turbidite facies and facies associations. In: *Examples of Turbidite Facies and Facies Associations of the Northern Apennines* (Ed. by E. Mutti, G.C. Parea, F. Ricci Lucchi, M. Sagki, G. Zanzucchi, G. Ghibaudo & S. Jaccarino), pp. 21–36. IX Int. Congr. Sedim. Nice, France, Field Trip A11.

MUTTI, E., SEGURET, M. & SGAVETTI, M. (1988) Sedimentation and deformation in the Tertiary sequences of the southern Pyrenees. *Field trip guide no. 7, Am. Ass. Petrol. Geol. Mediterranean Basins Conference*, Parma, Italy, 157 pp.

MUTTI, E., FONNESU, F., RAMPONE, G. & SONNINO, M. (1981) Channel-fill and associated overbank deposits in the Eocene Hecho Group, Ainsa-Boltaña region, (south-central Pyrenees). *Abstract 2nd Eur. Mtng int. Ass. Sediment.*, Bologna, Italy, 120–123.

MUTTI, E., LUTERBACHER, H.P., FERRER, J. & ROSELL, J. (1972) Schema stratigrafico e lineamenti di facies del Paleogene marino della zona centrale sud-pirinaica tra Tremp (Catalogna) e Pamplona (Navarra). *Memorie della Società Geologica Italiana*, **11**, 391–416.

MUTTI, E., REMACHA, E., SGAVETTI, M., ROSELL, J., VALLONI, R. & ZAMORANO, M. (1985) Stratigraphy and facies characteristics of the Eocene Hecho Group turbidite systems, south-central Pyrenees. In: *Excursion Guidebook 6th Eur. Mtng. int. Ass. Sediment.*, Lerida, Spain (Ed. by M.D. Mila & J. Rosell), 521–576.

NIJMAN, W. & NIO, S-.D. (1975) The Eocene Montañana Delta (Tremp-Graus Basin, provinces of Lerida and Huesca, southern Pyrenees, Spain). In: *The Sedimentary Evolution of the Paleogene South Pyrenean Basin, 11th Int. Congr. int. Ass. Sediment.*, Nice, France, part B, 1–20.

NORMARK, W.R. & PIPER, D.J.W. (1985) Navy Fan, Pacific Ocean. In: *Submarine Fans and Related Turbidite Systems* (Ed. by A.H. Bouma, W.R. Normark & N.E. Barnes), pp. 87–94. Springer-Verlag, New York.

RICCI LUCCHI, F. (1981) The Marnoso-arenacea: a migrating turbidite basin 'oversupplied' by a highly efficient dispersal system. *Excursion Guidebook 2nd Eur. Mtng int. Ass. Sediment.*, Bologna, 229–303.

RICCI LUCCHI, F. & VALMORI, E. (1980) Basin-wide turbidites in a Miocene, over-supplied deep-sea plain: a geometrical analysis. *Sedimentology*, **27**, 241–270.

RICCI LUCCHI, F., COLELLA, A., GABIANELLI, G., ROSSI, S. & NORMARK, W.R. (1985) Crati Fan, Mediterranean. In: *Submarine Fans and Related Turbidite Systems* (Ed. by A.H. Bouma, W.R. Normark & N.E. Barnes), pp. 51–57. Springer-Verlag, New York.

SCHUPPERS, J.D., DONSELAAR, M.E. & GEEL, C.R. (1990) Geological modelling of turbidite reservoirs. *Proc. Eur. Oil and Gas Conf.*, Palermo, Sicily, Italy, preprints vol. 1, 61–67.

SHANMUGAM, G. & MOIOLA, R.A. (1988) Submarine fans: characteristics, models, classification, and reservoir potential. *Earth-Sci. Rev.*, **24**, 383–428.

SOUQUET, P., PEYBERNES, B., BILOTTE, M. & DEBROAS, D. (1977) La chaine alpine des Pyrénées. *Geol. Alpine*, **53**, fasc. 2., 193–216.

TYLER, N., GALLOWAY, W.E., GARETT, C.M., JR. & EWING, T.E. (1984) Oil accumulation, production characteristics, and targets for additional recovery in major oil reservoirs of Texas. *The University of Texas at Austin, Bureau of Economic Geology, Geological Circular*, **84–2**, 31 pp.

TYLER, N. & GHOLSTON, J.C. (1988) Heterogeneous deep-sea fan reservoirs, Shackelford and Preston waterflood units, Spraberry Trend, West Texas. *The University of Texas at Austin, Bureau of Economic Geology, Report of Investigations*, **171**, 38 pp.

VAN LUNSEN, H.A. (1970) Geology of the Ara-Cinca region, Spanish Pyrenees, province of Huesca. *Geol. Ultraiectina*, **16**, 119 pp.

WALKER, R.G. (1978) Deep-water sandstone facies and ancient submarine fans: models for exploration for stratigraphic traps. *Bull. Am. Assoc. Petrol. Geol.*, **62**, 932–966.

WALKER, R.G. (1980) Modern and ancient fans—Reply. *Bull. Am. Assoc. Petrol. Geol.*, **64**, 1101–1108.

WEBER, K.J. (1982) Influence of common sedimentary structures on fluid flow in reservoir models. *J. Petrol. Technol.* **34**, 665–672.

WETZEL, A. (1984) Bioturbation in deep-sea, fine-grained sediments: influence of sediment texture, turbidite frequency and rates of environmental change. In: *Fine-Grained Sediments: Processes and Facies* (Ed. by D.A. Stow & D.J.W. Piper), Spec. Publ. Geol. Soc. London, Blackwell Scientific Publications, Oxford, **15**, 595–608.

WILDE, P., NORMARK, W.R., CHASE, T.E. & GUTMACHER, C.E. (1985) Potential petroleum reservoirs on deep-sea fans off central California. In: *Submarine Fans and Related Turbidite Systems* (Ed. by A.H. Bouma, W.R. Normark & N.E. Barnes), pp. 35–42. Springer-Verlag, New York.

ZEITO, G.A. (1965) Interbedding of shale breaks and reservoir heterogeneities. *J. Petrol. Technol.* **17**, 1223–1228.

Spec. Publs int. Ass. Sediment. (1993) **15**, 113–123

Outcrop studies of shale smears on fault surfaces

N.G. LINDSAY, F.C. MURPHY*, J.J. WALSH *and* J. WATTERSON

Fault Analysis Group, Dept of Earth Sciences, University of Liverpool, PO Box 147, Liverpool L69 3BX, UK

ABSTRACT

Shale smears occur on the surfaces of tectonic faults contained within lithified sequences of sandstones and shales. The smears occur on sandstone/sandstone fault surfaces and may extend from footwall to hangingwall cutoffs of the shale source layer. The nature and systematics of shale smears have been examined in working quarries exposing Westphalian sandstone/shale sequences, which are offset by faults of up to 15 m throw. The continuity of shale smears has been examined for a range of values of shale smear factor (SSF), which for these normal faults is the ratio of fault throw to vertical thickness of the offset shale layer. When this ratio is 7 or less, shale smears are continuous on sandstone/sandstone surfaces between base shale footwall cutoffs and top shale hangingwall cutoffs. Continuous smears, in 2-D sections, may occur at SSF values of up to 50. Shale smears are formed by one or a combination of three mechanisms: abrasion, shear and injection. Previous descriptions have been restricted to occurrences on growth faults in soft sediments where the shear mechanism is dominant.

In contrast to fault populations, which have well-defined scaling properties, the mechanisms of shale smearing are not sufficiently understood to establish their scaling laws and to extrapolate outcrop observations to larger scale faults. The shale smears on fault surfaces examined in outcrop are assumed to be equivalent to seals in the subsurface context and the observations are of most relevance to sealing of sub-seismic populations of intra-reservoir fault surfaces rather than to trap closures. The degree of sealing of fault surfaces due to shale smearing depends not only on the gross sandstone/shale ratio but also on the distribution of shale units within the sequence and on their individual thicknesses.

INTRODUCTION

The permeability of a reservoir volume varies with (i) the primary sediment architecture, expressed in terms of present day permeabilities, (ii) the fracture population and the hydraulic characteristics of the fractures, and (iii) the interaction between the fractures and primary architecture. Diagenesis spatially related to structural, as opposed to architectural, elements may be an additional variable.

Shale smears on fault surfaces are only one of the factors contributing to the hydraulic properties of faults and to the hydraulic characterization of a reservoir; the extent of their contribution varies with both the fault population and the sedimentary architecture. Shale smear is the general term used for a shale or clay layer intervening between sandstone units juxtaposed across a fault. Smith (1980) following Rettger (1935) refers to smears as being 'fault-plane fillings' to distinguish them from fault gouges. Previous descriptions of shale smears (Weber *et al.*, 1978; Smith, 1980; Weber, 1986) are of occurrences on growth faults. The examples described here are the result of tectonic faulting of lithified sediments. Data on shale smears are given below together with an outline of those aspects of fault populations and sedimentary architectures which are relevant to assessment of the effects of shale smears on permeability. It is emphasized that permeability is a scale-

* Present address: ERA, Level 8, 371 Queen Street, GPO Box 2927, Brisbane 4001, Australia.

dependent characteristic of a rock volume and that a quantitative understanding of the scale dependence is necessary if data acquired on one scale are to be applied on a different scale, as is usually required (Dubrule & Haldorsen, 1986). The scaling laws of faults are probably better known than those of other reservoir heterogeneities relevant to permeability and a brief review of fault scaling is given below to provide a context for description of shale smears on fault surfaces.

No measurement of permeability has been made on the fault surfaces with shale smears described here, so the assumption that they represent effective seals is untested. Fault sealing is significant in two different contexts and the effectiveness of a seal has to be assessed according to the context in which it occurs. For a sealed fault to contribute to the closure of a trap and be effective in preventing petroleum migration, the sealing requirement is rigorous in terms of the time scale for which the seal must remain effective and in terms of the pressure differential which it must sustain. In the production context, where the role of faults may be to compartmentalize the reservoir or perhaps may be limited to increasing the flow path tortuosity, the sealing requirements are much less stringent. The time scale on which the seal need be effective is perhaps seven orders of magnitude less than for trap seals, the pressure difference may be slight and even incomplete seals may have substantial effects on reservoir performance. The outcrop studies reported therefore have a different significance in each of these contexts and the results are more relevant in the intra-reservoir context. Compartmentalization of some reservoir units in Cormorant Block IV of the northern North Sea, has been accounted for by Sassi *et al.* (1992) by sub-seismic faults sealed by clay smears.

SCALE CHARACTERIZATION OF FAULTS

Both individual faults and fault populations have a systematic scale dependence over a scale range of at least six orders of magnitude of fault size (Walsh & Watterson, 1988a; Childs *et al.*, 1990; Heffer & Bevan, 1990). The scaling law for a fault population can be expressed as the fractal dimension of the population, which is determined by the relative abundances of faults of different sizes. The fractal dimension is the absolute value of the slope of the straight line obtained by plotting 'size' *vs.* cumulative number on a log/log plot. As fault size can be expressed in a variety of ways, the numerical value of the fractal dimension of a given population varies with the parameter used to express the size. The displacement (offset) on a fault varies systematically over the fault surface so fault size can be expressed in terms of the maximum displacement on each fault in the population. As there is a systematic relationship between maximum displacements and maximum dimensions of faults (Walsh & Watterson, 1988a; Marrett & Allmendinger, 1991), the fractal dimension of a population expressed in terms of maximum displacement can be converted to the fractal dimension of the population expressed in terms of maximum dimension. In practice, neither maximum displacement nor maximum dimension populations can be measured objectively from seismic data. Populations of *fault displacements* can be measured objectively from seismic data interpretations (Childs *et al.*, 1990), by measuring fault displacements on parallel sample lines usually oriented perpendicular to fault strike. Fault displacement populations can be used to derive practically useful quantities, such as fault surface area per unit volume of rock with area categorized by displacement; fault displacement populations are generally measured on parallel sample lines approximately perpendicular to the dominant fault strike (Walsh *et al.*, 1991). The smallest size of displacement which is adequately represented in a measured population is determined by the limit of seismic resolution of the data set. Data sets from outcrops and coal-mine plans, which have lower limits of resolution than seismic data, and theoretical modelling, show that the population of sub-seismic fault displacements can be estimated by extrapolating the straight segment of a measured population curve down to the smallest displacements of practical concern (Childs *et al.*, 1990). A single line displacement population plot of fault displacements measured in outcrop along a single horizon is shown in Fig. 1 and is described by a power law or fractal distribution from *c.* 10 m down to 1 mm. Fault populations vary not only in respect of the relative numbers of faults of different sizes, fractal dimensions, but also in the total number of faults, or the fault density. The population of sub-seismic faults in the Cormorant Field is estimated by Sassi *et al.* (1992) using a combination of core data and fault population data. The Cormorant fault population was derived by 2-D sampling of the maximum displacements on map fault traces, which

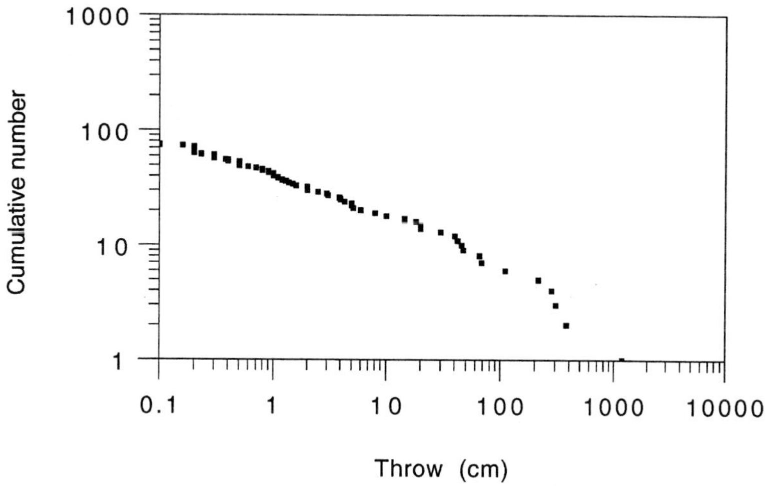

Fig. 1. Single line fault displacement population plot of fault displacements measured in outcrop along a profile across Round O Quarry, Lancashire. The population curve incorporates displacements of all faults encountered along a 157 m profile oriented approximately perpendicular to fault strike. Cumulative number refers to the number of faults with displacements greater than a given displacement value.

is less reliable than 1-D sampling of fault displacements (Walsh *et al.*, 1991). Using the 1-D sampling method, Walsh *et al.* (1991) have shown that for two North Sea fields displacement populations from seismic data are consistent with those measured from core data. Scaling relationships for the spatial distributions of faults are not sufficiently understood for the locations of sub-seismic faults to be predicted deterministically and stochastic methods are still required. The smallest displacements which are of interest will depend both on the hydraulic characteristics of the faults and on the spatial characteristics of the sediment architecture, which are discussed below. Given that fault populations are fractal, the minimum volume over which fault controlled permeability can be averaged, i.e. the representative elementary volume (REV), is the volume of the reservoir (Wheatcraft *et al.*, 1990).

The hydraulic effect of a given population of faults results from two separate factors. The first of these factors is the geometric alteration of pre-faulting architecture, and consequent change in continuity of permeable units, brought about by the faulting. In most cases the change is likely to be a reduction in continuity, although an increase is possible in those cases in which the permeable units initially are isolated lenses as opposed to laterally persistent layers. In the simplest general case, where lithological units approximate to laterally continuous layers of permeable sandstone and impermeable shale, the geometrical change in continuity brought about by a given fault population is primarily a function of the gross sandstone/shale ratio of the reservoir. The second factor by which a fault population changes the continuity of permeable reservoir units derives from the sealing or non-sealing nature of the fault surfaces. Where fault surface sealing is due to shale smears, the hydraulic effects on the reservoir depend not only on the net : gross ratio of sandstone but also on the thickness distribution and spacing of the shale layers, as is shown below.

SHALE SMEARS

The existence and potential hydraulic effects of shale smears have been described by several workers (Weber *et al.*, 1978; Weber, 1986; Speksnijder, 1987; Bouvier *et al.*, 1989). Where a shale layer is offset by a fault throw greater than the vertical thickness of the layer, a shale smear may occur on the sandstone/sandstone fault surface between the footwall and hangingwall cutoffs of the shale layer. The smear hydraulically separates the sandstone units in footwall and hangingwall which have been juxtaposed by the fault movement and which, in the absence of a shale smear, would be expected to be hydraulically connected across the fault surface. The circumstances which favour the formation and determine the extent of smears are not well established and an outcrop study was undertaken in order to acquire relevant quantitative data.

Outcrop studies

The outcrops studied occur at Round O Quarry, Lancashire, UK, a working quarry in lower West-

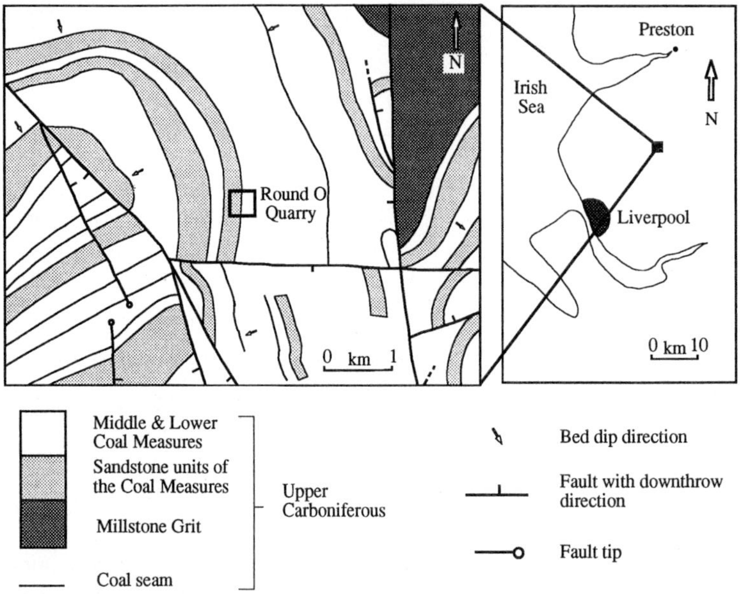

Fig. 2. Geological map showing location of Round O Quarry, Lancashire (modified from Geological Survey of England 1 inch Map—Sheet 84, Wigan).

phalian A (Pennsylvanian) fluvio-deltaic sandstones with shale interbeds (Fig. 2). Shales are interbedded on scales from centimetres to several metres. The thickest shale is approximately 10 m thick and contains subordinate interbedded layers (<0.5 m thick) of fine-grained sandstone/siltstone. From a distance, this shale layer, despite sandstone/siltstone interbeds, appears to be a single homogeneous unit and, where shale smearing is concerned, has behaved as such. The remainder of the sequence comprises thin shale units, typically <5 cm thick, which are interbedded with sandstone units of variable thickness (up to 3 m). The exposed sequence lies on the eastern limb of a SW plunging syncline of Late Carboniferous age and bed dips within the quarry are *c.* 25° west. The sequence is cut by an array of normal faults exposed on both the floor and sides of the quarry with fault throws ranging up to 15 m. Fault displacement populations measured on seven sample lines (up to 185 m long and 40–50 m apart), orientated perpendicular to the main fault strike (i.e. N–S), have fractal distributions over the measured range of fault displacements (1 mm–15 m, Fig. 1). Slopes of the displacement population curves range from −0.4 to −0.55, values which are at the limit of the range which is characteristic of coalfield and oilfield datasets (−0.5 to −1.0; Childs *et al.*, 1990). The faults are tectonic and post-date lithification and probably date from the time of maximum burial depth of *c.* 2 km, i.e. Late Carboniferous and pre-folding (Walsh & Watterson, 1988b); rotation of the predominantly eastward dipping faults by later folding has resulted in lower fault dips (*c.* 55°) than are usual for tectonic normal faults in the Coal Measures (mean dip *c.* 68°; Walsh & Watterson, 1988b). The mechanical properties of the rocks at the time of faulting were probably little different from now. The faults examined are analogues of a sub-seismic size population of a type which could be predicted from typical offshore seismic data on the basis of the fault population systematics described earlier. Small syn-sedimentary faults, with throws up to 15 cm, are restricted to a small area of the quarry and are easily distinguished from the tectonic faults.

Additional studies have been carried out in opencast coal workings of Westphalian B age in which some units, particularly seatearths, are still plastic clays.

Previous work (Smith, 1980; Weber, 1986) has emphasized the importance of shale smears in synsedimentary faults in unconsolidated sand/clay sequences, but the shale smears examined in this work were formed during tectonic faulting of a lithified sequence.

Types of shale smears

Shale smears occur on the surfaces of all faults which intersect shale layers and appear to be derived equally from both hangingwall and footwall of the source layer. Displacement on an individual fault is accommodated within a fault zone and is the cumulative displacement of numerous individual slip events, each occurring on a relatively smooth slip surface. Earlier slip surfaces may be intersected and distorted by later slip events and only the later slip surfaces retain their original features. The successive slip surfaces and the related fault rocks together form a fault zone which has significant thickness. Early slip surfaces are contained within a matrix of breccia, gouge and shale smears. Fault zone thickness generally increases with fault displacement and for small fault displacements (<*c.* 50 cm) there may be no clear distinction between individual slip surfaces. For larger fault displacements, individual slip surfaces can be distinguished within thicker fault zones (up to 1 m). We describe shale smears occurring on sandstone/sandstone slip surfaces, where throw is greater than the vertical thickness of the shale source layer, through to intra-shale slip zones in which highly polished slip surfaces occur.

Three distinct types of shale smear occur which are referred to as the abrasion, shear zone and injection types respectively. Abrasion smears are the most common and occur as uniform, wafer-thin (<1 mm) films or veneers on sandstone fracture surfaces (see Plate 1, facing p. 124); when pieces of the hangingwall and footwall sandstones are detached from the outcrop and separated, a shale veneer remains on both. Small pockets of shale occur in hollows on the veneered slip surface and surface markings such as striae and tool marks from embedded breccia clasts are common (Plate 1). Some shale smears have been partly eroded by further slip on sandstone/sandstone fault surfaces. The erosion is concentrated in slip-parallel zones, the positions and extents of which are dictated by the incompatibilities of the footwall and hangingwall slip surface topographies (see Plate 2, facing p. 124). As a fault zone may contain several individual slip surfaces, each with its shale veneer, a fault rock may include several thin shale veneers (see Plate 3, facing p. 124). No systematic data on thickness variation was obtained because individual smears are too thin to measure in the field; they appear to vary in thickness by factors of five or more, with the variation often controlled by the topography of the sandstone slip surfaces. In general, however, smear thickness is greatest when derived from thicker source layers and with small fault throw values. With this type of smear there is a clear distinction between the source layer and the shale smear.

The shear zone type of smear is less common and is associated with normal drag of the layers adjacent to the fault (see Plates 4, 5 and 6, facing p. 124). This is the type of smear described and figured by Weber *et al.* (1978) and Smith (1980). Normal drag is best developed within sequences with low sandstone/shale ratios and with high proportions of shale interbeds. Within individual fault zones increased attenuation of shale layers is accommodated either by displacement on closely spaced anastomosing slip surfaces or by distributed shear. Smear thicknesses are greatest near to the source layer cutoffs and reduce towards the region midway between hanging-wall and footwall cutoffs, where in some cases the attenuated shale layer is replaced by a shale veneer (Plate 5). Elements of both abrasion and shear zone types of smear often occur on the same slip surface and the two types are best regarded as end members of a continuous series.

Injection smears are usually associated with plastic clay layers and result from injection of shale or clay along the fault surface (see Plate 7, facing p. 124). Injection is a necessary interpretation in the rare cases in which the smear thickness is greater than the thickness of the source layer. Shale flowage is also required where displacement variations on individual faults are accommodated by thickness variations of shale interbeds.

Mechanisms

The naming of the three smear types reflects the mechanisms believed to be responsible for their formation, i.e. abrasion, shear and injection.

Abrasion

This is the simplest of the three mechanisms but is also the least quantifiable. Relative movement of shale past a sandstone fracture surface leaves a thin veneer of shale on the sandstone surface in much the same way as a film of abraded material builds up on, and smoothes, the surface of sandpaper or other abrasive. As soon as a continuous veneer is formed little or no further abrasion takes place and the thickness of the veneer is not progressively increased. When two sandstone surfaces are juxtaposed, each with a shale veneer, further movement will eventually erode the shale veneer. Whether or not a shale veneer is present will depend less on the thickness of the shale layer from which it is derived, than on the amount of subsequent movement on the juxtaposed sandstone surfaces and on their roughnesses. Formation and preservation of smears of this type will be favoured by sandstone fracture surfaces which are rough on the millimetre scale, aiding the formation of the shale film, but smooth on the centimetre and larger scales. Shale smears of this type frequently have a high polish similar to many slip surfaces contained entirely within shale units, which have mirror-like smoothness.

Shear

A shale smear can form by offset of a shale layer on a plastic shear zone comparable with the zones of simple shear in metamorphic rocks in which they are often the predominant displacement structures. The shear mechanism has been described by Weber *et al.* (1978) who simulated shear zones in a ring-shear apparatus. Shear zones are essentially faults in which continuity of layers is maintained within the shear zone, but individual layers are thinned, extended and rotated within the zone. Faults in soft sediments frequently show this continuous displacement geometry (see Plate 8, facing p. 124) which is largely determined by the rock strength. Displacement on tectonic faults in lithified rocks may be accommodated partly by continuous and partly by discontinuous displacement (Walsh & Watterson, 1987); in these cases the continuous element of displacement is usually referred to as normal fault drag (cf. reverse drag; Hamblin, 1965). The ideal single fracture (discontinuous) and the ideal shear zone (continuous) are simply the end members of a spectrum of displacement geometries in which the total displacement is the sum of the two components. The accommodation of displacement by a continuous fault geometry is favoured in sequences of low shear strength, typically with incomplete lithification of sandstones, a relatively high proportion of shale and a uniform distribution of shale layers within the sandstone. The geometry of such zones is schematically illustrated in Fig. 3 in which the relationship between displacement (d), zone thickness (t), layer thickness (T) and shear strain (γ) is shown. The thickness of a displaced layer within a shear zone, i.e. the smear thickness, decreases as a function of the shear strain (Ramsay & Huber, 1983) and is given by the expression

$$t_{(\text{smear})} = t_{(\text{original})} \cdot (\sin a^1/\sin a) \qquad (1)$$

where $t_{(\text{original})}$ and $t_{(\text{smear})}$ are the layer thicknesses outside and inside the shear zone and a and a^1 are the angles between the layer and the shear zone, outside and inside the shear zone respectively (Fig. 3). Beyond some minimum thickness individual shale layers, or shale smears, become discontinuous. A typical strain profile across a shear zone, with the maximum shear strain, and maximum layer thinning, at the centre of the shear zone is indicative of a strain softening mechanism by which strain is increasingly concentrated towards the centre of the shear zone. A feature of shale smears formed by this mechanism would be a systematic variation in shale thickness along the shear zone from footwall cutoff to hangingwall cutoff. When the shear zone throw exceeds the shale layer vertical thickness, i.e. when the shale smear is bounded on either side by sandstone, an ideal shear zone geometry is unlikely to be maintained if the sandstone is lithified.

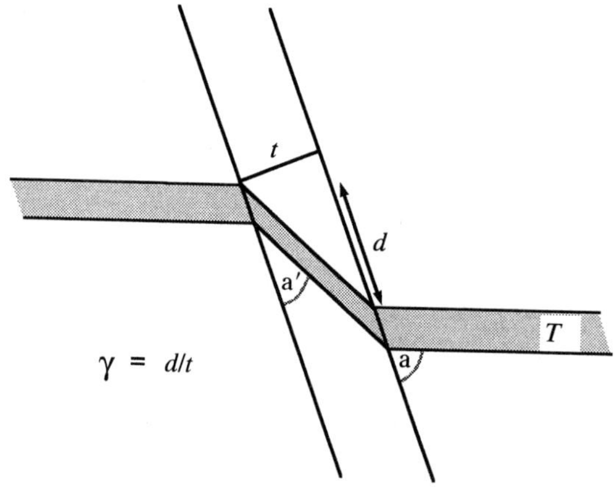

Fig. 3. Shear zone geometry. The relationship between displacement (d), shear zone thickness (t), layer thickness (T) and shear strain (γ) is shown. The ratio of layer thickness within the shear zone to original layer thickness is given by the ratio of sin a^1 to sin a.

Injection

Variable displacement along faults requires volume decrease adjacent to the fault on one side and dilatation on the other side (Fig. 4). Such changes are preferentially accommodated by shale layers, which are thinned on one side and thickened on the other side of the fault. These thickness changes are accomplished by flow of shale from one side of the fault to the other, similar to the flow of salt across faults (Petersen *et al.*, 1992). So long as the throw is less than the shale layer vertical thickness, flow of shale

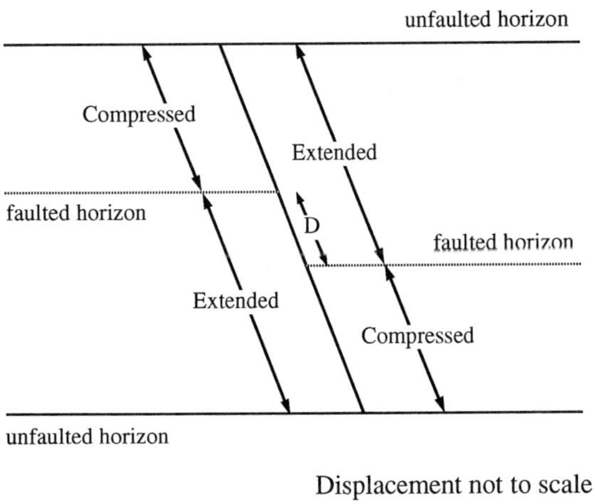

Fig. 4. Variable displacement along faults is accommodated by extension and compression on the opposite sides of the fault.

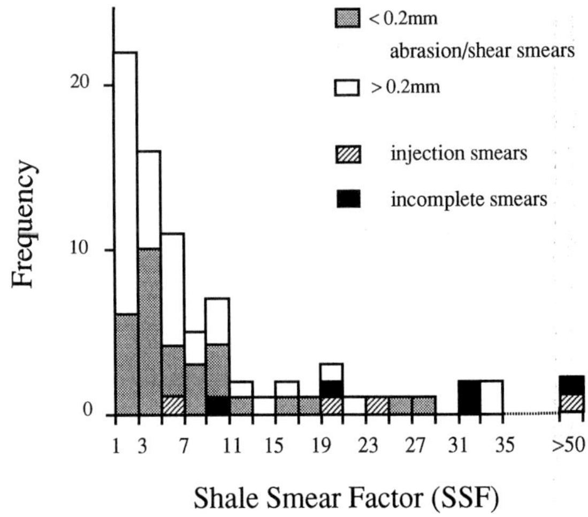

Fig. 5. Shale smear data for 80 slip surfaces at Round O Quarry. Histogram shows types of shale smear for the full range of shale smear factor (SSF) values. Smears are either continuous, as abrasion/shear smears (<0.2 mm and >0.2 mm) or injection smears, or incomplete.

is possible across the shale/shale interface; when footwall and hangingwall cutoffs of the shale layer are no longer juxtaposed, the same pressure gradient exists to drive shale from one side of the fault to the other along the fault surface. Shale thickness along the fault surface is not predictable and we have no data on the distances over which the mechanism is effective beyond the *c.* 1 m observed in outcrop. The mechanism is favoured by the presence of plastic clays and in the outcrops studied these characteristically occur as seatearths in coal-bearing sequences.

SHALE SMEAR CONTINUITY

The data presented in Fig. 5 represent measurements of fault zone smears collected from the interbedded sequences at Round O Quarry. Observations were mainly two-dimensional with only small areas of each fault surface, up to 1 m^2, available for examination. Smear thickness and its extent along the fault surface were recorded where possible, together with the true thicknesses of faulted shale layers and their displacements. Composite shale smears formed from closely-spaced thin shale layers (<10 cm) are not included in the data set. For fault zones containing closely-spaced multiple slip surfaces it is usually the combined smearing effect of the slip surfaces which has been recorded.

The data set includes shale smears on 80 faults with displacements ranging over four orders of magnitude; typically the displacement is <1 m. The thicknesses of shale beds from which measured shale smears are derived range up to 11.8 m, but most are <5 cm thick. Of the measured shale smears, half are of the abrasion type (i.e. <0.2 mm thick), over 75% of the total number are 1 cm thick or less and all are <5 cm thick except for one shale smear 40 cm thick which is derived from the 11.8 m thick shale layer.

Shale smear factor (SSF)

The ratio of fault displacement to shale layer apparent thickness, i.e. as measured along fault trace, is referred to as the shale smear factor (SSF). The data show that continuous smears are to be expected for SSF values of 9 or less but may also occur with SSF values of up to 50 (Fig. 5). A more accurate estimate of the lowest SSF value at which smears become incomplete is provided by restoration of individual incomplete smears to displacements at which they were continuous, by subtracting the length of eroded smear from the fault displacement. This method provides a minimum estimate of 7 for the SSF of incomplete smears, with 'restored' SSF values within the range 7–12 for four of the five incomplete smears restored. In the sequence examined, smears with SSF values greater than *c.* 30 are mostly of the composite type. The overlap in SSF values for incomplete and complete smears reflects the variety of controls. Although the statistical significance of the above systematics cannot easily be tested, it is likely that future work will reveal fault surfaces with incomplete smears at SSF values of less than 7. Incompletely smeared surfaces with only small proportions of unsmeared area would not significantly alter our conclusions regarding the sealing effects of faults on the production time scale.

The potential sealing effect of a particular SSF value on a complete fault surface is sensitive not only to the net:gross ratio, but also to the thicknesses and spacings of the shale layers in the sequence containing the fault. SSF values can be directly applied to examination of the sealing potential of fault surfaces on which the displacement variation is known (Bouvier *et al.*, 1989). These authors use the term 'clay smear potential' (CSP) but do not define or give the CSP value thought appropriate to a fault surface in the Nun River Field, Nigeria; examination of their diagrams (Bouvier *et al.*, 1989, figs 22, 23, 24) suggests that with the CSP classification used (high, medium, low), high CSP corresponds to an SSF value of 4 or 5.

Significance of the faulted sequence

The heterogeneity of the sequence affected by faulting is crucial to the hydraulic effects of a given fault population and to the fault surface characteristics. Quantitative description of lithological heterogeneities in 3-D are needed for quantitative assessment of the effect of a given fault population.

The effects of faulting on fluid flow in sandstones depends on several factors which include (i) the lithological sequence, (ii) the fault throw population, (iii) the sealing potential of the fault surfaces, with respect to the presence of shale smears or other fault products, and (iv) the 3-D geometry of faults, including the continuity of and displacement variations on individual fault surfaces. A single fault intersecting a sandstone–shale sequence will always produce a decrease in continuity of individual sandstone layers, but the effect at a particular point on an individual fault surface depends on the exact sequence of sandstones and shale intersected and on the fault throw at that point (Brand & Haldorsen, 1988). This effect can be illustrated by use of a simple model applied to a lithological sequence representative of the Round O Quarry (Fig. 6a), which incorporates the results of the outcrop observations on fault surface shale smearing. This 2-D model of sandstone continuity does not take account of the 3-D fault geometry and is not appropriate for 3-D analysis of sandstone continuity.

The modelled faults are assumed to be vertical and to have constant throw through the sequence. Results are given for faults with throws of up to 10 m (Fig. 6b). The effect of shale smearing is assessed by applying the empirical rules derived from analysis of shale smear data, i.e. with SSF < 7 continuous shale smears are expected, and with SSF $\leqslant 11$ continuous shale smears are probable. The proportions of sandstone/sandstone (without shale smearing), shale/shale, sandstone/shale and sandstone/sandstone (with shale smearing) contacts across the fault trace are illustrated for SSF values of 1, 6 and 11. With increase in fault throw there is a decrease in shale/shale and sandstone/sandstone contacts and a concomitant increase in sandstone/shale and smeared sandstone/sandstone contacts. For this particular example, sandstone continuity across the fault trace decreases from *c.* 70% of the fault trace length before faulting to *c.* 46%, 25% and 17% for SSF values of 1, 6 and 11 with 10 m fault throw (Fig. 6b). The contribution of shale smears to the reduction in sandstone continuity increases with increase in fault throw and, in the example illustrated, accounts for *c.* 33% of the decrease in continuity across faults with throws >5 m. With fault throws of >*c.* 20 m shale smears would be expected to cover the entire fault surface, whereas without shale smears zero continuity would occur only if fault throw is >*c.* 50 m, i.e. the thickness of the section. Where fault throw is greater than the vertical interval between successive shale layers, the shale smears from individual shale layers will partially overlap to form composite smears. In such cases the simple application of SSF values may underestimate the continuity of the composite smear.

Scaling

The outcrop observations are restricted to fault displacements up to *c.* 15 m and are directly applicable only to reservoir sequences in which the sandstone/shale layering is on a similar or smaller scale. Sassi *et al.* (1992) show that within the Cormorant Field, the Lower Ness sequence is sensitive to subseismic faults and appears to be compartmentalized by them. The Lower Ness reservoir sequence con-

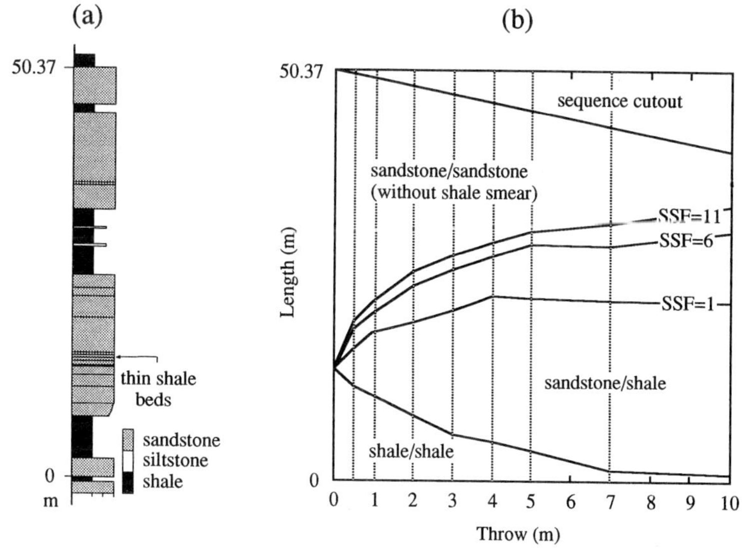

Fig. 6. (a) Representative stratigraphic sequence at Round O Quarry. (b) The effect of faults on sandstone continuity for the sequence shown in (a) with continuity represented by sandstone/sandstone without shale smear. Across-the-fault relationships are expressed as the lengths of fault trace (ordinate) occupied by sandstone/sandstone (without shale smear), sandstone/sandstone (with shale smear), sandstone/shale and shale/shale juxtaposition. Fault trace length occupied by sandstone/sandstone with shale smear is represented by the distance between SSF = 1 line (zero shale smear) and lines representing SSF values of 6 and 11. The model faults have throws from 0 to 10 m (abscissa) and are assumed to have constant throw through the sequence.

sists of thin (<3 m) sheet sands in a moderate net: gross sequence. Application of the observations to sequences with larger scale layering requires a knowledge of the scaling laws for each of the mechanisms concerned and these laws have not been established. An increase in displacement is accompanied by an increase in fault zone thickness such that the shale smearing is the product of slip on clearly multiple slip surfaces. The abrasion mechanism is clearly influenced by scaling of roughnesses and the rate of wear of shale veneers must be controlled by the amount of fault displacement subsequent to smear formation, so the process is likely to be strongly scale dependent. Characterization of the scale dependence of the shear zone mechanism requires a knowledge of the precise nature of shear zone growth. The shear zone mechanism is expressed mainly in terms of strain and might therefore be expected to be scale independent. However, the factors controlling shear zone width are not known and, as with faults, shear zones on different scales may not have self-similar geometries. The scale dependence of the shale injection mechanism is not known.

Application

Sassi *et al.* (1992) and Gauthier *et al.* (1991) have applied predictions of sub-seismic fault populations to the analysis of compartmentalization problems in Cormorant Block IV and Pelican Field respectively, both in the North Sea. In the case of Pelican, probabilistic fault patterns were generated. Sassi *et al.* (1992) report clay smears in core on faults with displacements in the 3–16 m range, presumably as measured at the core/fault intersections. Without further information on the lithological succession and position of the intersection on the fault surface the observations cannot be taken as evidence of a 16 m smear, but this extent is approximately double that of the largest smear we have observed, or could have observed, in the outcrops examined. Although the presence of clay smears and ductile behaviour of shale is taken by Sassi *et al.* (1992) to indicate faulting of sediments before consolidation the outcrop data which we report shows that this is not a necessary condition.

Routine application of fault sealing data requires developments in three areas: (i) the quantitative assessment of the effect of a population of sealed faults on connectivity; (ii) an understanding of scaling relationships of fault sealing mechanisms; and (iii) prediction of fault surface characteristics in the subsurface. The data are applicable to sealing in the intra-reservoir context rather than to sealing of traps.

On the 2-D sections examined in outcrop, the scales of sandstone/shale layering and of fault throws are such that shale smears appear to reduce lateral connectivity of sandstone to zero in the cross fault direction. When the 3-D geometry of discontinuous fault surfaces is taken into account, however, it is evident that connectivity is preserved. The preserved 3-D connectivity is associated with a greatly increased tortuosity relative to a similar sequence containing the same fault population but without sealing of the fault surfaces. The hydraulic differences between the two cases could be quantified for a given sequence and fault population but this has not yet been done. A secondary effect of the sealing of intra-reservoir fault surfaces is the formation of footwall traps by footwall uplift on faults of a wide range of size (Barnett *et al.*, 1987) with consequent reduction in potential recovery. Little progress has been made in the two other directions necessary for reliable practical application of fault sealing data, i.e. scaling relationships and prediction in the subsurface. Both these unknowns require investigation of all types of fault seal and not only shale smears.

CONCLUSIONS

On the scale of the outcrops studied, shale smears are continuous on fault surfaces where the SSF is 7 or less. In general, sandstone/sandstone fault surfaces are likely to have continuous shale smears if the vertical interval between shale layers is no greater than the fault displacements and the net : gross ratio of the sequence is no better than *c.* 6 : 7. Subsurface equivalents of the shale smears observed are likely to represent significant hydraulic barriers even where the shale smear is incomplete. Quantification of the effects of populations of intra-reservoir sealing faults is feasible. Other sealing processes in fault zones are likely to be significant and the discontinuous nature of fault surfaces in 3-D is likely to play a significant role in preventing compartmentalization of faulted reservoirs. Application of the results of outcrop studies on the larger scale will require an understanding of the scaling laws for each sealing process.

ACKNOWLEDGEMENTS

Acknowledgement is due to Statoil, Norway, and to Shell Research B.V., The Netherlands, for financial support of the Liverpool Fault Analysis Group and for permission to publish. C.A. Jourdan and R.J.H. Loosveld are thanked for their contructive reviews.

REFERENCES

Barnett, J.A.M., Mortimer, J., Rippon, J.H., Walsh, J.J. & Watterson, J. (1987) Displacement geometry in the volume containing a single normal fault. *Bull. Am. Assoc. Petrol. Geol.*, **71**, 925–937.

Bouvier, J.D., Kaars-Sijpesteijn, C.H., Kluesner, D.F., Onyejekwe, C.C. & Van Der Pal, R.C. (1989) Three-dimensional seismic interpretation and fault sealing investigations, Nun River Field, Nigeria. *Bull. Am. Assoc. Petrol. Geol.*, **73**, 1397–1414.

Brand, P.J. & Haldorsen, H.H. (1988) Modelling large-scale heterogeneities caused by faulting with a stochastic approach. *Revue de l'Institut Français du Petrole*, **43**, 647–657.

Childs, C.J., Walsh, J.J. & Watterson, J. (1990) A method for estimation of the density of fault displacements below the limits of seismic resolution. In: *North Sea Oil and Gas Reservoirs—II* (Ed. by A.T. Buller, E. Berg, O. Hjelmeland, J. Kleppe, O. Torsæter & J.O. Aasen), pp. 309–318. Graham and Trotman, London.

Dubrule, O. & Haldorsen, H.H. (1986) Geostatistics for permeability estimation. In: *Reservoir Characterisation* (Ed. by L.W. Lake & H.B. Carroll), pp. 223-247. Academic Press, London.

Gauthier, B.D.M., Lake, S. & Jensen, Y. (1991) Probabilistic modelling of subseismic faults in the Pelican Field (Block 211/26a), North Sea, Offshore U.K. *Bull. Am. Assoc. Petrol. Geol.*, **75**, 579. (Abstract.)

Hamblin, W.K. (1965) Origin of 'reverse drag' on the downthrown side of normal faults. *Bull. Geol. Soc. Am.*, **76**, 1145–1164.

Heffer, K.J. & Bevan, T.G. (1990) Scaling relationships in natural fractures—Data, theory and applications. *SPE*, reprint no. **20981**.

Marrett, R. & Allmendinger, R.W. (1991) Estimates of strain due to brittle faulting: sampling of fault populations. *J. Struct. Geol.*, **13**, 735–738.

Petersen, K., Clausen, O.R., & Korstgågd, J.A. (1992) The evolution of a salt-related listric growth fault near the D-1 Well, Block 5605, Danish North Sea: Displacement history and Salt Kinematics. *J. Struct. Geol.*, **14**, 565–577.

Ramsay, J.G. & Huber, M.I. (1983) *The Techniques of Modern Structural Geology, Volume 1: Strain Analysis*. Academic Press, London, 307 pp.

Rettger, R.E. (1935) Experiments on soft-rock deformation. *Bull. Am. Assoc. Petrol. Geol.*, **45**, 1759–1762.

Sassi, W., Livera, S.E. & Caline, B.P.R. (1992) Reservoir compartmentation by faults in Cormorant Block IV, U.K. northern North Sea. In: *Structural and Tectonic Modelling and its Application to Petroleum Geology*. (Ed. by R.M. Larsen), pp. 355–364. Norwegian Petroleum Society Special Publication no. 1. Elsevier, Amsterdam.

Smith, D.A. (1980) Sealing and non-sealing faults in Louisiana Gulf Coast salt basin. *Bull. Am. Assoc. Petrol. Geol.*, **64**, 145–172.

Speksnijder, A. (1987) The structural configuration of Cormorant Block IV in context of the northern Viking Graben structural framework. *Geologie en Mijnbouw*, **65**, 357–379.

Walsh, J.J. & Watterson, J. (1987) Distributions of cumulative displacement and seismic slip on a single normal fault surface. *J. Struct. Geol.*, **9**, 1039–1046.

Walsh, J.J. & Watterson, J. (1988a) Analysis of the relationship between displacements and dimensions. *J. Struct. Geol.*, **10**, 239–247.

Walsh, J.J. & Watterson, J. (1988b) Dips of normal faults in British Coal Measures and other sedimentary sequences. *J. Geol. Soc. London*, **145**, 859–873.

Walsh, J.J., Watterson, J. & Yielding, G. (1991) The importance of small scale faulting in regional extension. *Nature*, **351**, 391–393.

Weber, K.J. (1986) How heterogeneity affects oil recovery. In: *Reservoir Characterisation* (Ed. by L.W. Lake & H.B. Carroll), pp. 487–544. Academic Press, London, 659 pp.

Weber, K.J., Mandl, G., Pilaar, W.F., Lehner, F. & Precious, R.G. (1978) The role of faults in hydrocarbon migration and trapping in Nigerian growth fault structures. *10th Annual Offshore Technology Conference Proceedings, Society of Petroleum Engineers*, **4**, 2643–2653.

Wheatcraft, S.W., Sharp, G.A. & Tyler, S.W. (1990) Fluid flow and solute transport in fractal heterogeneous porous media. In: *Dynamics of Fluids in Hierarchical Porous Media* (Ed. by J.H. Cushman), pp. 305–326. Academic Press, London, 505 pp.

Plate 1. A shale veneered slip surface on a block of sandstone from the footwall of a normal fault with *c*. 2.5 m displacement, Round O Quarry. The veneered surface (1) overlies a 4 cm thick cataclastic fault zone (2) which itself overlies undeformed footwall sandstone (3). The thickness of the smear is no greater than 0.1 mm. Pronounced striae (4) on the smeared sandstone slip surface are associated with trails of gravel-sized clasts embedded in the footwall (5) with crescentic erosional hollows on their up-dip sides (6). No erosion of the shale smear is apparent because the matching hangingwall is a 9 m thick shale. Diameter of the coin is 2.5 cm.

Plate 2. An eroded shale veneer on a sandstone slip surface (1) from the footwall of a 15 m displacement fault, Round O Quarry. The erosion is along displacement-parallel zones (2) defined by irregularities on the respective incompatible slip surfaces. The slip surface is therefore incompletely sealed by the shale veneer. Note trails of debris (3) aligned parallel to the slip direction, which also have shale veneers. This slip surface is one of several within a fault zone and the combined effect of all slip surfaces is to provide a continuous smear along the zone. Photograph is of a 20 cm long section.

Plate 3. Outcrop with multiple shale smear veneers in fault zone, Round O Quarry. Five shale veneered slip surfaces (1–5) are visible in a fault zone which has *c*. 5 m aggregate displacement. A sandstone horse (6) and a shale horse (7) are isolated by shale smears, at least in 2-D.

[*facing page 124*]

Plate 4. Outcrop with mixed sandstone and shale forming a composite fault zone rock, Round O Quarry. A shale veneer on the slickenlined sandstone footwall (1) is juxtaposed against black shale in the hangingwall; the displacement is 2 m. Within the black shale layer, immediately adjacent to the shale veneer, a layer of fault rock (2) up to 10 cm thick consists mainly of highly deformed and contorted shale containing sandstone debris. Foliation in the deformed shale fault rock (3) consists of highly polished slip surfaces.

Plate 5. A 6 cm thick shale layer (1) within a thick sandstone unit displaced 24 cm along a normal fault (2), Round O Quarry. There is pronounced drag adjacent to the fault (3) and the thinned shale layer persists for some distance along the shear zone containing the fault discontinuity. The shale smear occupies about 80% of the distance between the footwall base shale cutoff and the hangingwall top shale cutoff, thinning towards the central region where it is replaced by a shale veneer.

Plate 6. Outcrop of a normal fault with displacement in excess of 10 m, at Potatopot opencast mine, Cumbria. A 2 m thick black shale layer (1) and the immediately underlying coal seam are dragged into the fault. A 10 m thick composite, coarse-grained sandstone unit overlying the shale shows much less drag folding than the shale but is fractured in response to the same fault movement. Drag of the shale layer (2) towards the slip surface is accommodated by slip on an anastomosing network of highly polished intersecting slip surfaces (3), sub-parallel to bedding. The individual slip surfaces have a mirror-like polish with slickensides and slickenlines as directional slip indicators. Ruler at left centre is 1 m.

Plate 7. Photograph showing a small-scale example of shale injection along a 25 cm fault displacing a 30 cm sandstone bed (1) within a 6 m shale unit (2) at Round O Quarry. The injected shale smear (3) connects the overlying shale in the hangingwall with the underlying shale in the footwall and is *c.* 1 cm thick. The shale smear has been injected upwards from the underlying shale and it is not intersected by the fault slip surface but cuts across it. The injected shale 'vein' has a strong foliation parallel to its margins.

Plate 8. Syn-sedimentary faults disrupting a finely laminated sandstone and shale sequence at Round O Quarry. Displacement is accommodated along shear zones in which continuity of layers is maintained. The beds were unlithified when faulting occurred, and are rotated and thinned within the shear zones. Diameter of the coin is 1.75 cm.

Spec. Publs int. Ass. Sediment. (1993) **15**, 125–142

Applications of the formation micro-scanner to modelling of Palaeozoic reservoirs in Oman

C.G.L. MERCADIER *and* S.E. LIVERA

Petroleum Development Oman, PO Box 81, Muscat, Sultanate of Oman

ABSTRACT

Formation micro-scanner (FMS) and core data have been acquired from a variety of Palaeozoic clastic and carbonate reservoirs in Oman in support of primary field development, water injection and enhanced oil recovery projects. The FMS is a four-pad downhole resistivity tool that provides orientated images of the borehole wall. The magnitude and alignment of structural dips, and the presence and orientation of cross-stratification and fractures, can be determined. Subtle features such as nodules, pebbles, clay laminae or stylolites may also be recognized. Core-FMS calibrations determined from master wells can be used to evaluate FMS data in other non-cored wells for correlation purposes.

In the siliciclastic sequences the FMS is used for specific facies recognition, correlation and reservoir modelling in the Cambro-Ordovician Haima-Haradh, -Amin and -Mahwis Formations and the Permo-Carboniferous Haushi Group. The Haima Group is particularly sand-rich and a target for horizontal drilling and enhanced oil recovery. These objectives require relatively detailed reservoir descriptions. The Haushi Group consists of a wide variety of facies with a moderate (30 – 60%) sand content deposited during a period of gradual climatic amelioration from the lowermost glacial sediments of the Al Khlata Formation to the uppermost fluvial redbeds of the Gharif Formation. FMS-derived palaeocurrent data are used to orientate small fluvial channel sandbodies in medium to low sand content alluvial plain systems (Gharif Formation) and to model large, truncated, high-angle aeolian cross-bed sets (Amin Formation). Palaeocurrent data are extracted either by direct measurements or by core orientation.

FMS applications in the carbonates of the Permo-Triassic Khuff Formation are concentrated on reservoir rock type discrimination, differentiating strongly leached and dolomitized ooidal and stromatolitic units from low permeability, heavily stylolitized mudstones with anhydrite nodules.

The FMS images are also used to study the tectonic overprint of the Palaeozoic sequence, particularly fracture orientation and frequency, in an attempt to quantify their effect on reservoir quality. Open fractures may enhance reservoir productivity but also cause rapid water and/or gas breakthrough in wells (Amin Formation siliciclastics and Khuff Formation carbonates). Compartmentalization by networks of cemented fractures reduces reservoir productivity (Gharif Formation siliciclastics).

INTRODUCTION

Petroleum habitat in Oman

Since the early days of oil exploration in Oman in the 1950s, geologists have recognized the complexity of the hydrocarbon habitats (Fig. 1). Oil and gas occur virtually throughout the stratigraphic column from Infracambrian to Tertiary rocks (Fig. 2), in structurally complex reservoirs that are often difficult to delineate because of poor seismic data. The major structural elements influencing oil accumulation are the highs of the Dhofar and Huqf axes and the major Infracambrian Ara salt basins lying to the north-west (Fig. 1). In the southern and central parts of Oman, salt removal has dominated structural evolution producing a variety of hydrocarbon trap types (Heward, 1990). In the north, structures are halokinetic, formed over basement highs and influenced by the Cretaceous–Tertiary emplacement

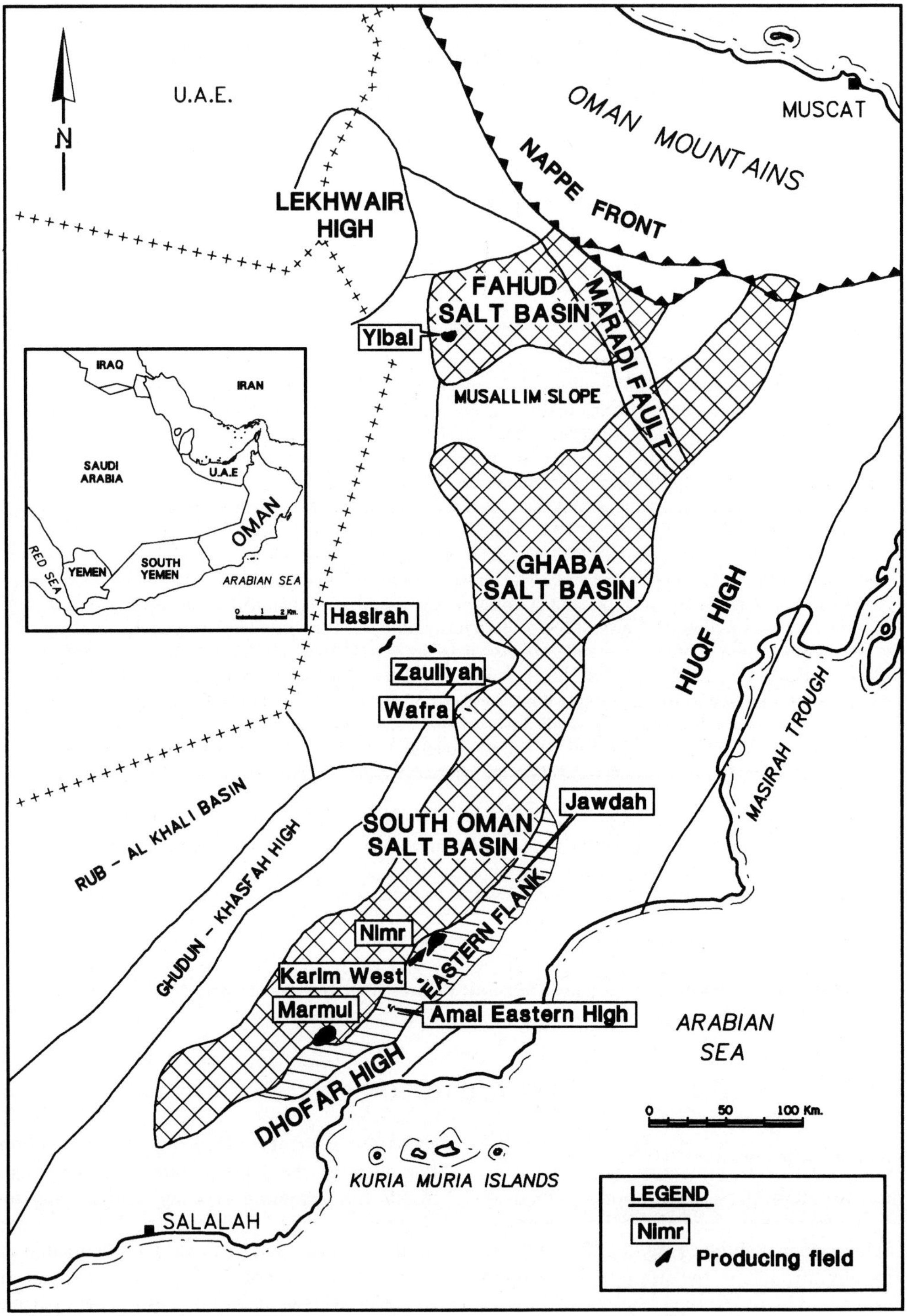

Fig. 1. Location map and structural elements of Oman.

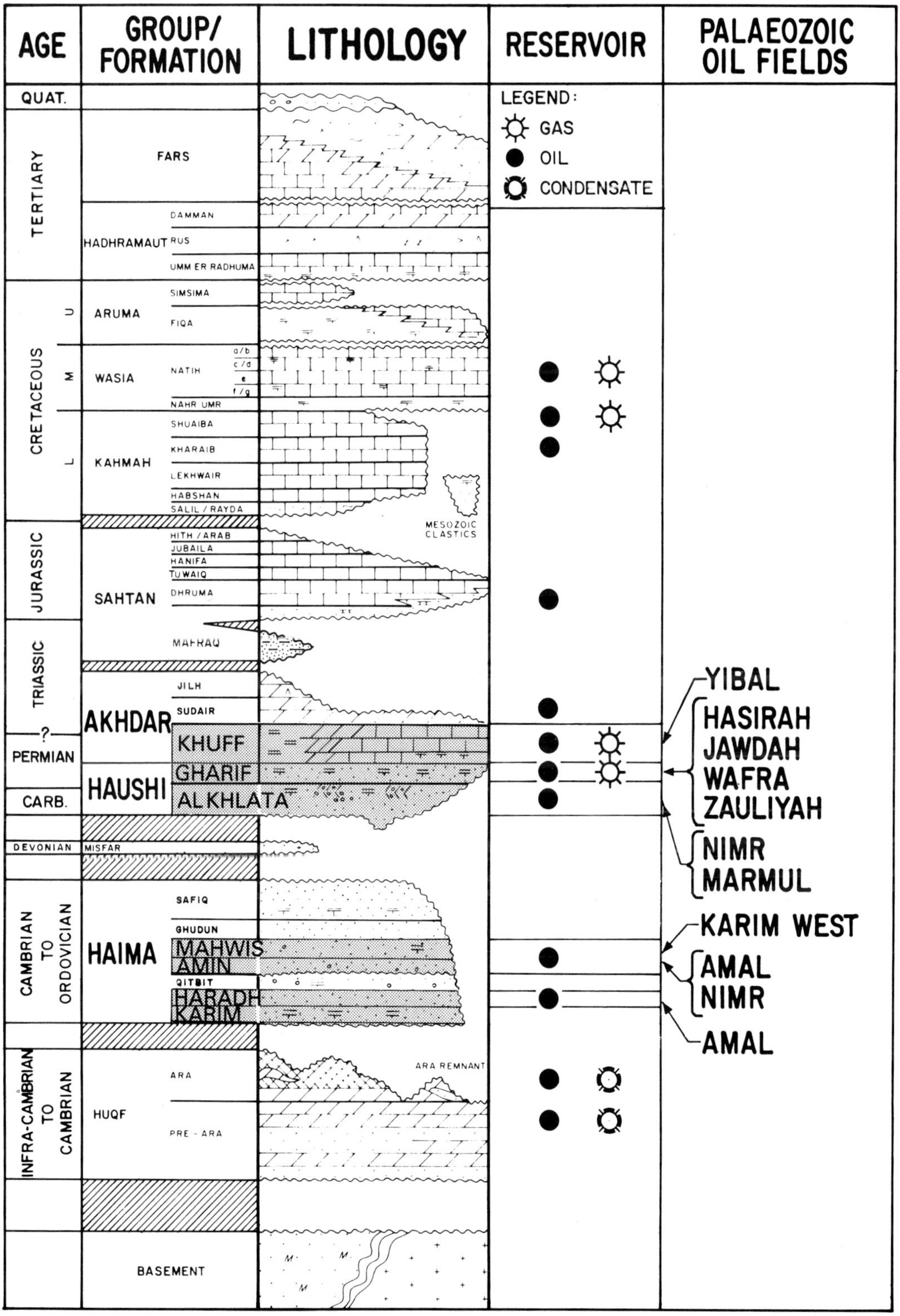

Fig. 2. Summary of the stratigraphic column of Oman.

of the Oman Mountains (Glennie *et al.*, 1974; Robertson *et al.*, 1990).

Oil derives mostly from Infracambrian sequences (Grantham *et al.*, 1988) but a history of re-migration, local biodegradation and secondary sources results in variable oil types in different fields. In the southern part of Oman, the 'Eastern Flank' reservoirs contain mostly heavy oil whereas the salt basin fields to the north contain lighter oil and gas (Fig. 1). The methods of oil exploitation vary in response to the reservoir configurations and oil type.

Because hydrocarbon production in Oman is land based, small fields can be brought onto production soon after discovery; thus there are presently over 60 oilfields connected to export facilities. A large portion of the production is primary, relying on reservoir pressure, gaslift or surface beam pumps to sustain flow. However, Oman is essentially a mature oil province and secondary and tertiary recovery mechanisms will play an increasing role in the future. Substantial thermal and chemical enhanced oil recovery (EOR) 'scouting' studies were carried out in the mid-1980s, aimed at the heavy oil fields of the Eastern Flank (Martin & Cooper, 1984; Watts *et al.*, 1986). More recently attention has been drawn to water injection schemes in the fields containing lighter oil. Although EOR plans have been delayed by the recent unstable oil prices, the scope for increasing ultimate recoveries in the longer term and the attractiveness of water injection schemes and horizontal drilling are placing an emphasis on more detailed reservoir description.

Stratigraphy

The Infracambrian Huqf Group contains source and reservoir rocks as well as the Ara Formation salt (and remnants of evaporitic rocks) that mostly controlled Palaeozoic and Mesozoic structuration. Unconformably overlying the Huqf are the main siliciclastic reservoir sequences in Oman, the Cambro-Ordovician Haima Group and the Permo-Carboniferous Haushi Group. The Permian Khuff Formation carbonates terminate the Palaeozoic, which is capped by thick sequences of Mesozoic carbonates (Fig. 2) that form important oil reservoirs in northern Oman.

Haima Group continental siliciclastic sediments contain over 4.4×10^9 bbls of oil in a sand-dominated sequence, the subdivision and correlation of which are based on lithofacies in the absence of biostratigraphic data (Heward, 1989). In the overlying Haushi Group, the spectacular and complex glacial sediments of the Al Khlata Formation contain over 5×10^9 bbls of oil in the Eastern Flank traps, usually in combination with the Haima Group (Levell *et al.*, 1988). Overlying the Al Khlata is the predominantly fluvial Gharif Formation. This interval was deposited during a significant period of climatic amelioration and is volumetrically the most important clastic reservoir sequence in Oman, containing over 7.5×10^9 bbls of oil. The Khuff Formation comprises cycles of shallow marine carbonates with reservoirs consisting of fractured, mouldic ooidal beds which alternate with mudstone. The reserves are relatively modest (0.5×10^9 bbls of oil), but the sequence is an exploration target in northern Oman and elsewhere in the Arabian Gulf region (Sykes & Abu Risheh, 1989). In total, the Palaeozoic sequence contains 55% of the remaining hydrocarbons in Oman. This is the main target for oil exploitation but, because of the complexity of the stratigraphic and structural setting outlined above, careful reservoir characterization is required on a field-by-field basis.

FMS applications

The FMS (formation micro-scanner) is a four-pad downhole tool that maps the resistivity variations across the formations through which it passes (Serra, 1989). Each pad is pressed against the borehole wall and comprises 16 electrode buttons which emit electrical current into the formation, producing resistivity curves. The tool is mainly run in holes filled with water-based drilling mud that provides a conductive medium around the electrodes. The resistivity curves may be displayed as raw traces ('wiggles') or processed to create an image of resistivity contrasts around the borehole wall. This image consists of four parallel traces that cover approximately 30% of the hole's surface in a 12¼ inch hole (Bourke *et al.*, 1989). The formation's resistivity variations are colour-coded on the displays, from white (resistive) through yellow and red to black (conductive), with a resolution of less than 1 cm. Image processing is in two ways: (i) 'static normalization' where the colour range displayed reflects the resistivity range of the entire logged interval; and (ii) 'dynamic normalization' where the colour range displayed is calculated from the resistivity range of a short 'window' (0.5 m long) sliding with depth. The static normalization processing produces an unorientated image whereas the dynamic normalization processing gives an image orientated relative to north (or 'BORMAP'). It must be emphasized that full interpretation of such images requires comparison with core material and integration with other wireline log data which allow the user to quality-control FMS outputs and to calibrate the tool's response to the specific characteristics of the various reservoirs (Adams *et al.*, 1990). The calibrations obtained from complementary core and FMS data in master wells can then be used to evaluate FMS data in other non-cored wells for correlation purposes within each field.

In Oman, the FMS is mostly run to obtain small-scale structural information which is then tentatively incorporated into the broader tectonic context by combination with seismic and regional data. The production of a reservoir model which predicts the geometry of the effective (open) fracture network at reservoir conditions in carbonates requires primarily the combination of data from: (i) the FMS, which provides fracture orientations, shows whether the fractures are closed or open at depth, and gives an indication of fracture aperture; and (ii) cores, which allow the recognition of the fracture type(s), overprinting relationship(s) and diagenesis. In clastic sequences, the FMS is also used to identify and model the influence of potential fault and fracture barriers or baffles to flow. However, the FMS is also a useful tool for sedimentological reservoir characterization where close well spacing, intensive drilling and lateral facies variability limit the representativeness of scattered cores in any single field.

HAIMA GROUP

The continental siliciclastic sediments of the Haima Group, together with the Al Khlata Formation, host the main heavy oil accumulations in the Eastern Flank area. Structural development of these fields was by salt dissolution and withdrawal, which resulted in sediment pods characterized by locally intense faulting and fracturing (Heward, 1990). Because the Haima Group is typically very sand rich (80–100%), these fractures, when open, interconnect and result in very rapid water cut development in new wells. A challenge with many Haima reservoirs, whether faulted or not, is minimizing the amount of formation water produced at the expense of oil as a direct result of coning (i.e. by-passing oil by inducing the rise of unwanted aquifer water through the oil column into a producing well). This phenomenon can be reduced by optimizing well completion intervals and by drilling horizontal wells, both of which require a knowledge of vertical permeability derived from sedimentological studies for adequate planning purposes. Additionally, the Haima Group as a whole is a major target for EOR.

Three of the six formations within the Haima Group on the Eastern Flank hold significant quantities of heavy oil (Haradh, Amin and Mahwis Formations; Fig. 3). The FMS is used to quantify open fracture frequencies, distributions and orientations and can assist in detailed correlations used in fault identification and structural interpretations. An additional significant planned use of the FMS is in detailed reservoir characterization studies in support of EOR and well completion policies, although the three main formations pose subtly different problems, as outlined below.

Haradh Formation

The Haradh Formation accumulations are typically thick (the Amal Eastern High Field has an oil column of over 250 m) and it is economically not feasible to core the whole stratigraphic interval in every well. Typically the Haradh Formation consists of dm-scale sets of trough cross-stratified, uniformly fine-grained sandstones, which are interpreted as braided stream deposits (Fig. 3 and Plate 1, facing p. 136). Characteristically the Haradh Formation is extremely sand rich (94% reservoir sand in the Amal Eastern High Field), but minor intercalated argillaceous intervals and cemented layers play a disproportionately important role on fluid flow. Plate 12 shows typical Haradh Formation sediments consisting of cross-stratified sandstones with abundant mudflake conglomerate, carbonate cementation and continuous, thin shale layers. These features dominate effective vertical permeability, a crucial factor in evaluating the economic attractiveness of EOR (Martin & Cooper, 1984). The carbonate cemented horizons and shale layers are impermeable and will control, for example, the rise of steam in a reservoir undergoing thermal EOR schemes. Unfortunately the majority of these features are well below the resolution of normal logging techniques and evaluations, but the FMS in comparison reveals important information. Plate 1 shows why it is impossible to identify many features accurately from conventional log data, which are essentially one-dimensional measurements averaged to a resolution limit of *c.* 25 cm. The FMS, in contrast, identifies continuous and

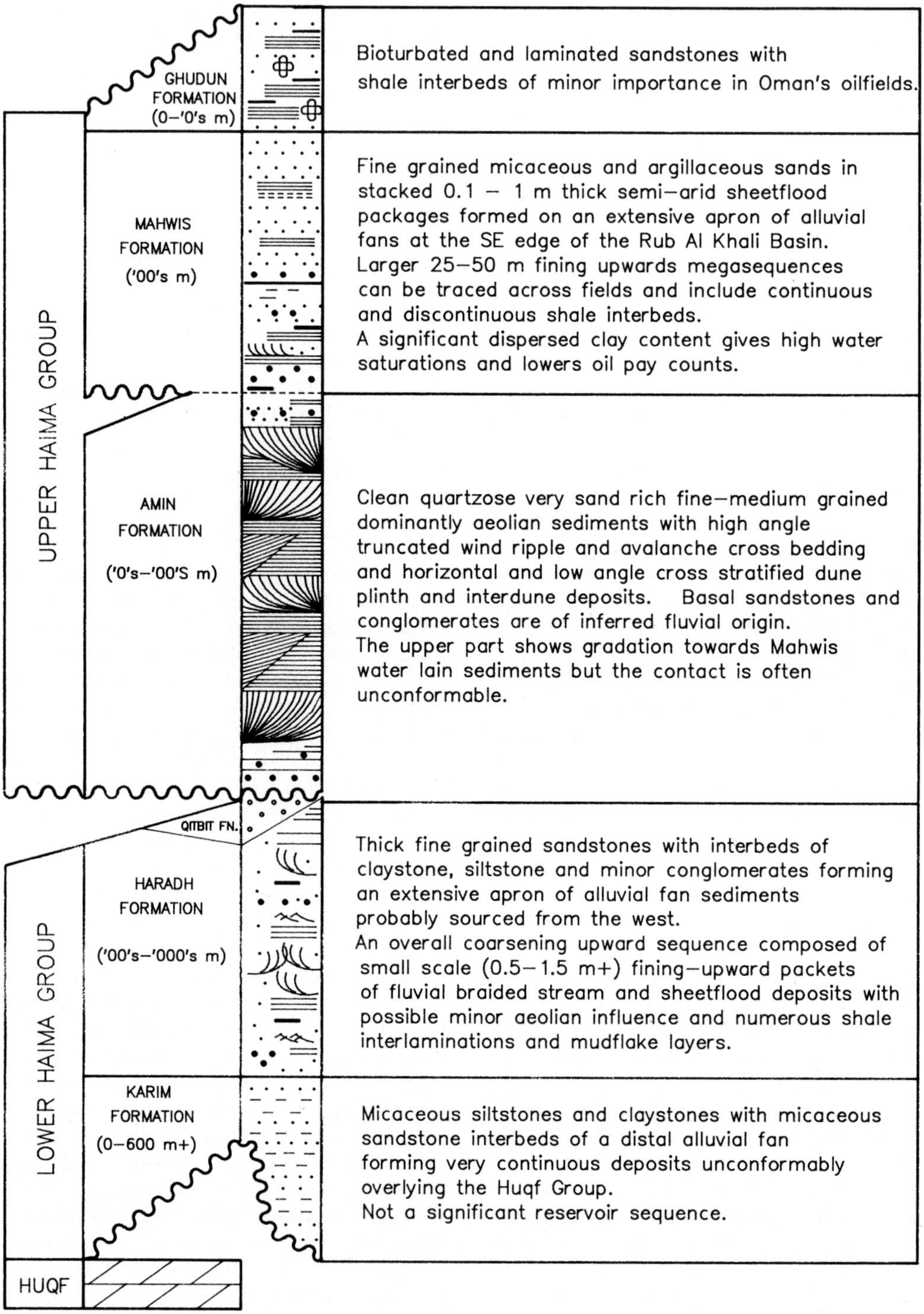

Fig. 3. Summary of the Cambro-Ordovician Haima Group in the Eastern Flank oilfields of Oman.

discontinuous mudstones, pervasive and non-pervasive carbonate cements in three dimensions and this example shows how 'static normalization' processing allows contrasts in resistivity (i.e. grain size/shaliness/oil saturation), and therefore reservoir quality, to be highlighted.

Amin Formation

Overlying the Haradh Formation, the dominantly aeolian Amin Formation provides spectacular examples of high-angle wind ripple cross-stratification in clay-free, high-permeability sandstones (Plate 2, facing p. 136). The excellent reservoir quality of the sands removes some of the reasons for detailed reservoir characterization; however, it is known that permeability anisotropy plays a significant role in hydrocarbon production from similar reservoirs (Weber, 1987; Lindquist, 1988). Horizontal permeability anisotropy results from the marked grain size variations within foresets with a minimum value parallel to palaeowind directions. Vertical permeability anisotropy will result from the superposition of differing aeolian stratification types and the presence of interdune deposits and thin fluvial intercalations. Such anisotropies have not yet been quantified in the Amin Formation, but any attempt to enhance recovery from Haima reservoirs, such as in the Nimr Field, will require evaluation of these effects. Marked differences in foreset laminae thickness and development/geometry of interdune sediments exist in the Amin Formation. These phenomena can be classified and quantified using the FMS (Plate 2). Cross-stratification orientations are particularly easy to determine on FMS images in the Amin Formation (Plate 2). The Amin Formation of the Nimr Field is becoming an important target for horizontal drilling and the ability to quantify vertical permeabilities is significant to assessing well productivities.

Mahwis Formation

The third important Haima Group Formation in the Eastern Flank is the Mahwis Formation. The characteristics of this sequence are summarized in Fig. 3, and further details are given in Heward (1989). Because of the fine-grained nature of the sands, their shaliness (including smectites) and variable carbonate cementation, it is difficult to distinguish between 'pay' (oil-bearing and oil-productive) and 'marginal' (containing non-producible oil due to high smectite clay and therefore high water content) zones. This problem is highlighted in Plate 3 (facing p. 136), taken from a well in the Karim West Field. Evaluations indicate that only 45.6 m can be considered pay in this well, as opposed to the 98 m thick oil-bearing section in a nearby well. Because the formation consists of small-scale (10–200 cm thick), fining-upward sheetflood units the pay zones are also thin, and scattered throughout non-pay intervals. In order to properly evaluate pay zones, micro-resistivity tools such as the FMS have to be used (Plate 3). The distinction between amount of pay and non-pay is also important for EOR evaluations and the smectites in the Mahwis Formation may severely limit steam injection by swelling, dislocation and migration, and possibly lead to polymer adsorption.

Most of the above comments on the Mahwis Formation relate to petrophysical (rock property) evaluations rather than directly to the sedimentology of the formation. However, any evaluation has to be placed in a geological framework in order to develop an overall 3-D model of a field such as Karim West (Fig. 1). Correlation of sediment packages (e.g. 25–50 m thick fining-upward sequences; Heward, 1989) is required for structural definition and to evaluate fault effects on reservoir continuity, which is potentially critical in such thinly-bedded, shaley reservoirs. In addition, the Mahwis Formation also contains discrete shale layers, mudflake conglomerate and calcite cement which is especially well developed in the cross-bedded lithofacies. As such, similar problems as in the Haradh Formation exist here, notably the definition of vertical permeability.

Much more can be made of the FMS as a potentially quantitative tool in the Mahwis Formation where permeability is found to be more closely linked to clay content and degree of cementation than porosity. Traditional methods of estimating permeability from porosity logs via a porosity–permeability relationship do not apply well to the Mahwis Formation.

HAUSHI GROUP

Al Khlata Formation

Reservoir setting

The sediments of the Al Khlata Formation have long been recognized to be of glacial and/or glacially-influenced origin (Braakman *et al.*, 1982; Hughes Clarke, 1988; Levell *et al.*, 1988). These sediments are dated as of Late Carboniferous (Late Westphalian) to Early Permian (Sakmarian) by sporomorphs (Besems & Schuurman, 1987). The main characteristics of the Al Khlata Formation are (i) a great variety of lithofacies, and (ii) an extreme lateral and vertical variability that results from deposition on the irregular Haima Group substratum, in unstable areas where salt movements were active during sedimentation (Levell *et al.*, 1988; Heward, 1990). The Al Khlata Formation thus consists of heterogeneous sedimentary packages that contain intraformational unconformities with abrupt, commonly fault-controlled, thickness variations (Fig. 4). Prediction of reservoir sand development in oil fields is therefore very difficult, even with well spacings of 300–400 m (e.g. Nimr Field, Fig. 5). Moreover, stacked reservoirs commonly have different fluid contents, properties and pressures probably due to their limited lateral extent.

The Al Khlata Formation, however, remains the largest EOR target in southern Oman and the need for more detailed reservoir description and modelling is being addressed via the integration of core and FMS data.

Sedimentology

Lithofacies comprise diamictites, conglomerates (Plate 3), gravels, pebbly and gravelly sands, cross-bedded and homogeneous sands, silts, silty shales and shales. Deposition was in the following two settings:

1 Deposition over Haima 'highs' (Figs 4, 5) with meltwater channels contained within narrow (a few hundreds of metres wide) palaeovalleys. These are filled with up to 200 m of both glacial and recycled eolian/sheet-flood Haima Group sediments. The dominant environment of deposition was, however, glacio-lacustrine with varve-like shaly deposits containing dropstones, interbedded with deltaic and/or subaqueous fan sands. This setting provides some lateral correlatability over a few square kilometres but rapid variations occur at the edges of the depositional bodies.

2 Accumulation directly over salt (sediments now lie directly on Huqf 'basement' or Haima 'lows' as a result of salt removal) consist of few sandbodies interbedded with thick packages of claystone, siltstone and diamictite (Fig. 4). These sediments were deposited as deltas, subaqueous fans and locally turbidites, within a basinal glacio-lacustrine environment.

FMS applications

The dominant problem for reservoir modelling of Al Khlata reservoirs is their lateral variability. This variability can only be quantified with dense drilling patterns because (i) the major boundaries and internal faults of the Al Khlata Formation cannot be resolved from seismic data, and (ii) there are no established models that quantitatively predict facies changes in glacial environments. The palynological subdivision obtainable from sidewall samples in wells assists correlation and differentiation of reservoir from poor to non-reservoir rock types. To obtain data needed to construct detailed simulation reservoir models a full core would be required in virtually every well but this procedure would not be economically viable.

A potential use of the FMS is in good reservoir sands that are selected for EOR purposes. In the Marmul Field (Fig. 1), the target intervals for both water and polymer injection consist of poorly consolidated, cross-bedded, glacio-fluvial sand with permeabilities locally up to several darcies. These sands overlie a sequence of diamictites with irregularly distributed thin sands (Plate 4, facing p. 136). Although the main depositional sequences may be interpreted from conventional log response (Levell *et al.*, 1988), the FMS is needed to recognize intervals with subtle reservoir and non-reservoir features, after calibration with core data (Plate 4). This information may then be compared with oil saturation and porosity profiles to better define the internal characteristics of pay intervals and achieve detailed correlation with neighbouring wells.

FMS imagery is also used to detect and analyse faults and fractures in the Al Khlata Formation. A survey in the Nimr Field (Fig. 1) has demonstrated that (sub)vertical, closely-spaced (a few centimetres), small-scale (a few tens of centimetres) fractures and faults are located in the sand-rich intervals and do not extend across the impermeable, shaly/silty diamictites and conglomerates (Fig. 4). However, this lithofacies-controlled fracture distribution appears to be insignificant on a larger scale since production data to date (over a period of less than 10 years) indicate homogeneous depletion of the field.

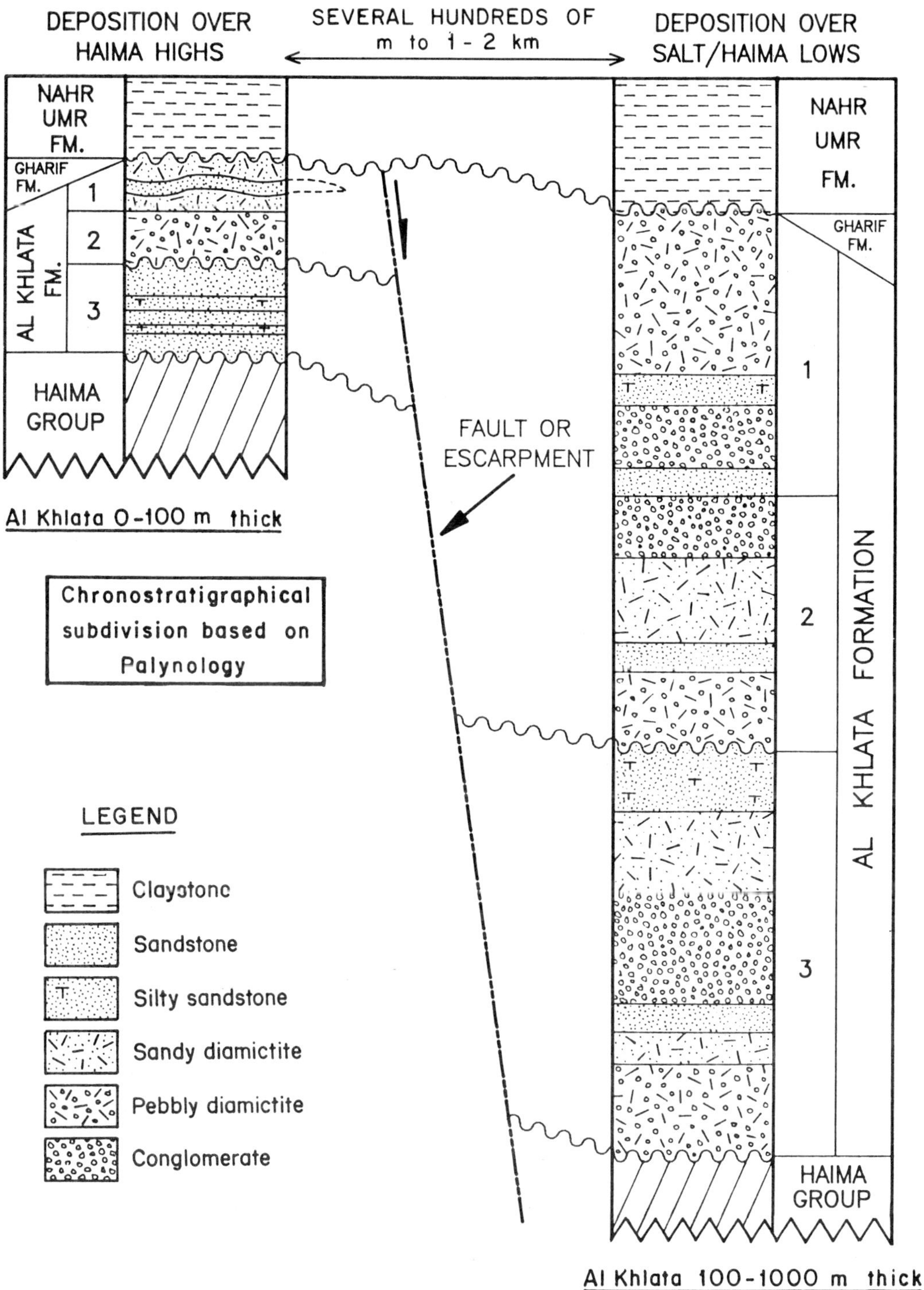

Fig. 4. Summary of the Permo-Carboniferous Al Khlata Formation in the Eastern Flank. The stratigraphical subdivision into Al Khlata Units 1, 2 and 3 is based on palynological assemblages.

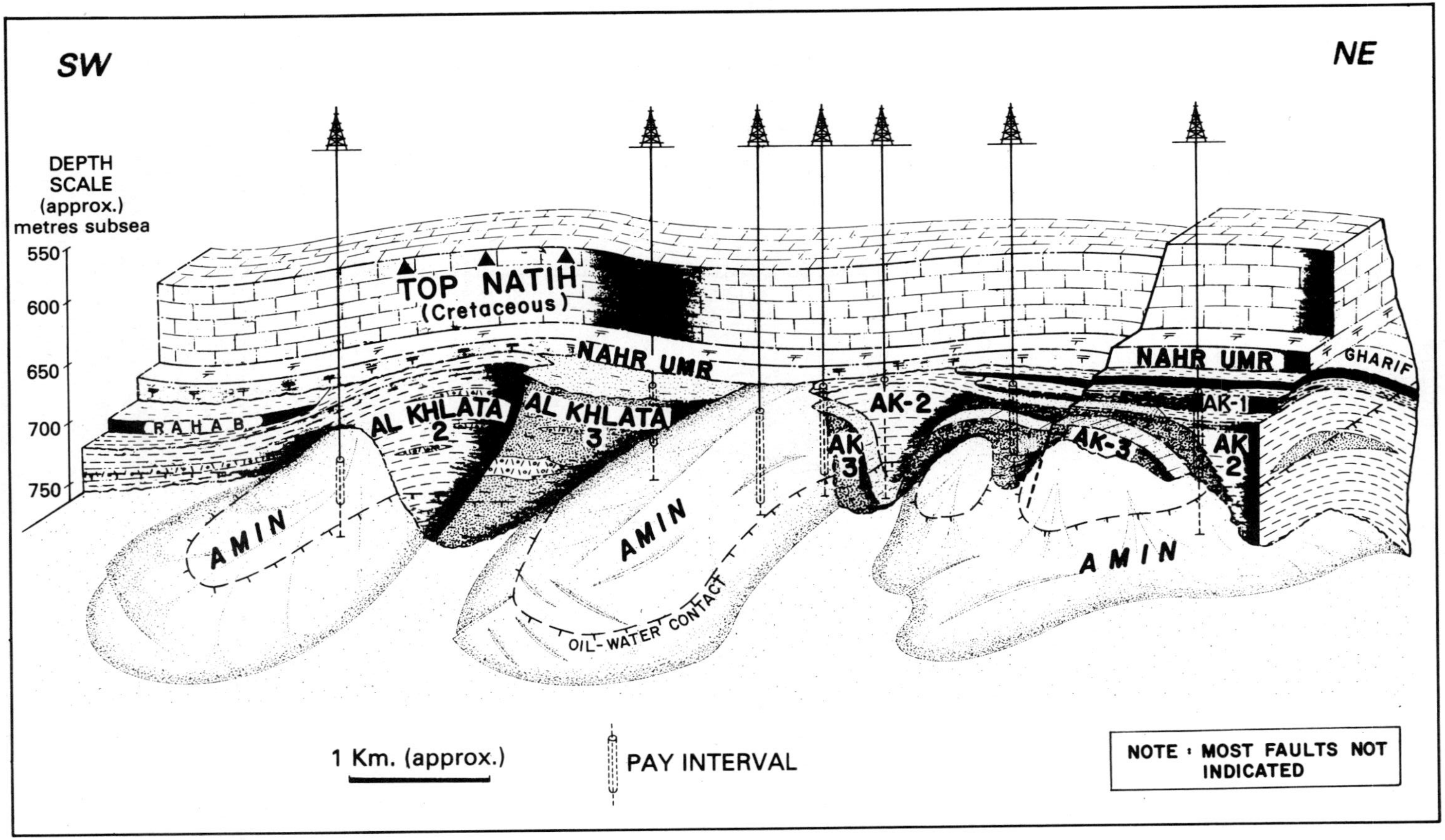

Fig. 5. Block diagram of the Nimr Field showing the relationships between the Amin Formation (Haima Group) and the Al Khlata Formation Units 1, 2 and 3 (Haushi Group).

Gharif Formation

Reservoir setting

The Gharif Formation (Fig. 2) contains hydrocarbons throughout Oman and has been subdivided, based on lithostratigraphy, into Lower, Middle and Upper Members (Hughes Clarke, 1988; Focke & Popta, 1989). On the Eastern Flank (Figs 1, 6), the Lower Gharif Member includes glacial sediments that mark a late phase of Gondwana glaciation and is the main Gharif reservoir in this area. To the north, in the salt basins (Figs 1, 6), the mainly continental red beds of the Middle and Upper Gharif Members are the dominant reservoirs. These sequences were deposited during a period of climatic amelioration in a stable intracratonic setting on the margin of the Rub Al Khali Basin (Fig. 1).

Sediment source was from the south and east and included feldspathic material reworked from the underlying Al Khlata Formation. Periods of source-area regeneration and denudation, and a major marine incursion that formed the Haushi Limestone, controlled the distribution of reservoirs and seals in the formation (Fig. 6). A very generalized palaeogeographic reconstruction indicates sand-rich fluvial systems to the south and east, draining from the Huqf and Dhofar axes into increasingly red bed, mudstone dominated distal sequences of a mature alluvial plain (Fig. 7). Within this setting, reservoir characterization has to address a number of different problems, generally on a field-by-field basis. A major use of the FMS is in lithofacies recognition, both for general correlation purposes and for more detailed reservoir architecture definition.

Correlation in the Gharif Formation is complicated by the poor gamma ray distinction between mudrocks and various types of sandy lithofacies due to the high feldspar content of the latter and low radioactivity of the shales. In many fields intense calcite cementation affects the density and sonic wireline log responses. Continuing salt withdrawal and compactional effects give marked Gharif thickness variations between and within fields on the Eastern Flank. Accurate correlation is required to evaluate reservoir sandbody interconnections, to define fault locations and amounts of throw, and to predict the distribution of hydrocarbons in the stacked reservoir sequences created by intra-Gharif sealing mudstones.

Sedimentology

As in the Al Khlata Formation, diamictites play a role as inter-reservoir seals in a few fields in the Eastern Flank but only the massive, silty diamictites act as permeability barriers. However, the sandy, stratified, conglomeratic diamictites share similar wireline log responses but are not barriers to fluid flow. Thus core material or FMS images are required to distinguish the two facies associations.

In the Lower Gharif Member, major datum horizons are defined by the top of the Al Khlata Formation/Rahab Shale and a transgressive event that is the time equivalent of the Haushi Limestone of the salt basins (Fig. 6). The transgressive event is recognized by the coarsening-upward offshore-shoreface regression, the 'bioturbated interval' of earlier authors (Heward, 1990). In both core and on the FMS images this thin (3–15 m thick) sequence is clearly recognizable (Plate 5, facing p. 136) and the shaly lower part is distinct from other mudrocks in the Lower Gharif Member. The transgressive surface acts as a datum that is used to subdivide the Lower Gharif Member into two submembers, but the shale is often ambiguous on routine wireline logs, as in the Jawdah Field (Plate 5).

In the Middle and Upper Gharif Members, thickness variations are small in the salt basins where these stratigraphic horizons contain most oil. Here the correlation problem is one of defining individual reservoir drainage units based on stratigraphic subdivision rather than defining fault cut-outs or identifying salt or compactional effects as in the Eastern Flank. In fields such as Hasirah, Zauliyah, or Wafra (Fig. 1), multiple oil-water contacts occur within a stratigraphy of varying sand content, from low (25%), poorly connected, often sub-economic net-to-gross intervals to well-connected channel sandstone belts that are potential targets for water injection schemes. A meaningful reservoir subdivision relies on the integration of all available data, including the recognition of specific floodplain environments from cores and FMS images. Within mudstone-dominated packets, well-drained floodplain is marked by calcrete horizons, which form ideal correlation markers. Unfortunately these calcretes 'wash out' in boreholes due to the high smectite content of the palaeosols and thus give very poor FMS recordings. However, in sandier sequences or under good borehole conditions, root-bearing horizons can be clearly identified (Plate 6, facing p. 136). Localized lacustrine sequences have specific FMS

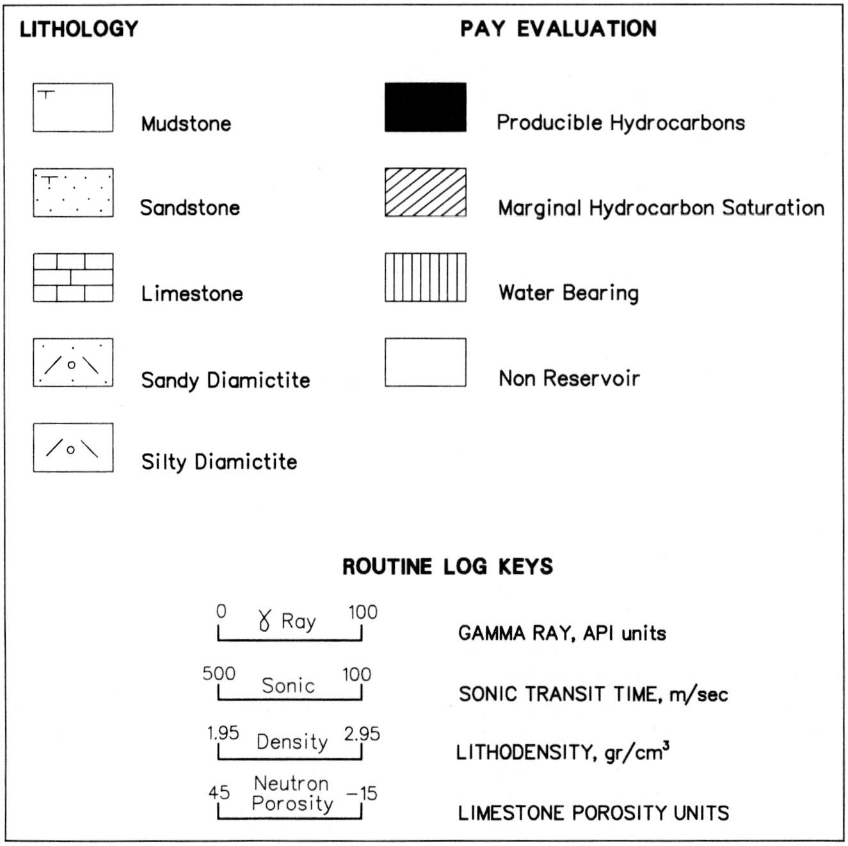

Key to colour plates 1–12

signatures (Plate 7, facing p. 136) that can be calibrated with other wireline logs and correlated throughout the field. The lacustrine facies contain silty sandstones with moderate reservoir quality that interconnect the main channel sand reservoir belts over tens of square kilometres in fields such as Hasirah (Fig. 7 and Plate 7).

FMS applications

In all moderate to low net-sand fluvial reservoirs with discontinuous channel sands, knowledge of sandbody orientation is needed to model reservoir geometry and optimize well spacing. Two examples of clear palaeocurrent indicators are given in Plates 8 and 9 (facing p. 136). In one case the resistivity contrasts of the foresets are clearly picked out by the FMS in a minor distributary channel in coastal plain sediments overlying the Haushi Limestone (Plate 8). This feature would probably be distinguished with a standard dipmeter (SHDT) log (Hocker *et al.*, 1990). The great advantage of the FMS tool is that it allows the geologist to interpret poor quality data, as in Plate 9. Here the foresets are only visible on two traces of the FMS image because the sedimentary signature is heavily masked by cementation.

In the shallow marine sediments of the Lower Gharif Member, Central Oman, high sand contents and light oil ensure good oil recovery. However, facies variation is marked, with both small and large-scale heterogeneities present that must be included in numerical simulation studies, as for similar depositional environments elsewhere (e.g. Harzog Draw Field; Hearn *et al.*, 1984). FMS images in the Lower Gharif Member show how cross-bedded, clay-free sands with high permeabilities can easily be recognized (Plate 10, facing p. 136).

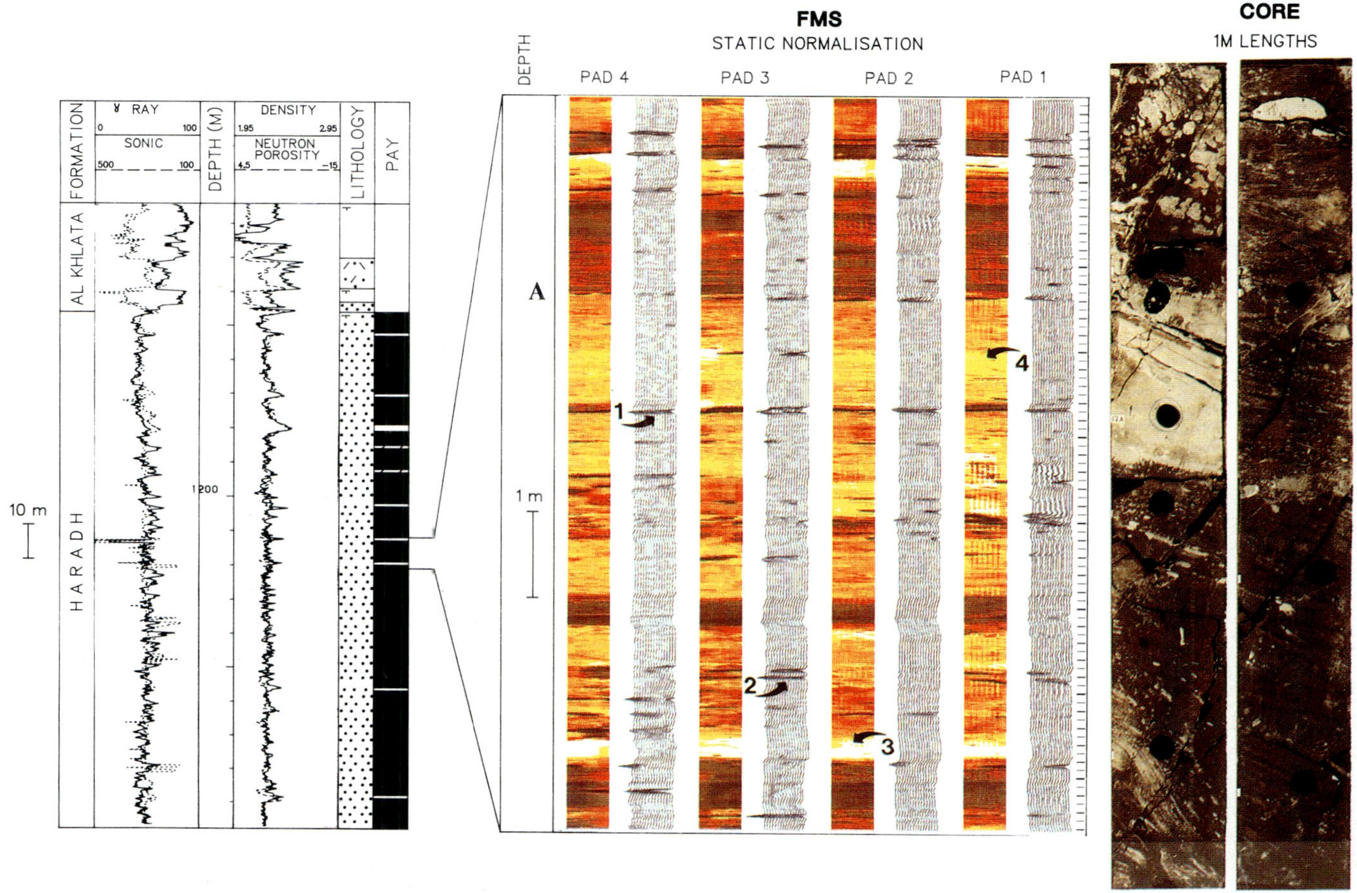

Plate 1. Core of mostly oil-stained cross-stratified sandstones with abundant mudflake conglomerate, carbonate cementation and thin continuous shale layers from the Haradh Formation. The FMS image from a nearby well shows the resolution of continuous (1) and discontinuous (2) shale layers, carbonate cement (3) and excellent high-resistivity sands (4). A comparison with petrophysical evaluations shows the poor thin-bed resolution of the conventional logs, and highlights the variation in reservoir quality (porosity and hydrocarbon saturations) above and below (A) although both zones are evaluated as 'pay'*. A realistic evaluation of the effects of shales on fluid flow requires that all impermeable beds be recognized and their lateral continuity quantified from direct measurement (discontinuous) or by outcrop analogue (continuous). (**Note*: Pay = thickness of strata calculated having producible hydrocarbons).

[*facing page 136*]

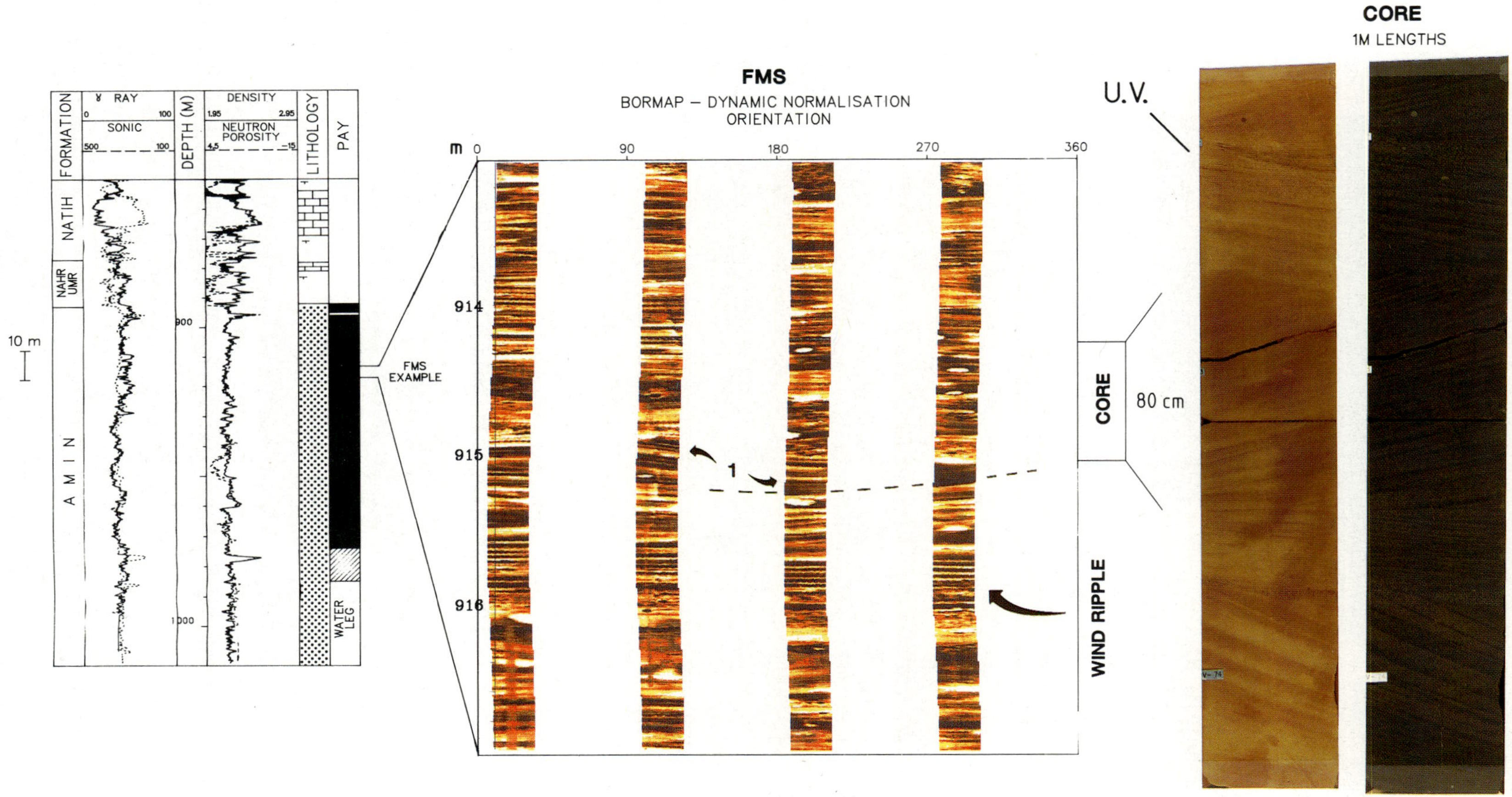

Plate 2. Core of wind ripple lamination with marked grain size, porosity and permeability variations and the corresponding vertically condensed BORMAP from the Amin Formation. The laminae orientations (1) and thickness (a possible indication of grainflow deposits) in the Amin Formation can clearly be seen on FMS traces. Both horizontal and vertical permeability anisotropies will result from the grain size variations, which require careful core plug measurement and orientation before they can be quantified (Weber, 1987). (*Note*: 'U.V.' indicates an ultra violet light photograph).

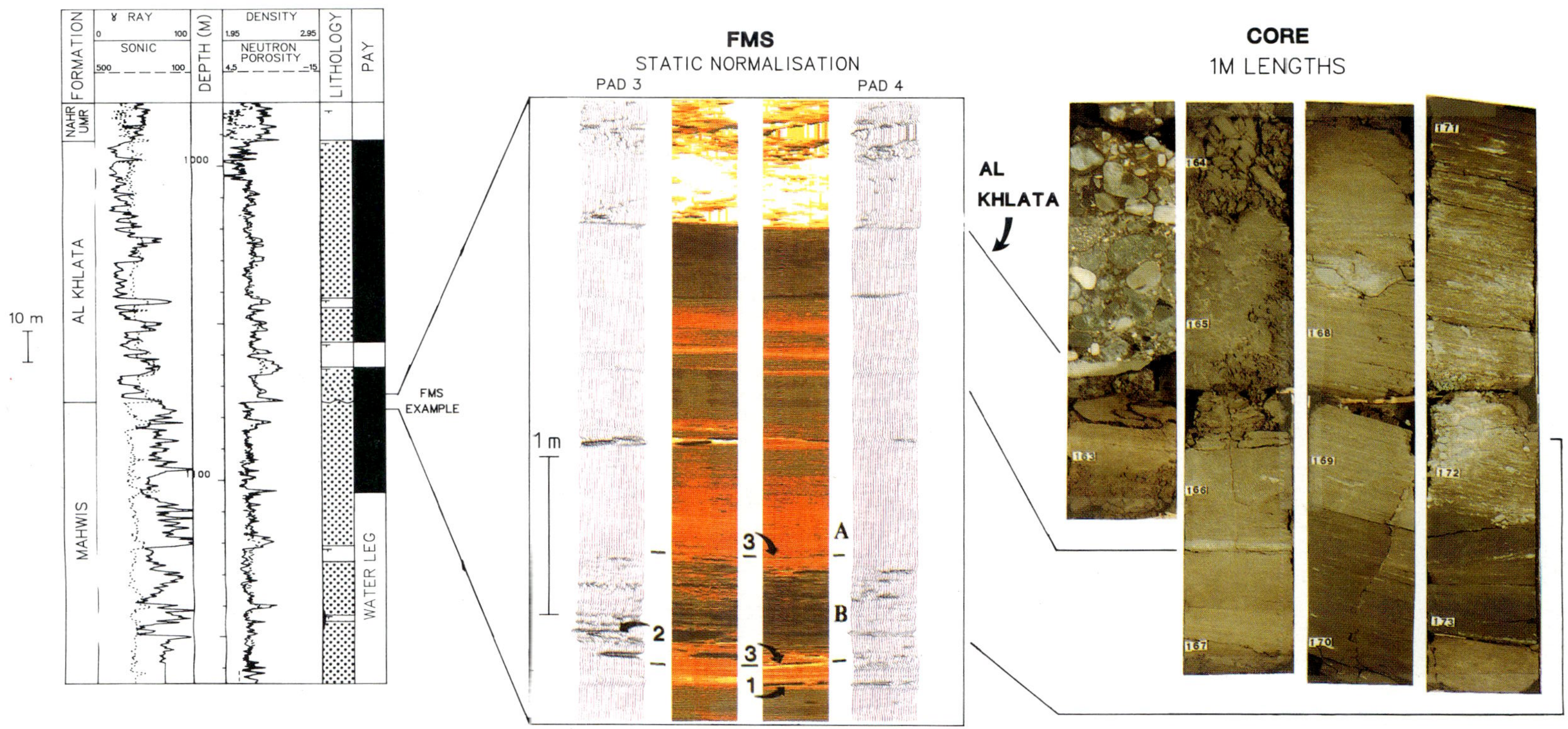

Plate 3. Fine-grained, shaley and partly cemented sandstones of the Mahwis Formation. Note how very subtle the distinction between pay (A) and non-pay (B) is on the core. Routine logs evaluate these differences very poorly. The FMS static image highlights the variable sand quality as well as illustrating minor shale layers (1), mudflake conglomerate (2) and cross-stratification (3), the latter partly masked by the vertical condensation of the FMS image.

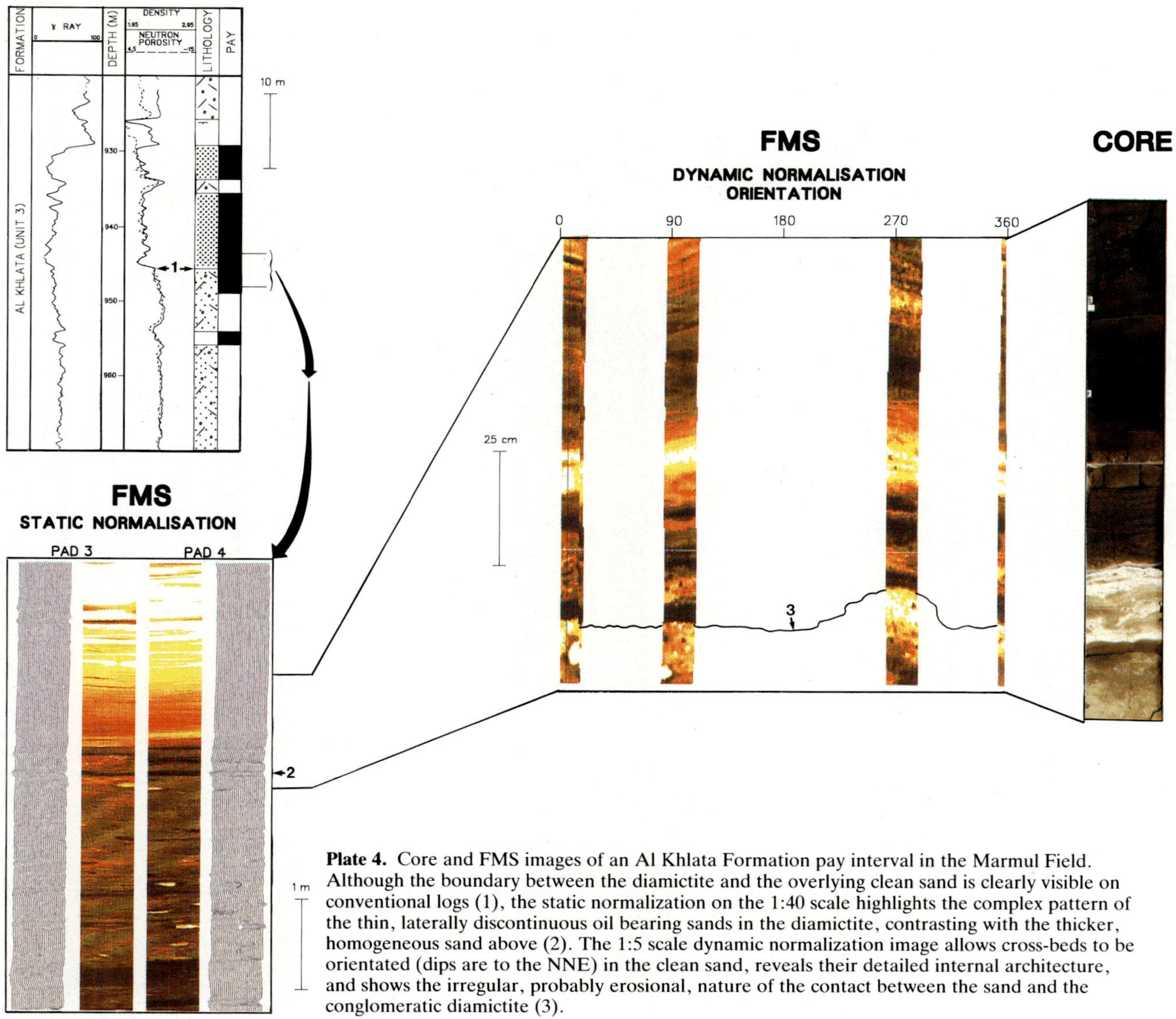

Plate 4. Core and FMS images of an Al Khlata Formation pay interval in the Marmul Field. Although the boundary between the diamictite and the overlying clean sand is clearly visible on conventional logs (1), the static normalization on the 1:40 scale highlights the complex pattern of the thin, laterally discontinuous oil bearing sands in the diamictite, contrasting with the thicker, homogeneous sand above (2). The 1:5 scale dynamic normalization image allows cross-beds to be orientated (dips are to the NNE) in the clean sand, reveals their detailed internal architecture, and shows the irregular, probably erosional, nature of the contact between the sand and the conglomeratic diamictite (3).

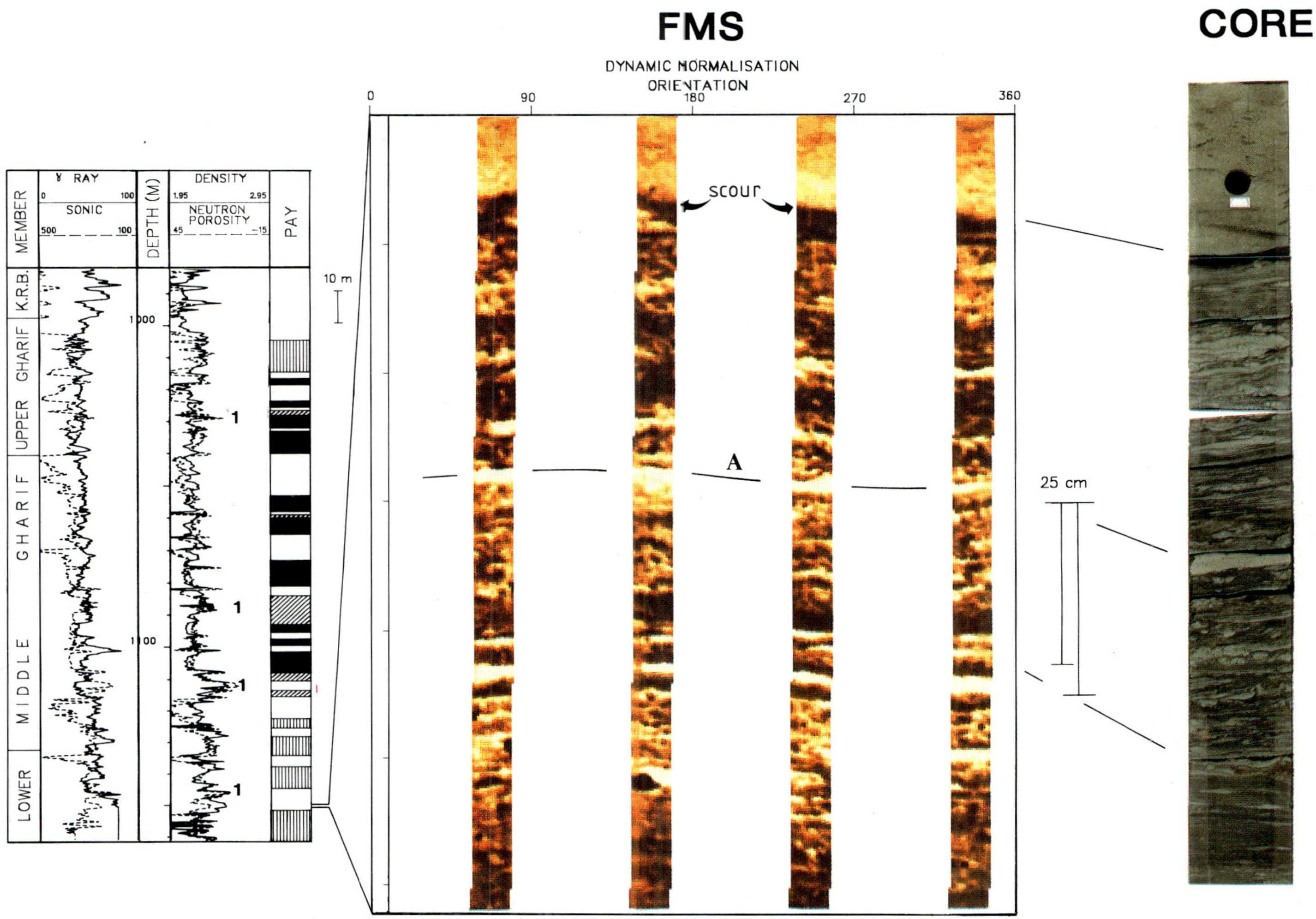

Plate 5. Distinctive bioturbated and lenticular bedded silty mudstones from the Lower Gharif in the Jawdah Field. This lithofacies occurs at the base of a shoreface horizon, which can be traced throughout the Eastern Flank, and is used as a datum for correlation purposes. The horizon is very ambiguous on the standard wireline log responses due to the high feldspar content of the sands (masking the gamma ray contrasts with shale) and the widespread carbonate cementation (1). This facies provides excellent structural dip definition with the FMS (A).

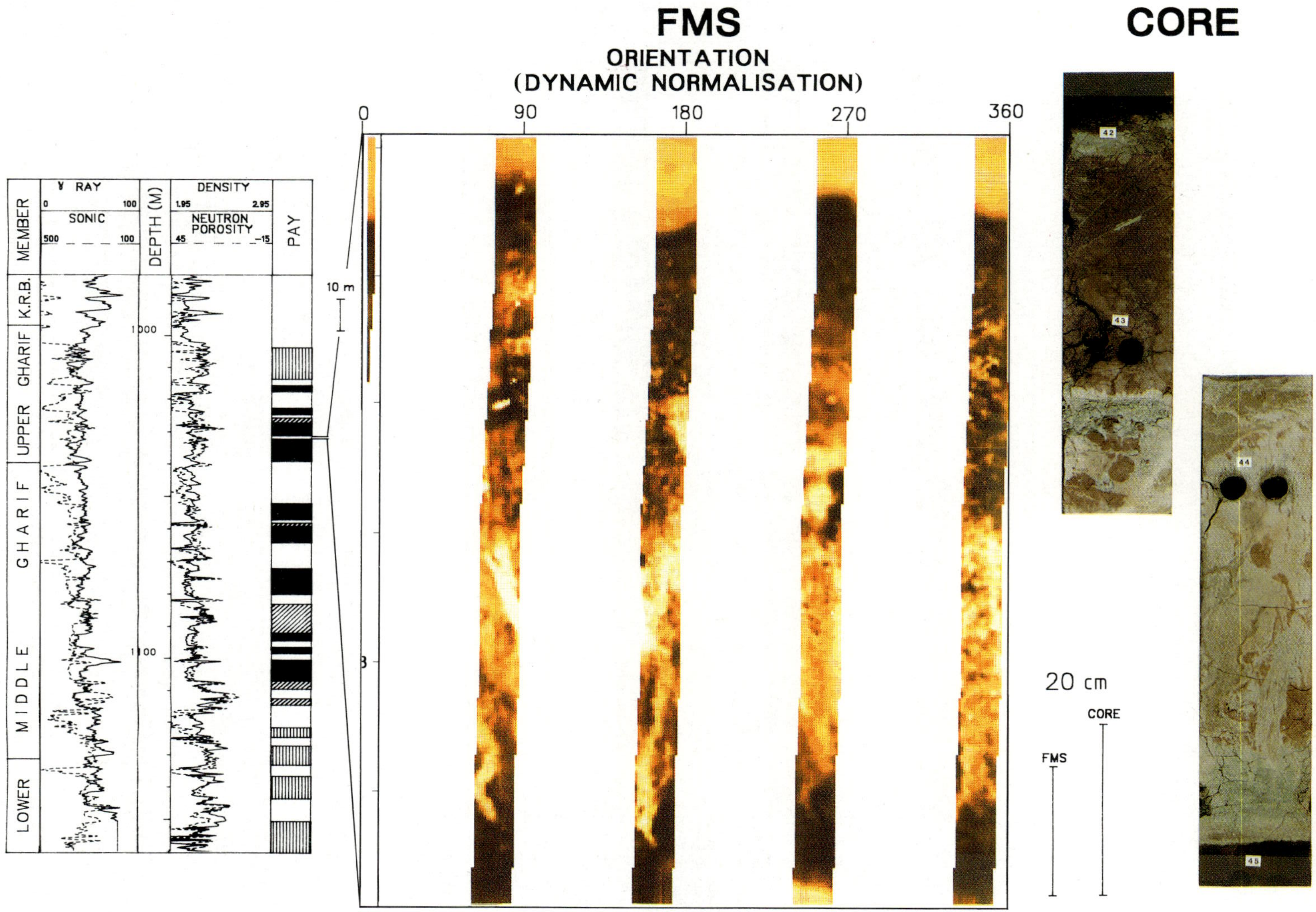

Plate 6. Root-bearing floodplain siltstone with minor carbonate cementation and reddening from oxidative diagenetic processes. The FMS image can pick out such detail in good borehole conditions.

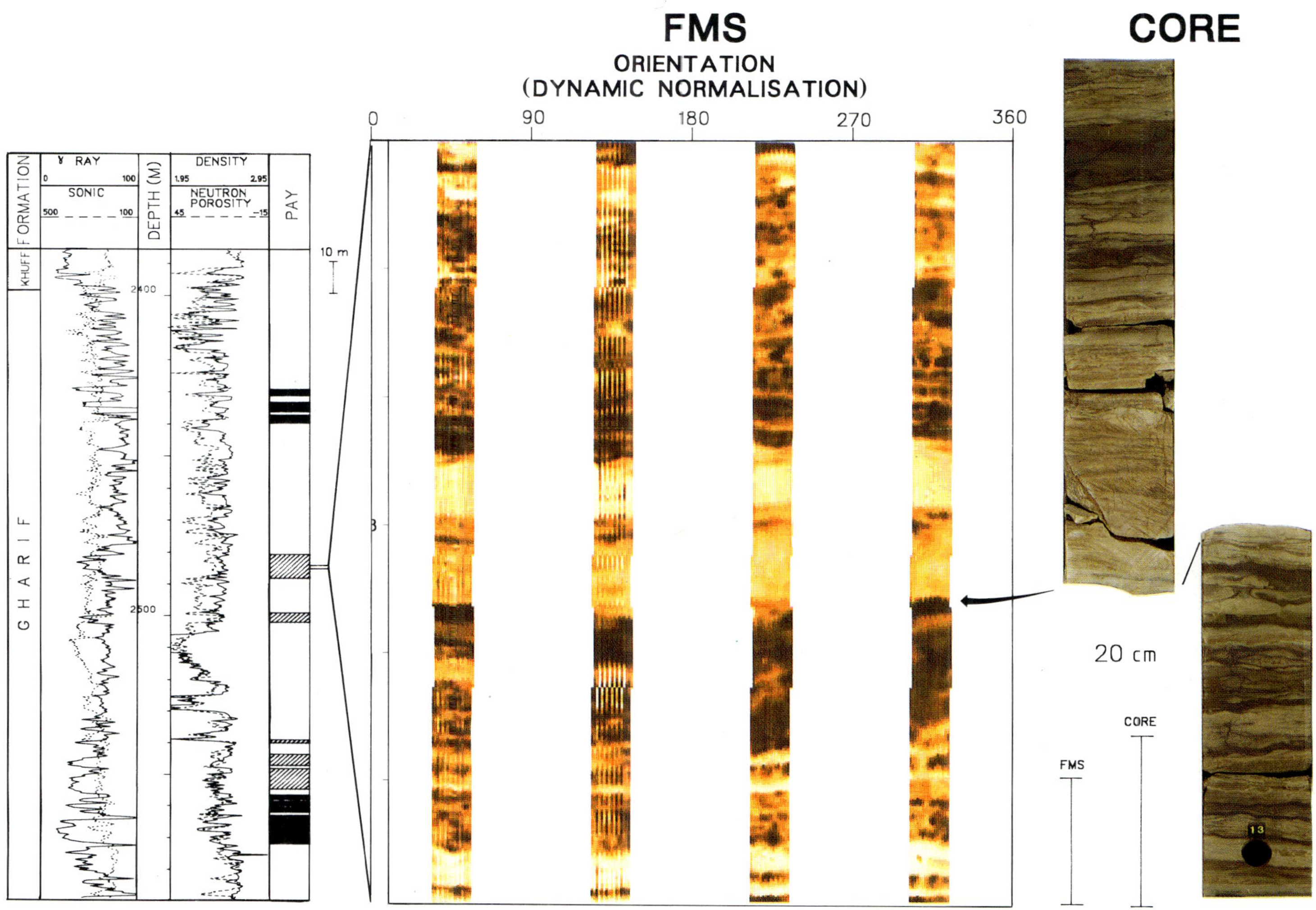

Plate 7. Interlaminated rippled and bioturbated shallow lacustrine siltstones and very fine-grained sandstone from the Upper Gharif in Hasirah. These moderate reservoir quality sandstones are continuous at present well spacings (1 km) and provide pressure support between wells. The distinctive response on the FMS allows periods of subaqueous deposition to be distinguished from the predominant semi-arid red bed calcrete bearing floodplain mudstones. The former can be correlated as a method for defining reservoir units in stacked reservoir sequences.

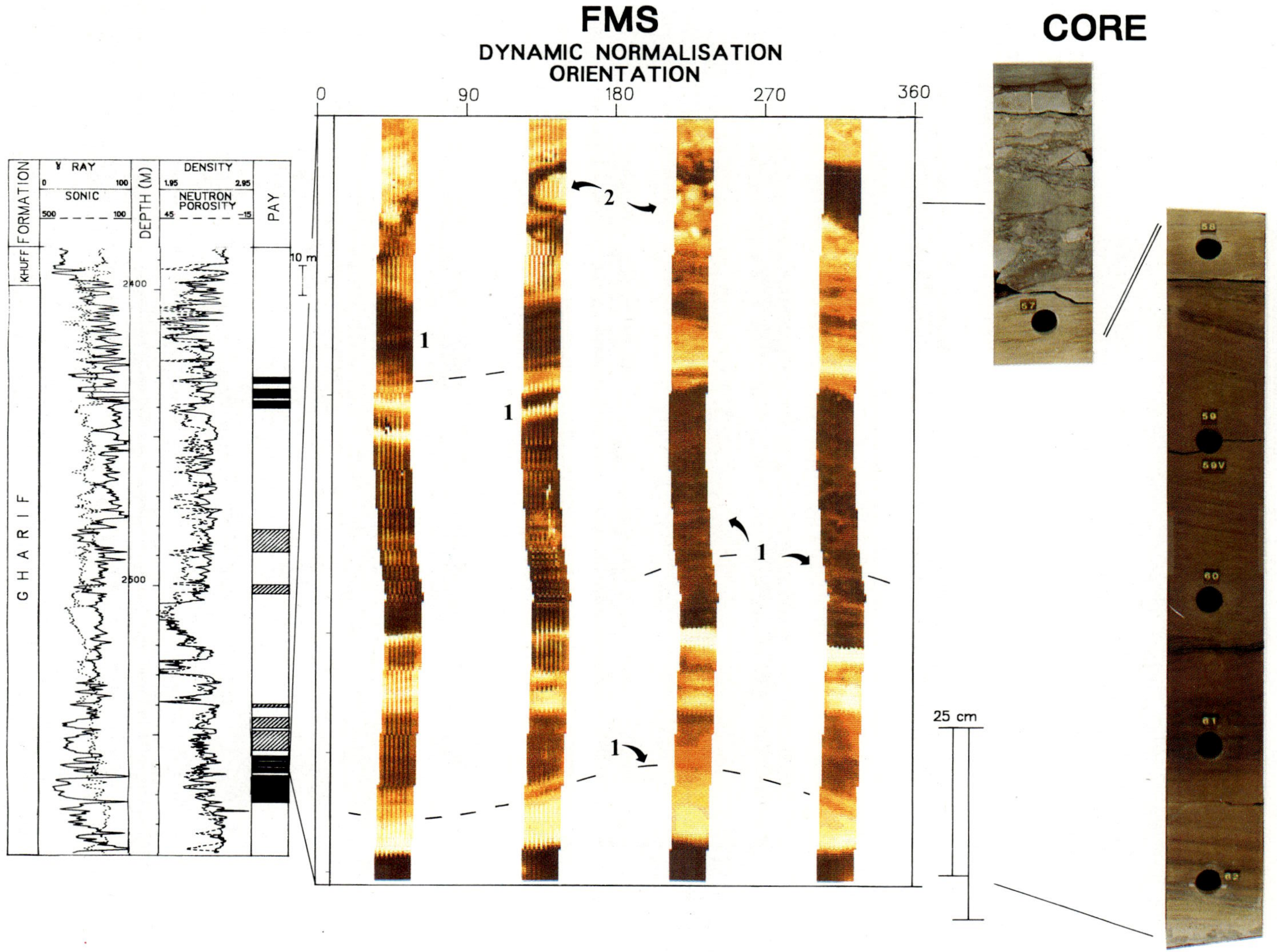

Plate 8. Cross-bedded sandstone from a small distributary channel in a regressive basal Middle Gharif sequence overlying the Haushi Limestone. Palaeocurrents can clearly be measured on the FMS (1). Note the conglomerates at the base and top of the stacked channel sequence (2).

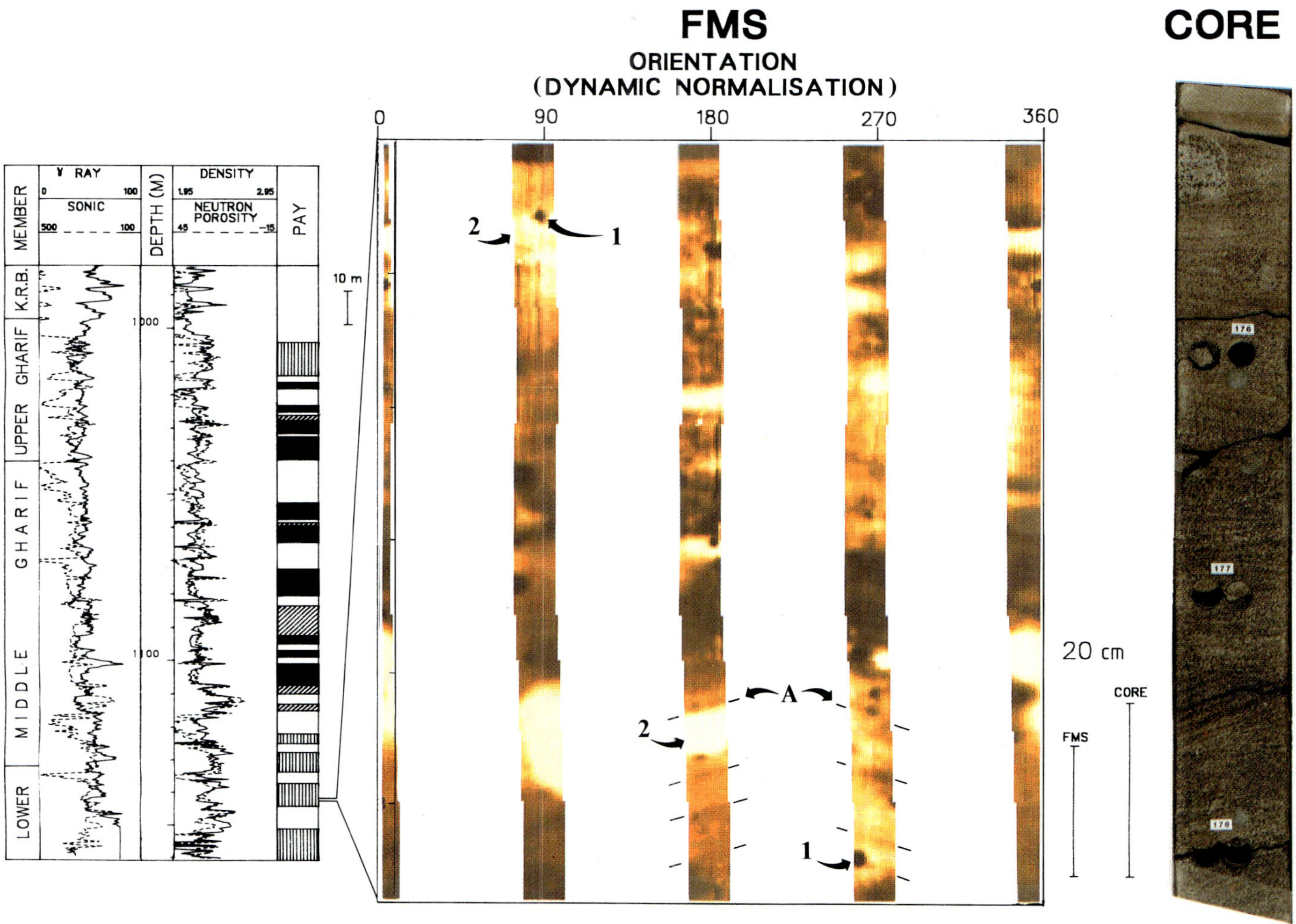

Plate 9. Cross-bedded channel sandstone in the Lower Gharif. The foreset images are obscured by pyrite (1) and carbonate cement (2) but are still visible on two of the FMS pads (A). This would normally be an impossible evaluation on an SHDT but the FMS allows a geologist to interactively determine a palaeocurrent direction even in poor-quality data.

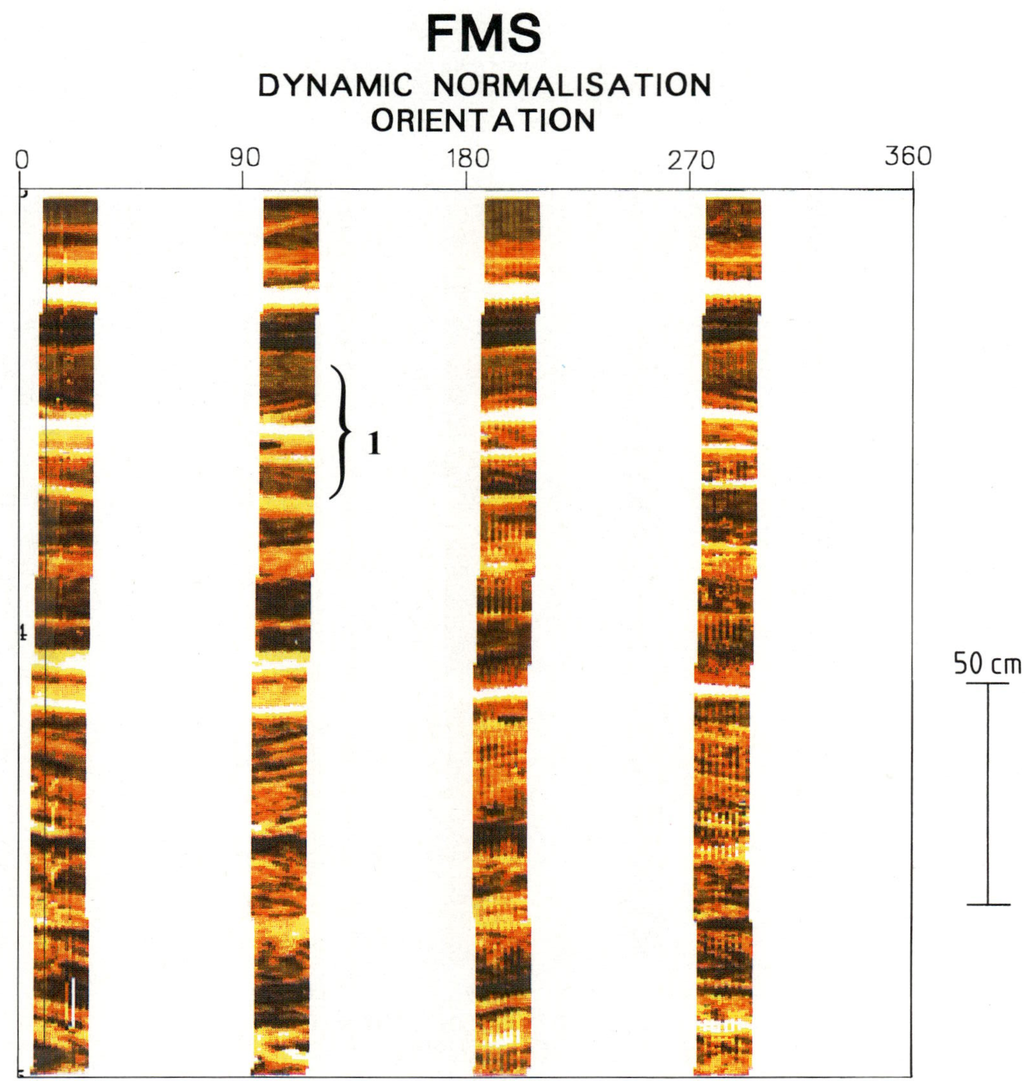

Plate 10. Cross-stratified shallow marine sandstones in a water-bearing interval of the Lower Gharif in Hasirah showing how palaeocurrents can easily be evaluated in ideal conditions. These sets are very thin and would not be evaluated in large diameter (12¼ inch) boreholes with an SHDT due to variable resistivity contrasts at scour surfaces. Note the hint of [illegible]

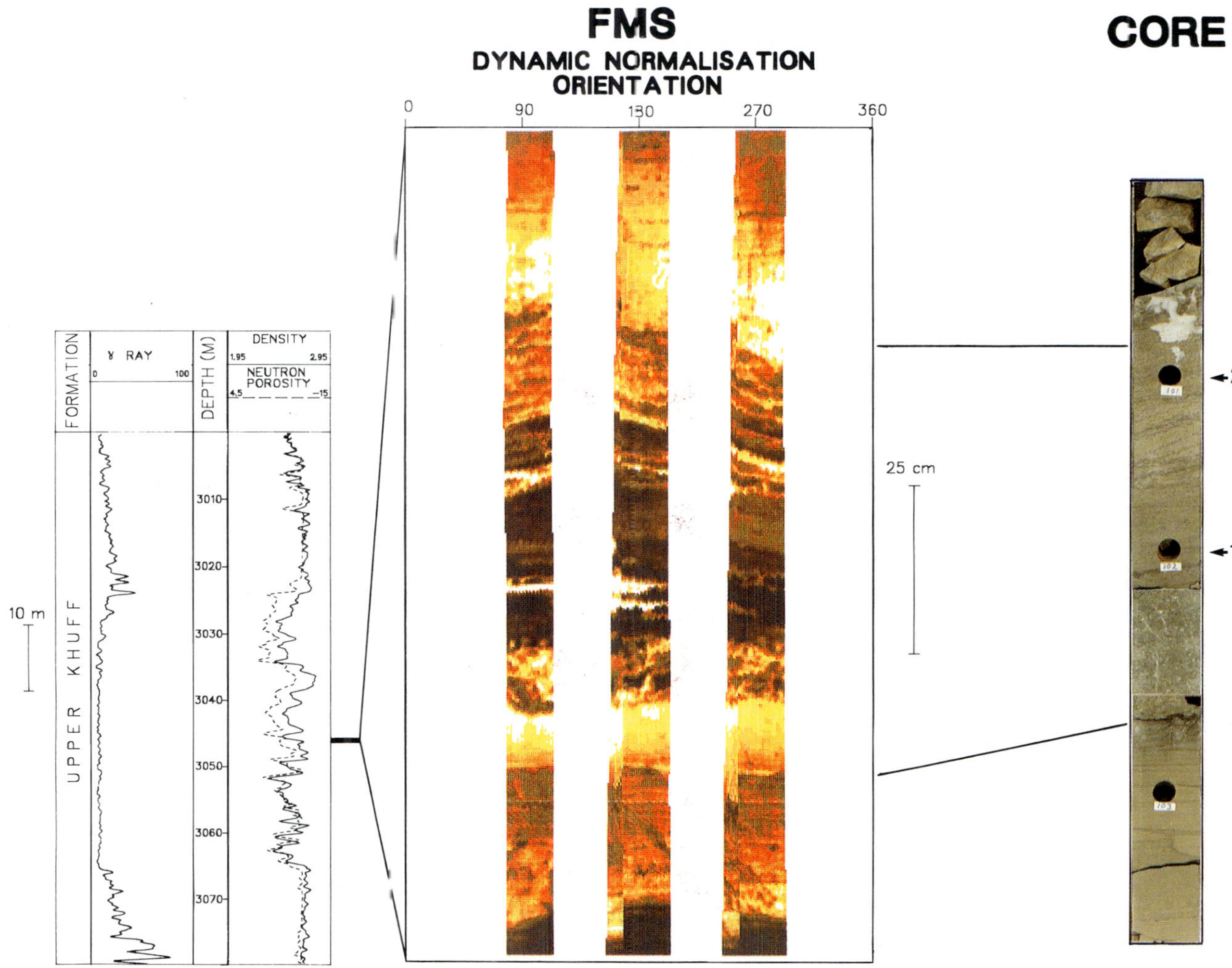

Plate 11. Small (metre thick) fining-up depositional cycles with an irregular, gravelly transgressive lag at the base (mottled pattern on the FMS image) and ooidal, cross-bedded, dolomitized sand with patchy anhydrite cement at the top. The FMS image clearly highlights the better reservoir interval that is invaded by the drilling mud (1: porosity 14%, permeability 280 mD), contrasting with the overlying, poorer reservoir interval (2: porosity 10%, permeability 25 mD).

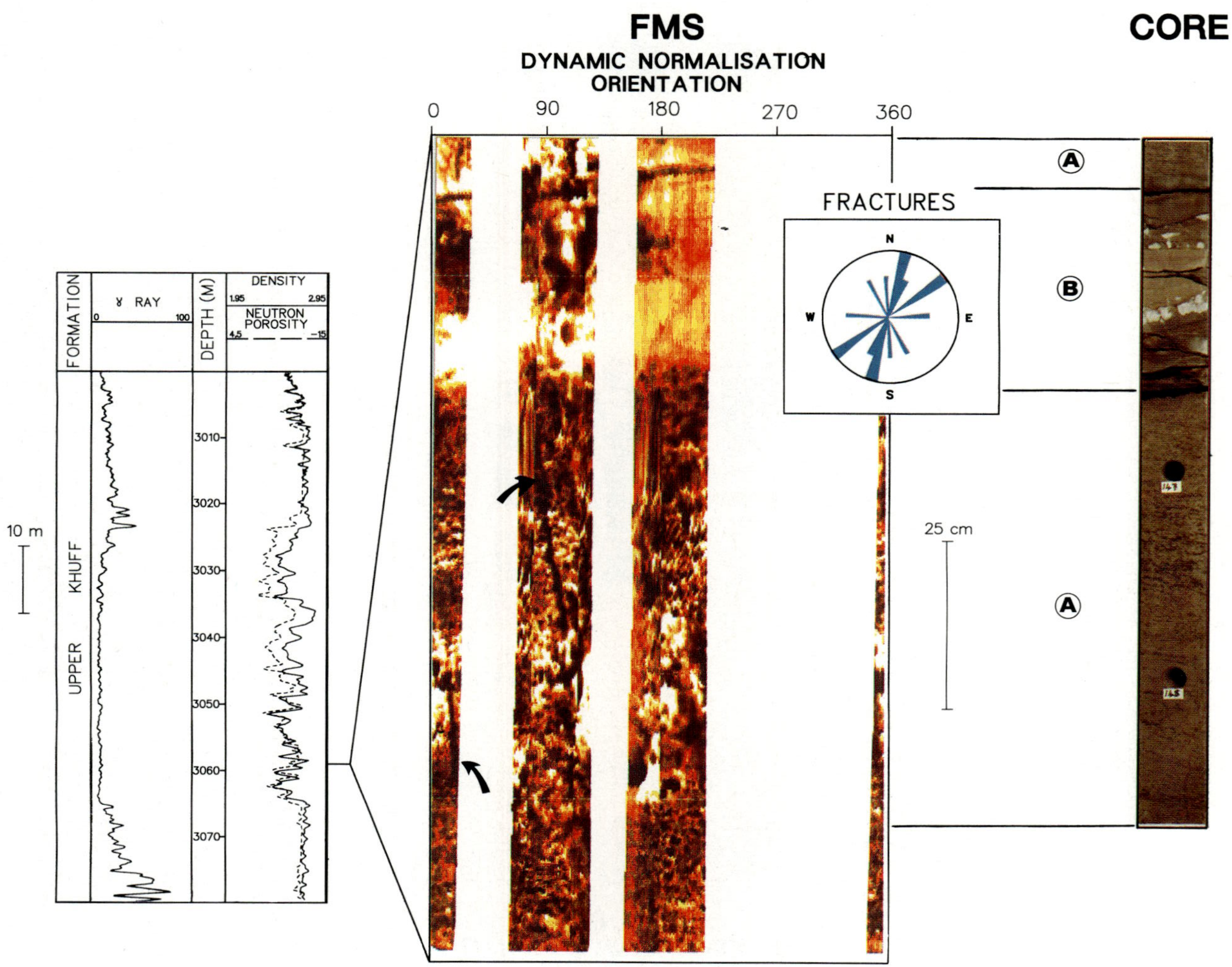

Plate 12. Cross-bedded, oomouldic limestone (A) and intercalation of mudstone with anhydrite nodules (B). Note the characteristic mottled pattern on the FMS image across the mouldic reservoir zone which indicates a poorly connected pore network (permeability ranges between 0.01 and 0.1 mD) in spite of porosities ranging between 15 and 18%. Open, (sub)vertical, extensional fractures are visible throughout the interval (arrows). Fracture orientation (insert) is from the entire cored interval in the well.

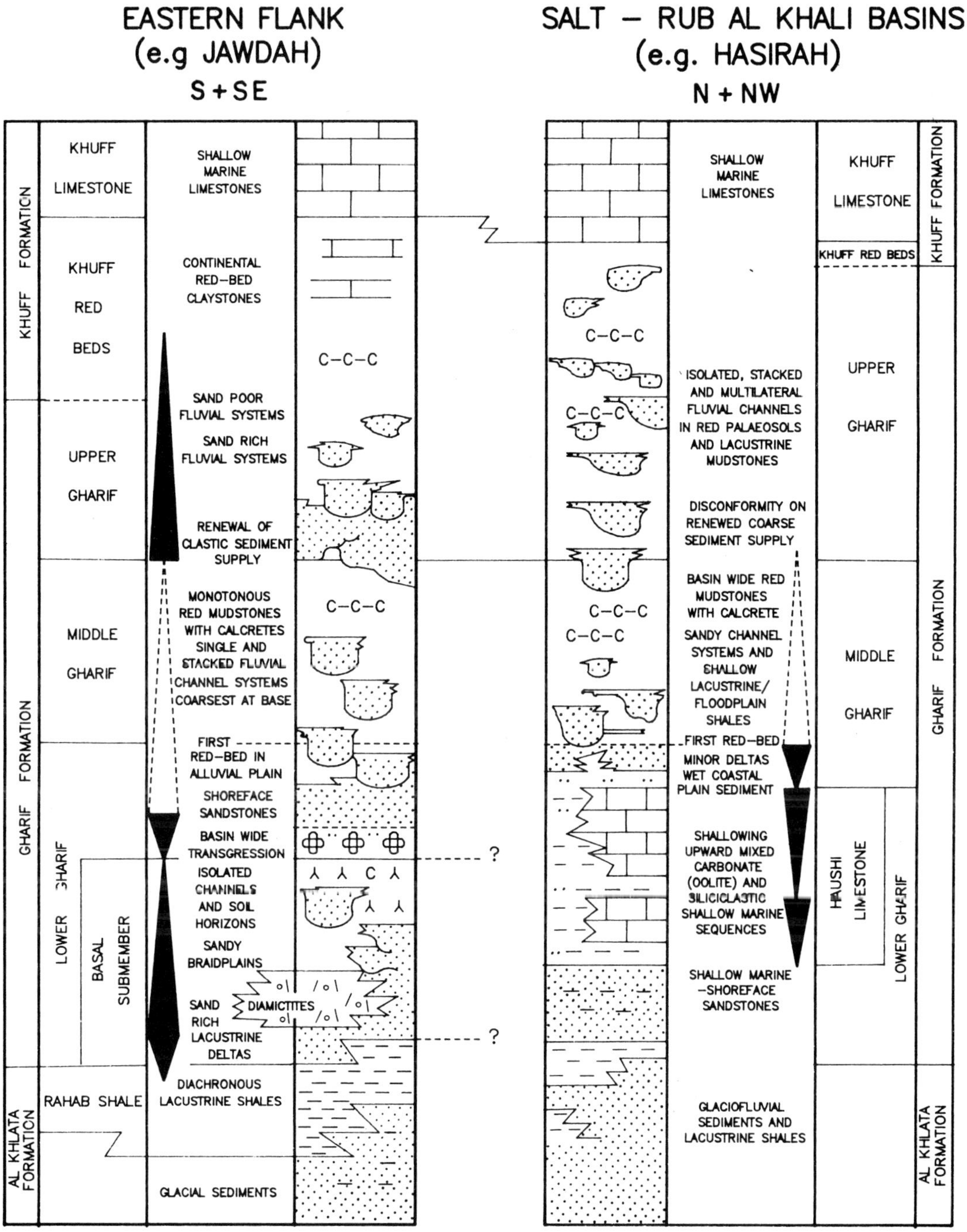

Fig. 6. Summary of the stratigraphy and sedimentology of the Gharif Formation in Oman. In the Eastern Flank oil is found mainly in the Lower and Middle Gharif, whilst in the Ghaba Salt Basin oil is found mainly in the Middle and Upper Gharif. The location of the example fields is shown in Fig. 1.

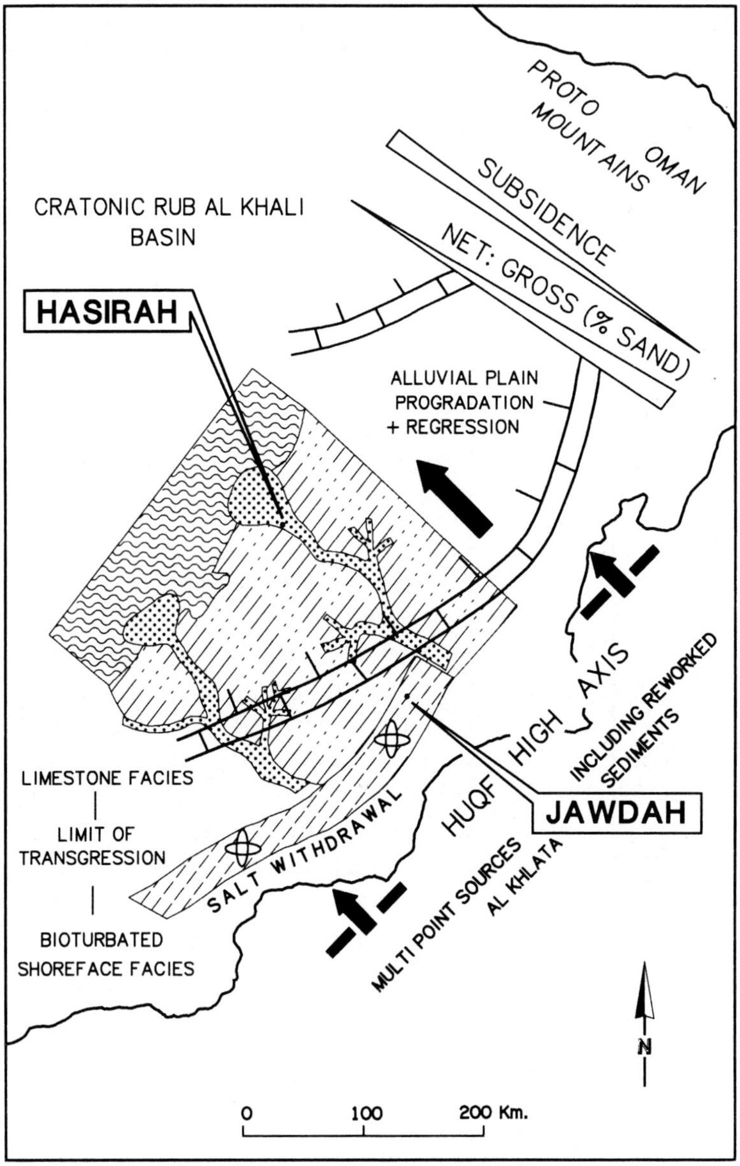

Fig. 7. Palaeogeographic summary of the Lower Gharif Formation in Oman.

AKHDAR GROUP: KHUFF FORMATION

Reservoir setting

The Khuff Formation (Fig. 2) is a major gas reservoir in the Middle-East province, but oil rates of up to 1200 m^3/day (7550 bbls/day) were tested in 1986 from around 3000 m depth in the Yibal Field, North Oman (Fig. 1; Bos, 1989). The expected oil-in-place and producible reserves are relatively modest (95×10^6 and 15×10^6 m^3 respectively) compared with other plays, but the Khuff Formation is a significant exploration target in North Oman (Sykes & Abu Risheh, 1989). The Khuff Formation marks the onset of dominantly marine carbonate deposition that prevailed over northern Oman during the Mesozoic, contrasting with the predominantly continental clastic deposition of Palaeozoic times. The Upper Khuff Formation, which lies at the transition between Permian and Triassic intervals, comprises laterally persistent limestone and dolomite sequences which were deposited on a

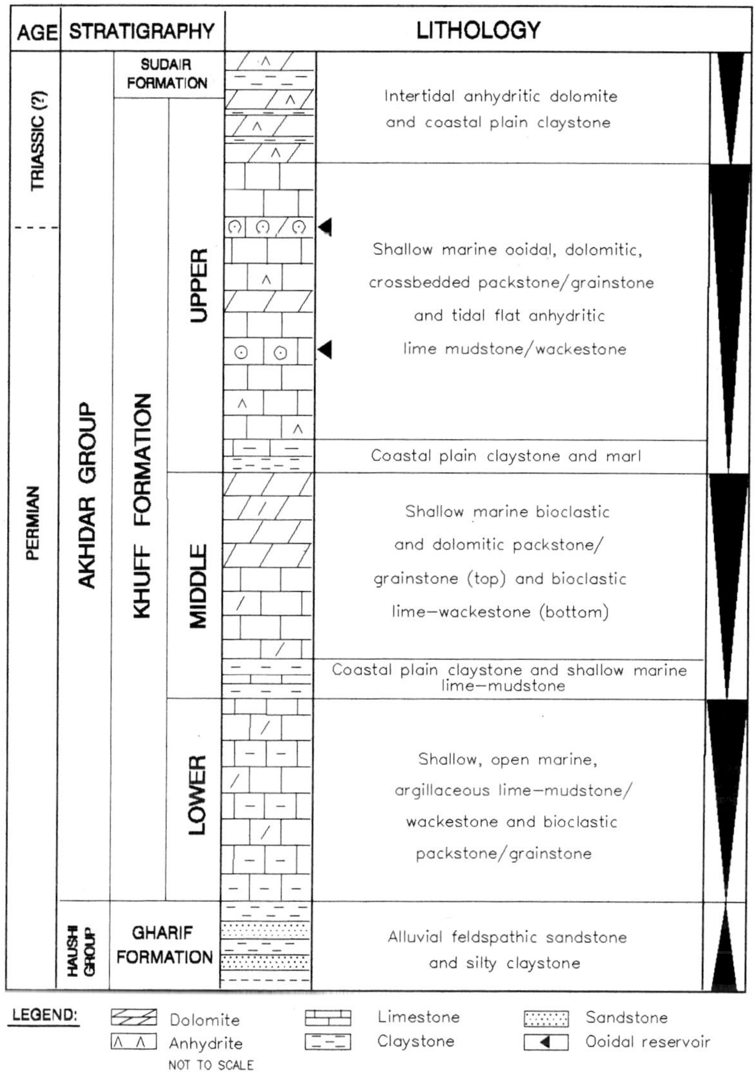

Fig. 8. Summary of the Khuff Formation in Northern Oman.

broad, shallow shelf in North Oman (Fig. 8). The shelf margin was roughly along the axis of the Oman Mountains (Fig. 1; Blendinger, 1988). To the south, the carbonates grade laterally into clastic red-bed lithofacies, which dominate in South Oman where the chronostratigraphic boundary with the underlying Gharif Formation is unclear.

Sedimentology and diagenesis

Each cycle of deposition comprises centimetres to decimetres-thick mudstone beds, with anhydrite nodules at the base, overlain by metres-thick, locally cross-bedded ooidal grainstone and stromatolitic boundstone with anhydrite nodules (Fig. 8). These cycles reflect deposition in coastal, normal to semi-restricted, marine environments and may be traced laterally over tens of kilometres. However, the Khuff cycles include rock types that have variable reservoir quality and differing lateral extent as a result of strong, poly-phase diagenetic and tectonic overprints. These post-depositional overprints modified the initial porosity distribution, principally in lime-grainstones with the formation of oomouldic porosity. Such modifications occurred through dolomitization, cementation, leaching and fracturing events. Open fractures, although accountable for high initial production rates, may cause rapid gas

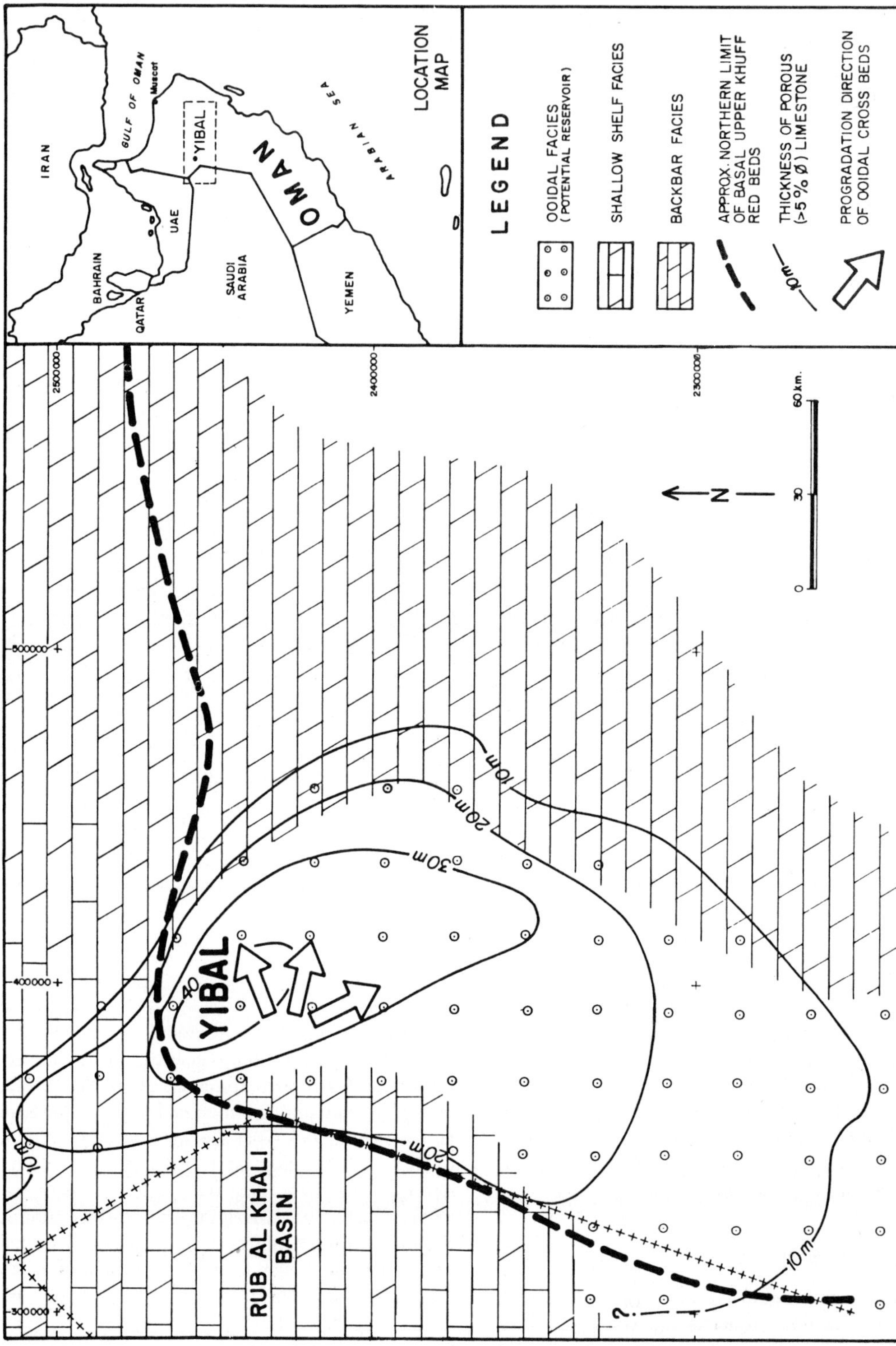

Fig. 9. Upper Khuff Palaeogeographic map showing the inferred distribution and progradation direction of ooidal lithofacies in North Oman.

and/or water cut to occur in wells. As a result, areal prediction of economic hydrocarbon production from Khuff Formation reservoirs is presently impossible.

FMS applications

The FMS is used in the Khuff Formation in combination with other open hole logs to assist in the discrimination of reservoir rock types. From core data, five different reservoir rock types have been identified, namely, dolomite with intergranular and/or intercrystalline porosity (1), micro-intercrystalline (2) or mouldic (3) porosity, and limestone with intergranular (4) or mouldic (5) porosity. Calibration of the FMS images with cores shows that reservoir rock types with intergranular and/or intercrystalline porosity (types 1, 2 and 4) can be differentiated from rock types with mouldic porosity (3 and 5). The FMS image across intervals with intergranular and/or intercrystalline porosity reflects a continuous pore network with smooth transitions between zones of different permeabilities except for low-permeability (white colour), well-delineated anhydrite nodules (Plate 11, facing p. 136). By contrast, intervals with mouldic porosity appear mottled on FMS images because the pore network is only locally connected ('black dot' appearance; Plate 12, facing p. 136). Reservoir rock types (1/2), (3), (4) and (5) can be further discriminated by the combination of FMS images and the neutron and density log curves which allow the identification of limestone and dolomite intervals.

An important diagnostic feature detected by the FMS is cross-bedding in good reservoir quality, dolomitized, ooidal grainstone. The cross-bedding shows dips of 10–20° towards the south, the east or the ENE in general, or locally to other directions (Plate 11). This observation is in line with the palaeogeographical distribution of ooidal sandbodies, which forms an important lead in the exploration for Khuff reservoir lithofacies (Fig. 9). The FMS also allows recognition of impermeable, burrowed horizons at the top of metre-scale, regressive depositional cycles overlain by transgressive layers rich in mudstone pebbles. These transgressive layers form potentially good correlation markers on a field scale (i.e. tens of kilometres).

The FMS has also been used to study fractures within the Khuff Formation. These consist of mostly (sub)vertical, NNE–SSW to NE–SW striking, shear and extensional fractures that formed during early Tertiary folding and probably subsequent uplift of the Oman Mountains (Fig. 1 and Plate 12). This set of fractures postdates low-angle, N–S to NW–SE trending shear fractures that may have been generated during a Late Cretaceous overthrusting event (e.g. emplacement of the North Oman Nappes). Core and FMS data show that fracture type, frequency and dimensions are dependent on lithology. The cemented, decimetre-long, (sub)vertical, extensional and shear fractures develop in low-permeability (mudstone) or other poor reservoir lithofacies. The open, often metre-long, (sub)vertical extensional fractures and the decimetre-long low-angle shear fractures form in porous dolomite and limestone (Plate 12).

CONCLUSIONS

Although the FMS tool is run in Oman essentially for structural information, it has considerable use for sedimentological reservoir characterization and modelling purposes. Oman is a mature oil producing area that comprises a large number of oilfields and complex hydrocarbon habitats. Extensive coring campaigns would be needed to collect the detailed reservoir descriptions that are increasingly required to support secondary and tertiary oil production and horizontal well planning in the Palaeozoic sequences. The amount of coring can be reduced by careful core-log calibrations and a wider use of micro-resistivity tools such as the FMS.

In sand-rich (80–100%) sequences, such as the Cambro-Ordovician Haima Group, the FMS is used to assist in determining effective vertical permeabilities to aid horizontal well planning and completion policies. In marginal fine-grained siliciclastics, such as the Mahwis Formation, the tool can additionally assist in the qualitative definition of thin pay zones although more absolute values of reservoir quality will be required in the future.

The relatively less sand-rich (20–60%) Permo-Carboniferous Haushi Group poses contrasting problems to the Haima Group. This sedimentologically more complex siliciclastic sequence is characterized by poor correlatability which makes waterflood project planning difficult. In the glacial Al Khlata Formation the FMS improves the definition of the internal characteristics of pay intervals that are selected for EOR purposes. Such detailed information may then be used to assist correlation with neighbouring wells, and to analyse faults and fracture distribution within the formation. In the fluvial Gharif Formation the FMS is used to orientate channel sandstone bodies where the amount of data frequently allows the geologist to extract information from moderate quality logs. Lithofacies recognition and correlation is also possible from FMS logs in the Gharif Formation in conditions where other logs provide insufficient resolution. Palaeoenvironmental determination from the FMS is also possible in the carbonate Khuff Formation, both as an aid to exploration and as a framework to model diagenetic modifications and fracture development at the production scale.

ACKNOWLEDGEMENTS

The authors wish to thank the Ministry of Petroleum and Minerals of Oman and the management of Petroleum Development Oman for their permission to publish this paper. Continued technical support of Schlumberger-Oman and Roy Nurmi of Schlumberger-Middle East for this project is gratefully acknowledged.

REFERENCES

Adams, J., Bourke, L.T. & Buck, S. (1990) Integrating Formation Microscanner images and cores. *Oilfield Review*, **2**, 52–65.

Besems, R.E. & Schuurman, W.M.L. (1987) Palynostratigraphy of Late Palaeozoic glacial deposits of the Arabian peninsula with special reference to Oman. *Palynology*, **11**, 37–53.

Blendinger, W. (1988) Permian to Jurassic deep water sediments of the Eastern Oman Mountains: their significance for the evolution of the Arabian margin of the South Tethys. *Facies*, **19**, 1–32.

Bos, C.F.M. (1989) Planning of an appraisal/development program for the complex Khuff carbonate reservoir in the Yibal Field, North Oman. *7th Soc. Petr. Eng. Middle East Oil Tech. Conf. Exhib.*, Manama, Bahrain, 11–14 March 1989, Ref. SPE 17988, 631–640.

Bourke, L., Delfiner, P., Trouiller, J.C., Fett, T., Grace, M., Luthi, S., Serra, O. & Standen, E. (1989) Using Formation Microscanner images. *The Technical Review*, **37**, 16–40.

Braakman, J.H., Levell, B.K., Martin, J.H., Potter, T.L. & Vliet, A. van (1982) Late Palaeozoic Gondowana glaciation in Oman. *Nature*, **299**, 48–50.

Focke, J.W. & Popta, J. van (1989) Reservoir evaluation of the Permian Gharif Formation, Sultanate of Oman. *7th Soc. Petr. Eng. Middle East Oil Techn. Conf. Exhib.*, Manama, Bahrain, 11–14 March 1989, Ref. SPE 17978, 517–528.

Glennie, K.W., Boeuf, M.G.A., Hughes Clarke, M.W., Moody Stuart, M., Pilaar, W.H.F. & Reinhardt, B.M.D. (1974) Geology of Oman Mountains. *Verhandelingen van het Koninklijk Nederlands Geologisch Mijnbouwkundig Genootschap*, Delft, **31**, 423 pp.

Grantham, P.G., Lijmbach, G.W.M., Posthuma, J., Hughes Clarke, M.W. & Willink, R.J. (1988) Origin of crude oils in Oman. *J. Petrol. Geol.*, **11**, 61–80.

Hearn, C.L., Ebanks, W.J., Tye, R.S. & Ranganathan, V. (1984) Geological factors influencing reservoir performance of the Hartzog Draw Field, Wyoming. *J. Petrol. Technol.*, **36**, 1335–1344.

Heward, A.P. (1989) Early Ordovician alluvial fan deposits of the Marmul Oil Field, South Oman. *J. geol. Soc. London*, **146**, 557–565.

Heward, A.P. (1990) Salt removal and sedimentation in Southern Oman. In: *The Geology and Tectonics of Oman* (Ed. by A.H.F. Robertson, M.P. Searle & A.C. Ries), *Spec. Publ. geol. Soc. London*, **49**, 637–652.

Hocker, C., Eastwood, K.M., Herweijer, J.C. & Adams, J.T. (1990) Use of dipmeter data in clastic sedimentological studies. *Bull. Am. Assoc. Petrol. Geol.*, **74**, 105–118.

Hughes Clarke, M.W. (1988) Stratigraphy and rock units nomenclature in the oil-producing area of interior Oman. *J. Petrol. Geol.*, **11**, 5–60.

Levell, B.K., Braakman, J.H. & Rutten, K.W. (1988) Oil-bearing sediments of Gondwana glaciation in Oman. *Bull. Am. Assoc. Petrol. Geol.*, **72**, 775–796.

Lindquist, S.J. (1988) Practical characterisation of aeolian reservoirs for development: Nugget Sandstone, Utah–Wyoming thrust belt. *Sedim. Geol.*, **56**, 315–339.

Martin, J.H. & Cooper, J.A. (1984) An integrated approach to the modelling of permeability barrier distribution in a sedimentologically complex reservoir. *59th Soc. Petr. Eng. Ann. Techn. Conf. Exhib.*, Houston, Sept. 16–19, Ref. SPE 13051.

Robertson, A.H.F., Searle, M.P. & Ries, A.C. (eds) (1990) The geology and tectonics of the Oman region. *Spec. Publ. geol. Soc. London*, **49**, 845 pp.

Serra, O. (1989) *Formation Microscanner Image Interpretation*. Schlumberger Educational Services, Houston, 117 pp.

Sykes, R.M. & Abu Risheh, A.K. (1989) Exploration of deep Palaeozoic and Pre-Cambrian plays in the Sultanate of Oman. *Proc. O.A.P.E.C. Sem. 'Deep Formations in the Arab Countries: Hydrocarbon Potential and Exploration Techniques'*, Abu Dhabi, U.A.E., 8–11 October 1989, 71–113.

Watts, N.L., Ellis, D. & Heward, A.P. (1986) Geologic input to enhanced oil recovery project planning in South Oman. *Bull. Am. Assoc. Petrol. Geol.*, **70**, p. 661. (Abstract.)

Weber, K.J. (1987) Computation of initial well productivities in aeolian sandstone on the basis of a geological model, Leman gas field, U.K. In: *Reservoir Sedimentology* (Ed. by R.W. Tillman & K.J. Weber). *Spec. Publ. Soc. econ. Paleont. Miner.*, Tulsa, **40**, 333–354.

Spec. Publs int. Ass. Sediment. (1993) **15**, 143–156

Predicting reservoir sandbody orientation from dipmeter data: the use of sedimentary dip profiles from outcrop studies

H. WILLIAMS *and* H.F. SOEK*

Koninklijke/Shell Exploratie en Produktie Laboratorium, 2280 AB Rijswijk, The Netherlands

ABSTRACT

Dipmeter data obtained from the sub-surface contain important sedimentary information concerning the origin and directional trend of sandbodies. Such parameters are an essential input to the optimal assessment of reservoir orientation, connectivity and directional permeability. Although problems associated with dipmeter processing can be minimized through the application of sophisticated software, manipulating and interpreting both raw and processed dipmeter data remains problematic.

Vertical sedimentary dip profiles have been collected from outcrops where the environment of deposition and orientation can clearly be identified. These data have been used to create sedimentary dip 'fingerprints' characteristic of low sinuosity channel and shoreface sandbody types, and to design dip filters which leave only those dips which point along these sandbody axes. Such studies provide analogues which can be compared directly with dipmeter data obtained from the sub-surface in order to optimize our use of dipmeter data in reservoir geological models.

INTRODUCTION

Recent developments in computer-based reservoir geological modelling techniques have made it possible to determine the dominant parameters that govern 3-D reservoir connectivity. Such studies appear to show that in low sand/shale ratio reservoirs of fluvio-deltaic origin, the orientation of elongate or tabular sandbodies, such as those deposited by river channels, is one of the dominant parameters (Bryant & Flint, this volume; Dubrule *et al.*, in press). In addition, several workers have shown that cross-bedding orientation can significantly effect directional permeability and residual oil saturation (Emmett *et al.*, 1971; Weber, 1982; Kortekaas, 1985). Predicting such orientations in the sub-surface relies on the correct interpretation of cross-bedding foreset dip azimuths.

Dipmeter data are the most important source of true dip information concerning both structural and sedimentological features in the sub-surface. Two main factors have previously limited the application of dipmeter data to predicting reservoir sandbody orientation:

1 Only relatively recently has it been recognized that dipmeter processing parameters must be carefully selected to derive realistic sedimentary dips (Hocker *et al.*, 1990; Herweijer *et al.*, 1990)

2 Very little seems to be known about the frequency, distribution and reliability of sedimentary dips in sandbodies with respect to the overall orientation of the sandbody (Steinmetz, 1972; Rider, 1978; Cameron, 1986).

To improve our ability to analyse and interpret sedimentary dips, and to predict sandbody orientations from dipmeter data, vertical dip profiles have been collected from outcrops where the sandbody orientation and depositional environment are unambiguous.

Such studies provide analogues which can be used comparatively with dipmeter data from the subsurface to aid the recognition and interpretation of sedimentary dips. The resultant outcrop dip profiles can be used for two specific purposes. Firstly, the vertical arrangement of sedimentary

* Present address: Petroleum Development Oman, Muscat, Sultanate of Oman.

dips (magnitude and azimuth) constitute a genetic 'fingerprint', characteristic of the depositional environment. Secondly, the dip profile data collected from elongate or tabular sandbodies can be used to design statistical filters which leave only those dips which point along the sandbody axis.

Similarly, dipmeter data have recently been used by Muwais & Smith (1990), both to distinguish vertical from lateral accretion channel deposits, and to map channel trends in the McMurray Formation, northeast Alberta.

This contribution describes dip profiles collected from a Carboniferous fluvial sandbody exposed on the west coast of Ireland, and from a Cretaceous shoreface sandbody exposed in Utah, USA. These vertical dip profiles may be used comparatively with dipmeter data from analogous reservoir sandbodies to reliably determine their orientations.

DIP PROFILES FROM OUTCROP

In this study, only surfaces that have the potential to be recorded by a high resolution dipmeter tool (i.e. successive surfaces characterized by a high resistivity contrast) were measured from outcrop (Cameron, 1986). True dip magnitude and azimuth readings were taken at a vertical spacing of 10 cm to provide the best analogues to dipmeter output. A description was made of the surface causing the textural or compositional contrast (e.g. F = foreset) and the quality of that contrast (e.g. 1 = excellent, 3 = moderate). Tectonic dip magnitude and azimuth were measured from shales that immediately underlie or overlie the channel sequences. Realistic tectonic dip data could not be collected from intra-sandbody shales since some of these were found to be draping bedforms and therefore possessed a component of sedimentary dip.

Carboniferous fluvial sandbodies

Geological background

Central County Clare contains a *c.* 1600 m thick sequence of Namurian clastic sediments which conformably overlie Dinantian shelf limestones (Fig. 1). The uppermost Namurian Central Clare Group contains five regressive cycles, each approximately

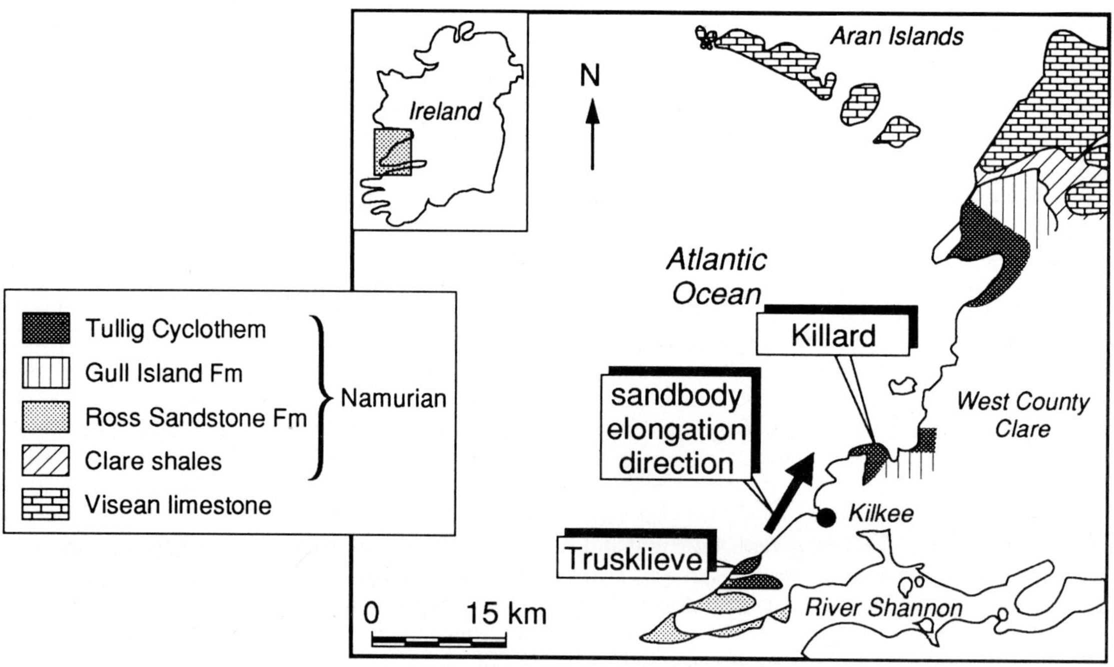

Fig. 1. Geological map and key localities of the study area in SW County Clare, Ireland. (Modified after Rider, 1969.)

100–200 m thick, which represent deposition in a fluvial-dominated but wave-influenced deltaic environment (Rider, 1969, 1974; Pulham, 1988, 1989).

This study documents some reservoir characteristics of a large sandbody contained in the lowest regressive (Tullig) cycle, interpreted as the deposits of a low-sinuosity, multistorey, multilateral channel complex which flowed towards the NNE (Pulham, 1988, 1989) (Fig. 1). Two separate outcrop sections through the sandbody are described in this report: *Trusklieve*, where the deposits comprise at least four vertically stacked channels (Figs 2, 3); and *Killard*, some 25 km down palaeocurrent, where the same unit comprises only three vertical stacks (Fig. 4).

Detailed sedimentology

The sandbody comprises several stacks of fine- to medium-grained channel deposits whose bases are usually marked by erosion surfaces and breccio-conglomerates. Internally, each channel storey is characterized by gently undulating or dipping bounding surfaces that separate sets and co-sets of trough cross-beds, parallel-laminated and massive beds (Figs 2, 3, 4). The dips obtained from cross-bed foresets are of most relevance to sandbody orientation studies, since these are the features that consistently point down-palaeocurrent, along the length of the channel.

Statistical manipulation of dip profile data

Statistical analysis of the dip data presented in Figs 2 and 4 was carried out after structural rotation of the dips to ensure that only sedimentary dips were included in the study. For each outcrop section, the total sequence of dip magnitude data was successively filtered to leave only dips between 5 and 35°, 9 and 35°, 13 and 35°, etc. (Fig. 5a), to assess an optimum range of magnitudes, whilst still retaining a significant number of dips whose azimuths are directed along the direction of the channel elongation direction (SSW–NNE). This filtering procedure was also applied to individual channel stacks at both localities, as shown, for example, in Fig. 5b. The results show that dip magnitudes should be filtered to leave only the range between 17 and 35°, because this process preferentially sorts for cross-bed foreset dips that point down the palaeocurrent and along the length of the channel sandbody.

Each set of sorted dip magnitude data was then systematically assessed for the number of remaining high quality dips whose azimuths occurred within the defined channel azimuth range of N000°–045°E (Fig. 6). The results of these analyses show that after optimum dip magnitude filtering in the range of 17–35° for the total sequence of dips at each outcrop, 48% of the remaining dips from Killard were within the specified azimuth range. At Trusklieve, 53% of the remaining dips fell within the specified range.

However, when only dips from the upper two channel stacks are considered, the remaining dips within range rise to 71% and 89% respectively (Fig. 6a,b). These results show that, although the number of dips has been drastically reduced, the remaining dips in the uppermost stacks of multistorey channel sequences are the most reliable for orientation studies. This is probably because:

1 the uppermost stack is the only complete, fully preserved channel sequence without an erosion surface at its upper boundary;

2 the lower stacks are more likely to be compactionally deformed by successive channel deposits above them;

3 the lowest stack comprises the deposits of channel initiation/avulsion events, or very minor channels that preceded the development of the main channel belt, and therefore contain anomalous dip azimuths.

The basic assumption made when using cross-bedding for palaeocurrent studies is that the current flowed in the direction of maximum foreset inclination (High & Picard, 1974).

By comparing our results with the original dip profile data, it can be seen that the more reliable dips are generally associated with the upper part of the thickest cross-bed sets. This presumably reflects the fact that larger-scale bedforms are more likely to move directly downstream, parallel to the channel axis, during high-stage fluvial discharge events. A model for the possible development and preservation of steeper foreset dips only in the axial zone of lunate trough cross-bedforms is depicted in Fig. 7.

Cretaceous shoreface sandbody

Geological background

The Cretaceous (Campanian) Blackhawk Formation, exposed in the Book Cliffs of Utah, was deposited as a series of regressive shoreface sandbodies, laterally equivalent to the marine, pro-delta Mancos Shale, and unconformably overlain by stacked, low-sinuosity channel deposits of the Castlegate

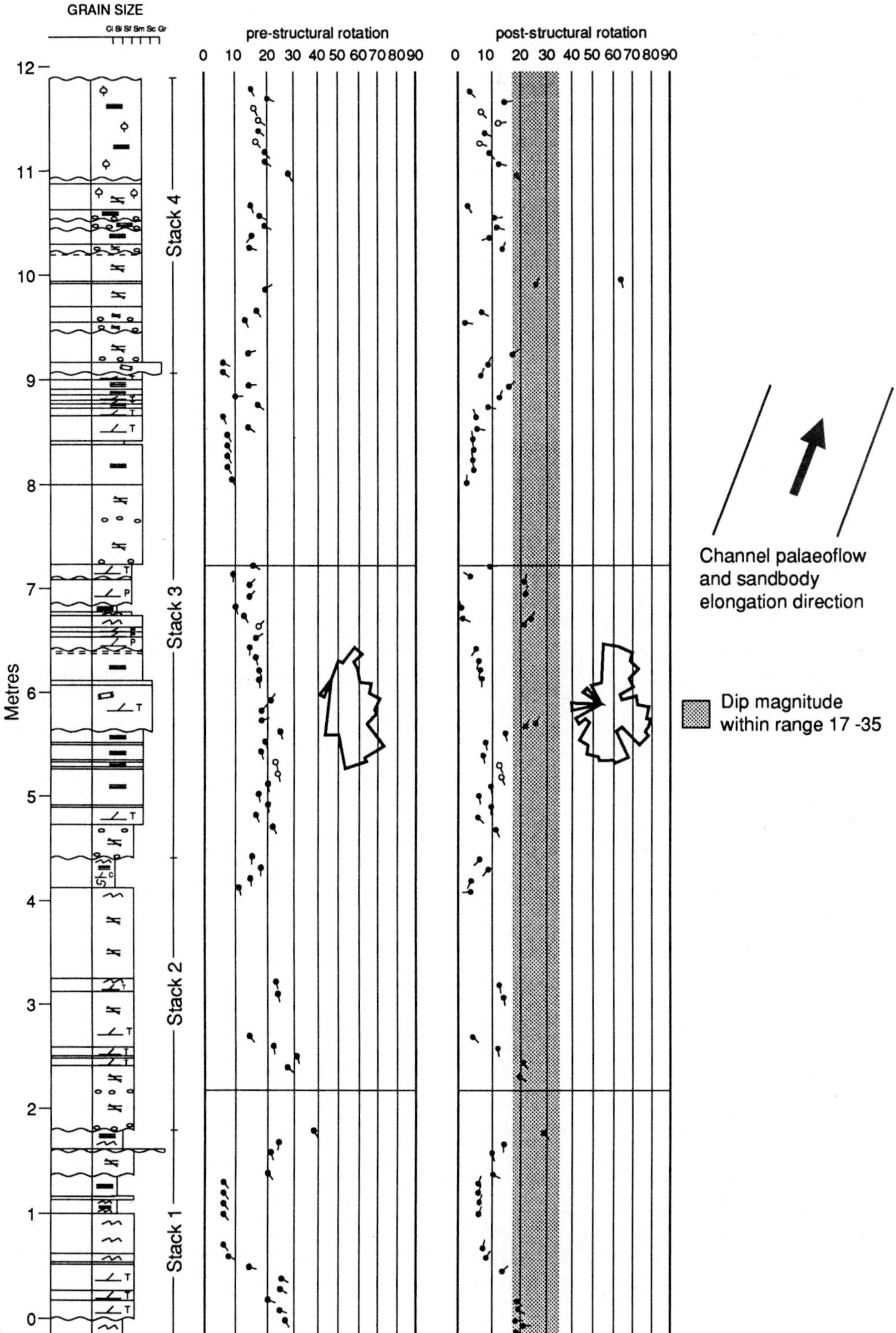

Fig. 2. Sedimentary log and dip profile through the Tullig Cycle stacked, low-sinuosity channel deposit at Trusklieve.

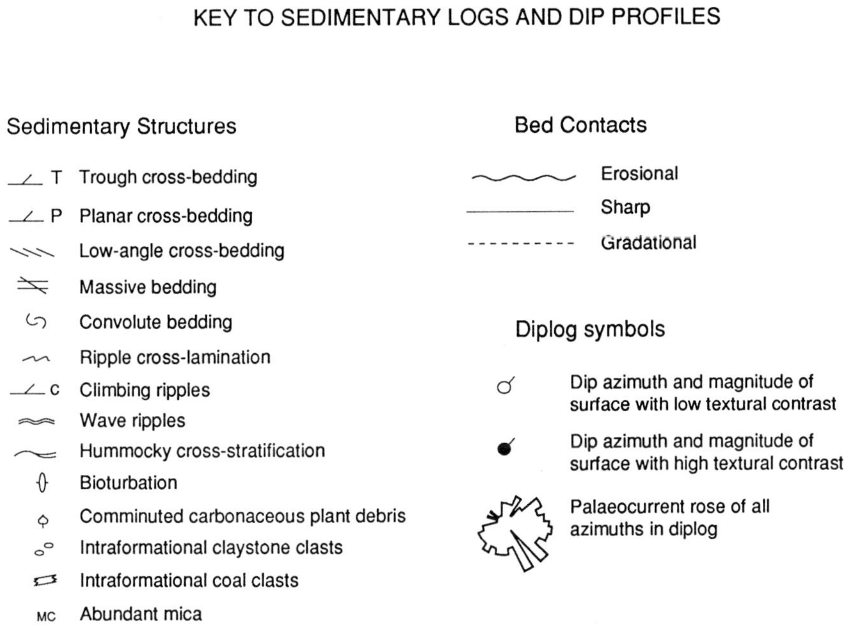

Fig. 2 (*continued*). Key.

Sandstone (Balsley, 1982; Van Wagoner *et al.*, 1990). Internal sedimentary structures and pinch-out directions within these shoreface deposits indicate that the external tabular, sheet-like geometry and internal directional permeability of these sandbodies is dominated by a NE–SW trend, parallel to the regional depositional strike (Fig. 8). Palaeochannels oriented normal to this trend support these observations.

Detailed sedimentology

The base of the study section comprises a 60 cm thick *in situ* coal seam, overlain by a thin, poorly-laminated siltstone (10 cm thick). This is overlain by a sharp-based, fine- to medium-grained, well-sorted sandstone which attains a thickness of some 24 m (Figs 9, 10). Internally, the sandbody comprises two similar, vertically-stacked units, each approximately 12 m thick. The detailed description, interpretation and sedimentary dip profile presented in this summary refers only to the lower interval.

The lower 8 m of the section is interpreted to represent the deposits of a middle shoreface environment, with storm-generated sheet-flow, symmetric, asymmetric ripples and hummocky cross-stratification, interspersed with fair weather burrowing (Clifton, 1976; Kumar & Sanders, 1976). The uppermost 4 m represent deposition in the upper shoreface environment, with high-angle trough cross-beds deposited by shoaling and breaking waves, surf, and rip currents (Clifton, 1976; Reineck & Singh, 1973) (Fig. 11).

This 12 m thick unit is interpreted to represent a single regressive episode of shoreline progradation. Following the first episode of deltaic sedimentation (represented by the lower 12 m of sediment), a period of non-deposition and landward transgression of the shoreline must have taken place across the subsiding delta platform. This is represented by the erosive surface at the top of the upper shoreface deposits which has removed the foreshore deposits, indicating that significant subaqueous erosion and/or subsidence must have occurred (Balsley, 1982) (Fig. 9).

Dip profile data

Analysis of the dip data presented in Fig. 9 was carried out after structural rotation of the dips, using a dipmeter processing package, to ensure that only sedimentary dips were included in the study. The

Fig. 3. Outcrop photograph showing position of log section and dip profile measured at Trusklieve.

Fig. 4 (*facing page*). Sedimentary log and dip profile through the Tullig Cycle stacked, low-sinuosity channel deposit at Killard.

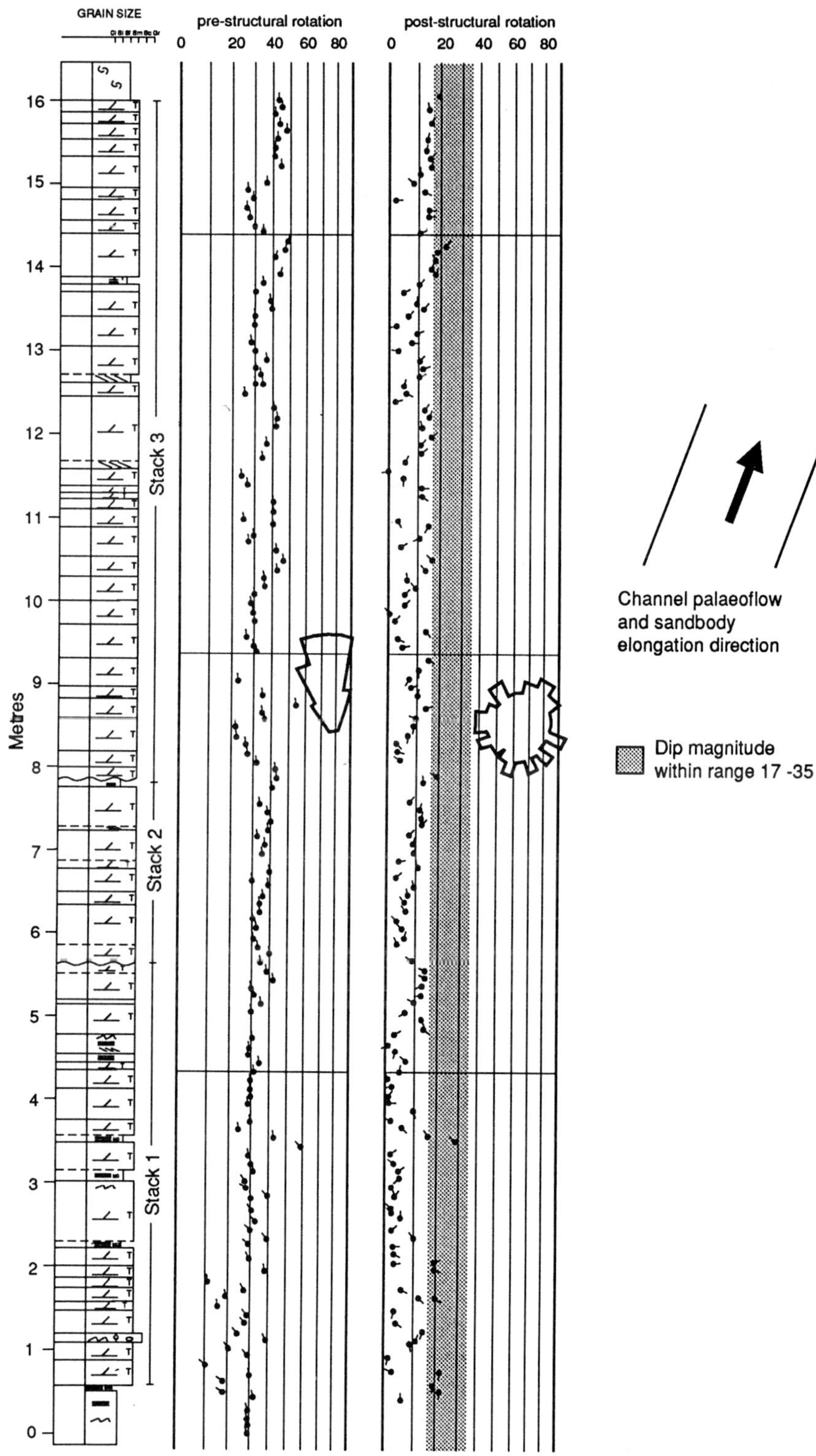
GRAIN SIZE
pre-structural rotation
0 20 40 60 80
post-structural rotation
0 20 40 60 80
Metres
Stack 1
Stack 2
Stack 3
Channel palaeoflow and sandbody elongation direction
Dip magnitude within range 17 -35

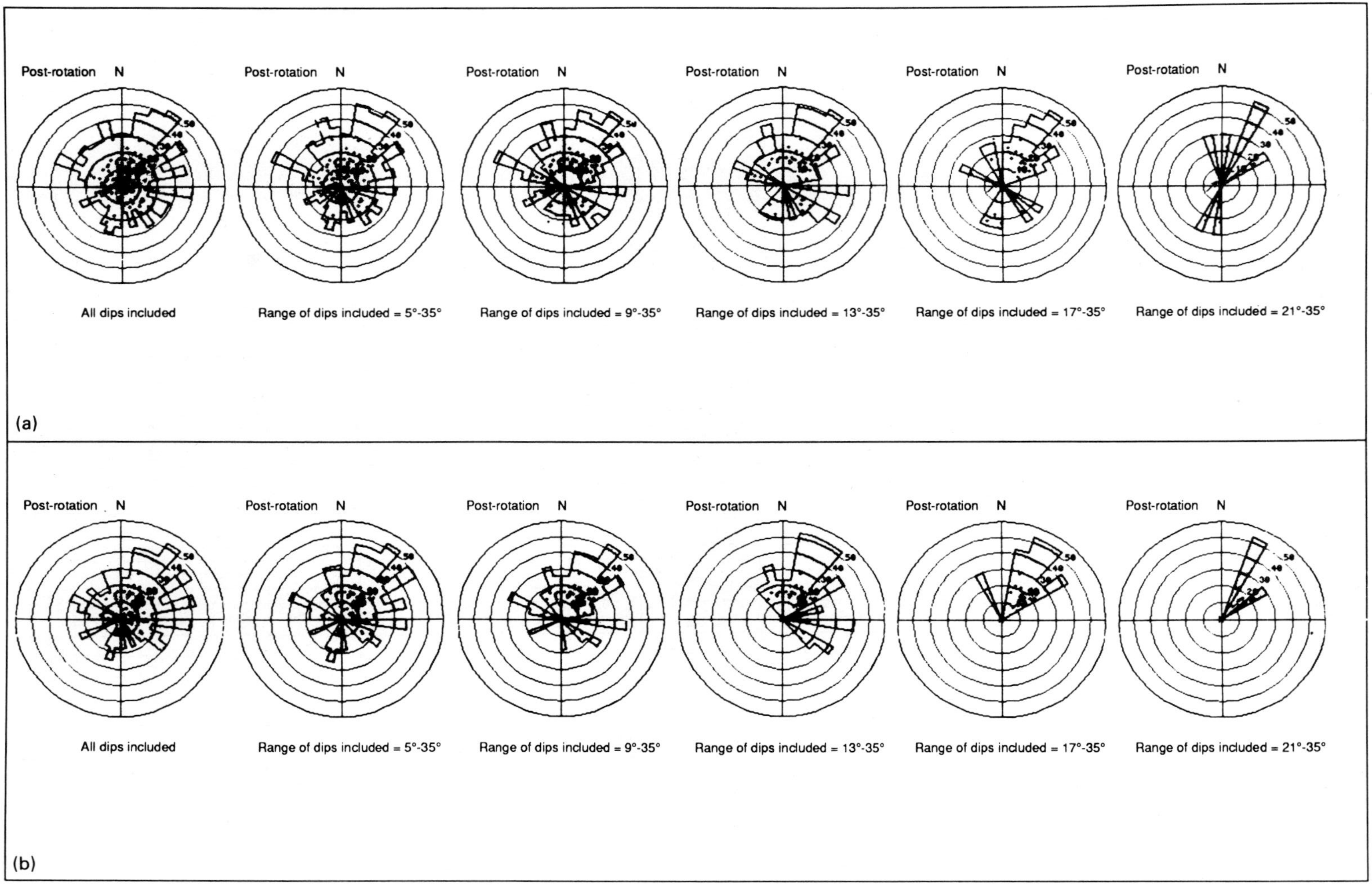

Fig. 5. (a) Killard distributary channel: Schmidt plots of the total sequence (stack numbers 1, 2 and 3) after different result filters have been applied to the post-structurally rotated dips (structural dip = 30° to N341°E). (b) Killard distributary channel: Schmidt plots of the top stack (number 3) after different result filters have been applied to the post-structurally rotated dips (structural dip = 30° to N341°E).

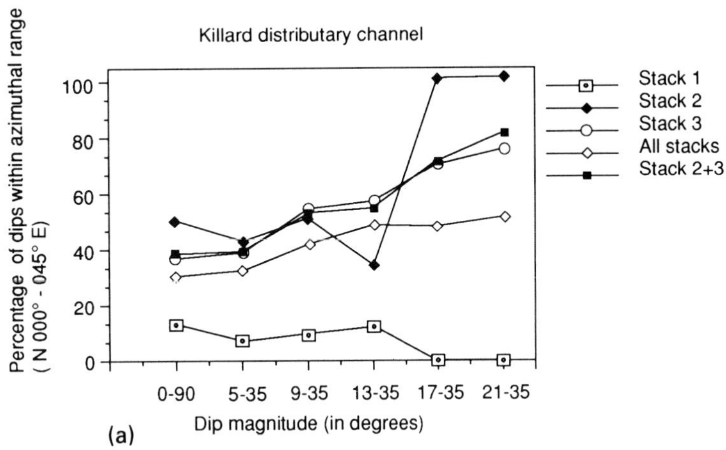

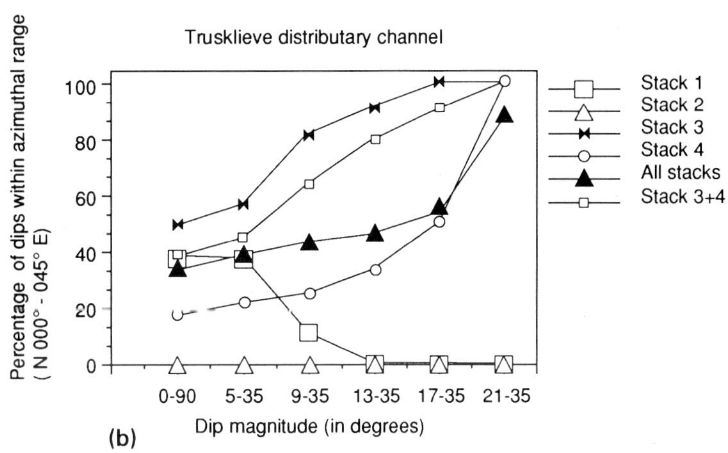

Fig. 6. (a) Derivation of optimum range of dip magnitude and azimuth for Killard distributary channel deposits. Selective dip magnitude filtering to a smaller range shows that in all cases except stack 1, an increasing percentage of dips fall within the known azimuth range. (b) Derivation of optimum range of dip magnitude and azimuth for Truskleve distributary channel deposits. Selective dip magnitude filtering to a smaller range shows that in stacks 3 and 4, an increasing percentage of dips fall within the known azimuth range.

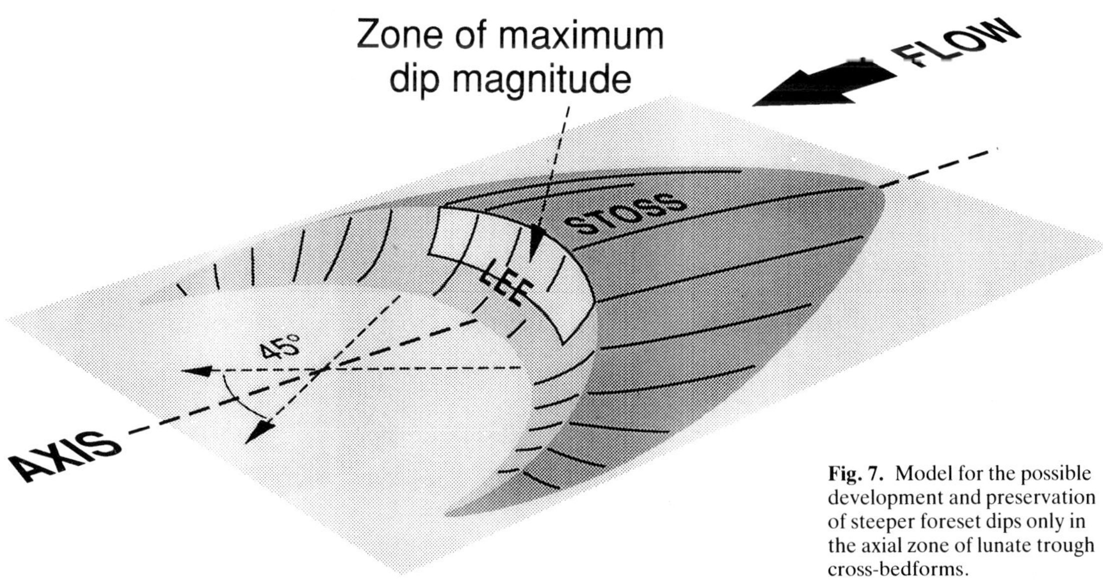

Fig. 7. Model for the possible development and preservation of steeper foreset dips only in the axial zone of lunate trough cross-bedforms.

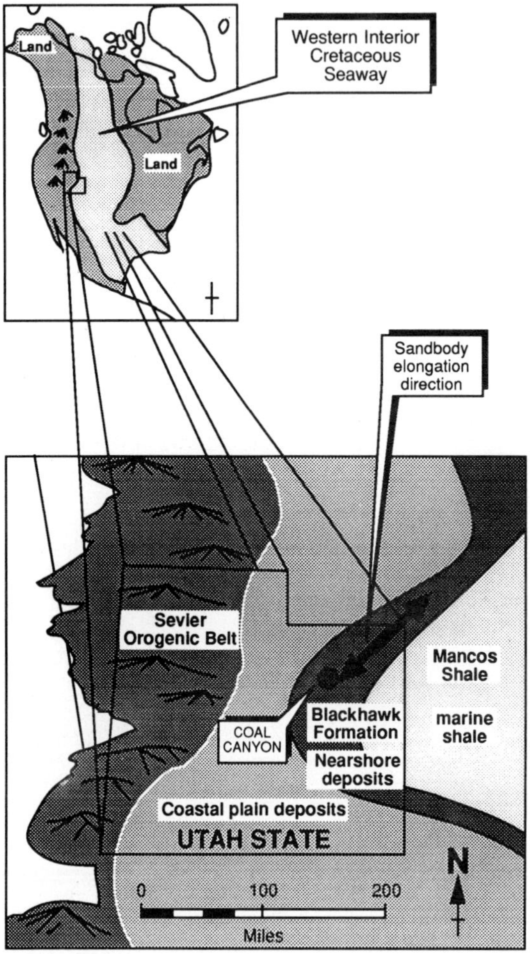

Fig. 8. Location map showing Western Interior Cretaceous Seaway and palaeogeographical position of the Blackhawk Formation study area in NE Utah, USA. (Modified after Balsley, 1982.)

post-rotation dip results show an azimuth dispersal pattern which is chiefly directed to the NE, SE and SW with only a minor component directed to the NW (Fig. 9). This shows that the majority of sedimentary dips have a bi-modal direction alongshore, parallel to the sandbody elongation direction (NE–SW), or offshore to the SE. Comparison of the dip profile and sedimentary log presented in Fig. 9 reveals that the majority of high-angle (10–35°) dips are from trough cross-beds which are directed offshore or alongshore to the SW. Conversely, the majority of low-angle (< 15°) dips are measured from hummocky cross-strata which is directed alongshore toward the NE.

These results are in sharp contrast with those reported by Campbell (1968), Serra (1985) and Hocker *et al.*, (1990), who claim that in beach and barrier bar sandbodies, tabular cross-beds will have low-angle dips toward the sea and higher angle dips directed onshore. The results are, however, consistent with measurements made from present-day surf zone and near-bottom shelf currents on the Texas coastal margin, and also with palaeocurrent measurements made from shoreline to shelf sequences in Cretaceous strata of New Mexico (Snedden & Nummedal, 1990).

A model to explain the development of sedimentary structures and dips likely to be identified by a high resolution dipmeter in shoreface sandbodies is presented in Fig. 11.

CONCLUSIONS

Statistical analyses of dip profiles collected from outcrops through multistorey, low-sinuosity distributary-channel deposits have shown that in order to optimize the orientation of sandbodies in reservoir geological models

1 vertical profiles of sedimentary dips give reliable data on the directional trends of distributary channels;

2 the optimum range of dip magnitude filtering that should be applied to derive the most meaningful sedimentary dips from SHDT results is 17–35°;

3 more reliable dip azimuth data occur towards the top of each storey, but particularly within the best-preserved, uppermost storey of stacked channel sequences.

Analysis of the dip profile collected from outcrop of a tabular-shaped, shoreface sandbody has shown that in order to optimize the orientation of analogous sandbodies in geological reservoir models, using the results of dipmeter data:

1 vertical profiles of sedimentary dips can be used to reliably obtain directional trends of shoreface sandbodies;

2 dip magnitude filtering is unnecessary to derive

Fig. 9 (*facing page*). Sedimentary log and dip profile from part of the Blackhawk Formation, Coal Canyon, Utah, USA.

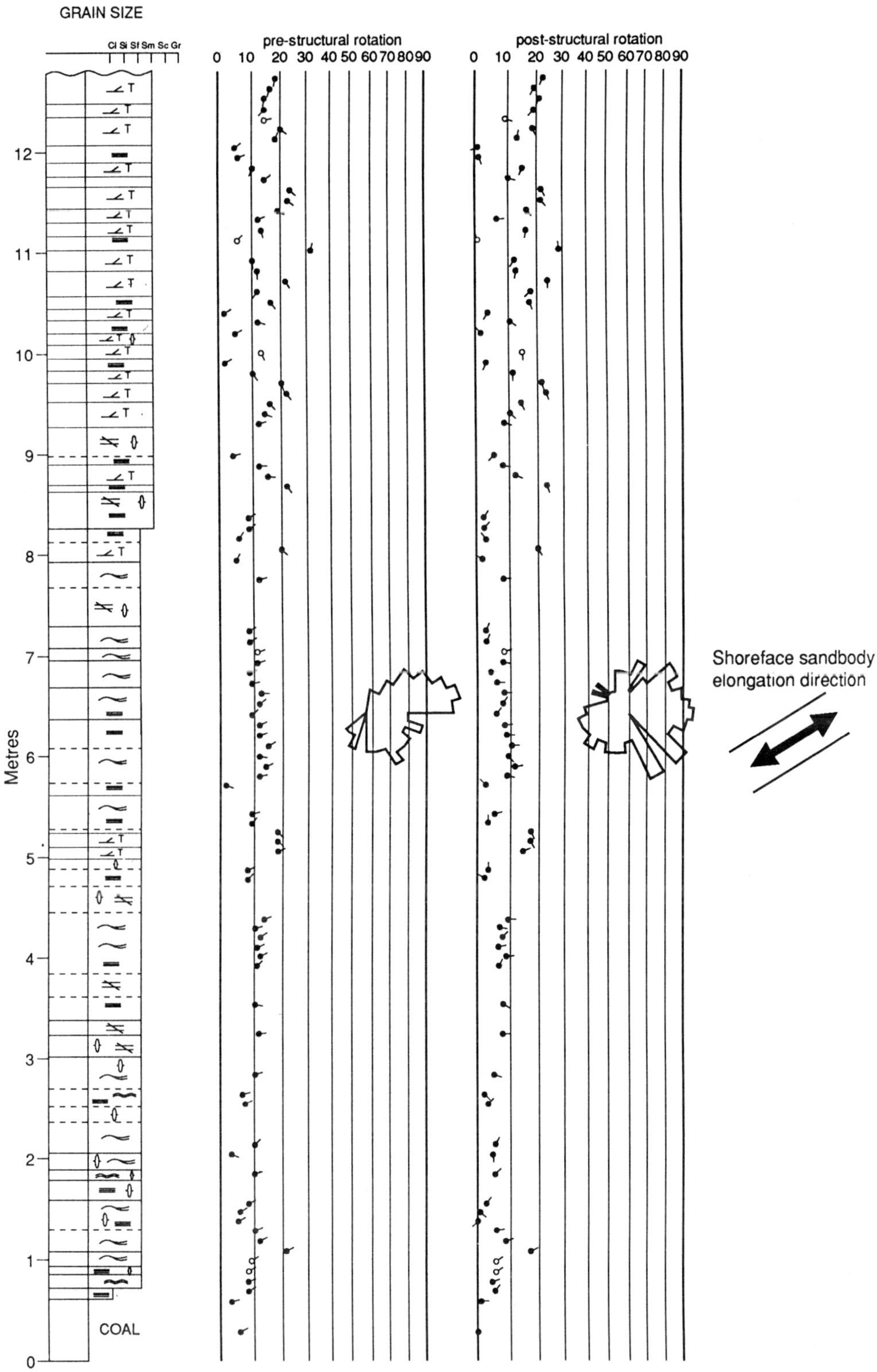
GRAIN SIZE
Cl Si Sf Sm Sc Gr
pre-structural rotation
0 10 20 30 40 50 60 70 80 90
post-structural rotation
0 10 20 30 40 50 60 70 80 90
Metres
0
1
2
3
4
5
6
7
8
9
10
11
12
COAL
Shoreface sandbody elongation direction

Fig. 10. Outcrop photograph showing position of log section and dip profile measured in Coal Canyon.

meaningful sedimentary dips from dipmeter data in shoreface sandbodies;

3 the elongation direction of tabular, sheet-like shoreface sandbodies, and the trend of directional permeability can be reliably assessed by extracting sedimentary dips from trough cross-beds (10–35°) and hummocky cross-strata (<15°) which dominantly point both alongshore and offshore.

ACKNOWLEDGEMENTS

The ideas contained in this paper are further developments to the original work of Steinmetz, Rider and Cameron. We wish to thank John Kantorowicz, Ian Bryant and Marc Budding for valuable comments on earlier versions of this manuscript. The fieldwork was carried out with the aid of Steve Flint,

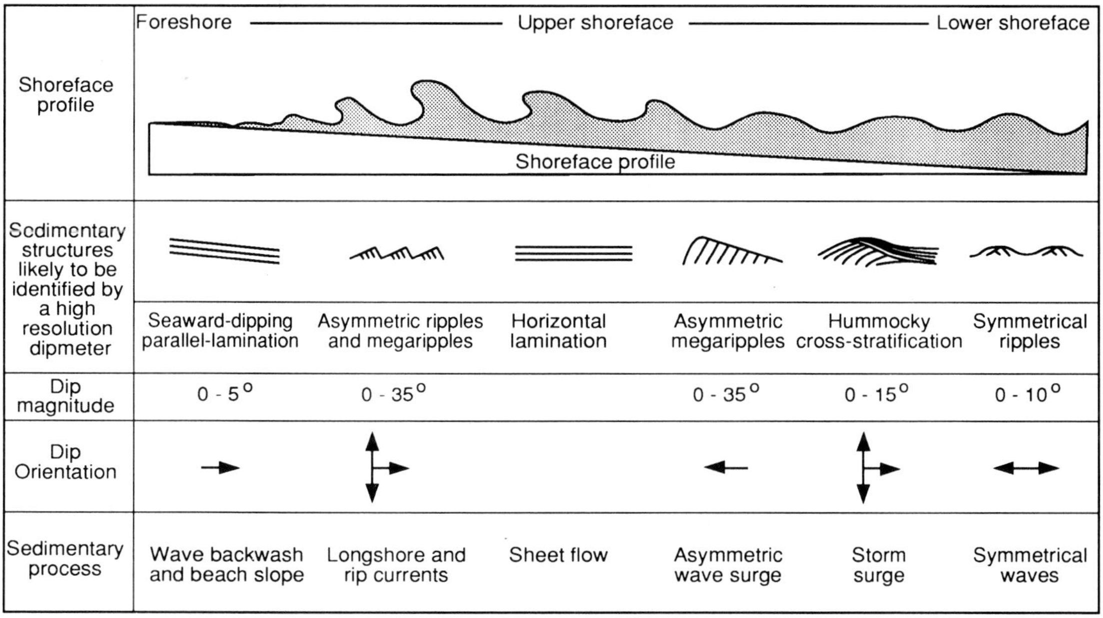

Fig. 11. Sedimentary structures, dips and processes in shoreface deposits. (Modified after Clifton, 1976, and Balsley, 1982.)

now at Liverpool University, and Katy Williams. The paper has also benefitted from the constructive comments of the referees, especially those of Malcolm Rider. This paper is published with the permission of Shell International Research Maatschaapij.

REFERENCES

BALSLEY J.K.. (1982) *Cretaceous Wave-dominated Delta Systems: Book Cliffs East Central Utah*. Amoco Production Company Guidebook, 219 pp.

CAMPBELL, R.L. JR. (1968) Stratigraphic applications of dipmeter data in Mid-Continent. *Bull. Am. Assoc. Petrol. Geol.*, **52**, 1700–1719.

CAMERON, G.I.F. (1986) Confidence and the identification of foresets in stratigraphic dipmeter surveys. *Society of Professional Well Log Analysts, 10th European Formation Evaluation Symposium*, Paper **0**, 1–16.

CLIFTON, H.E. (1976) Wave-formed sedimentary structures a conceptual model. In: *Beach and Nearshore Sedimentation*. (Ed. by R.A. Davies, Jr. & R.L. Ethington). Spec. Publ. Soc. econ. Paleont. Miner., Tulsa, **24**, 126–148.

DUBRULE, O.R.F., FLINT, S.S., PAARDEKAM, A.H.M. & BUDDING, M.C. (1992) Three-dimensional reservoir geological modelling of the Caño Limon field, Colombia. *Bull. Am. Assoc. Petrol. Geol.* (in press).

EMMETT, W.R., BEAVER, K.W. & MCCALEB, J.A. (1971) Little Buffalo Basin Tensleep heterogeneity—its influence on infill drilling and secondary recovery. *J. Petrol. Technol.*, Feb 1971, 161–168.

HERWEIJER, J.C., HOCKER, C.F.W., WILLIAMS, H. & EASTWOOD, K.M. (1990) The relevance of dip profiles from outcrops as reference for the interpretation of SHDT dips. In: *Geological Applications of Wireline Logs* (Ed. by A. Hurst, M.A. Lovell & A.C. Morton). Spec. Publ. geol. Soc. London, **48**, 39–43.

HIGH, L.R. JR. & PICARD, M.D. (1974) Reliability of cross-stratification types as palaeocurrent indicators in fluvial rocks. *J. sedim. Petrol.*, **44**, 158–168.

HOCKER, C.F.W., EASTWOOD, K.M., HERWEIJER, J.C. & ADAMS, J.T. (1990) Use of dipmeter data in clastic sedimentological studies. *Bull. Am. Assoc. Petrol. Geol.*, **74**, 105–118.

KORTEKAAS, TH.F.M. (1985) Water/oil displacement characteristics in cross-bedded reservoir zones. *SPE* Paper **12112**.

KUMAR, N. & SANDERS, J.E. (1976) Characteristics of shoreface storm deposits, modern and ancient examples. *J. sedim Petrol.*, **46**, 145–162.

MUWAIS, W. & SMITH, D.G. (1990) Types of channel-fills interpreted from dipmeter logs in the McMurray Formation, northeast Alberta. *Bull. Can. Petrol. Geol.*, **38**, 53–63.

PULHAM, A.J. (1988) Depositional and syn-sedimentary deformation processes in Namurian deltaic sequences of West County Clare, Ireland. PhD thesis, University of Wales.

Pulham, A.J. (1989) Controls on internal structure and architecture of sandstone bodies within Upper Carboniferous, fluvially-dominated deltas, County Clare, western Ireland. In: *Sites and Traps for Fossil Fuels*. (Ed. by M.K.G. Whateley & K.T. Pickering). Spec. Publ. geol. Soc. London, Blackwell Scientific Publications, Oxford, **41**, 179–203.

Reineck, H.E. & Singh, I.B. (1973) *Depositional Sedimentary Environments*. Springer-Verlag, New York, 439 pp.

Rider, M.H. (1969) Sedimentological studies in the West Clare Namurian and the Mississippi River delta. PhD thesis, Imperial College, London.

Rider, M.H. (1974) The Namurian of West County Clare. *Proc. Roy. Irish Acad.*, **74**, 125–142.

Rider, M.H. (1978) Dipmeter log analysis: an essay. *Society of Professional Well Log Analysts 19th Annual Symposium*, El Paso, Texas, June 13–16. Paper G, 1–18.

Serra, O. (1985) *Sedimentary Environments from Wireline Logs*. Paris, Schlumberger, M-081030/SMP, 211 pp.

Snedden, J.W. & Nummedal, D. (1990) Coherence of surf zone and shelf current flow on the Texas (U.S.A.) coastal margin: implications for interpretation of paleocurrent measurements in ancient coastal sequences. *Sedim. Geol.*, **67**, 221–236.

Steinmetz, R. (1972) Sedimentation of an Arkansas river sand bar in Oklahoma: a cautionary note on dipmeter interpretation. *Shale Shaker*, **23/1**, 32–38.

Van Wagoner, J.C., Mitchum, R.M., Campion, K.M. & Rahmanian, V.D. (1990) *Siliciclastic Sequence Stratigraphy in Well Logs, Cores and Outcrops: Concepts for High-resolution Correlation of Time and Facies*. Am. Ass. Petrol. Geol. Methods in Exploration Series, No. **7**.

Weber, K.J. (1982) Influence of common sedimentary structures on fluid flow in reservoir models. *J. Petrol. Technol.*, **34**, 665–672.

Spec. Publs int. Ass. Sediment. (1993) **15**, 157–168

Permeability patterns in point bar deposits: Tertiary Loranca Basin, central Spain*

C.A. HARTKAMP-BAKKER *and* M.E. DONSELAAR

Faculty of Mining and Petroleum Engineering, Delft University of Technology, PO Box 5028, 2600 GA Delft, The Netherlands

ABSTRACT

A composite point bar sediment body in the Tertiary Loranca Basin of central Spain was the subject of a detailed analysis of the relation between sedimentary facies heterogeneities and permeability contrasts.

Comparison of field-measured permeability data with grain size, sorting and diagenetic features shows that a strong decrease in permeability (from 10 D to 0.1 D) over the contact zone of two stacked point bar units is related to the occurrence of a cemented, rootletted mud layer which formed during a channel abandonment stage. This type of low-permeability horizon is observed to extend from the adjacent floodplain downslope onto the abandonment layer. In the topographically lowest part of the abandonment slope, however, the mud layers usually are eroded during a subsequent channel reactivation stage, and stacking of point bar units is characterized by sand-on-sand contacts with normal, high-permeable transitions.

Statistical processing of permeability data from cross-bedded sets within a point bar unit reveals that the average foreset permeability is significantly higher than the average bottomset value (total permeability range from 0.5 to 17 D), and that the foreset values are highly variable. Both observations are directly related to grain size and sorting differences.

In reservoir equivalents of the composite point bar sediment body, the effect of the permeability contrasts on fluid flow is two-fold:

1 channel abandonment layers, when preserved after channel reactivation, cause partial compartmentalization of the reservoir unit;

2 in water-flooded oil-reservoirs, permeability contrast between adjacent foreset laminae and bottomsets may cause a capillary pressure-determined water fingering effect.

INTRODUCTION

The low recovery efficiency of fluvial reservoirs in particular, and of all heterogeneous reservoirs in general, highlights the need for better reservoir models based on detailed quantitative knowledge of the sedimentological parameters which determine the reservoir properties. Quantifiable sedimentological data include reservoir characteristics on all relevant scales, such as width–thickness ratios of sediment bodies, interconnectedness of the reservoir units, internal facies distribution, grain size and sorting, bioturbation rate, shale length distribution, diagenesis and burial-decompaction history.

If data density is low, as is the case during the early stages of field development, probabilistic reservoir modelling can be employed and many possible realizations will be generated by the modelling exercise. Increasing the knowledge about the sediment-

* Research Center for Geological Reservoir Characterization (GeoRes) publ. no. 004.

ological parameters (the deterministic model) improves and validates the stochastic procedures and reduces the number of stochastic realizations. The upgraded stochastic models then form the basis for flow simulation studies, which help to outline the oil production strategy.

Research of permeability distribution on a wide range of scales and in a variety of sedimentary environments has been carried out over the last 20 years. Scales of reservoir heterogeneities based on outcrop and/or sub-surface data have been defined by Morrow (1971), Haldorsen & Lake (1984), Weber (1986) and Lewis (1988). Studies of permeability distributions in fluvial deposits are presented by Weber *et al.* (1972), Pryor (1973), Jacobsen & Rendall (1991) and Dreyer *et al.* (1990). The majority of the studies illustrate permeability characteristics on a large scale (reservoir and reservoir-unit scale; see Weber, 1986). Permeability values are generally averaged using statistical methods (e.g. Dodge *et al.*, 1971; Pryor, 1973; Stalkup, 1986; Jacobsen & Rendall, 1991). Weber *et al.* (1972), Pryor (1973) and Chandler *et al.* (1989) focus on small-scale heterogeneities in the permeability distribution. A flow simulation study has stressed the influence of permeability anisotropy and capillary forces on fluid flow in cross-stratified sets. This study has clearly illustrated the need to incorporate small-scale (cross-set and lamina scale) permeability variations in flow simulation studies (Kortekaas, 1985).

Fluvial hydrocarbon reservoirs with a low sand/shale ratio typically consist of a spatial arrangement of single or stacked lenticular sandbodies embedded in shale: the so-called 'labyrinth' reservoirs (Weber & van Geuns, 1989). The spatial distribution of the fluvial sandbodies as well as the geometry and internal facies arrangement contribute to the complex heterogeneous nature and result in compartmentalization of the reservoir. Oil entrapment in compartments and low-sweep efficiency through cross-bedded sedimentary structures (Kortekaas, 1985) are considered to be important causes for the low recovery efficiency typical of fluvial reservoirs. Recovery rates may be as low as 25–40% of oil in place, depending on the degree of reservoir complexity (Barwis *et al.*, 1990). Injection programmes intended to enhance recovery are often unsuccessful if knowledge of the spatial arrangement, geometry and internal permeability barriers in this type of reservoir is incomplete (van Veen, 1977).

The present study is concerned with the analysis of a detailed outcrop data set describing permeability contrasts in fluvial point bar deposits. Data acquisition was carried out in the Loranca Basin of central Spain. The object of this study is a composite point bar sandstone body which forms part of well-exposed distal fluvial fan deposits of late Oligocene to early Miocene age. Two types of depositional permeability heterogeneities have been studied: (i) the contact area of two stacked point bar units; and (ii) the internal permeability distribution in a cross-bedded point bar. Data acquisition comprised conventional sedimentological procedures as well as on-site permeability measurements with the use of a portable permeability meter: the 'minipermeameter'.

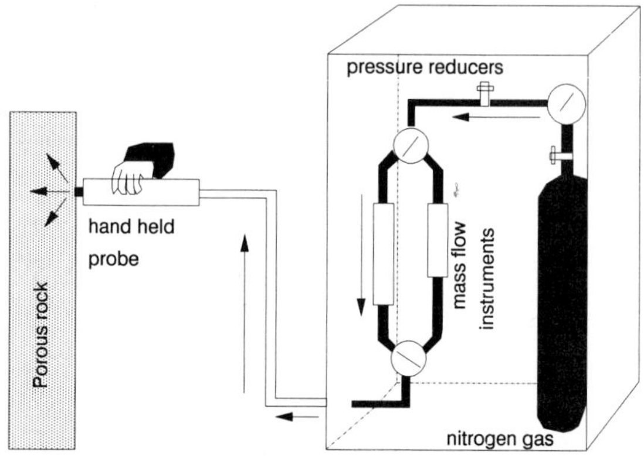

Fig. 1. Schematic drawing of the portable permeability apparatus, the 'minipermeameter'.

METHODOLOGY

With the portable permeability recording apparatus, the 'minipermeameter' (Fig. 1), nitrogen gas is injected into the porous rock, using a hand-held probe, with a 2 mm diameter tip-seal opening. Both the gas-flow induced pressure differential (ΔP) and the injection pressure (P_{inj}) are recorded by a micromanometer, and recalculated to gas flow rate, using a generalized form of Darcy's law for minipermeameter experiments:

$$q_t = \frac{C\Delta P}{P_{inj}} \tag{1}$$

where q_t is the flow rate in cm^3/min, C the specific calibration constant per flow tube, ΔP the pressure differential, and P_{inj} the injection pressure.

The permeability values are subsequently calcu-

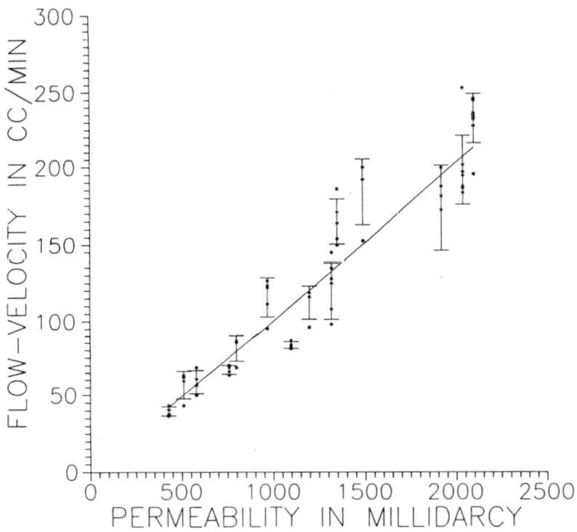

Fig. 2. Calibration curve of the relationship between flow velocity and permeability, based on 65 measurements (marked with*) of 14 cores with laboratory-derived permeability values.

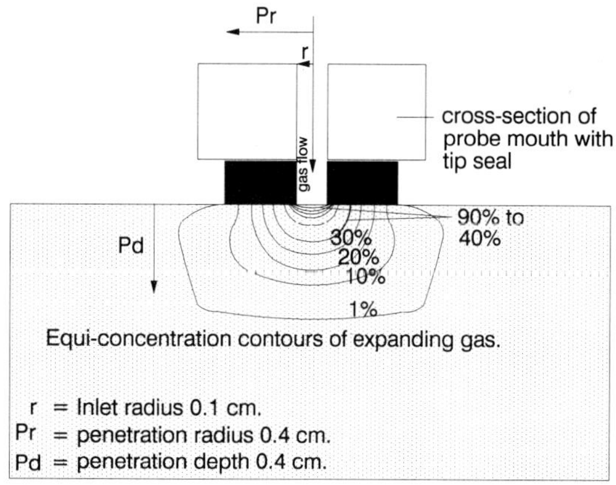

Fig. 3. Penetration radius and depth of gas invasion in a porous medium. Modified from Goggin *et al.* (1988).

lated from a series of previously established calibration curves (Fig. 2). Nitrogen gas is chosen because of its dry character and stable viscosity and flow performance in a relatively wide temperature range (Goggin *et al.*, 1988). It should be noted, however, that a high temperature contrast between the nitrogen gas in the minipermeameter and the surrounding environment causes strongly varying, non-reproducible pressure readings (contrast of 10°C and higher). Measurements in highly permeable rock were expected to be problematic due to turbulent flow of the nitrogen gas (Goggin *et al.*, 1988), but micromanometer readings were usually stable and reproducible.

The effective penetration depth of the gas is equivalent to a core plug sample with a radius and length four times the internal tip seal radius (Goggin *et al.*, 1988) (Fig. 3). In our case the minipermeameter measurements affect an area with a radius and depth of 4 mm. Eighty to ninety per cent of the measurements record the permeability properties of the first few cubic millimetres of rock. Reliability of the measurements is enhanced by removing the weathered surface of the rock. The fluvial sediments in the Loranca Basin have a low degree of cementation and hence a friable surface. Removal of the weathered surface was achieved with a steel brush. Dust was subsequently removed with a soft brush. This cleaning method enhanced the visibility of sedimentary structures on the measured surface.

The limited penetration depth (Fig. 3) and radius of the minipermeameter measurements allow the use of a closely-spaced grid without overlap of flow-velocity values in successive grid points. For this study various closely-spaced grids were applied to characterize the permeability patterns of the point bar sediment body. A grid with a vertical to horizontal spacing of 2.5 cm to 5 cm respectively (number of grid points: 69) was used to measure permeability contrasts across the boundary of two stacked point bar units. A grid with a 1 cm vertical and 10 cm horizontal spacing (total number of grid points: 1096) was measured to define the permeability patterns within a single point bar unit. Additional grid points were measured on key horizons inbetween the grid rows. Data were processed and correlated in the field in order to have direct control on the measurements and to re-record doubtful grid point readings. Selected core samples were taken with an electrical drill. Several samples were impregnated with blue-coloured Araldite for microscopic analysis of the porosity distribution and petrology of the rock.

GEOLOGICAL SETTING

The depositional history of the Iberian Peninsula during the Tertiary was strongly influenced by plate tectonic movements. The Iberian plate was squeezed between the northward moving African plate and the rigid Eurasian plate in a compressive period that lasted from the Late Cretaceous to the Middle Miocene (Choukroune *et al.*, 1973). On the Iberian

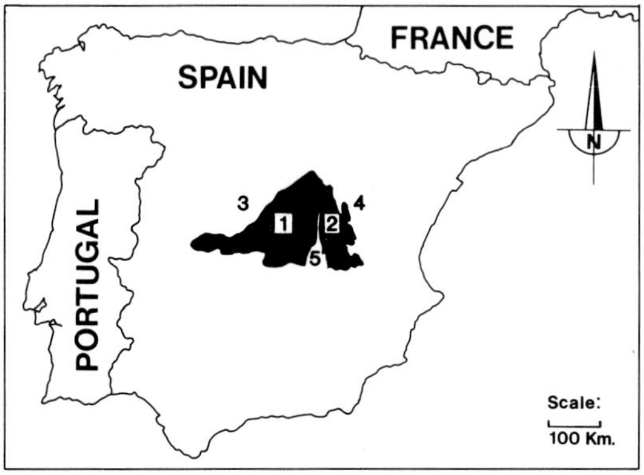

Fig. 4. Tertiary cratonic basins (1–2) in central Spain surrounded by mountain belts (3–5). 1 = Madrid Basin; 2 = Loranca Basin; 3 = Central System; 4 = Iberian Range; 5 = Sierra de Altomira. After Diaz Molina *et al.* (1989).

plate the compression generated a number of cratonic depressions separated by mountain belts. The depressions were subsequently filled with molasse-type sediments.

The Loranca Basin of central Spain (Fig. 4) is one of these cratonic basins. The basin comprises a 100 km long and 80 km wide north–south trending depression which is limited to the east by uplifted, folded and faulted Mesozoic rocks of the Iberian Range. This mountain range acted as principal sediment source area to the Loranca Basin (Diaz Molina *et al.*, 1985). The western basin margin consists of a narrow, north–south trending thrust belt, the Sierra de Altomira. Subsidence in the Palaeogene was strongest adjacent to the Sierra de Altomira and resulted in an asymmetric basin fill (Fig. 5) with a maximum cumulative thickness in the western part of the basin of over 1200 m of Palaeocene to Miocene clastic sediments (Diaz Molina *et al.*, 1985). Basin fill took place by successive alluvial fan systems which deposited suites of fluvial fan conglomerates and sandstones and associated fine-grained clay flat, playa lake and lacustrine deposits.

The deposits discussed in the present paper comprise the upper part of the Tórtola fan of Late Oligocene–Early Miocene age (Diaz Molina *et al.*, 1989). The Tórtola fan has its source area in the Iberian Range, its maximum length and width are 94 km and 40 km respectively, and it covers an area of 2500 km^2 (Fig. 6; Diaz Molina *et al.*, 1989). The fan is of the wet fan type (Schumm, 1977) and has a radial fluvial distributary pattern. Fan progradation in the western direction was restricted by the Sierra

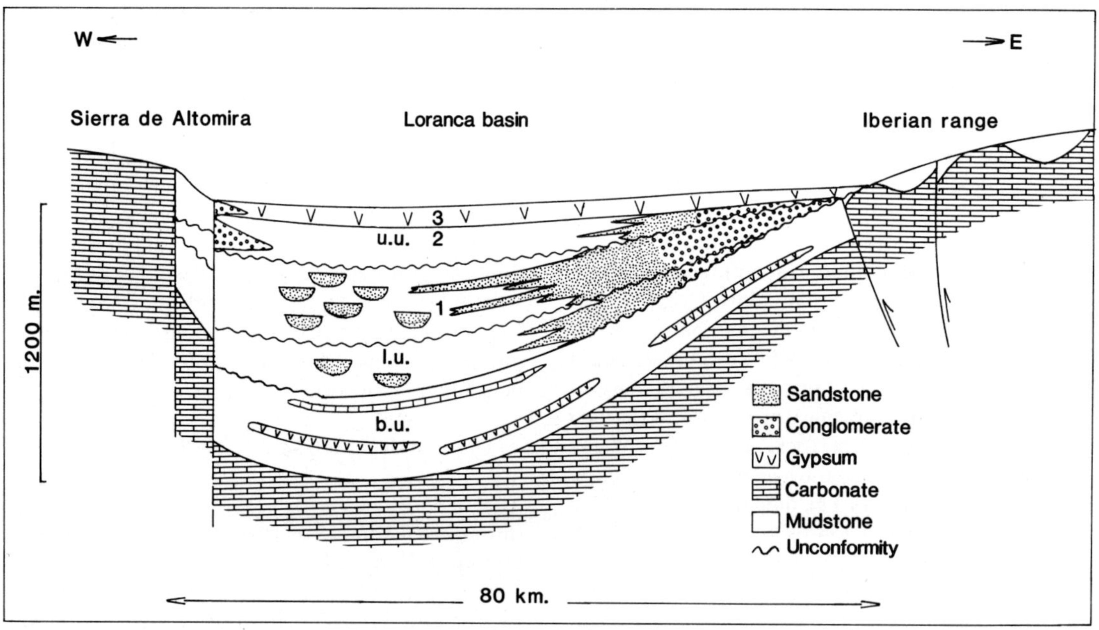

Fig. 5. Schematic cross-section through the Loranca Basin. b.u. = basal unit; l.u. = lower unit; u.u. 1–3 = upper unit with sub-units 1 to 3. Thrust faults in the Sierra de Altomira have a sinistral strike-slip component (Diaz Molina *et al.*, 1985).

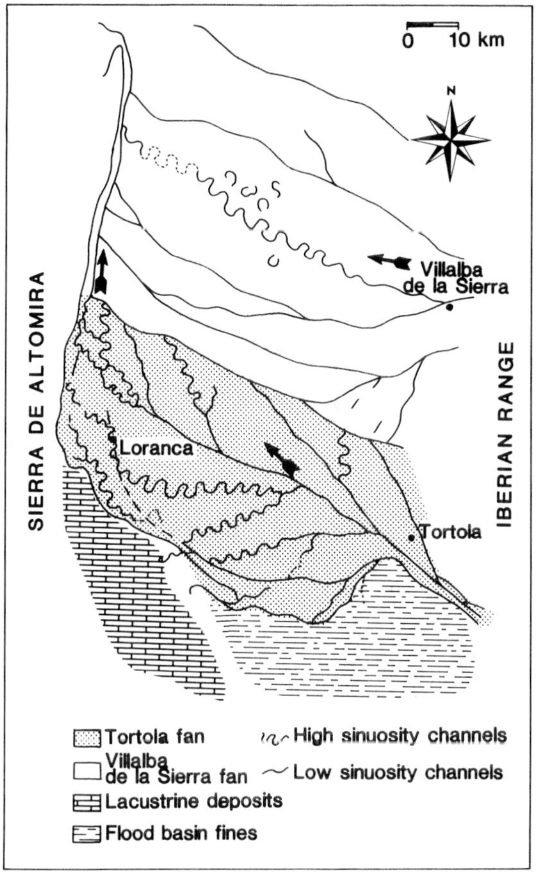

Fig. 6. Palaeogeographic reconstruction of the Tórtola and adjacent Villalba de la Sierra fan systems. Arrows indicate general flow direction of the fluvial channels. After Diaz Molina *et al.* (1989).

de Altomira thrust belt which forced the fluvial channels to flow to the north, parallel to the axis of the thrust belt. The northern part of the fan coalesced with an adjacent fan system (Fig. 6). Distal fan sedimentation produced vertically- and laterally-stacked sandstone bodies embedded in fine-grained deposits. Sandstone bodies were formed by meandering and braided fluvial channels.

SEDIMENTARY FACIES DESCRIPTION

The studied fluvial sediment body has a lenticular geometry and contains seven vertically and laterally stacked point bar units (Fig. 7). The sediment body has a length and thickness of 100 m and 6 m respectively.

Each point bar unit consists of lateral accretion deposits. During stages of channel abandonment the accretion surface was draped with mud. Partial preservation of the mud drape depends on the rate of erosion during subsequent channel reactivation, and on the orientation of the next meander loop (Diaz Molina, in press). In the studied composite point bar complex the bounding surfaces between stacked point bar units show partially-preserved mud drapes (Fig. 7C). The drapes are restricted to the topographically higher part of the accretion slope. Downslope the mud drape is eroded and sand-on-sand contact between successive point bar units is a common feature.

The preserved mud drape measured in detail in this study (Fig. 8) consists of a 10–15 cm thick, 7 m long very-fine sand and silt layer. The mud layer is

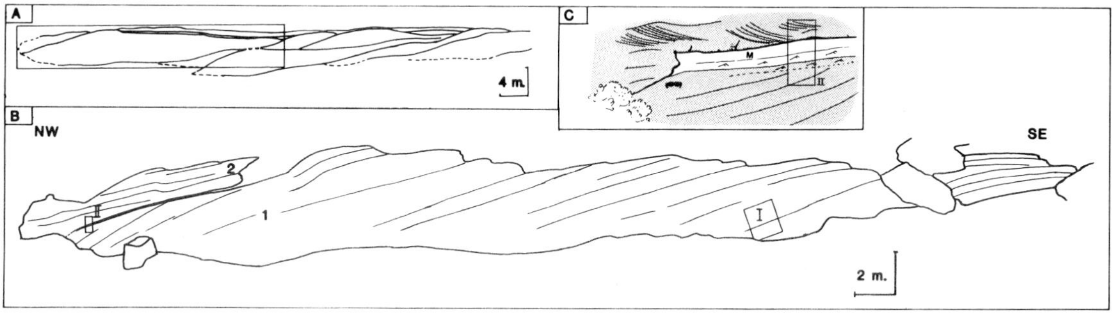

Fig. 7. (A) Outcrop drawing of stacked point bar units (after Diaz Molina *et al.*, 1989). Box indicates the area of Fig. 7B. (B) Drawing (from photo-panel) of the studied point bar units 1 and 2. Thin lines represent the lateral accretion surfaces. Thick line between units 1 and 2 is the very fine-grained, cemented horizon. Boxes I and II are the grid locations (see Figs 8 and 9). (C) Drawing of the channel abandonment mud drape between two stacked point bar units. Down the accretion slope, the mud drape is eroded and a sand-on-sand contact is present. Height of box II is 30 cm.

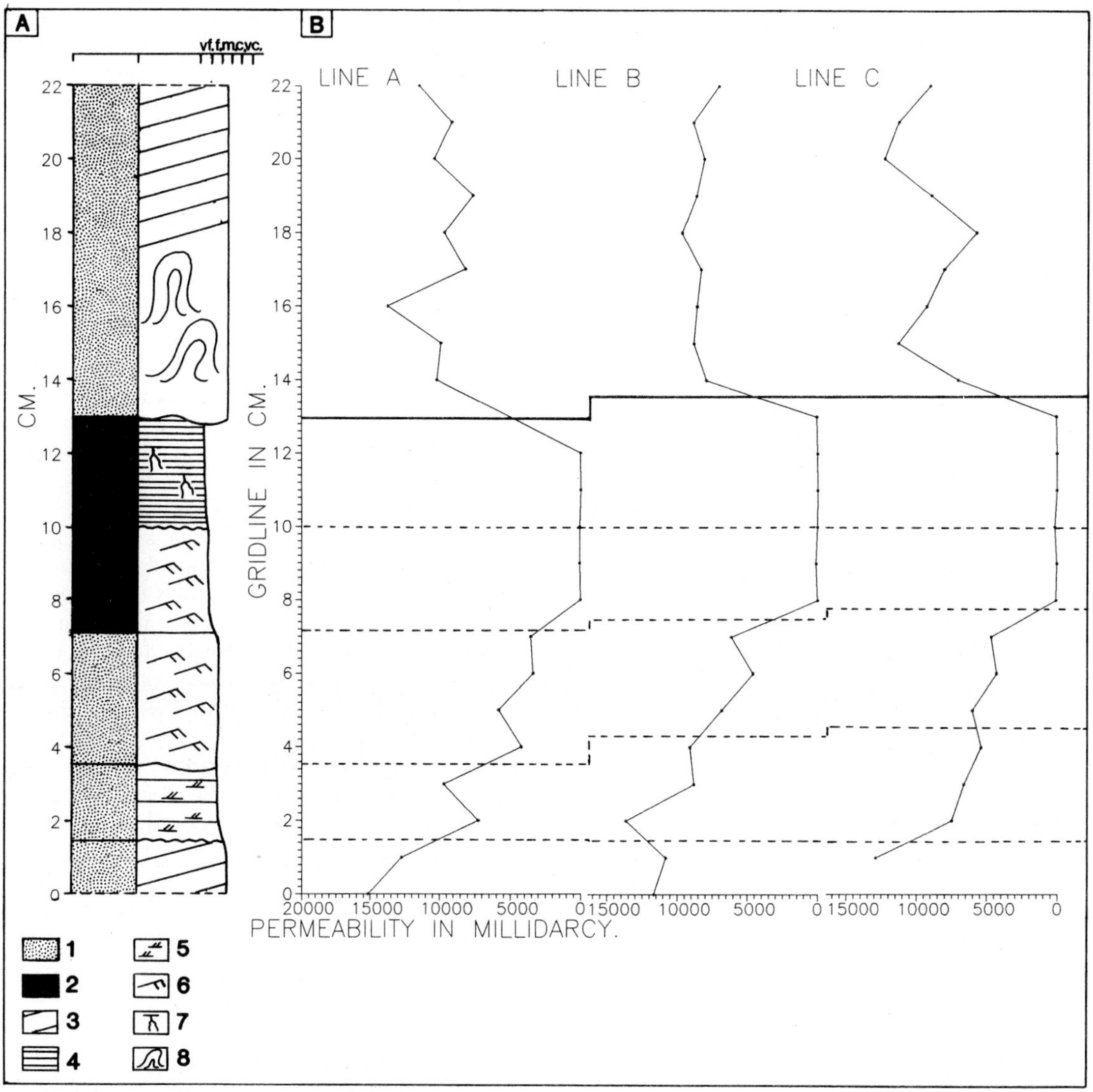

Fig. 8. (A) Vertical log across the boundary of two superimposed point bar units. 1 = sand; 2 = very fine sand; 3 = lateral accretion; 4 = horizontal lamination; 5 = small-scale ripples; 6 = climbing ripples; 7 = rootlets; 8 = contorted bedding. (B) Permeability pattern across the point bar transition zone. Continuous line: contact horizon between the successive point bar units. Broken line: correlation horizons. Spacing of the grid columns is 5 cm.

rootletted and thoroughly cemented with calcite of early-diagenetic origin. The high degree of cementation contrasts with the over- and underlying point bar sands, where calcite cementation is restricted to the grain contacts.

The lower 70–80% of the point bar units consist of trough cross-bedded, medium- to coarse-grained sand. The upper 20–30% is characterized by small-scale rippled, even-laminated to non-laminated, structureless (bioturbated) fine sand. Internally each point bar unit shows lateral accretion units. These units are sandy throughout; no mud drapes are observed on the accretion surfaces. Trough cross-bed sets climb up along the lateral accretion surfaces

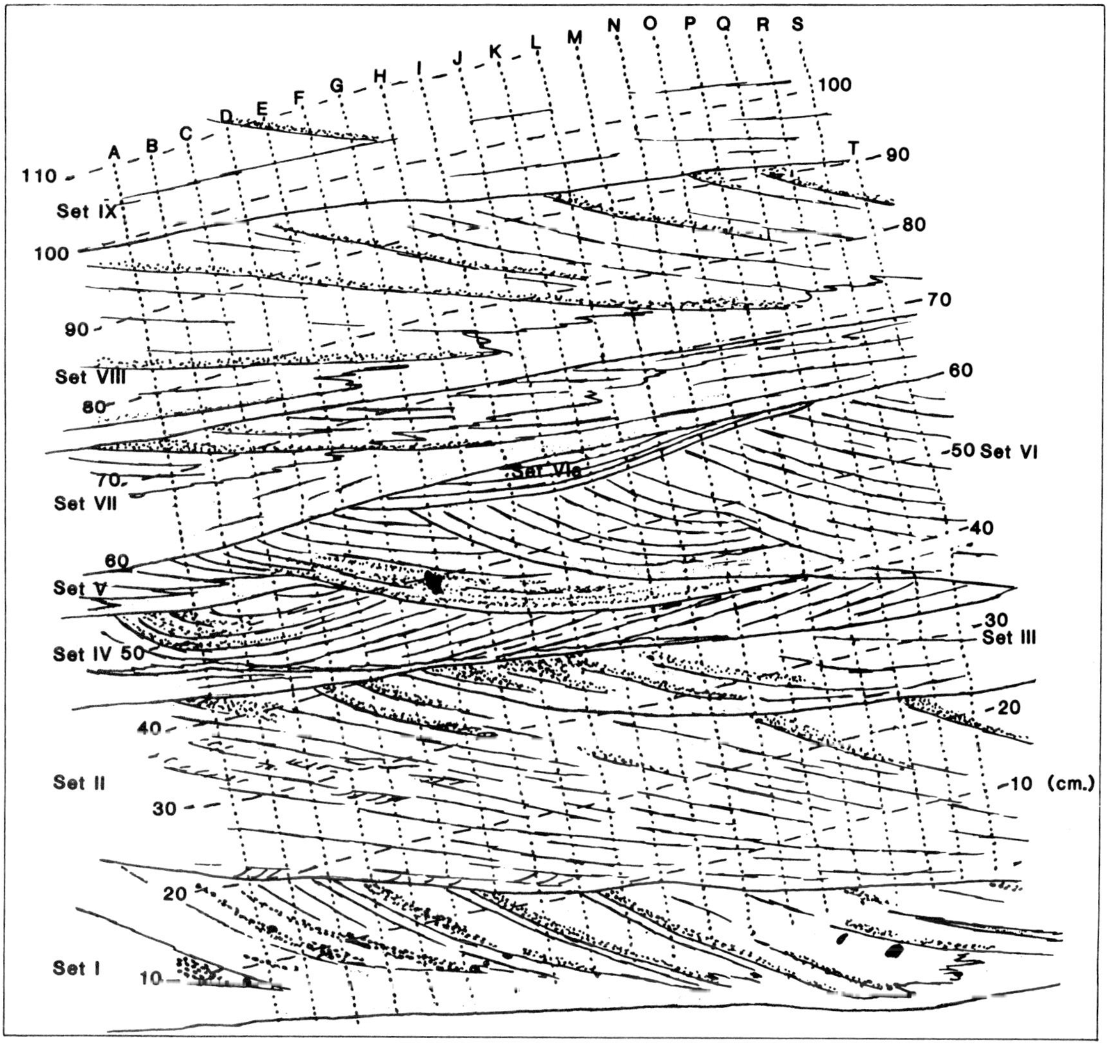

Fig. 9. Drawing (from photograph) of the measured grid over the trough cross-bed sets (set I to IX) that make up the lateral accretion units. See Fig. 7B, box I for location of the grid. Spacing of the grid columns A–B, B–C, etc. is 5 cm; grid row points are 1 cm apart.

(Fig. 9), such that bottomsets are parallel to the dip of the accretion surfaces. The foresets consist of alternating laminae of fine- and medium- to coarse-grain size. Lamina thickness is 1–10 mm. Individual foreset laminae are well sorted. By contrast, the bottomsets are poorly sorted and range in grain size from very fine to coarse sand. Sorting differences typically reflect the flow conditions during sedimentation: foreset laminae are sorted by size during avalanching along the leeside of the mega-ripple, while bottomsets result from vortex mixing of fine-grained suspension fall-out with coarser-grained avalanche sand and coarse-grained topset sands of the underlying mega-ripple. Bottomsets make up 20–30% of the total volume of the preserved part of the cross-sets.

The sandstone consists of 75% quartz, 20% carbonate and 5% feldspar and gypsum clasts, with a primary porosity of 25–30%. Dissolution of the gypsum and gypsum clasts is responsible for a secondary porosity and the friable nature of the sandstone. The calcite cement in the very fine-grained abandonment facies is probably formed as an early-diagenetic cement due to palaeosol formation.

PERMEABILITY MEASUREMENTS

Permeability distribution over the contact area of two point bar units

The decrease of grain size and the increased degree of cementation at the transition between two stacked point bar units are reflected in the permeability pattern across the boundary (Fig. 8). A strong upward decrease in permeability (from 10 D to 0.1 D) characterizes the upper part of the lower point bar unit, and is followed by a rapid increase in permeability in the upper point bar unit to values comparable with the cross-bedded parts of the lower one. This low permeability zone between the two highly permeable units is discontinuous and extends from the adjacent floodplain down onto the abandonment layer (the preserved part of the abandonment mudstone drape). Further downslope, stacking of the point bar units is characterized by sand-on-sand contacts with normal, high-permeable transitions (mudstone drape eroded) (Fig. 7C).

Fluvial oil reservoirs of meandering river deposits are often compartmented by similar low-permeability abandonment layers, which result in a marked permeability anisotropy that can influence sweep efficiency.

Small-scale permeability heterogeneities

Permeability distribution in the trough cross-bedded part of the point bar is directly proportional to grain size and sorting of the cross-bed sets (Beard & Weyl, 1973). This results in a permeability contrast between the poorly-sorted bottomsets (values 0.5–7 Darcy) and the moderately- to well-sorted foresets (values 5–17 Darcy) (Figs 9, 10, 11A). Foresets are characterized by a larger permeability variation because of the varying grain sizes of the individual laminae. The highly variable permeability values in set 2 (Fig. 10) are caused by the occurrence of back flow ripples. In some sets the coarse-grained foreset laminae grade into the bottomsets and are characterized by a vertical alternation of higher (foreset) and lower (bottomset) values.

In order to test the statistical difference in the mean of both sets of measurements (foreset and bottomset) the Wilcoxon test was done. The total number of datapoints (1096) has been subdivided into foreset data and bottomset data. The data sets represent stochastically normal populations. The nine average bottomset values of the nine cross-sets were tested against the nine average foreset values. The test showed that the foreset population was characteristically more than 2 Darcy higher than the bottomset population (confidence limit 99%) (Fig. 11). This test case confirms the expected result of average foreset permeability values to be higher than the average bottomset values. In Table 1, average permeability values show an upward decrease and subsequent increase (from set 1 to set 9). This trend is related to the average grain size trend from medium-coarse sand in set 1, to fine-medium sand in sets 4–8, and back to medium sand in set 9.

The standard deviation of permeability values is higher in the foreset than in the bottomset. This can be explained by the higher variability of foreset permeability values as compared to the bottomset.

Discussion

Permeability heterogeneity in cross-bedded fluvial reservoirs is principally caused by depositional grain size and sorting characteristics. The permeability contrast is often enhanced during burial and diagenesis. A study by Kortekaas (1985) indicated the likelihood of significant oil trapping in foresets as the result of different imbibition speed of water into laminae of contrasting permeability and capillary pressure characteristics (Beard & Weyl, 1973). The

Table 1. Average bottomset permeability values and average foreset permeability values of the nine stacked cross-bed sets. Refer to Figs 9 and 10 for position of cross-bed sets 1–9

	Foreset			Bottomset		
	Average	SD	Number	Average	SD	Number
Set 1	7.13	4.02	65	5.18	3.64	31
Set 2	7.86	3.12	176	6.61	3.4	51
Set 3	7.15	2.75	34	5.32	2.11	14
Set 4	6.83	3.99	41	5.29	2.76	26
Set 5	7.54	3.82	33	5.04	2.2	19
Set 6	6.91	2.28	79	4.9	1.94	17
Set 7	6.57	2.19	49	4.44	2.29	65
Set 8	7.84	2.77	174	6.75	2.78	62
Set 9	9.94	2.53	96	7.83	1.76	59

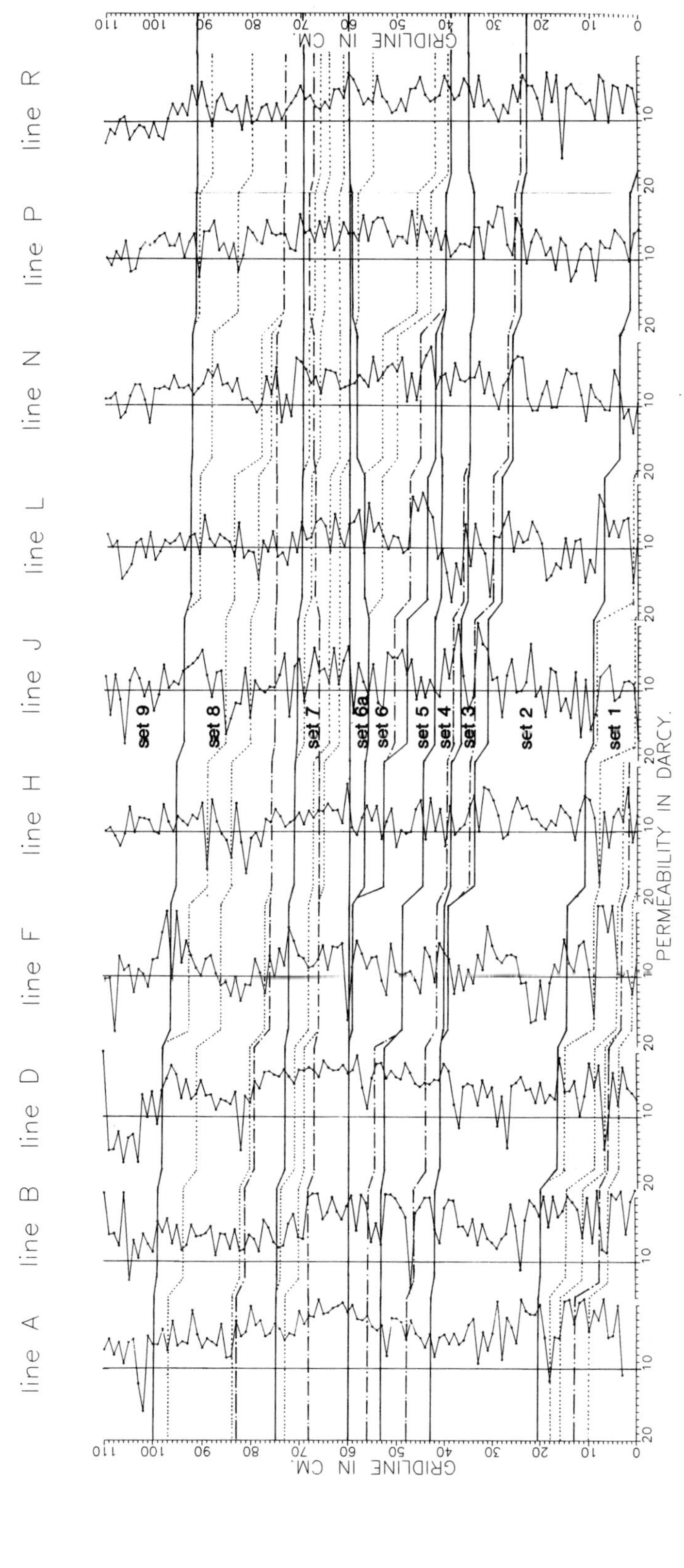

Fig. 10. Permeability pattern of trough cross-bed sets. Continuous line: base of cross-beds. Dashed–dotted line: boundaries between foreset and bottomset. Dotted line: inclined foreset laminae. Spacing between vertical grid lines is 10 cm, except for Lines A and B, which are 5 cm apart. Refer to Fig. 9 for exact location of grid columns.

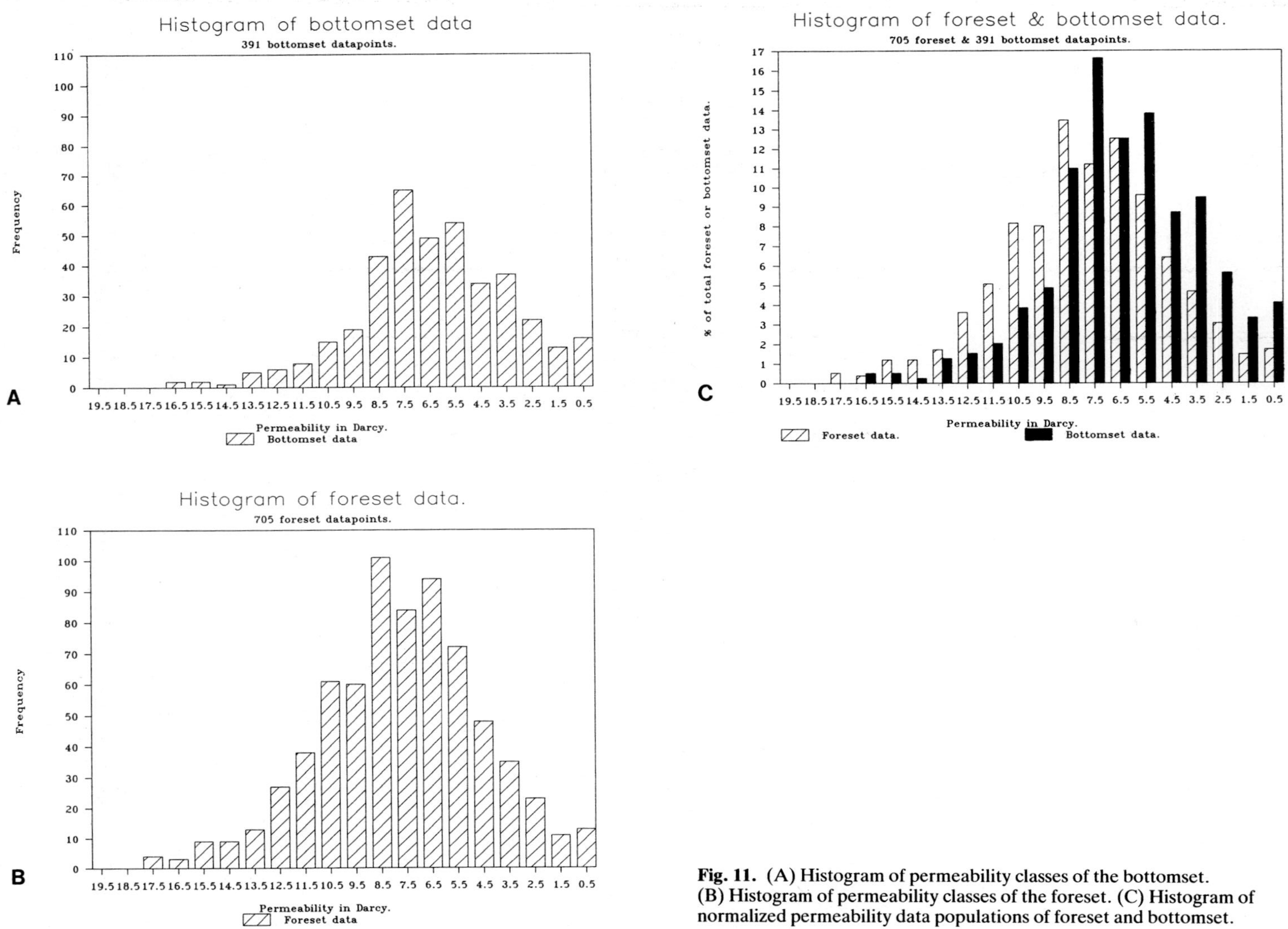

Fig. 11. (A) Histogram of permeability classes of the bottomset. (B) Histogram of permeability classes of the foreset. (C) Histogram of normalized permeability data populations of foreset and bottomset.

permeability in the bottomsets is generally lower than in the foresets, and consequently the capillary pressure is higher in the former. Fast imbibition of water and subsequent water tonguing will occur in the bottomsets and in the lower-permeable foreset laminae. When maximum water saturation is reached, a non-flow boundary is formed. The remaining mobile oil in the highly-permeable foreset laminae is trapped by the non-flow boundaries (Weber, 1986). The percentage of trapped oil is a function of the ratio of permeability of adjacent laminae and the mobility ratio of the displacing and displaced fluids. Permeability heterogeneity on the scale of cross-set laminae is likely to have a marked influence on fluid flow in oil reservoirs with a fluvial architecture similar to the studied outcrop. In this type of fluvial reservoirs suitable enhanced oil recovery methods may be developed that decrease the interfacial tension of the driving fluid, e.g. low concentration surfactant flooding.

CONCLUSIONS

Detailed permeability measurements of a composite point bar sediment body reveal a direct relationship between permeability heterogeneity and grain size, sorting and diagenesis, and allow the definition of two types of permeability heterogeneity for this group of fluvial deposits. The two types are:

1 The bounding layer between stacked point bar units may form a low-permeable zone. The layer originates from channel abandonment and in this example is 10–15 cm thick, very-fine grained, early-cemented and rootletted. Preservation of the low-permeability zone and lateral extent of the abandonment surface are a function of the subsequent channel reactivation. In the studied outcrop this type of low-permeability zone commonly extends from the adjacent low-permeability floodplain down onto the abandonment surface over a distance of about 7 m. Measured permeability ranges from 0.1 Darcy (abandonment layer) to 10 Darcy (over- and underlying point bar sand).

2 The cross-bedded part, which makes up 70–80% of the bulk volume of the point bar sediment body, shows a marked permeability contrast between poorly-sorted bottomsets (permeability 0.5–7.0 Darcy) and well-sorted foresets (5.0–17.0 Darcy). In reservoir equivalents of the composite point bar, both types of permeability heterogeneity are considered to influence fluid flow:

(a) The low-permeability channel abandonment surfaces may act as permeability baffles to fluid flow, and cause partial compartmentalization of the reservoir.

(b) In water-flooded oil-reservoirs, permeability contrast between adjacent foreset laminae and bottomsets may cause water fingering caused by capillary pressure effects.

ACKNOWLEDGEMENTS

We are indebted to M. Diaz Molina for introducing us to the sedimentology of the Loranca Basin, and for the valuable discussions in the field. J. Arribas Mocoroa, A. Tortosa Lopez and M. Boering are thanked for their guidance and help in the field. Thanks are due to the members of the Research Center for Geological Reservoir Characterization (GeoRes) at TU Delft, especially to K.J. Weber and M.C. Cuevas Gozalo, for their constructive comments on the manuscript. We are grateful for the careful reviews of I. Bryant and T. Dreyer, which improved the final version of the manuscript. The field research was carried out with financial support of the Royal Dutch Shell Exploration and Production Laboratory (KSEPL), Rijswijk, The Netherlands.

REFERENCES

Barwis, J.H., McPherson, J.G. & Studlick, J.R.J. (eds) (1990) *Sandstone Petroleum Reservoirs*. Springer-Verlag, New York, 583 pp.

Beard, D.C. & Weyl, P.K. (1973) Influence of texture on porosity and permeability of unconsolidated sand. *Bull. Am. Assoc. Petrol. Geol.*, **57**, 349–369.

Chandler, M.A., Kocurek, G., Goggin, D.J. & Lake, L.W. (1989) Effect of stratigraphic heterogeneity on permeability in eolian sandstone sequence, Page Sandstone, Northern Arizona. *Bull. Am. Assoc. Petrol. Geol.*, **73**, 658–668.

Choukroune, P., Le Pichon, X. & Seguret, M. (1973) Bay of Biscay and Pyrenees. *Earth Planet. Sc. Lett.*, **18**, 109–118.

Diaz Molina, M., Arribas Mocoroa, J. & Bustillo Revuelta, A. (1989) The Tórtola and Villalba de la Sierra fluvial fans: late Oligocene–early Miocene, Loranca Basin, Central Spain. *Guidebook for field trip 7. 4th Int. Conf. on Fluvial Sedimentology, Barcelona-Sitges, Spain*. Servei Geologic de Catalunya, Spain, 74 pp.

Diaz Molina, M., Capote, R. & Lopez Martinez, N. (1985) Wet fluvial fans of the Loranca Basin (Central Spain), channel models and distal bioturbated gypsum with chert. In: *6th I.A.S. Eur. Reg. Meeting, Excursion*

Guidebook (Ed. by Mila, M.D. & Rosell, J.), Institut d'Estudis Illerdencs (Lleida), pp. 149–185.

Diaz Molina, M. (1993) Geometry and lateral accretion patterns in ancient meander loops, Loranca Basin, Spain. In: *Alluvial Sedimentation* (Ed. by M. Marzo & C. Puigdefabregas) *Spec. publs int. Assoc. Sediment.*, **17**, Blackwell Scientific Publications, Oxford (in press).

Dodge, C.F., Holler, D.P. & Meyer, R.L. (1971) Reservoir heterogeneities of some Cretaceous sandstones. *Bull. Am. Assoc. Petrol. Geol.*, **55**, 1814–1828.

Dreyer, T., Scheie, Å & Walderhaug, O. (1990) Minipermeameter-based study of permeability trends in channel sand bodies. *Bull. Am. Assoc. Petrol. Geol.*, **74**, 359–374.

Goggin, D.J., Thrasher, R.L. & Lake, L.W. (1988) A theoretical and experimental analysis of minipermeameter response including gas slippage and high velocity flow effects. *In Situ*, **12**, 79–116.

Haldorsen, H.H. & Lake, L.W. (1984) A new approach to shale management in fieldscale models. *SPE Journal*, **24**, 447–457.

Jacobsen, T. & Rendall, H. (1991) Permeability patterns in some fluvial sandstones. An outcrop study from Yorkshire, North East England. In: *Reservoir Characterization—II* (Ed. by C.W. Lake, H.B. Carroll Jr & T.C. Wesson), pp. 315–339. Academic Press, London.

Kortekaas, T.F.M. (1985) Water/oil displacement characteristics in crossbedded reservoir zones. *SPE Journal*, December, 917–926.

Lewis, J.J.M. (1988) Outcrop-derived quantitative models of permeability heterogeneity for genetically different sand bodies. *SPE Paper* **18153**.

Morrow, N.R. (1971) Small-scale packing heterogeneities in porous sedimentary rocks. *Bull. Am. Assoc. Petrol. Geol.*, **55**, 514–522.

Pryor, W.A. (1973) Permeability-porosity patterns and variations in some Holocene sand bodies. *Bull. Am. Assoc. Petrol. Geol.*, **57**, 162–189.

Schumm, S.A. (1977) *The Fluvial System*. Wiley, New York, 338 pp.

Stalkup, F.I. (1986) Permeability variations observed at faces of crossbedded sandstone outcrops. In: *Reservoir Characterization—I* (Ed. by L.W. Lake & H.B. Carroll Jr.), pp. 141–175, Academic Press, London, 650 pp.

Van Veen, F.R. (1977) Prediction of permeability trends for water-injection in a channel-type reservoir, Lake Maracaibo, Venezuela, *SPE Paper* **6703**.

Weber, K.J. (1986) How heterogeneities affects oil recovery. In: *Reservoir Characterization—I* (Ed. by L.W. Lake & H.B. Carroll Jr.), pp. 487–544, Academic Press, London, 650 pp.

Weber, K.J., Eijpe, R., Leijnse, D. & Moens, C. (1972) Permeability distribution in a Holocene distributary channel-fill near Leerdam (The Netherlands). *Geologie en Mijnbouw*, **51**, 53–62.

Weber, K.J. & Van Geuns, L.C. (1989) Framework for constructing clastic reservoir simulation models. *SPE Paper* **19582**.

Spec. Publs int. Ass. Sediment. (1993) **15**, 169–180

Knowledge base development for the estimation of reservoir rock properties in the interwell area: examples from the Texas Gulf coast

F.C.J. MIJNSSEN*, N. TYLER† *and* K.J. WEBER**

* *TNO—Institute of Applied Geoscience, PO Box 6012, 2600 JA Delft, The Netherlands;*
† *Bureau of Economic Geology, The University of Texas at Austin, University Station, Box X, Austin, Texas 78713-7508, USA;*
** *Department of Mining and Petroleum Engineering, Delft University of Technology, PO Box 5028, 2600 AG Delft, The Netherlands*

ABSTRACT

In hydrocarbon reservoirs deterministic interpolation of rock properties between wells is hampered by the high spatial variability of these properties. The modelling techniques presently used frequently yield unreliable results because they are only marginally conditioned to geology. This can be attributed, in part, to the lack of quantitative knowledge on geological parameters, and relationships between these parameters.

Quantification of geological variation is possible where the rate of variation may be adequately sampled. Such opportunities are presented by outcrops, densely-drilled fields, and modern environments. Data from outcrops and modern environments give continuous information on distribution of rock properties, but the collection of these data is relatively expensive and time consuming. Data from densely-drilled fields are easier to obtain. However, these data are not continuous.

Three case studies of data collection are presented from densely-drilled fields. The case studies show that sedimentological interpretations are often speculative due to the limited resolution of the logging tools. However, quantification of large-scale phenomena is possible in depositional settings characterized by laterally continuous facies units.

INTRODUCTION

Numerical simulations of fluid flow are used to predict the production behaviour of hydrocarbon reservoirs. These reservoir simulations have three ingredients: fluid flow equations, fluid properties and rock properties. Fluid flow equations are universal. Fluid properties may show variation with depth as a function of pressure; laterally these properties show limited variation. However, rock properties can vary abruptly in space and can therefore often be the main cause of uncertainty in reservoir simulation studies. For this reason it is important to describe this variation as accurately as possible.

In this paper various modelling techniques are analysed and an approach is given for the development of a knowledge base containing quantitative geological information which can aid in the description of reservoir architecture. The potential of densely-drilled fields for the development of such a knowledge base is illustrated with data from three reservoirs in the Texas Gulf coast.

DESCRIPTION OF RESERVOIR ARCHITECTURE

Reservoir data

For the description of reservoir architecture four types of data from the reservoir are available: seismic data, well test data, wire line log data and core data. Well test data and seismic data give three-dimensional information on the interwell area. However, although a number of papers have been published recently on the improvement of these techniques (e.g. Ehlig-Economides & Joseph, 1987; Cramer, 1988), these techniques seldom yield information that is detailed enough to be used in a reliable construction of the reservoir architecture. Well logs and cores give detailed information on the rock properties, but this data is only valid in the near surroundings of the well and cannot usually be easily interpolated into the interwell area.

Modelling techniques

Because reservoir data usually do not provide enough information at the interwell scale, modelling techniques have to be used that incorporate data from the reservoir with geological knowledge. In general terms three groups of modelling techniques can be differentiated: deterministic interpolation, geological process simulation (forward modelling) and geostatistical techniques. This paper is concerned with data preparation for geostatistical techniques.

The most promising results are probably obtained if object-based and sequence-based geostatistical techniques are combined (Dubrule, 1989). The distribution of genetic units could then be modelled with an object-based technique (Damsleth *et al.*, 1990; Gundeso & Egeland, 1990). The resulting model should, however, be conditioned to sequential data, i.e. the chance that a particular genetic unit type is succeeded by another particular genetic unit type. At the moment there are only a few reservoir modelling techniques based on this so-called 'conditioned random modelling' approach. The best known are the models for the calculation of vertical permeabilities in sandy reservoirs containing discontinuous shales. The first model of this kind was presented by Haldorsen & Lake (1984). The position of the shales are randomly defined while the lengths of the shales are determined from shale distribution curves obtained from outcrops as published by Weber (1982). A major problem with conditioned random modelling techniques is a lack of quantitative geological knowledge for the conditioning of the models to geology. For example, the earlier mentioned shale models do not take into account the three-dimensional geometry of the shales because information on length/width ratios of shales is not available. Therefore these methods could be improved by development of a knowledge base containing the necessary quantitative data.

KNOWLEDGE BASE DEVELOPMENT

For the development of a knowledge base containing quantitative geological data, firstly relevant geological relations must be defined. Geological variation on three scales can be differentiated: large scale (variation between genetic units), medium scale (variation within genetic units), and small scale (variation due to textural and compositional features of the rocks). Variation on a large scale is mainly concerned with geometries and internal configuration of sedimentary environments. Variation on an intermediate scale is concerned with geometries and internal configuration of genetic units. Examples are width/thickness ratios of genetic units, and sizes of baffles to flow. Variation on a small scale is concerned with variation in rock properties caused by the texture and composition of the rocks. Relations on this scale incorporated in the knowledge base are mainly concerned with porosity and permeability trends, for example those caused by sedimentary structures.

Quantification of geological variation is possible where the rate of variation may be adequately sampled. Such opportunities are presented by outcrops, recent environments and densely-drilled fields. Outcrops give continuous information on the distribution of rock properties in a cross-sectional view. However, outcrop studies are relatively expensive and time consuming, the orientation of the sections is invariate and information on the areal distribution of the genetic units is limited. Data from recent environments yield continuous information on the areal distribution of rock properties, but for the distribution of rock properties in a cross-sectional view shallow wells have to be drilled (e.g. van Heerden & Roberts, 1988), or high resolution seismic data have to be available (e.g. Corsmit *et al.*, 1987). A major problem in using analogue data from recent environments for reservoir modelling is that many depositional environments are reworked and not preserved in the same geometry, e.g. tidal channels rework barrier islands. Also compactional and diagenetic features cannot be assessed from recent deposits. The gathering of data is again relatively time consuming and expensive. Studies on densely-drilled fields are relatively less expensive and time consuming than outcrop studies or studies on recent

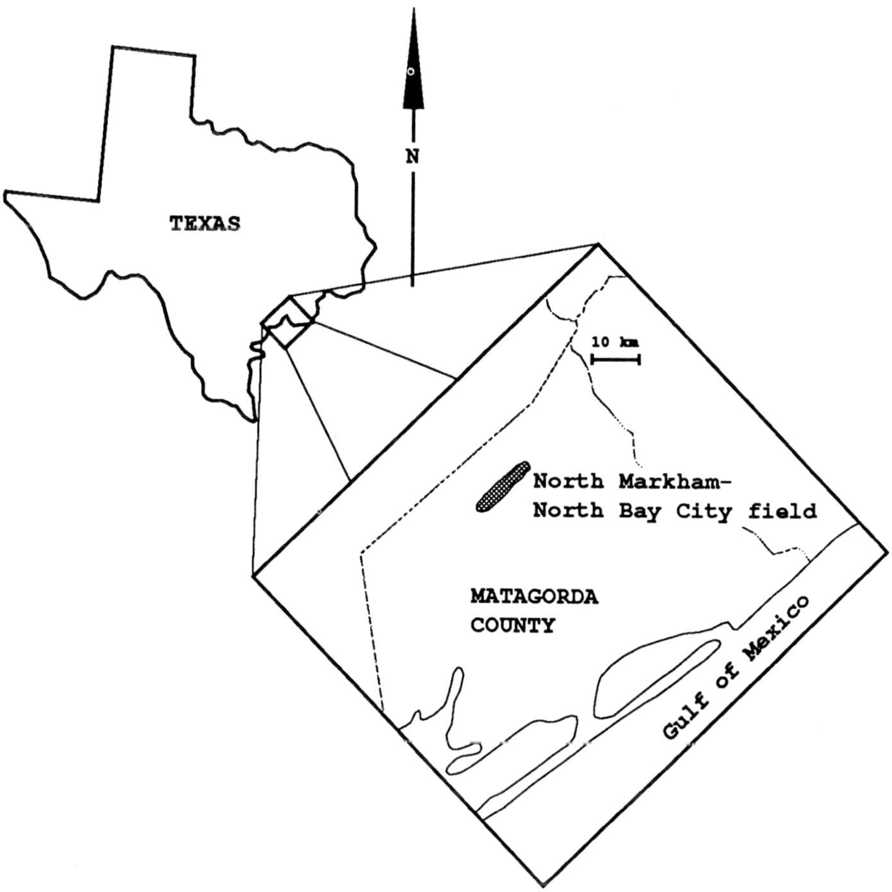

Fig. 1. Location of the North Markham–North Bay City field.

environments. Also, the direction of the sections is less restricted and diagenetic and compactional features can be taken into account. On the other hand, a major problem is the limited continuity of the data. This causes uncertainties in the sedimentological interpretation of the genetic units, as well as in their three-dimensional distribution.

Several examples of the use of outcrop studies for the development of a knowledge base have been published (e.g. Ravenne *et al.*, 1989; Flint *et al.*, 1989). Studies of recent environments or densely-drilled fields for the construction of a knowledge base are not generally available. In this paper an evaluation is given of the possibilities for knowledge-base development based on information from densely-drilled fields. This evaluation is based on three reservoirs from the North Markham–North Bay City field in the Texas Gulf coast.

Methodology

For the extraction of quantitative geological data from the three reservoirs under consideration detailed sedimentological models were constructed based on all available data (wireline logs and production data). These models were represented by a number of closely spaced cross-sections along stratigraphic dip and strike, and a palaeogeographic map.

Dimensions of genetic units were obtained by measuring widths from the palaeogeographic maps, and thicknesses from the cross-sections. Relations between genetic units were obtained by overlaying a

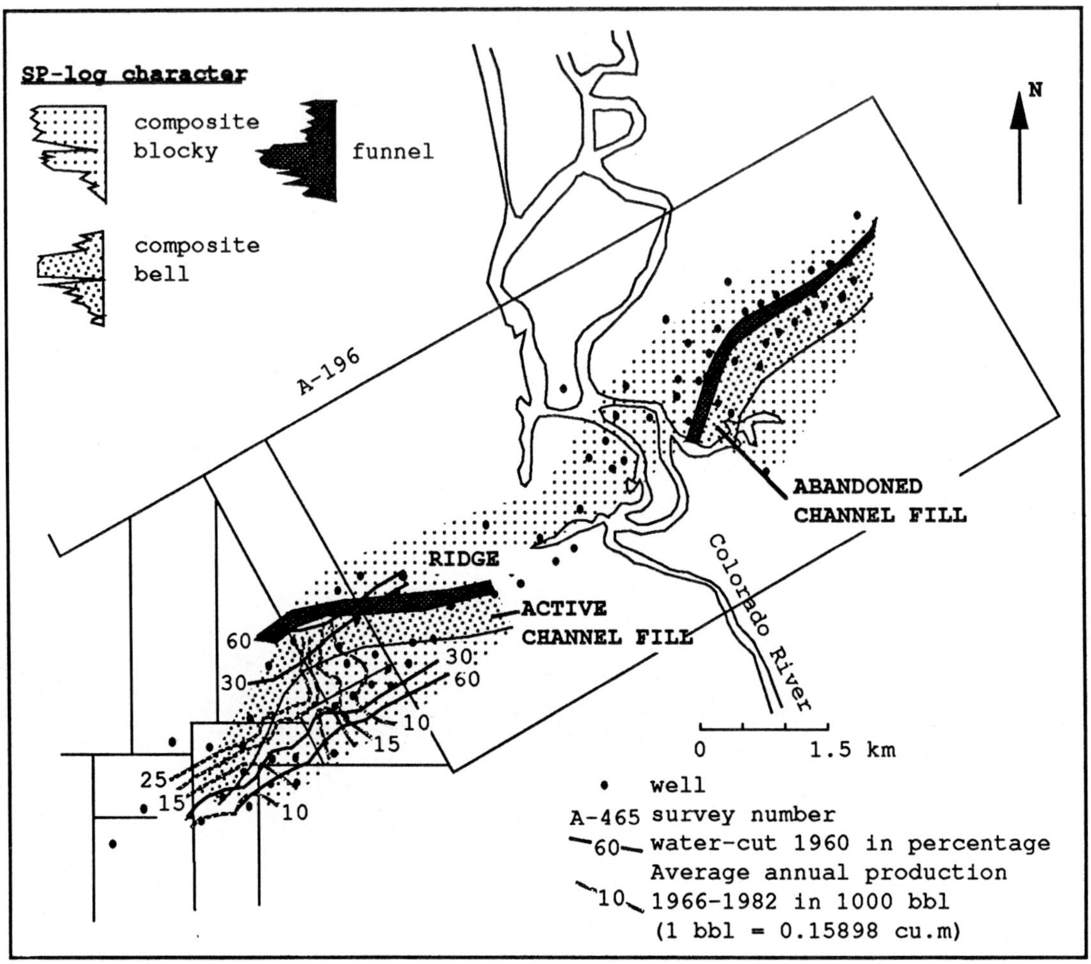

Fig. 2. Palaeogeographic map of transgressed strandplain sandstones of the Carlson reservoir showing log-shape distribution, water-cut data and productivity data (modified from Tyler & Ambrose, 1983).

grid on the cross-sections and palaeogeographic maps. Counting of vertical and horizontal pixel transitions gives information on the mutual distribution of the genetic units.

North Markham–North Bay City field

The North Markham–North Bay City field is located in the eastern part of the Texas Gulf coast, USA (Fig. 1), and it produces oil and gas from multiple stacked shoreline sandstones of the Oligocene Frio Formation (Tyler & Ambrose, 1983). Contemporaneous growth faulting was the dominant structural process during deposition of the North Markham–North Bay City reservoirs. The faulting produced strike-elongated roll-over anticlines, which form the hydrocarbon traps (Tyler & Ambrose, 1983). By 1969 the Railroad Commission of Texas recognized 20 separate hydrocarbon reservoirs in the field, 9 are principally oil reservoirs and the remaining 11 are gas reservoirs. Cumulative oil production from the field by the end of 1982 was 7.1 × 10^6 m^3 and total gas production was 2.8 × 10^6 m^3 (Railroad Commission of Texas, 1983).

The three main oil reservoirs of the field are the Carlson, Cornelius and Cayce reservoirs. Data avail-

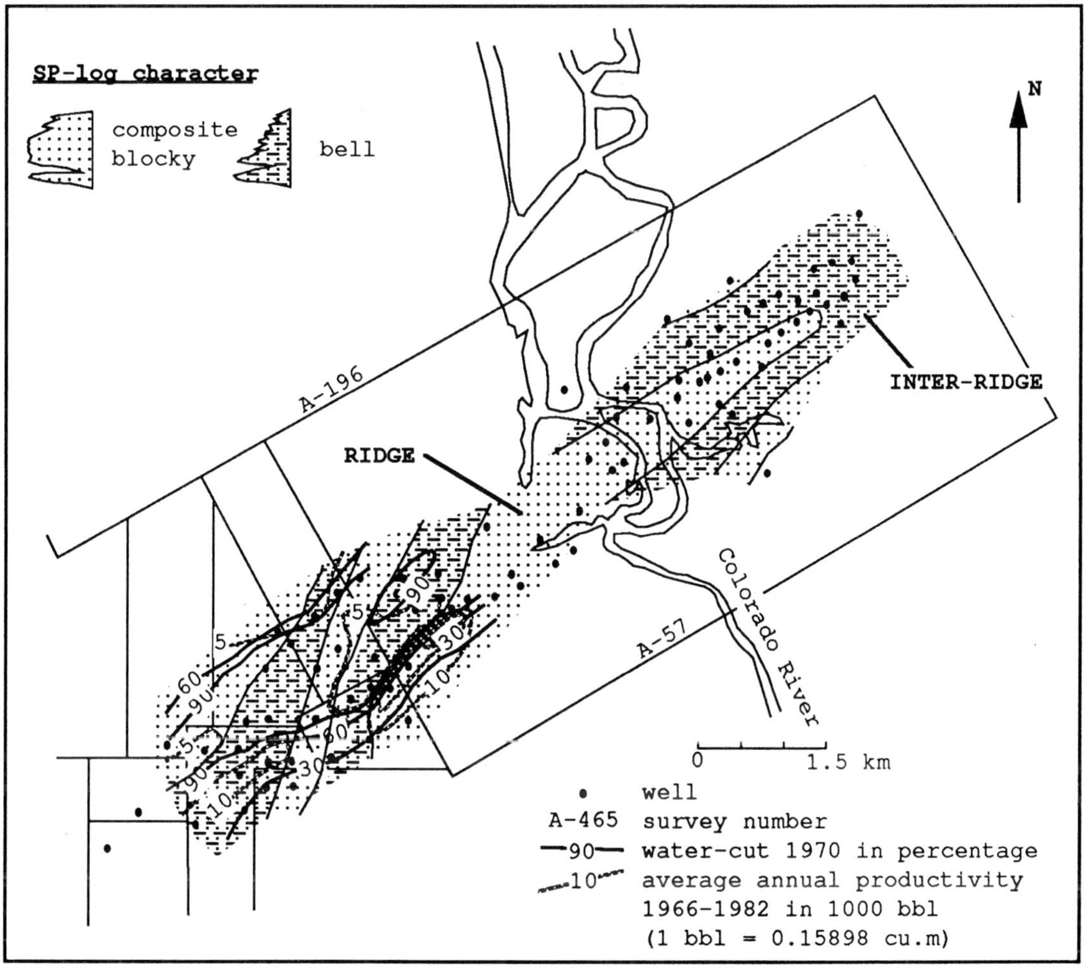

Fig. 3. Palaeogeographic map of mud-rich strandplain deposits of the Cornelius reservoir showing log-shape distribution, water-cut data and productivity data (modified from Tyler & Ambrose, 1983).

able for these reservoirs are spontaneous potential (SP) logs, resistivity logs, water-cut data and well productivity data. Average well spacing in the North Markham–North Bay City field varies between 100 and 350 m and most wells were drilled in the early 1940s. SP- and resistivity logs have limited vertical resolution and only allow speculative sedimentological interpretation. However, based on regional geological information, available production data and log shape mapping, Tyler & Ambrose (1983) were able to interpret the Carlson, Cornelius and Cayce reservoirs as different types of strandplain deposits (Figs 2, 3, 4).

Data collection

For the purpose of knowledge-base development the geological models of Tyler & Ambrose (1983) were quantified. As much as possible of the details incorporated in the original maps of Tyler & Ambrose (Figs 2, 3, 4) were utilized and additional cross-sections were constructed. The grid of sections that was used in this study is shown in Fig. 5. An example of a cross-section is given in Fig. 6. Orientation of the sections is parallel or perpendicular to depositional dip as inferred from the geological models. A grid was overlain on the maps and cross-

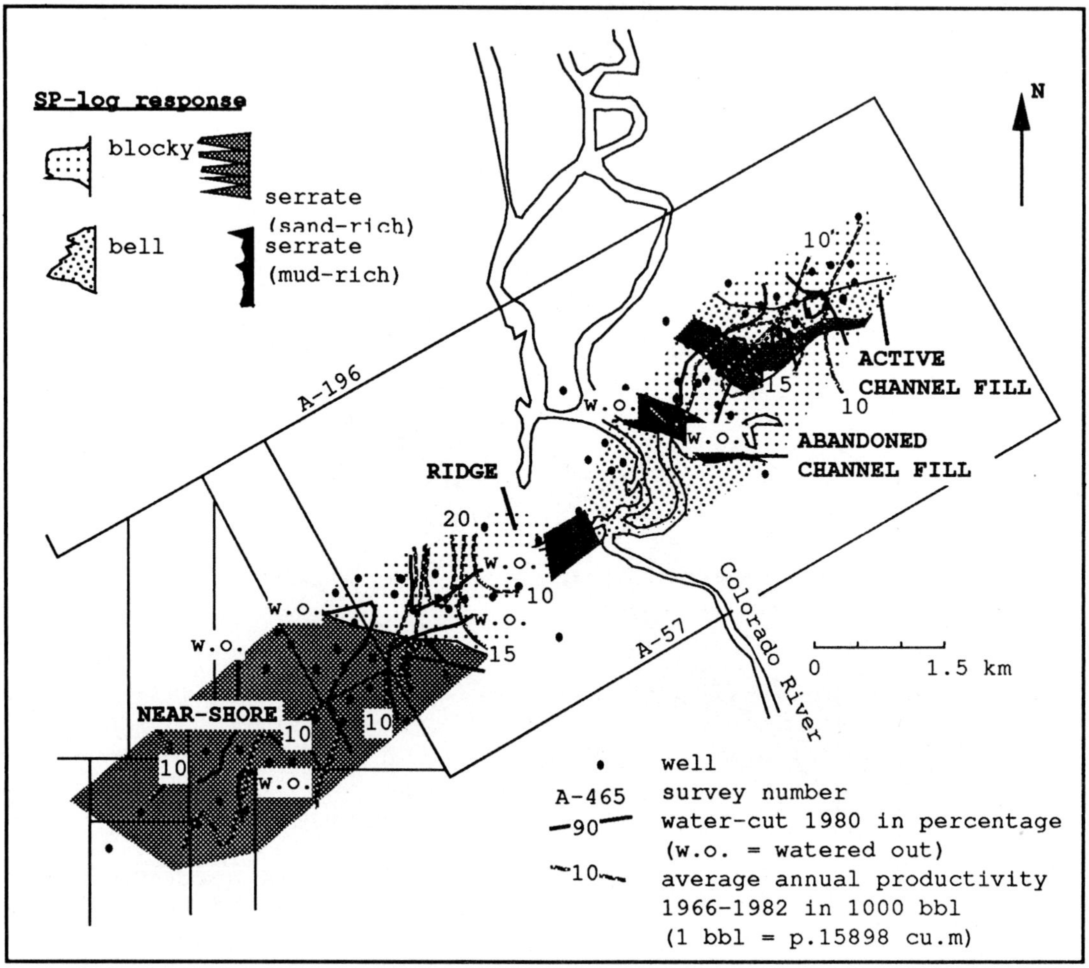

Fig. 4. Palaeogeographic map of sand-rich strandplain sandstones of the Cayce reservoir showing log-shape distribution, water-cut data and productivity data (modified from Tyler & Ambrose, 1983).

sections to obtain quantitative information on upward- and lateral transition probabilities between genetic units and on relative sizes of genetic units (Fig. 6). Lateral and vertical transition probabilities can be used in reservoir modelling to condition the position of genetic units to each other. The relative size of genetic units can be used for conditioning of the relative abundance of various genetic units. In addition to relative sizes and transition probabilities, dimensions of individual genetic units were also determined. These geometrical data can be used to generate realistic dimensions for individual genetic units within a reservoir model.

The results of the quantification are shown in Fig. 7. Results are plotted separately for strike and dip lines to investigate if the results are direction dependent. There is no major difference in upward-transition probability between dip and strike sections in the transgressed sand-rich strandplain sandstones of the Carlson reservoir. Figure 7 shows that the largest variability in upward transition probabilities occurs in the upper part of the sequence. This can be explained by the limited lateral continuity of the genetic units there (Fig. 6). In the Cornelius reservoir there are minor differences in upward transition probabilities between strike and dip lines

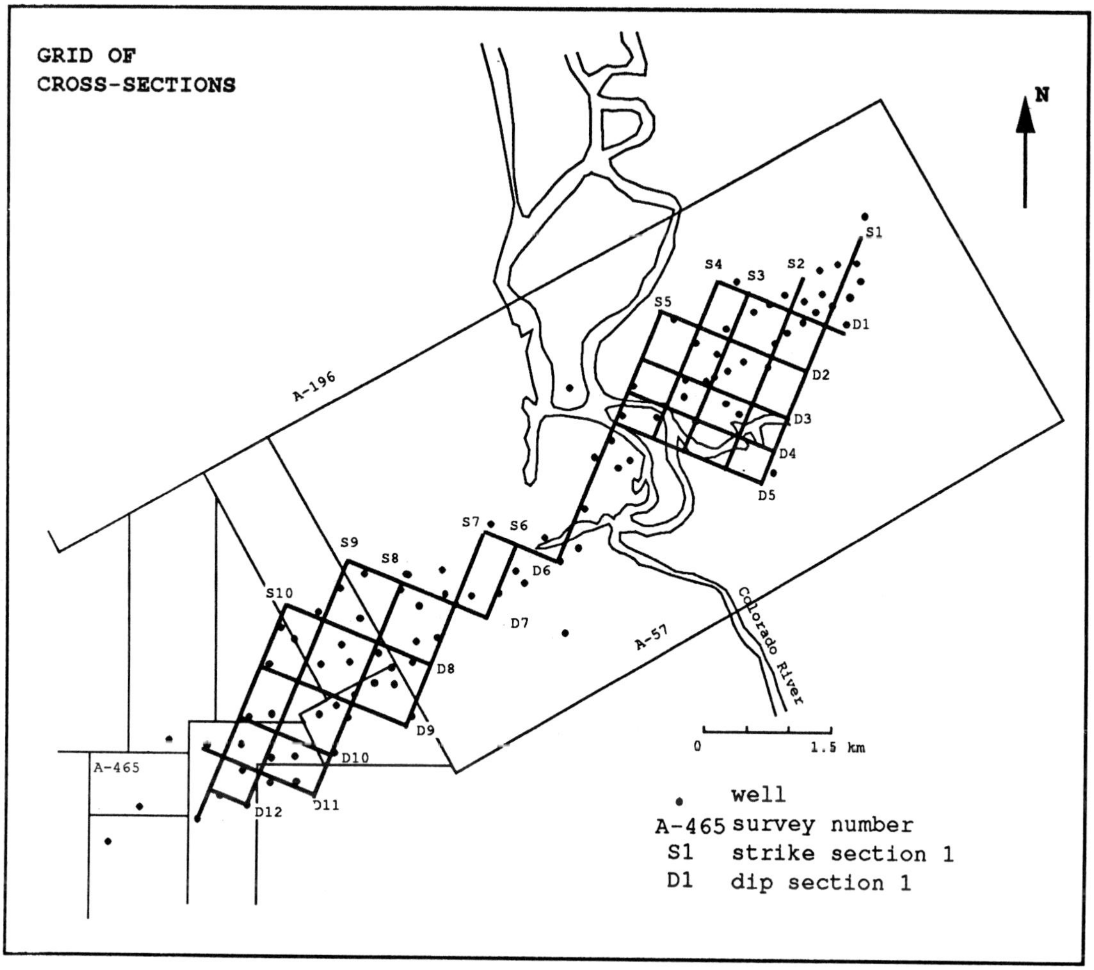

Fig. 5. Base-map of the North Markham–North Bay City field showing grid of sections that was used for the quantification of the sedimentological models of Tyler & Ambrose (1983).

(Fig. 7). The somewhat higher transition probability from foreshore to inter-ridge sediments in the strike sections is caused by the fact that the orientation of the genetic units is more or less parallel to the strike lines, and the strike lines used in this study run predominantly through inter-ridge material. There is also no major difference in upward-transition probability between dip and strike sections in the Cayce reservoir (Fig. 7).

Lateral transition probabilities in the Carlson reservoir show variation between dip and strike lines (Fig. 8). Most remarkable is the higher transition probability between active channel and ridge sediments in the dip sections. This is caused by the orientation of the channel which is more or less parallel to the strike lines. Lateral transition probabilities in the Cornelius reservoir also show some differences between dip and strike sections (Fig. 8). This is partly caused by the limited number of transitions (n = 227) and partly by the orientation of the genetic units within the Cornelius reservoir. For example, the relatively large number of inter-ridge to ridge transitions in the dip sections is caused by the fact that the genetic units are oriented perpendicular to the dip sections. Lateral transition probabilities in the Cayce reservoir show a major

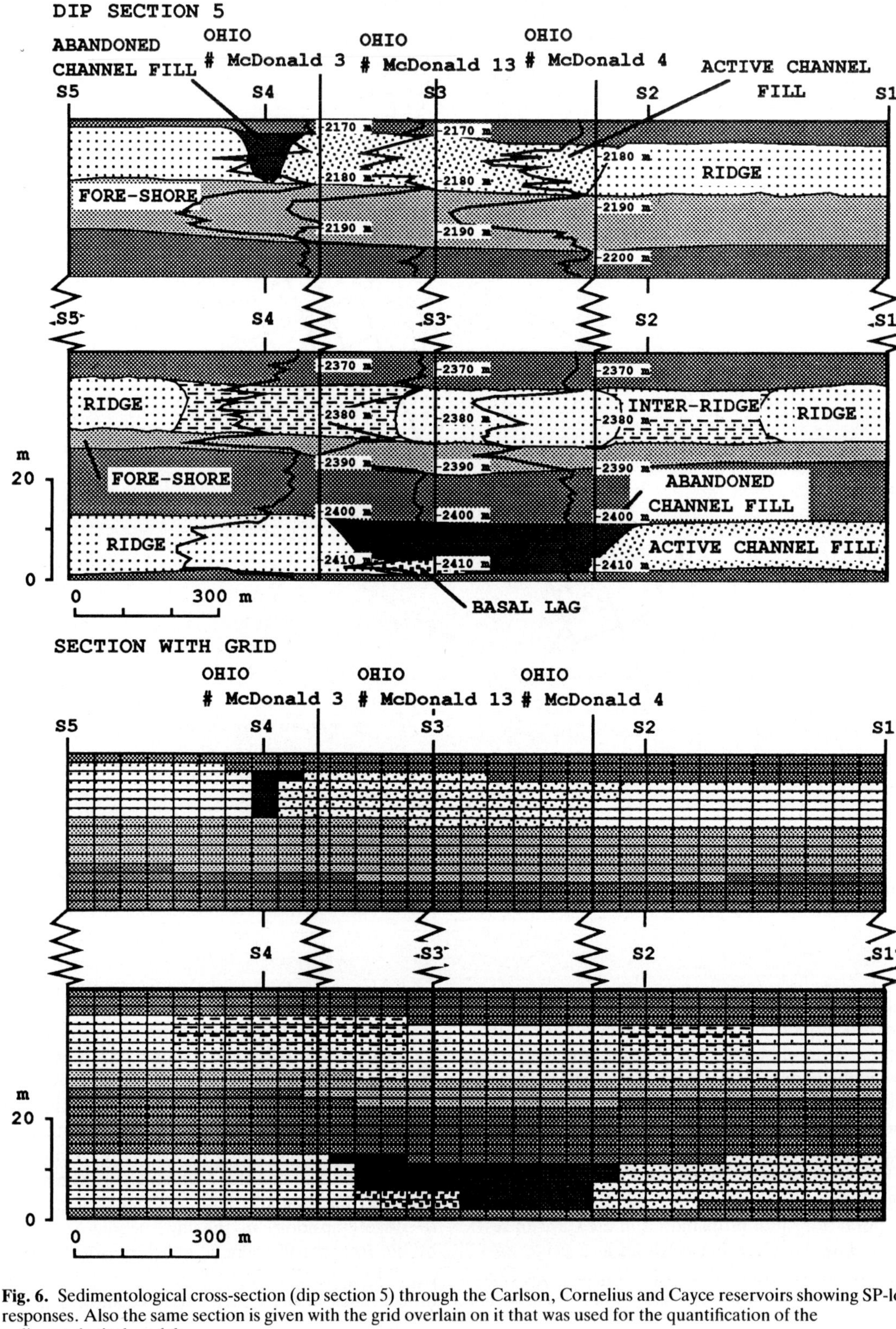

Fig. 6. Sedimentological cross-section (dip section 5) through the Carlson, Cornelius and Cayce reservoirs showing SP-log responses. Also the same section is given with the grid overlain on it that was used for the quantification of the sedimentological models.

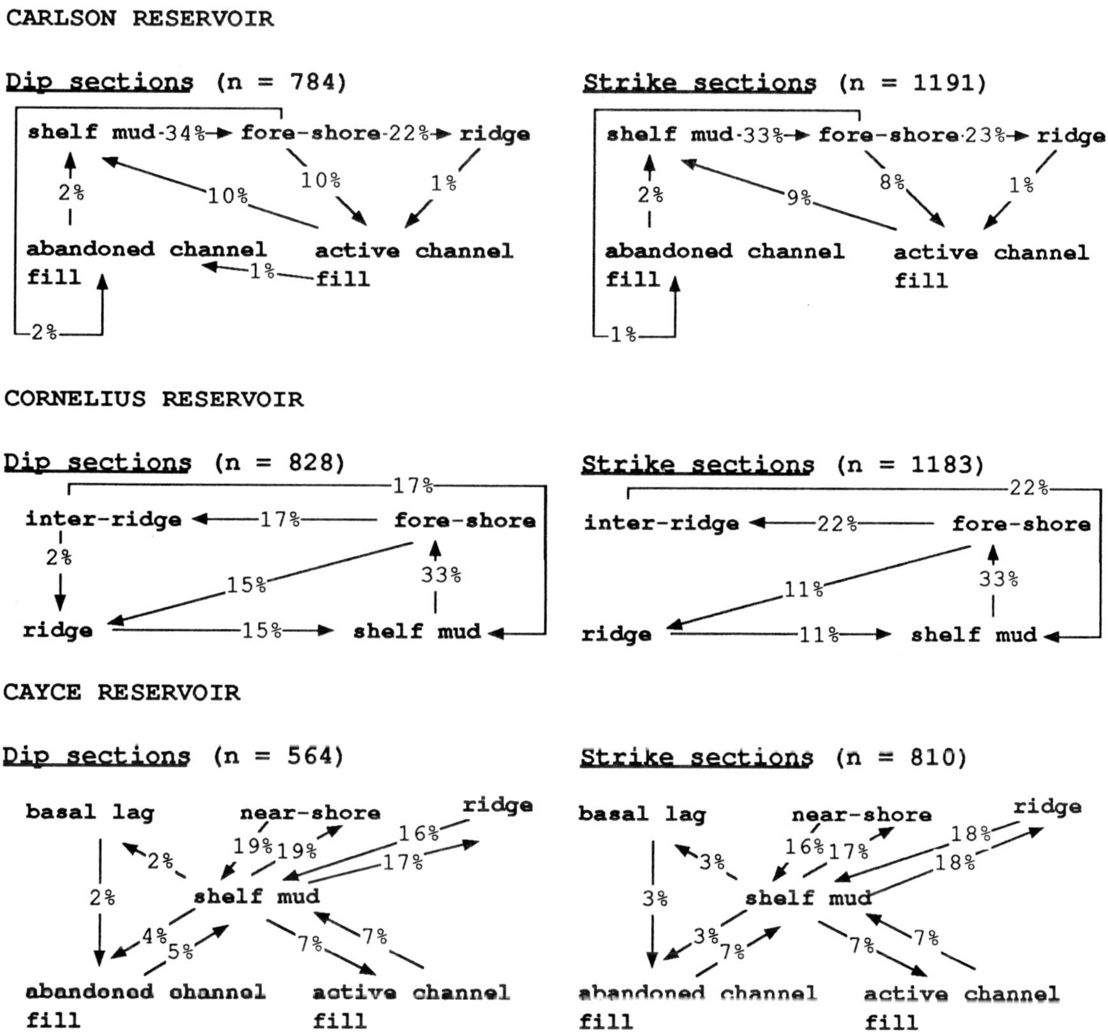

Fig. 7. Upward transition probabilities in the Carlson, Cornelius and Cayce reservoirs.

difference between dip and strike lines (Fig. 8). This is probably caused by the relatively limited number of considered transitions ($n = 133$) coupled with a high variability in sediment types within the Cayce reservoir.

Comparison of relative sizes of genetic units in the Carlson reservoir again shows no major difference between strike and dip lines (Fig. 9). The relatively large amount of foreshore deposits is caused by the occurrence of these deposits everywhere in the reservoir as a basal zone (Fig. 6). From the genetic units in the upper part of the sequence the ridge sediments are most abundant. Comparison of relative sizes of genetic units in the Cornelius reservoir shows that in the dip sections the ridge sediments are relatively more abundant than in the strike sections (Fig. 9). This is explained by the orientation of the genetic units parallel to the strike sections, and because the strike sections run predominantly through inter-ridge sediments. Comparison of relative sizes of genetic units in the Cayce reservoir shows again no major difference between strike and dip lines (Fig. 9).

Relative sizes of the genetic units in the horizontal

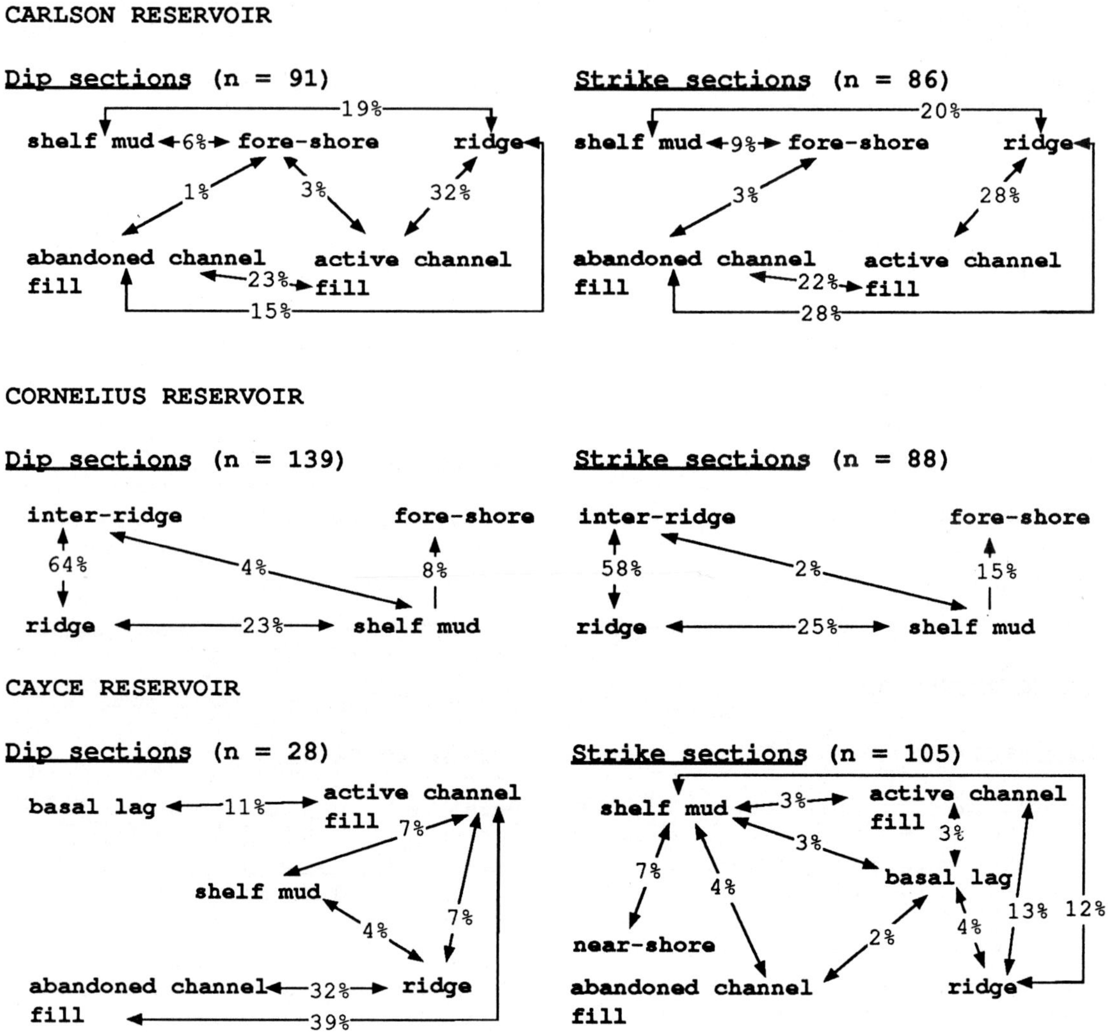

Fig. 8. Lateral transition probabilities in the Carlson, Cornelius and Cayce reservoirs.

direction were determined by overlying a grid on the palaeogeographic maps. The results are shown in Fig. 9. The majority of the area in the Carlson reservoir is occupied by ridge sediments. Abandoned channel-fill sediments occur about one-third as frequently as active channel-fill sediments. In the Cornelius reservoir inter-ridge and ridge sediments occur in more or less the same amounts, and in the Cayce reservoir strandplain and near-shore sediments occur in more or less the same amounts. The significance of the data on relative sizes of genetic units both vertically and laterally as presented in Fig. 9 is probably of limited use because the study area is relatively small in relation to the size of individual genetic units and is defined geologically in an arbitrary way by the field boundaries.

Width and thickness were measured for individual genetic units within the three reservoirs under study. Width is defined as the width perpendicular to the orientation of the genetic unit under consideration (as inferred from the geological model). Thickness is defined as the maximum thickness of the genetic unit at the location of the width measurement. In the Carlson and Cayce reservoirs channel dimen-

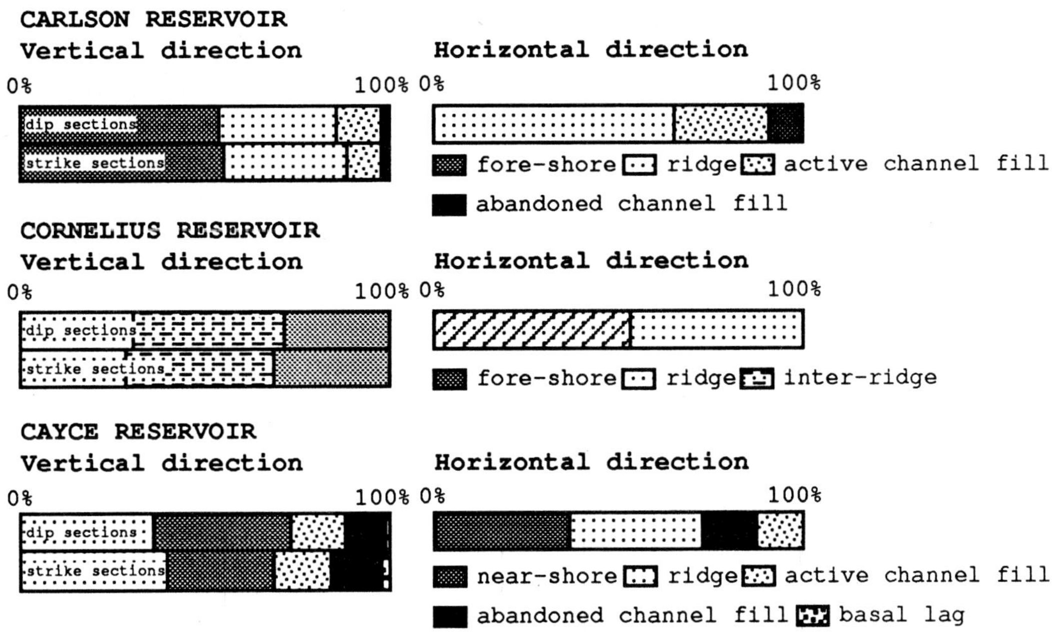

Fig. 9. Relative sizes of genetic units in the Carlson, Cornelius and Cayce reservoirs.

Table 1. Channel dimensions in the Carlson and Cayce reservoirs

	Average width (m)	Average thickness (m)	Width/ thickness ratio
Carlson reservoir			
Total channel fill	633	9	70
Active channel fill	516	9	57
Abandoned channel fill	117	9	13
Cayce reservoir			
Total channel fill	882	12	69
Active channel fill	373	12	31
Abandoned channel fill	366	12	28

Table 2. Ridge dimensions in the Cornelius reservoir

Average width (m)	Average thickness (m)	Width/thickness ratio
390	14	28
438	12	37
323	11	29

sions were measured (Table 1). In the Cornelius reservoir ridge dimensions were measured (Table 2). A remark must be made with respect to the dimensions of the channels measured in the Carlson and Cayce reservoirs. MacPherson & Miller (1990) reported that in the Bell Ridge Field, California (USA), individual channels were not correlatable, despite a 10 acre well spacing. The data on channel dimensions from the Carlson and Cayce reservoirs might therefore reflect composite rather than single sediment bodies. Also, dimensions could be measured for only a small number of genetic units because the area under study is small in relation to the sizes of individual genetic units. Additional data from other studies in similar sedimentological settings are required to obtain a statistically significant data set.

CONCLUSIONS

From three densely-drilled reservoirs in the Texas Gulf coast (USA) quantitative sedimentological parameters are derived to demonstrate the use of densely-drilled fields as analogues for the conditioning of reservoir models. If the sedimentological setting is characterized by laterally continuous sediment bodies (e.g. strandplain environment), data from densely-drilled fields have significant use in addition to data from outcrops and recent environments. Data from densely-drilled fields give three-dimensional information on the distribution of genetic units, while outcrops usually give only two-dimensional information. Recent environments cannot easily be sampled vertically and do not give information on the effects of burial.

ACKNOWLEDGEMENTS

The work presented in this paper forms part of the PhD research of F.C.J. Mijnssen carried out at the TNO Institute of Applied Geoscience. The authors thank three anonymous reviewers for their constructive comments on the original manuscript.

REFERENCES

Corsmit, J., Versteeg, W.H., Brouwer, J.H. & Helbig, K. (1987) High resolution 3D reflection seismics on a tidal flat: acquisition, processing and interpretation. *First Break*, **6**, 9–23.

Cramer, P.W. (1988) Reservoir development using offset VSP techniques in the Denver-Julesberg basin. *J. Petrol. Technol.*, **40**, 197–205.

Damsleth, E., Tjolsen, C.B., Omre, K.H. & Haldorsen, H.H. (1990) A two-stage stochastic model applied to a North Sea reservoir. *SPE Paper* **20605**, 12 pp.

Dubrule, O. (1989) A review of stochastic models for petroleum reservoirs. In: *Geostatistics, Volume 2* (Ed. by M. Amstrong), pp. 493–506. Kluwer Academic, Dordrecht.

Ehlig-Economides, C.A. & Joseph, J. (1987) A new test for determination of individual layer properties in a multi-layered model. *SPE Formation Evaluation*, **3**, 261–283.

Flint, S.S., van Rossem, S.J. & Williams, H. (1989) Quantitative outcrop analysis of coastal plain sequences from eastern Kentucky: Example of high quality data input to three-dimensional reservoir modelling. *Bull. Am. Assoc. Petrol. Geol.*, **73**, 356. (Abstract.)

Gundeso, R. & Egeland, O. (1990) SESIMIRA—A new geological tool for 3D modelling of heterogeneous reservoirs. In: *North Sea Oil and Gas Reservoirs*—II (Ed. By A.T. Buller, O. Hjelmeland, J. Kleppe, O. Torsaeter & J.O. Aasen), pp. 229–282. Graham and Trotman, London.

Haldorsen, H.H. & Lake, L.W. (1984) A new approach to shale management in field-scale models. *SPE Journal*, **24**, 447–457.

MacPherson, J.G. & Miller, D.D. (1990) A subsurface study of fluvial sandbody geometries and connectedness. In: *13th International Sedimentological Congress*: Abstracts Papers, pp. 349–350. International Association of Sedimentologists, Utrecht.

Railroad Commission of Texas (1983) *Annual Report of the Oil and Gas Division for 1982*. Railroad Commission of Texas, Austin, 733 pp.

Ravenne, C., Eschard, R., Galli, A., Mathieu, Y., Montadert, L. & Rudkiewicz, J.L. (1989) Heterogeneities and geometry of sedimentary bodies in a fluvio-deltaic reservoir. *SPE Formation Evaluation*, **5**, 239–246.

Tyler, N. & Ambrose, W.A. (1983) Facies architecture and production characteristics of strandplain reservoirs in the Frio Formation, Texas. *Bureau of Economic Geology Report of Investigations No. 146*. The University of Texas at Austin, Austin, 42 pp.

Van Heerden, I.L. & Roberts, H.H. (1988) Facies development of Atchafalaya delta, Louisiana: a modern bay-head delta. *Bull. Am. Assoc. Petrol. Geol.*, **72**, 439–453.

Weber, K.J. (1982) Influence of common sedimentary structures on fluid flow in reservoir models. *J. Petrol. Technol.*, **34**, 665–672.

Weber, K.J. & van Geuns, L.C. (1990) Framework for constructing clastic reservoir simulation models. *J. Petrol. Technol.*, **42**, 1248–1253.

Spec. Publs int. Ass. Sediment. (1993) **15**, 181–188

The use of 3-D seismic in reservoir geological modelling

K.J. WEBER

Shell Internationale Petroleum Maatschappij B.V.,
The Hague, The Netherlands

ABSTRACT

High-resolution 3-D seismic has become a key tool in hydrocarbon field appraisal and development planning. So far the emphasis has been on structural analysis. However, in many cases sedimentary features can also be detected. Large-scale features have already been described for some time using 2-D seismic data and seismic stratigraphy is widely applied. The use of 3-D seismic has further refined techniques but has also resulted, in some cases, in a remarkable delineation of individual sandbodies and of reservoir architecture. The major reason for this is the improved resolution resulting from the increased subsurface sampling provided by a closely spaced 3-D grid and the advantages of 3-D migration. In addition, detailed analysis at reservoir level is greatly assisted by the introduction of powerful, user-friendly, interactive interpretation workstations.

Even with high-quality 3-D seismic it is in general still very difficult to resolve internal reservoir configuration because of insufficient acoustic contrast or insufficient thickness of the individual beds. Favourable characteristics for seismic delineation are reservoirs consisting of thick, relatively homogeneous sandstone or carbonate rock bodies, separated by thick shales or other rocks with a strongly contrasting acoustic impedance. The presence of gas in porous reservoir bodies can significantly enhance acoustic impedance contrasts. Morphological features that can lead to the delineation of reservoir bodies are irregular lower or upper surfaces and sharp lateral boundaries.

INTRODUCTION

Large-scale sedimentary features can often be detected by conventional 2-D seismic, especially in coastal and marine settings (e.g. Berg, 1982). Carbonate reefs are also frequently observed, and seismic sections sometimes even show a good definition of marginal reef facies and internal bedding (Fontaine *et al.*, 1987). Deltaic wedges can be outlined in sections perpendicular to coastlines, and global correlations of sea level changes have been inferred from seismically traced transgressive and regressive units (Vail *et al.*, 1977).

In this paper, the contributions of the 3-D seismic method and modern interactive interpretation workstations to the delineation of clastic sedimentary units will be discussed. It will be shown that the enhanced horizontal and vertical resolution provided by closely spaced 3-D grids and 3-D migration can sometimes enable individual sandbodies to be imaged.

Resolution is a function of the frequency content of the reflected signal, the acoustic impedance contrast across the reflection interface, sandbody thickness and the separation between the sandbodies (Kallweit & Wood, 1982).

Large-scale sedimentary bodies such as turbidite fans can be observed in outline on moderate-quality 2-D seismic lines, but high resolution seismic is required to detect and interpret features such as individual turbidite lobes or distributary channels. In practice, this means that we can only image specific types of turbidites, shallow marine sand bars and fluvial deposits such as distributary channels and point bars embedded in clay.

Modern interactive workstations provide a quick and flexible means of interpreting the much larger volumes of data inherent in the 3-D method. One can make use of the reflection seismic profiles or of profiles integrated to acoustic impedance (Neidell,

1985). Workstations can be used to map amplitude variations along successive levels in the reservoir sequence (e.g. Ruijtenberg *et al.*, 1990), which is called the horizon seiscrop method (Brown, 1986). Amplitude variations may then be associated with variations in reservoir, lithology, mineralogy, porosity or fluid content. This paper is intended as a brief guide to the detection of sedimentary features on seismic data.

LARGE-SCALE SEDIMENTARY FEATURES

There are numerous publications describing large-scale sedimentary features based on seismic data. The practice of unravelling a sedimentary sequence in terms of general environment of deposition and sea level changes is called 'seismic stratigraphy'. The subject is dealt with extensively in the well-known AAPG Memoir 26 (Payton, 1977).

For a sandbody to be delineated by seismic, it must be thick enough in relation to the signal frequency and produce a sufficiently large acoustic contrast with its surroundings. This usually means that the sandbody must have a minimum thickness of about 15–25 m and be enveloped by shale in order to be imaged individually.

Without special techniques, only the larger sandbodies can be seen. Most prominent are turbidite fans because they have an undulating upper surface and are often covered and underlain by clay. A good example is provided by 2-D seismic from the North Sea Frigg Field (Skaug & Gundesø, 1986), in which only the top of the fan can be mapped with some confidence (Fig. 1). A more detailed delineation of both the upper surface and an internal sand-prone distributary system was achieved for the Eocene

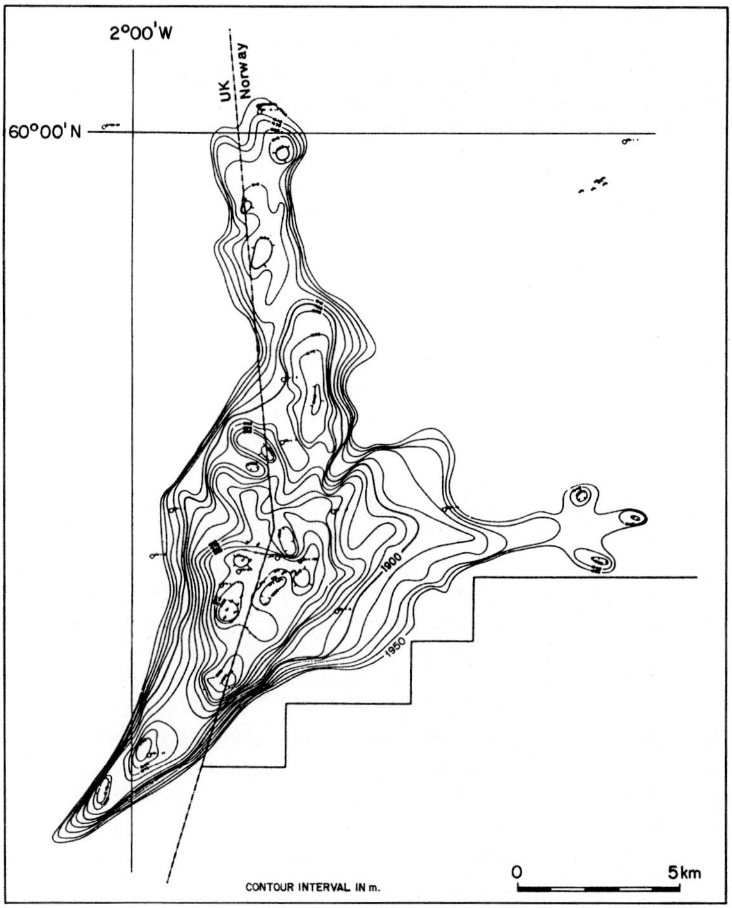

Fig. 1. Top surface of a turbidite fan from 2-D seismic, Frigg Field, Norwegian North Sea (from Skaug & Gundesø, 1986).

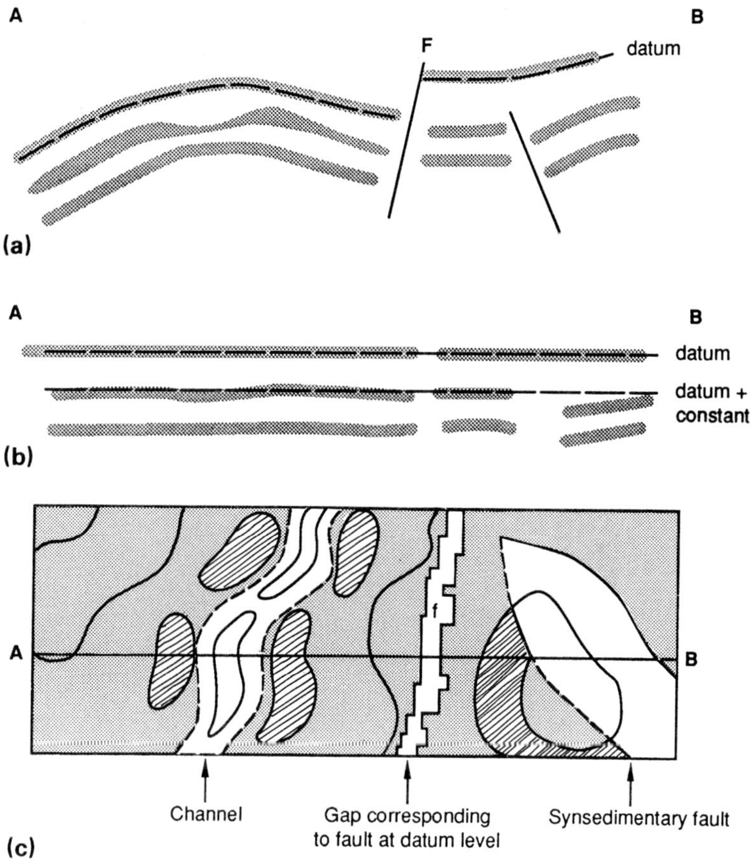

Fig. 2. The basis of horizon amplitude analysis (after Ruijtenberg *et al.*, 1990). (a) Schematic seismic cross-section with datum horizon. Some irregularities in thickness distribution are apparent in the underlying interval. (b) The datum horizon is flattened and the amplitude of a specific time interval underneath is extracted to analyse stratigraphic and lithologic information. (c) The mapped amplitude data show a channel fill and various recognizable structural elements.

Clontarf submarine fan in the Porcupine Basin off the SW coast of Ireland (Macurda & Nelson, 1989). Typical for this facies are the hummocky clinoforms within the sequence. Although the study was based on a 2-D survey consisting of only 12 seismic lines, modern workstation techniques made it a simple task to construct isochron maps of the whole fan and of a number of sequences representing stages of fan development. In the basinward direction, the fan sediments are progressively more readily differentiated from surrounding basinal sediments. In consequence, even relatively thin turbidite-filled troughs along the main fan development could be distinguished.

In upper fan sequences, with large single channels, it is possible in many cases to map the channels. In the middle and lower fan environments it may be possible to separate individual sand lobes if the shale intercalations are sufficiently thick (Harris *et al.*, 1982). Particularly detailed seismic images have been obtained over recent turbidite fan complexes which have been buried to shallow depths, such as the Rhône deep-sea fan (Droz & Bellaiche, 1985).

Large shallow marine sand bars like the Draugen field reservoir, offshore Norway, can also be detected by 2-D seismic (Ellenor & Mozetic, 1986). These sand bars have a flat lower surface on top of marine shales. The grain size coarsens upwards and a pronounced acoustic contrast is present along the crest by the abrupt transition from clean sand to marine shale. The convex upper surface of the bars can be observed quite clearly as a result.

RESERVOIR ARCHITECTURE FROM SEISMIC

The true value of 3-D seismic for reservoir geological modelling is demonstrated by the delineation

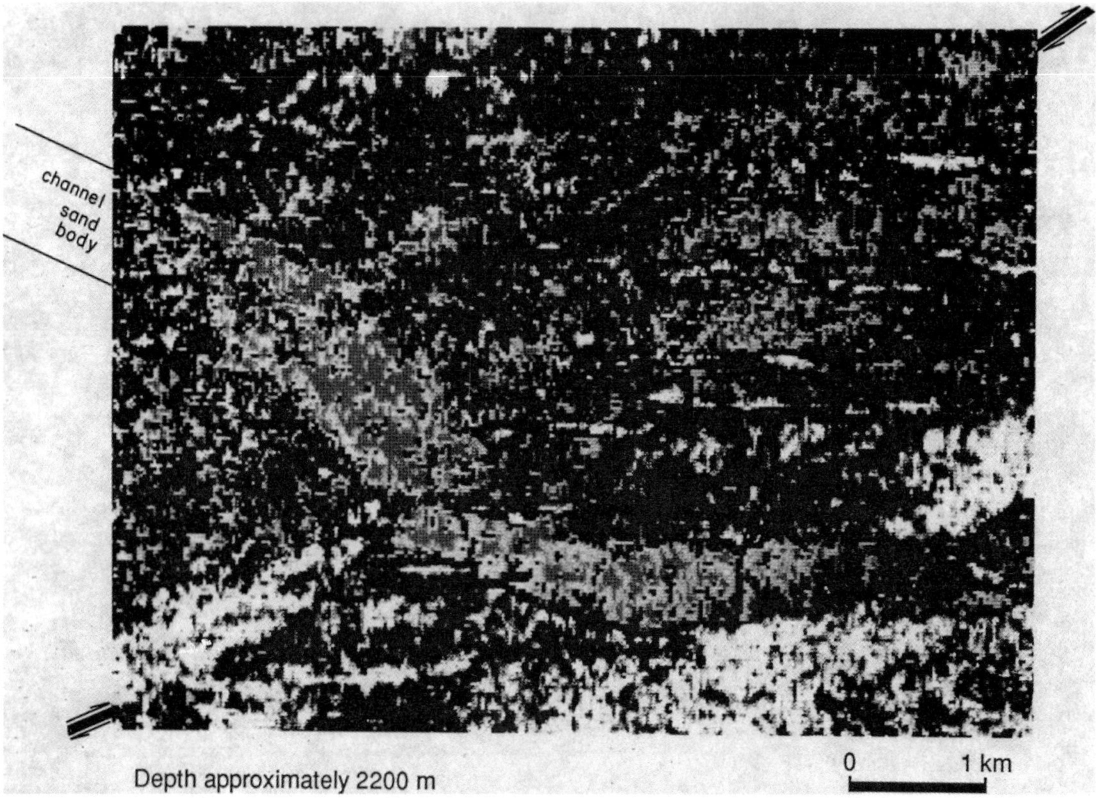

Fig. 3. Amplitude map showing a channel sandbody offset by a strike-slip fault (after Ruijtenberg *et al.*, 1990).

of small-scale features within reservoirs. With 2-D seismic it is usually very difficult to resolve individual sandbodies with thicknesses of less than 20–30 m. However, various 3-D mapping procedures can help to reveal details of reservoir architecture that are at the limits of seismic resolution (Neff, 1990a, b).

Modern interactive interpretation workstations provide automatic volume tracking facilities which can be used to efficiently map seismic horizons throughout an entire 3-D data set. Amplitude maps can then be produced for any tracked horizon such as the cap rock/reservoir interface (Ruijtenberg *et al.*, 1990). The amplitudes can be mapped either directly on the reflection seismic data or on acoustic impedance traces (Fig. 2). The strength of the seismic reflection can be enhanced if one sandbody contains gas instead of water. Results are also strongly influenced by structural deformations.

The great filtering capability of the human eye is exploited by modern workstations. Seismic amplitudes are displayed in colour using suitable colour schemes. Subtle differences in displayed colours corresponding to sandbodies may then be picked up clearly by the eye. Figure 3 demonstrates this principle in black and white, where a sinuous distributary channel sandbody is clearly outlined because it is surrounded by clay overbank deposits. An interesting feature is the small offset of the channel along a post-depositional strike-slip fault.

On the basis of a sedimentological review of clastic depositional environments one can predict which types of sandbodies should be detectable and in what settings. Obviously rather sharp transitions from sand to shale are required and individual sandbodies should be enclosed by an envelope of shales. In addition, both sands and shales should be sufficiently thick in relation to the seismic frequency and acoustic velocity. The distinction between sandstone and shale in seismic is strongly dependent on

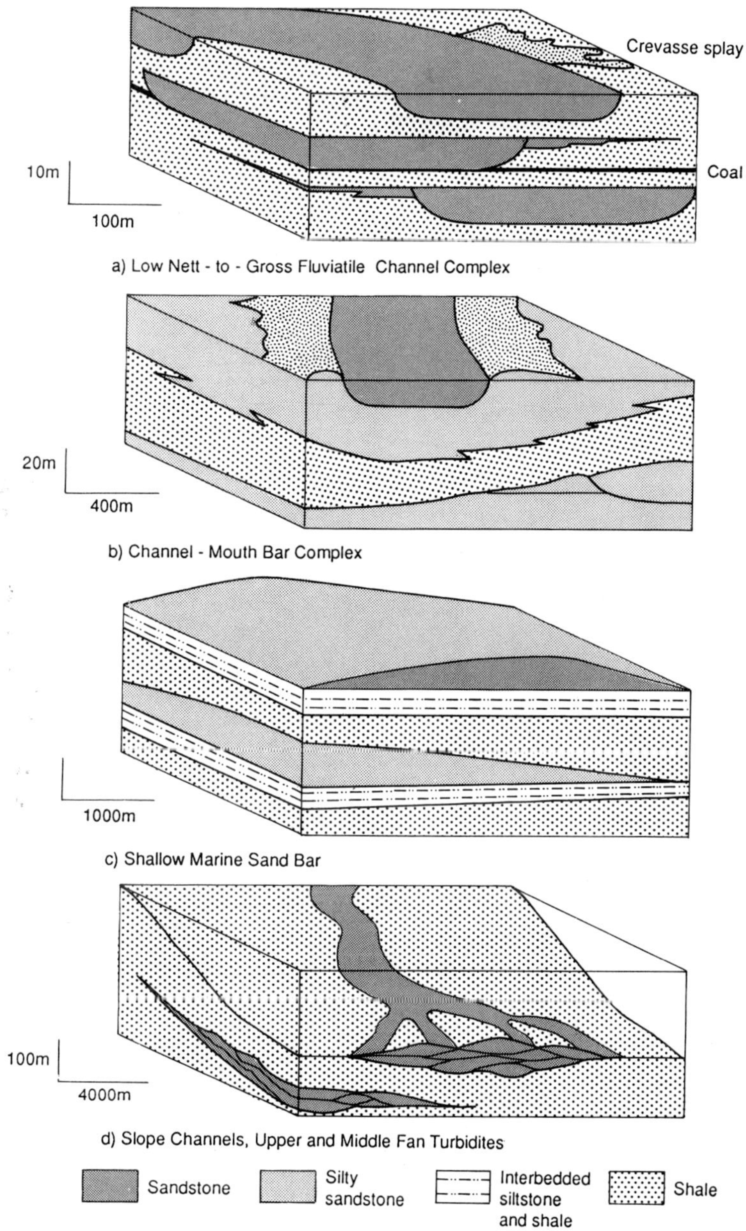

Fig. 4. Sandbodies that may be delineated with high quality 3-D seismic.

the type of shale. Flood-plain silty sands have a much lower acoustic contrast with sandstones than organic-rich lacustrine or marine shales. Such circumstances are present in the cases shown in Fig. 4. Within the large sandbodies mentioned in the previous section one may often be able to map some of the constituent elements. On a smaller scale it is mainly distributary channels and point bars in low sand/shale ratio reservoirs that may be delineated (Ruijtenberg *et al.*, 1990).

Figure 5 schematically illustrates how the method of plotting the amplitudes below a stratigraphic datum level could be used to map the amplitude structure of turbidite distributary channels, from

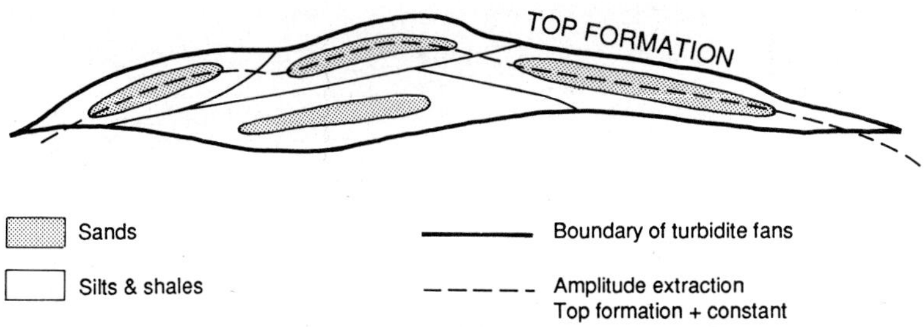

a) Cross-section showing the seiscrop horizon

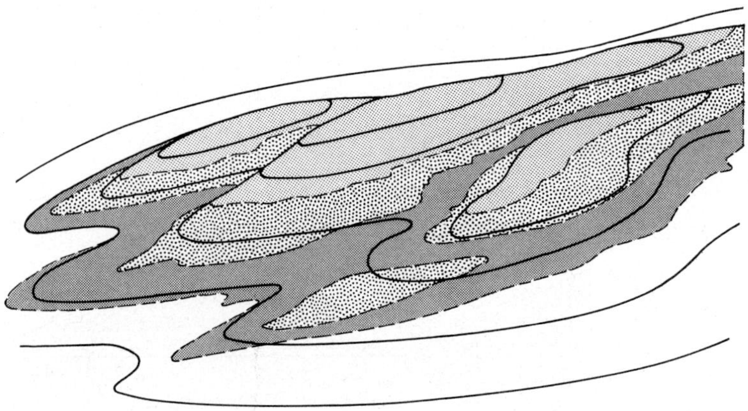

b) 3-D perspective display of amplitude and structure

0 1 km

Contour spacing 20 m

Fig. 5. Mapping and sandbody of a turbidite fan using the seiscrop method. Especially along the upper zone of the fan, sandy lobes can be distinguished from the much less sandy inter-lobe areas. Because of differential compaction the amplitude contrasts are more or less conformable with the structural contours along the top of a tectonically underformed fan.

which reservoir quality may be inferred. In a similar manner, the rough shapes and trends of lacustrine mouth bars have been mapped (Flint *et al.*, 1988). A technique to predict the detectability of sandbodies or shales is forward modelling of the seismic response. For a given seismic wavelet and frequency range one can model the seismic response of a reservoir sequence with a known acoustic impedance distribution. Examples are shown in the AAPG Memoir 26 (Payton, 1977) and numerous other papers. The use of this method is mostly restricted to 2-D sections, but 3-D models can also be made. Special circumstances can sometimes allow the delineation of sandbodies. For example, it is not uncommon to have distributary channels or point bars erode away part of an extensive coal layer. Although the sands may be too thin to be picked up individually, the absence of a strong coal bed reflection shows their position. Coals may also be present along the upper part of abandoned channel fills and oxbow lake fills. In that case the channels themselves show up on amplitude maps. The presence of gas in high-porosity sandbodies is a fairly common reason for increasing the impedance contrast between a sandbody and its surrounding sediments. (Fig. 6).

CONCLUSIONS

Macroscopic sedimentological features can be identified very well on 2-D and 3-D seismic. However, the delineation of individual sandbodies is often difficult because of lack of sufficient acoustic contrast and inadequate frequency content to resolve the sandbody thickness. Thus, special attention to

Fig. 6. A horizon seiscrop showing amplitude distributions. A channel shaped feature is present. The structurally highest part of the channel above a time level of 805 msec has a brighter amplitude, which is interpreted to be gas filled (from Brown, 1986).

high-frequency conservation in processing is crucial if sandbody analysis is contemplated.

A number of clastic sedimentological configurations are favourable for seismic detection particularly via horizon tracking of 3-D data sets. These are characterized by lens-shaped sandbodies and an envelope of shales. Certain turbidites, shallow marine sand bars, mouth bars, some coastal barrier bars, distributary channel fills and point bars can fit these requirements.

Although the scope for sandbody delineation is thus somewhat limited, it is nevertheless an important technique, especially in low sand/shale ratio reservoirs, where the location and trend of the sandbodies is of crucial importance in field development planning.

REFERENCES

BERG, O.R. (1982) Seismic detection and evaluation of delta and turbidite sequences; their application to exploration for the subtle trap. *Bull. Am. Assoc. Petrol. Geol.*, **66**, 1271–1288.

BROWN, A.R. (1986) Interpretation of three-dimensional seismic data. *Am. Assoc. Petrol. Geol. Memoir*, **42**, 194.

DROZ, L. & BELLAICHE, G. (1985) Rhône deep-sea fan; morphostructure and growth pattern. *Bull. Am. Assoc. Petrol. Geol.*, **69**, 460–479.

ELLENOR, D.W. & MOZETIC, A. (1986) The Draugen oil discovery. In: *Habitat of Hydrocarbons on the Norwegian Continental Shelf* (Ed. by A.M. Spencer *et al.*), pp. 313–316. Graham and Trotman, London.

FLINT, S., STEWART, D.J., HYDE, T., GEVERS, E.C.A., DUBRULE, O.R.F. & VAN RIESSEN, E.D. (1988) Aspects of reservoir geology and production behaviour of Sirikit oil field; an integrated study using well and 3-D seismic data. *Bull. Am. Assoc. Petrol. Geol.*, **72**, 1254–1269.

FONTAINE, J.M., CUSSEY, R., LACAZE, J., LANAUD, R. & YAPAUDJAN, L. (1987) Seismic interpretation of carbonate depositional environments. *Bull. Am. Assoc. Petrol. Geol.*, **71**, 281–297.

HARRIS, D.G., SARG, J.F., BRANISA, F. & VAIL, P.R. (1982) Trace inversion for interpreting reservoir continuity and lithology. *SPE Paper* **10013**.

KALLWEIT, R.S. & WOOD, L.C. (1982) The limits of resolution of zero-phase wavelets. *Geophysics*, **47**, 1035–1046.

MACURDA, D.B. JR. & NELSON, H.R. JR. (1989) Interactive interpretation of a submarine fan, offshore Ireland: a case history. In: *Seismic Interpretation Series, 1, Soc. Expl. Geophys.*, 59–65.

NEFF, D.B. (1990a) Incremental pay thickness modelling of hydrocarbon reservoirs. *Geophysics*, **55**, 556–566.

NEFF, D.B. (1990b) Estimated pay mapping using three-dimensional seismic and incremental pay thickness modelling. *Geophysics*, **55**, 567–575.

NEIDELL, N.S. (1985) Stratigraphic modelling and interpretation; geophysical principles and techniques. *Education course note series 13*, AAPG, Tulsa.

PAYTON, C.E. (ed.) (1977) Seismic stratigraphy—applications to hydrocarbon exploration. *AAPG Memoir*, **26**, 516 pp.

RUIJTENBERG, P.A., BUCHANAN, R. & MARKE, P. (1990) Three-dimensional data improve reservoir mapping. *J. Petrol. Technol.*, **42**, 22–61.

SKAUG, M. & GUNDESØ, R. (1986) Geological modelling of the Frigg field with special emphasis on shale mapping. *SPE Paper* **15859**.

VAIL, P.R., MITCHUM, R.M. JR., TODD, R.G., WIDMIER, J.M., THOMSON, S. III, SANGREE, J.B., BUBB, J.N. & HATLELID, W.G. (1977) Seismic stratigraphy and global changes of sea level. In: *AAPG Memoir 26* (Ed. by C.E. Payton), 49–212.

Modelling Methods

Spec. Publs int. Ass. Sediment. (1993) **15**, 191–204

Sedimentary flow units in hydrocarbon reservoirs: some shortcomings and a case for high-resolution permeability data

A. HURST

Production Geoscience Research Unit, Department of Geology and Petroleum Geology, University of Aberdeen, Aberdeen AB9 2UE, UK

ABSTRACT

Accurate definition of sedimentologically-defined flow units is often central in the successful modelling of reservoir performance and reliable economic predictions. Inappropriate reservoir simulation models may be generated because the dimensions and characteristics of flow units are defined inadequately. Laboratory probe permeametry provides data which may alleviate such problems, since it identifies and quantifies sedimentary heterogeneities which are unresolved by conventional data. For example, net sand is defined directly using permeability data, and a closer relationship is made between sedimentary characteristics and quantitative data.

INTRODUCTION

Hearn *et al.* (1984) give an example of the successful implementation of sedimentologically-defined flow units made for the reservoir characterization of the Hartzog Draw Field. The flow units were designated by assuming that they were continuous laterally and vertically, and had the following characteristics:

1 similar porosity;

2 similar permeability;

3 similar bedding characteristics.

The reservoir model evolved using these characteristics was subsequently used for numerical simulation of oil recovery and is shown in simplified section in Fig. 1. The concept of flow units as defined by Hearn *et al.* (1984) can potentially release geologists from the layer-cake rigidity often imposed by reservoir engineering constraints on reservoir modelling for the purpose of numerical simulation of reservoir behaviour. It also allows geologists to shift their focus of attention from 'static' volumetric models, which are largely based on mapping stratigraphic horizons, porosity, net/gross ratios and saturation, to consideration of 'dynamic' characteristics of reservoirs (e.g. using permeability and pressure data).

Many examples exist where the sedimentological characteristics of reservoirs have proven to be important when determining reservoir performance (Weber, 1987; Jensen, 1990). Field experience, and the availability of sedimentary facies models, provide a basis for wide-scale application of sedimentology to definition of reservoir architecture (Galloway, 1986). Prior to establishing production, permeability data, together with sub-surface pressure test data and analogues from previous experience, are normally used to define the flow characteristics in reservoir models. There may, however, be an element of confusion, caused by overestimation of the influence of sedimentological features, if the permeability data used to define the flow characteristics of the flow units are inadequate. Once production is established, pressure and fluid data from the reservoir itself are available to augment definition of flow units.

AIM

As reservoir volumes, flow units may be used as input for volumetric calculations, although their ultimate role is to provide a description for modelling the dynamic behaviour of a reservoir during production. Dynamic behaviour is usually modelled numeri-

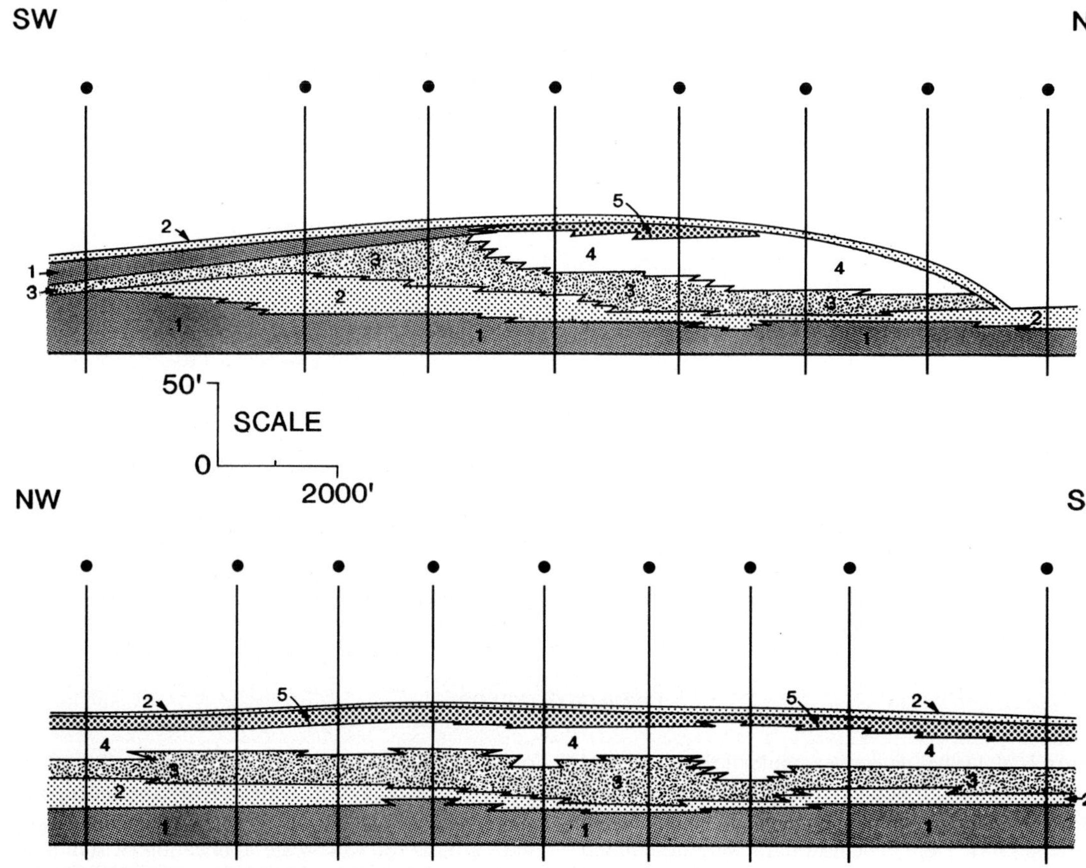

Fig. 1. Flow unit distribution in the Hartzog Draw Field where five flow units are recognized. Modified from Hearn *et al.* (1984). Vertical lines represent well positions.

cally with reservoir simulation models (Odeh & Heinemann, 1988). Simulations are used to assess factors such as extraction mechanisms, well performance, optimal well positioning, and to project the lifetime and economics of fields (Mattax & Dalton, 1990). The constraints of simulation modelling, for example the shape, size and number of grid cells available to construct a model, place many restrictions on the amount of geological heterogeneity which may be input. However, the role of the geologist when defining flow units is not to furnish reservoir engineers with 'convenient' input for modelling, but rather to identify, and quantify, the heterogeneity present. When defining 'ten golden rules for simulation engineers', Aziz (1989) begins by stating categorically in rule 1, 'before you do any simulation, understand the geological characteristics of your reservoir . . .'.

The extreme complexity of grain-to-grain relationships in sandstone reservoirs precludes description, and averaging, of their geometry for practical use in flow equations (Navier-Stokes equations; Yuan, 1967). To circumvent the problem of microscopic reservoir description, reservoir engineers commonly apply a coarser scale of averaging, the macroscopic scale (e.g. core plug measurements, Fig. 2), when describing flow in reservoir units (Haldorsen, 1986). Fundamental to this approach, the 'continuum approach', is to decide how great a volume of rock needs to be averaged to give a representative value for a rock property. This volume, a representative elementary volume (REV; Bear, 1972), is essential to the concept of a continuum, and the engineer's task is to determine how and if, from any point within a reservoir, rock properties vary with changes in the dimensions of REVs. Clearly, sedimento-

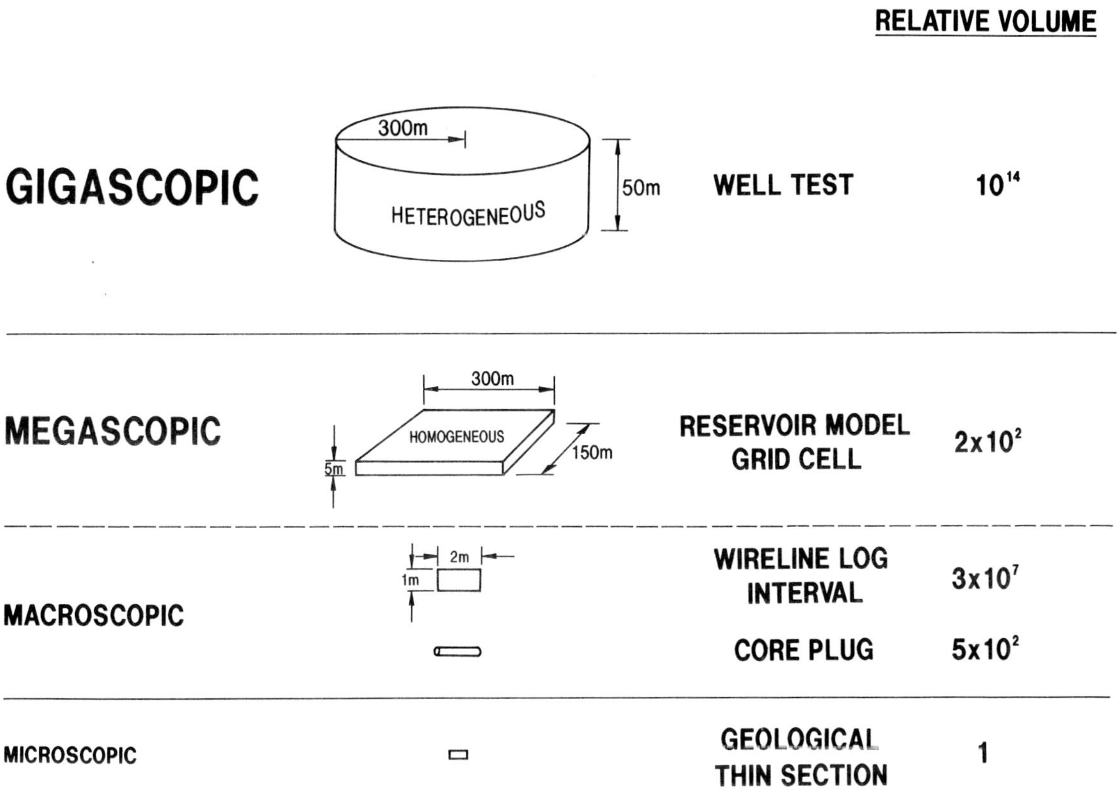

Fig. 2. Scales of investigation used in reservoir characterization. Modified from Archer & Hurst (1987) and Haldorsen (1986).

logical units are potentially useful building blocks which may have value as REVs, dependent on the rock property concerned.

A hitch in the process of describing flow units is that fluid flow equations in practice are formulated at the macroscale, but are solved at the megascale (Archer, 1983) — the scale of reservoir simulation grid blocks (Fig. 2). Thus, similarly severe simplifications are made both from the microscale (a common scale of geological observation) to the (REV) megascale, and from the megascale to the macroscale (reservoir simulation and its many applications). Properties of porous media are averages of measurements made on heterogeneous volumes and, because of the heterogeneity, measurements at one scale are not directly applicable at another scale unless an appropriate averaging is made over the new larger volume (Haldorsen, 1986).

In the recent past, reservoir engineers have shown an increasing awareness of the importance of obtaining accurate numerical geological descriptions of reservoirs. The advent of 'stochastic' and other statistical methods for modelling reservoir heterogeneity, and so evaluating uncertainty, have presented geologists with the problem of, and opportunity for, collecting 2-D and 3-D data from outcrops. These data can be used as the input data for calculating parameters such as vertical permeability (Haldorsen & Lake, 1984; Begg & King, 1985) and sandbody continuity (Clementsen *et al.*, 1990). Concurrently, sedimentologists have made the significant change from describing sequences as series of laterally-correlated 1-D vertical sections, to 2-D descriptions which identify the internal hierarchy of bedding surfaces (Allen, 1980). These advances, together, have created a surge of activity focusing on the collection of outcrop (analogue) data, often with a view to applying the data for definition of flow units as part of a specific reservoir characterization. Clearly, considerable responsibility should lie in the hands of geologists from whom reservoir engineers now await quantitative analogue data or, at least, uncertainty factors associated with geological interpretations.

MANIFESTATION

When the economic expectations based on the results of reservoir simulation are not met, re-analysis of reservoir models is requisite. In reality, and when modelled in simulation studies, key economic factors, such as timing of production peak and peak water production, projected production rates and field lifetime, are all extremely sensitive to uncertainties in the initial geological model (Aziz, 1989). Inappropriate flow units and geological zonation of reservoirs need not be the immediate cause of erroneous simulation modelling; however, if the user of the simulator is familiar with the physics of the processes being modelled, other errors should be minimal. Inaccuracies in geological input to simulation models may often be caused by a failure to recognize the significance of a particular heterogeneity when zoning the reservoir, and may necessitate building highly contrived models which include 'geological' permeability barriers which are not readily substantiated geologically (van Oort, 1988).

Possibly the single greatest problem in reservoir characterization is description and definition of vertical permeability (K_V). The heterogeneous character of all porous media, particularly in relation to grain size and diagenetic variations, makes both measurement and estimation of K_V from core and borehole data problematic. In sandstone reservoirs with interbedded shales, the sedimentological origin of the shales is usually readily identifiable and useful for estimating shale (barrier) geometry, which then may be incorporated into statistical models for estimation of K_V (Haldorsen & Lake, 1984; Hazeu *et al.*, 1988). Similarly, where carbonate-cemented horizons are believed to be the main constraints to K_V, sedimentological models are invoked to explain their distribution (van der Helm *et al.*, 1990). More lithologically homogeneous reservoirs, where the lithological definition (both in cores and from wireline logs) of K_V barriers is less clear than for shale and carbonate layers, present more subtle problems. In these reservoirs, permeability variations are often related to the changes of grain size and shape associated with sedimentary structures (Hurst & Rosvoll, 1991).

Procedure for reservoir simulation

Recent practice for reservoir simulation studies (Nybråten *et al.*, 1990; Wills & Peattie, 1990) may involve several scales of analysis (Fig. 3). At each stage of scaling-up or averaging, the descriptors of flow through porous media are described by pseudofunctions (Kyte & Berry, 1975), the role of which have recently been reviewed (Lake *et al.*, 1990).

2-D analysis allows considerable geological information to be incorporated, and the sensitivity of recovery efficiency to various geological parameters can be tested with reference to defined well points (Braithwaite *et al.*, 1989; Nybråten *et al.*, 1990). 2-D (cross-sectional) models typically investigate interwell distances, e.g. the distance between an injector and a producer (Fig. 3a).

Expansion to 3-D elements involves scaling-up information from preceding 2-D analyses, including what are considered to be important geological data, and will normally focus on areas of particular significance to evaluation of commerciality. For example, up to 18 000 cells (areal size 75 m × 75 m, thickness 3 m) were used in 3-D segment models (Fig. 3b) to model the behaviour of 5–12 wells in volumes containing 8–80 Sm^3 of oil in place in the Snorre Field (Nybråten *et al.*, 1990).

Full-field models require a further averaging of data from smaller scales of investigation, but aim to retain the reservoir characteristics critical to production. The number of grid cells in modern full-field simulations is often very large, e.g. 13 000 in the Forties Field (Wills & Peattie, 1990).

Geological description of flow units and their attributes will, potentially, supply the data to describe the grid cells in simulation models. Therefore, an important goal of geological reservoir characterization is to describe lateral variations in reservoir characteristics, particularly with respect to vertical and horizontal permeabilities, K_V and K_H respectively.

Lower Brent Group

The Lower Brent Group (Fig. 4) is a prolific reservoir in many North Sea Fields (Johnson & Stewart, 1985). It comprises a prograding, coarsening-upward sequence of lower shoreface sandstones (Rannoch Formation) overlain by channelized upper shoreface sandstones (Etive Formation). Despite excellent reservoir characteristics, timing of water breakthrough

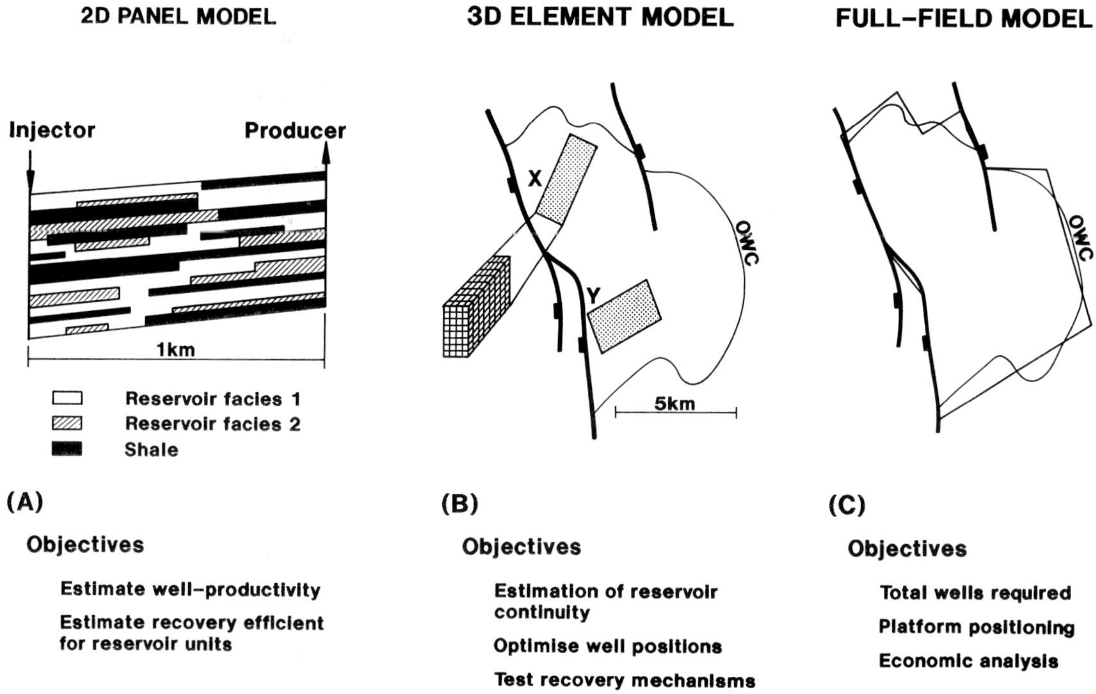

Fig. 3. Examples of some typical numerical simulation models: (A) two-dimensional, cross-sectional model; (B) three-dimensional element model for investigating recovery in specific, well-controlled or critical volumes of the reservoir; (C) three-dimensional, full-field model for modelling gross recovery characteristics.

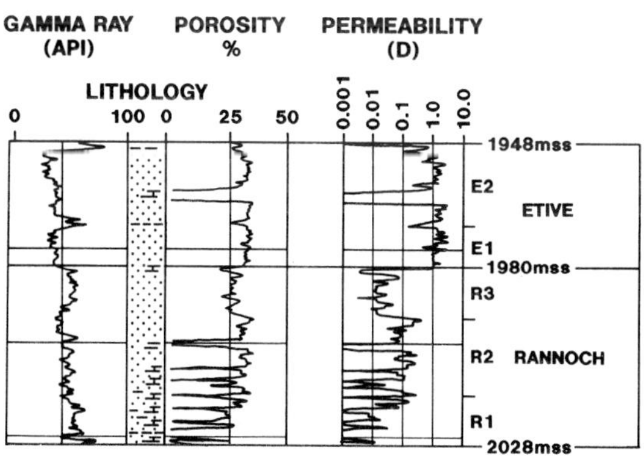

Fig. 4. A typical section from the Lower Brent Group of the Gullfaks Field (from Erichsen *et al.*, 1987). Reservoir zones are marked, E2, E1, R3, R2 and R1.

and cumulative water production frequently exceed predictions (e.g. Bayat & Tehrani, 1985; Petterson *et al.*, 1990). If a flow unit has been assigned inappropriate permeability characteristics when integrating geological and petrophysical data at the macroscopic scale (Fig. 2), successive scaling-up of data (Fig. 3) compounds the error.

A cartoon reservoir model demonstrates the type of misconception possible when predicting a water flood (Fig. 5). In reality, the simulation model will be more complex than illustrated, perhaps with several layers of grid cells averaging approximately 5 m in thickness, and with a range of values for K_H and K_V derived from averages of core plug data. If a water flood is simulated that assumes a general increase in K_H and K_V, and K_V/K_H upwards, an 'ideal' scenario develops with successive time steps (t_1 to t_5, Fig. 5b), which in the simulation encourages propagation of a stable flood front. In practice, Lower Brent water flooding probably rarely produces a stable flood front but, rather, tends to exploit relatively narrow 'thief zones' (Fig. 5c). Consequently, water breakthrough occurs earlier than predicted, and the percentage of water produced in the total fluid produced (water cut) attains high levels early in the production life of a field (Bayat &

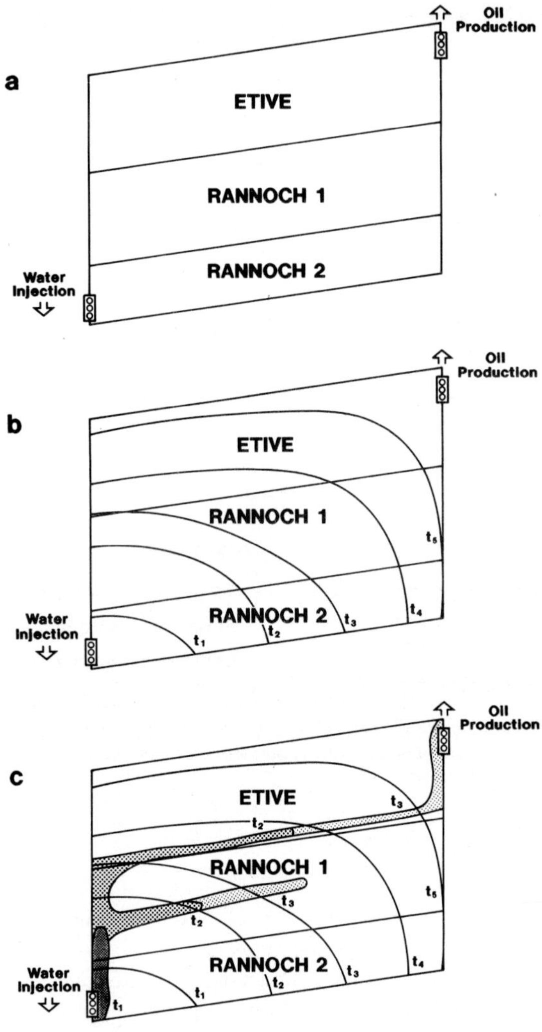

Fig. 5. (a) Layer-cake zonation of a Lower Brent Group reservoir for use in numerical simulation modelling. The layers are assumed to have decreasing average permeabilities downwards. (b) Prediction of water flooding through time (t_1–t_5). (c) Actual behaviour of water flood during production. Important to note is that the rapid water movement vertically along the injection well may not be attributable only to geological heterogeneity, but also to thermal fracturing of the formation and cement adjacent to the well caused by the cooling influence of the injected water.

Tehrani, 1985; Petterson *et al.*, 1990). Typically, in a reservoir simulation the Etive Formation is eventually almost entirely flushed by water, whereas the majority of the oil in the Rannoch Formation is bypassed (e.g. 76% of the oil in place in the Rannoch Formation of the Dunlin Field would have been bypassed by the year 2000 if a programme of improved recovery was not instigated; Braithwaite *et al.*, 1989). If the inaccuracy in a simulation model is attributable to definition of flow units, the initial characterization of permeability distribution is the prime potential source of error.

When water breakthrough has occurred, the flow unit transmitting injected water directly to the production well can be identified, and its geological and petrophysical characteristics examined. A well example (Fig. 6) shows the location of water breakthrough in interval 'X' near the Etive–Rannoch boundary, together with the permeability data available for zonation of the reservoir. On the basis of the correlation between the core plug (KELH) and log-derived (KLOGH) permeability data, the petrophysical formation evaluation of the interval was considered successful. Interval 'X' has an average permeability of 2116 mD, which is intermediate between the average permeabilities of the Etive (V and W, Fig. 6) above and the Rannoch (Y) below. From the histogram presentation of the data, it is apparent that interval 'X' has a markedly skew population distribution, the bulk of the permeabilities lower than 2116 mD, but with several more permeable exceptions. It is also apparent visually that the match between core data (KELH) and the log-derived permeability (KLOGH) is poorer in interval 'X' than elsewhere.

Core from this 5–6 m interval is typified by interbedded, fine-grained, micaceous sandstones and coarse-grained, quartzose sandstones, where the individual lithological units average from 10 to 25 cm in thickness. The bedding heterogeneities are not resolved by the density (RHOB) log (the measurement used to derive KLOGH) and, when the core is examined, are seen to be only partially sampled by the core plugs. These wireline log measurements are not designed to evaluate permeability (see Rider, 1986); they do, however, give accurate estimates of porosity (particularly the density—RHOB—log) from which empirical evaluations of permeability are often made (Coates & Dumanoir, 1973). It is recognized that the lack of an established log-derivable permeability indicator places severe limitations on the input to dynamic modelling of hydrocarbon recovery (Worthington, 1991).

Identification of permeability heterogeneities is largely the task of the geologist when describing core. However, quantification of permeability heterogeneity is normally limited to core plug data, which usually sample K_H and K_V at 30 cm (1 foot)

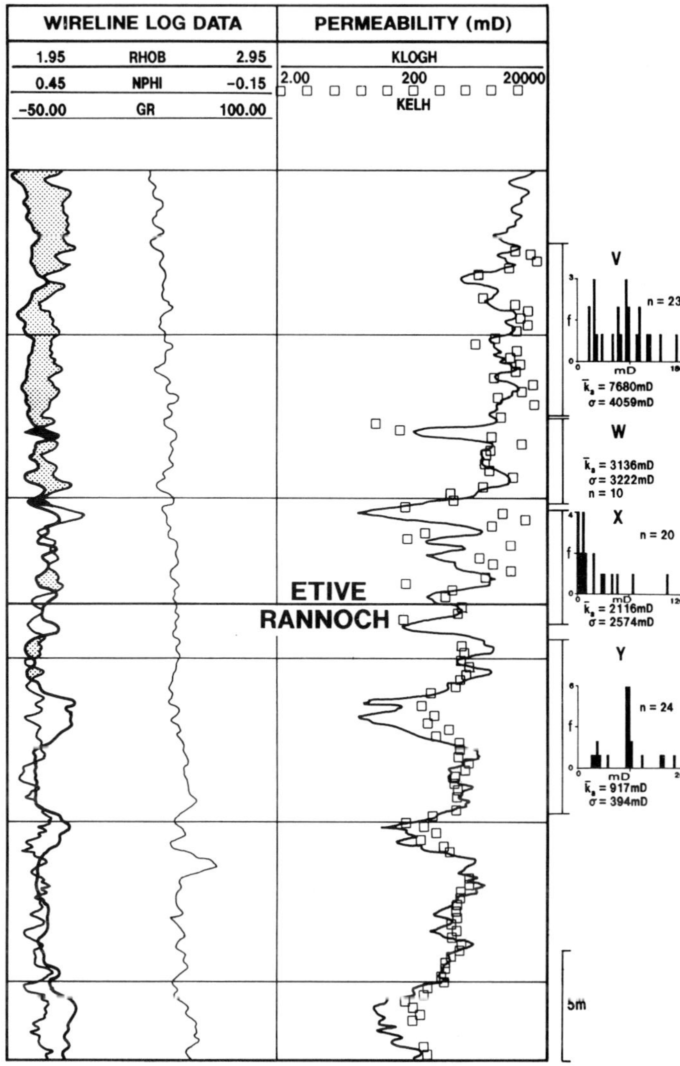

Fig. 6. An example of a Lower Brent Group well showing the location of the interval 'X' where water breakthrough occurred. Estimated best quality reservoir is stippled between the RHOB and NPHI log traces. Summary statistics for selected intervals V, W, X, and Y are as follows: n = number of samples, k_a = arithmetic average, σ = standard deviation, f = frequency. All permeabilities are in millidarcies (mD). Log and core data are as follows: RHOB = formation density (g.cm^{-3}); NPHI = neutron porosity (porosity units); GR = gamma ray (API units); KLOGH = log-derived horizontal permeability; KELH = Klinkenberg corrected horizontal permeability to air.

vertical intervals, a convenient rather than an appropriate sampling interval. Loss of core plug data may occur from highly permeable intervals which are frequently not sampled because of their poorly consolidated character, or, when sampled, are frequently disturbed, both by coring and extraction of core plugs, thus giving unrepresentative data (Hurst, 1987). Core plugs in interval 'X' sampled very few of the high-permeability sandstones within the interval, known from subsequent work to often have permeabilities between 15 D and 30 D. Clearly, the average permeability of 2116 mD (Fig. 6) represents neither the permeability of the low permeable, micaceous zones, nor the highly permeable coarse sandstones.

Assuming that interval 'X' had been identified as a sedimentological flow unit, its average permeability does not reflect the behaviour of the unit as a thief zone during production. Indeed, even the highest permeabilities within interval 'X' are similar or lower than the highest permeabilities measured in the overlying section. On the basis of the available permeability data (KELH and KLOGH), interval 'X' cannot be described as a flow unit, even if the geological heterogeneity is identified. Thus, the flow unit known to exist from the evidence of water breakthrough can be identified but not quantified. As has been demonstrated from the same formations elsewhere (Corbett & Jensen, 1991), core plug data undersample the permeability heterogeneity present, frequently missing high and low permeability intervals and giving an incomplete picture of the permeability distribution. Clearly, a case exists for acquiring data which can resolve the permeability heterogeneity present.

HIGH-RESOLUTION DATA

Ultimately, quantification of the flow units used in reservoir prediction will depend in part upon the availability of permeability data. Most often, these data will be obtained either from measurements made directly on core samples, or indirectly from wireline log measurements which can be correlated with actual permeability data. However, routinely available data are not always adequate to identify flow units (Fig. 6). A source of permeability data which allows significant improvements to be made in the geological characterization of reservoirs is probe (mini) permeametry (Eijpe & Weber, 1971; Weber, 1986; Halvorsen & Hurst, 1990).

Probe permeameter data

The advantage of obtaining several probe permeameter measurements rather than a single core plug measurement is known to be of considerable advantage when characterizing the permeability of heterogeneous reservoirs (Weber & van Geuns, 1990). Furthermore, it is apparent from recent work (Hurst & Rosvoll, 1991) that the permeability distribution of specific bedding types, the building blocks of sedimentary flow units, can be more effectively defined using a probe permeameter. Fully-automated probe permeameter instrumentation allows accurate, non-destructive measurements to be made rapidly; it is commonplace in a routine (even 'quick-look') study to collect 2–300 measurements on orthogonal sampling grids per metre length of core (Halvorsen & Hurst, 1990). At this level of sampling density, permeability heterogeneities, such as those present in the Lower Brent, are readily identified and quantified (high permeabilities in the range 15–30 D are known from similar intervals; Corbett & Jensen, 1991). Acquisition of dense data sets helps evaluation of the statistical significance of permeability variations within centimetre-scale sedimentary structures (Hurst & Rosvoll, 1991) and, thereby, permeability assessment of sedimentary flow units is possible.

Example 1

Core plug permeability data from a sandstone sequence show a clear correspondence between the coarsest grained part of the sequence and the highest permeabilities (Fig. 7a). Comparison of the permeability data with the core description allows the net pay to be defined and an average permeability of 180 mD is estimated for that section. Although the sandstone is characterized by the highest permeabilities, there is no particular relationship between core plug permeability and grain size variation, i.e. the permeability distribution appears to be random.

Subsequently, probe permeameter data were obtained from the same interval (Fig. 7b). It is comforting, especially from an experimental petrophysicist's standpoint, that the core plug data and the probe permeameter data lie on approximately the same trend. Far more significant, however, is that the permeability distribution is clearly not random, but corresponds to the depositional trend of increasing grain size and scale of sedimentary structure upwards. Permeability distribution is closely related to sedimentological variations, and only the lower 20 cm of the net sand in Fig. 7a has an average permeability close to the 180 mD estimated from the plug data. In the top part of the interval, the average permeability varies between 700 and 800 mD, with the exception of a thin, partly carbonate-cemented horizon from which a core plug measurement was made. Furthermore, the assumption that the average permeability of the net pay can be approximated by using a geometric mean, in which it is implicit that permeability distribution has no strong systematic pattern of variation (Jensen, 1991), is erroneous.

Although the probe permeameter data have similar values at any vertical depth increment, or between neighbouring depths, some spread occurs—up to around half an order of magnitude difference in some cases. These variations are unrelated to experimental error (Halvorsen & Hurst, 1990), and reflect the actual small-scale permeability variations present. Thus, small-scale permeability variations are identified which the geologist can now attempt to explain by making more detailed examination of the core and, possibly, by obtaining petrographic data. In so doing, the geological characterization of the core may be more

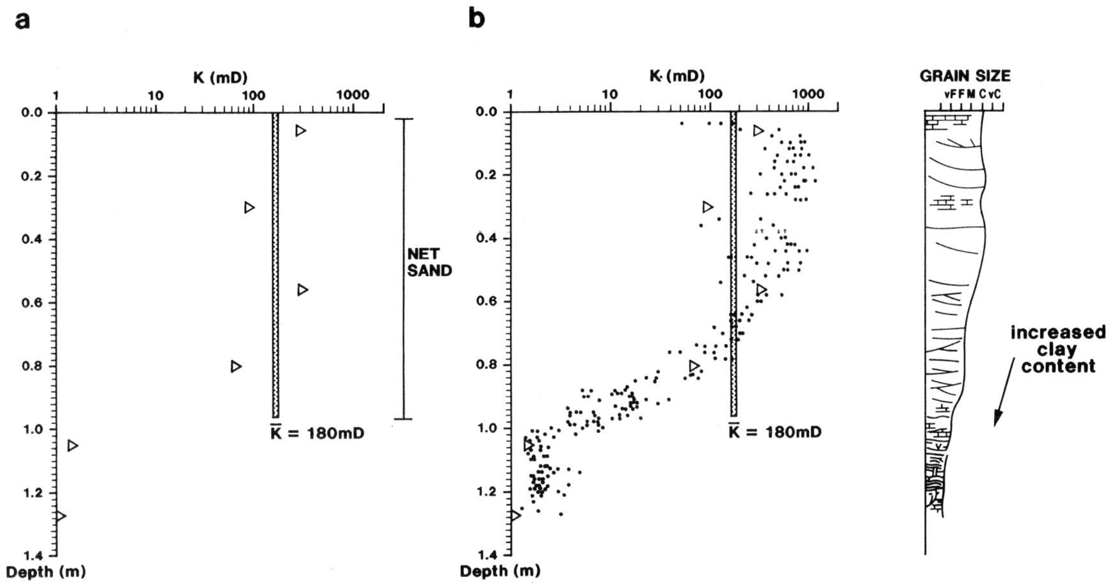

Fig. 7. (a) Core plug permeability data and mean permeability value (180 mD) from a shallow marine, coarsening-upwards, sandstone unit. Triangles are core measured permeabilities (K_H). Net sand is defined by the length of the mean permeability curve. (b) Minipermeameter data (dots) and geological core description from slabbed core. Triangles are the same data points as in (a). The minipermeameter data confirm the permeability trend, allow precise correlation with sedimentary features, and show that the average permeability at the top of the net sand is higher than shown by the core plugs.

closely linked with the petrophysical characterization. An important point made by Halvorsen & Hurst (1990), and confirmed by Corbett & Jensen (1991), is that the existing routines for plugging, slabbing and description of cores are not optimal for reservoir description. The non-destructive nature of probe permeametry allows petrophysical and geological characterization to be made on *exactly* the same materials. In this example, too few permeability measurements are available from the core plug data, either to accurately define the net sand and its average permeability, or to characterize the permeability heterogeneity.

Example 2

A 'shaly' sandstone reservoir with typically high gamma log response, limited neutron/density log cross-over, and poor correlation between core- and log-derived permeabilities (Fig. 8a) presents an obvious task for application of a laboratory probe permeameter. Definition of the net/gross ratio, average effective horizontal permeability (K_H), and K_V are targets for particular attention. Core reveals the heterolithic character of the formation (Fig. 9); interbedded sandstones and mudstones with sandstone beds vary in thickness from a few millimetres up to approximately 20 cm. Probe permeameter data were obtained from five parallel vertical transects with measurements taken at 5 cm vertical spacings (Fig. 8b).

The petrophysical implications of these data are discussed elsewhere (Halvorsen & Hurst, 1990) but in summary are as follows. First, neither core plugs nor the available wireline logs are able to resolve the permeability or lithological variations present. Second, approximately two orders of magnitude of permeability variation occur about the mean permeability derived from the core plug and log data. Third, using the conventional data the laminated character of the reservoir is 'lost', thus giving a homogeneously (low) average value for the permeability.

The data allow net sand, the thickness from which hydrocarbons are deemed to be produceable, to be defined using permeability (a dynamic characteristic) rather than porosity (a static characteristic). Once permeability is characterized in this way in cored intervals, it is possible that calibration of wireline logs for permeability estimation will allow more

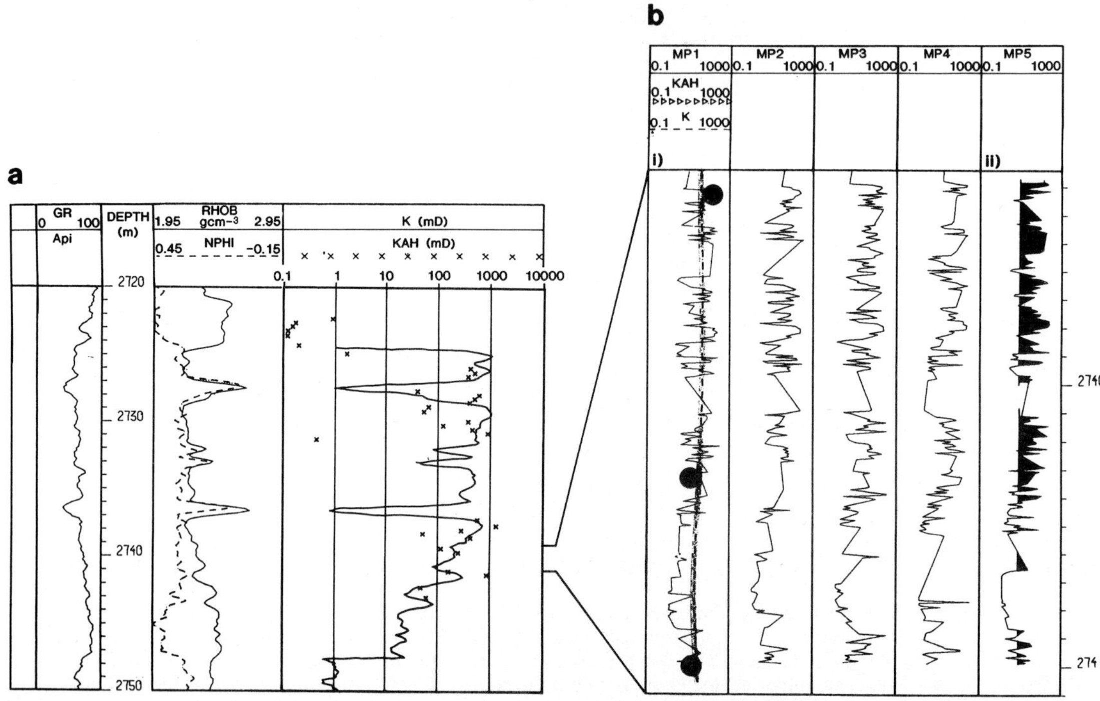

Fig. 8. (a) Wireline and permeability data from a 'shaly' sandstone interval. K = log-derived permeability (calibrated from the density log); KAH = horizontal core plug permeabilities; RHOB is formation density ($g.cm^{-3}$); NPHI is neutron porosity (porosity units); GR is natural gamma radiation. (b) Probe permeameter data from interval 2739 m to 2741 m. Column (i) includes three core plug data points and the trace of the log-derived permeability (K). Column (ii) shows a net sand cut-off at 20 mD defined by the probe permeameter data.

accurate permeability estimation in geologically similar uncored intervals. The average K_H of approximately 200 mD is similar for both permeability data sets. In this case, a net sand cut-off of 20 mD is applied, and the net/gross ratio for the interval is estimated as 66%. As permeability data are available from all the lithologies present, K_V can be evaluated by building a three-dimensional, numerical simulation model, where the sedimentological model would define the architecture of the heterogeneities, and the probe permeameter would provide facies-defined permeabilities (cf. Braithwaite *et al.*, 1989). In such a model, appropriate analogue data for describing the sedimentary architecture are fundamental. The main advantage of the probe permeameter data is that a greater range of permeability variation is identified than from the core plugs, which reflects the sedimentological heterogeneity.

DISCUSSION

Definition of reservoir flow units ultimately has a single, clear goal: to enhance our understanding of reservoir behaviour in order to reduce the uncertainties of economic prediction for field development. The acquisition of high-resolution data is important, not only for describing geological flow units, but also for quantifying them. Routine acquisition of probe permeameter data makes an important contribution towards achieving that goal. Considerable impact on the understanding of recovery efficiency is possible even when limited additional (probe permeameter) permeability data are used together with accurate sedimentological description (Braithwaite *et al.*, 1989; Weber & van Geuns, 1990), and in general the correlation between core data and log data will be improved (Jensen, 1990). The common petrophysical origin of the log and probe permeameter data, and the application of the

Fig. 9. Laminated sandstone and shale typical of the section presented in Fig. 6. Diameter of hole left by core plug is 1 inch (2.54 cm) in diameter. Note that the core plug has sampled a mixture of sand, shale and silt, and thus will not give an accurate measure of the permeability of any of these lithologies.

data in a geologic context, ensures better integration and quantification of geological core description. There are undoubtedly other measurements that could give similar enhancement of reservoir zonation. Indeed, the time is ripe for introduction of experimental methods which make automatic, continuous measurements of grain size variation and sorting, thus superseding the imprecise, time-consuming laboratory practices (hand lenses and comparators) used at present.

Although several examples exist where reservoir modelling is enhanced by using probe permeameter data (e.g. Martin & Cooper, 1984; Giordano *et al.*, 1985; Braithwaite *et al.*, 1989), none have had the benefit of automated instrumentation (Halvorsen & Hurst, 1990) and, thus, have not generated the type of high density data presented here. In general, routine acquisition of high density data in cored wells, as in both examples, allows net sand to be defined on the basis of permeability distribution, thus avoiding the convolutions of normal practice where empirical relationships are made between porosity and permeability (Coates & Dumanoir, 1973). In turn, this would allow geologically more realistic estimation of the thickness of permeable intervals from the '$k_a h$' (mD feet) products derived from the flow rates of drill-stem tests. To determine the average permeability (k_a) of a drill-stem tested interval, the thickness (h, height) of the interval from which fluids are flowing must be known. Thus, if core is available, probe permeameter data could be better used to determine h.

In the examples, the probe permeameter data identify both higher and lower permeabilities than those identified with conventional data (Figs 7, 8). With these additional data, it is clearly possible to model reservoir performance in detail, by depicting reservoir heterogeneity more accurately, as is often required if reliable predictions are to be made (Journel & Alabert, 1988; Aziz, 1989). One can at least define a greater variety of flow units, for which the relative effects on recovery efficiency can be modelled. Without high-resolution data, intervals with marked permeability characteristics, such as in the Lower Brent Group (Fig. 6), may not be identified until water breakthrough has occurred. After breakthrough, irreversible changes have occurred within the reservoir. Injected water will be channelled by direct, unfavourable routes ('thief zones') between injection and producing wells, thus reducing the effect (sweep efficiency) of the water front elsewhere in the reservoir (Bayat & Tehrani, 1985; van Oort, 1988; Petterson *et al.*, 1990). Equally, well-head facilities may not have sufficient capacity to handle the larger than expected volumes of water produced, and will either operate below expectation, or require modification. Conceivably, the uncertainty associated with these serious economic risks can be greatly reduced by improving the definition of sedimentary flow units during field appraisal and production planning. As shown in practice (Braithwaite *et al.*, 1989), similar attention to detail can be invaluable when evaluating improved hydrocarbon recovery from poorly-swept formations in fields with established production history.

An improved understanding of small-scale permeability heterogeneity is important to obtain a better understanding of viscous-dominated flow, specifically in reservoirs where waterflooding is used. Sedimentary heterogeneities and small-scale permeability variations are known to affect residual oil saturations (Kortekaas, 1985), the pseudo-permeability to water (Muggeridge, 1991), and to increase the effective dispersivity (Kelkar & Gupta, 1988). Jensen (1990) states that conventional core plugs and special core samples have been found to capture poorly the presence of small-scale permeability variations; probe permeameter data have been applied to enhance interpretation of the results of special core analysis (Giordano *et al.*, 1985).

From a geological standpoint, probe permeameter data allow the relationships between grain size, sorting and permeability, and specific sedimentary structures or facies units to be examined (Hurst & Rosvoll, 1991). Probe permeameter data are also objective physical measurements which can be applied in a variety of ways to help characterize and quantify the significance of sedimentary features in terms of their reservoir heterogeneity. Given probe permeameter data similar to those in Figs 7 and 8, it may be possible to rationalize sedimentological core description by focusing on those intervals perceived to be significant in terms of permeability. The probe permeameter data can be conceived as an approximately continuous log, which augments the traditional grain size log. In the case of heterolithic formations, where grain size variations are very time-consuming to resolve, the permeability trends are an alternative measurement with which sedimentary description can be correlated and quantified. The possibility of calibrating high-resolution wireline logs with cores for the purpose of facies analysis (Luthi & Banavar, 1988) can then be extended to permeability imaging by integrating borehole images with probe permeameter data from cores (Bourke *et al.*, in press).

It is imperative that sedimentologists do not forsake the imterpretational aspects of core description which provide the basis for estimating, either subjectively or statistically, the three-dimensional geometry of sedimentary units. Acquisition of analogue data from outcrops plays an important role in validating such modelling. However, if creation of a three-dimensional reservoir model is not preceded by the identification of the location, and context of, sedimentary heterogeneities, the significance of the models as flow units during production may be spurious (cf. Aziz, 1989). Acquiring satisfactory, quantitative, geological data, and ensuring a willingness to accept the inherent uncertainties with these data, are major goals of sedimentological reservoir characterization. The limitations of routinely available core and log data have been demonstrated, in particular with respect to their lack of resolution. Though observations of small-scale variations have been made (Giordano *et al.*, 1985; Weber, 1987), many hundreds of measurements per metre of core will often be required if permeability heterogeneity is to be accurately described and correlated to sedimentological characteristics.

CONCLUSIONS

The basic definition of reservoir flow units combines sedimentological and petrophysical reservoir characteristics. However, there is a need to acquire high-resolution data so that permeability distribution may be mapped, and subsequently modelled, more accurately. Failure to identify permeability contrasts with routinely available data may lead to inappropriate numerical simulation of reservoir behaviour, and unreliable economic predictions.

Probe permeameter data are demonstrated to enhance greatly the task of identifying permeability heterogeneity and allow close integration of geological and petrophysical data. Definition and quantification of sedimentological flow units can be made more accurately, and at higher resolution, than is possible with routine data. Thus, description of the sedimentary architecture in reservoir models enables testing of the sensitivity of hydrocarbon recovery to various flow unit configurations.

High-resolution data, whether from a probe permeameter or from wireline logs, enhances the integration of sedimentological data into engineering models. If the permeability distribution of a reservoir interval is characterized rigorously, collection of outcrop data can focus on the sedimentological features recognized to be of most significance as flow units.

ACKNOWLEDGEMENTS

Christian Halvorsen and Charles Jourdan are thanked for their co-operation, support and criticism. Referees Steve Livera and Ian Bryant made many constructive comments, as did Jerry Jensen and Pat Corbett. All statements made are, however, entirely the opinions of the author. Den norske Stats

Oljeselskap (Statoil) and Unocal UK are acknowledged for supporting publication of the paper.

REFERENCES

ALLEN, J.R.L. (1980) Sand waves: a model of origin and internal structure. *Sedim. Geol.*, **26**, 281–328.

ARCHER, J.S. (1983) Reservoir definition and characterisation for analysis and simulation. *Proc. 11th World Petrol. Cong.*, London, PD 6(1).

ARCHER, J.S. & HURST, A. (1987) Evaluation of formation damage in heterogeneous reservoirs. *Marine Petrol. Geol.*, **4**, 119–126.

AZIZ, K. (1989) Dialog: ten golden rules for simulation engineers. *J. Petrol. Technol.*, **41**, 1157.

BAYAT, M.G. & TEHRANI, D.M. (1985) The Thistle Field—analysis of its past performance and optimization of its future development. *SPE Paper* **13989/1**.

BEAR, J. (1972) *Dynamics of Fluids in Porous Media.* Elsevier, Amsterdam.

BEGG, S.H. & KING, P.R. (1985) Modelling the effects of shales on reservoir performance: calculation of effective vertical permeability. *SPE Paper* **13529**.

BOURKE, L.T., CORBIN, N., BUCK, S.G. & HUDSON, G. (in press) Permeability images—a new approach to reservoir characterisation. In: *Advances in Reservoir Geology* (Ed. by M. Ashton). Spec. Publ. geol. Soc. London.

BRAITHWAITE, C.I.M., MARSHALL, J.D. & HOLLAND, T.C. (1989) Improving recovery from the Dunlin Field, UK northern North Sea. *SPE Paper* **19878**.

CLEMENTSEN, R., HURST, A., KNARUD, R. & OMRE, H. (1990) A computer program for evaluation of fluvial reservoirs. In: *North Sea Oil and Gas Reservoirs—II* (Ed. by A.T. Buller, E. Berg, O. Hjelmeland, J. Kleppe, O. Torsæter & J.O. Aasen), pp. 373–385, Graham and Trotman, London.

COATES, G. & DUMANOIR, J. (1973) A new approach to improved log-derived permeability. *SPWLA 14th Ann. Logg. Symp.*, **R**, 28 pp.

CORBETT, P.W.M. & JENSEN, J.L. (1991) A comparison of small-scale permeability measurement methods for reservoir characterisation. Paper presented at *Advances in Reservoir Technology*, 21–22 February, 1991, Edinburgh.

EIJPE, R. & WEBER, K.J. (1971) Mini-permeameters for consolidated and unconsolidated sands. *Bull. Am. Assoc. Petrol. Geol.*, **55**, 307–309.

ERICHSEN, T., HELLE, M., HENDEN, J. & ROGNEBAKKEN, A. (1987) Gullfaks. In: *Geology of Norwegian Oil and Gas Fields* (Ed. by A.M. Spencer, *et al.*), pp. 273–286, Graham and Trotman, London.

GALLOWAY, W.E. (1986) Reservoir facies architecture of microtidal barrier systems. *Bull. Am. Assoc. Petrol. Geol.*, **70**, 787–808.

GIORDANO, R.M., SALTER, S.J. & MOHANTY, K.K. (1985) The effects of permeability variations on flow in porous media. *SPE Paper* **14365**.

HALDORSEN, H.H. (1986) Simulator parameter assignment and the problem of scale in reservoir engineering. In: *Reservoir Characterisation* (Ed. by L.W. Lake & H.B. Carroll Jr.), pp. 487–544, Academic Press, New York.

HALDORSEN, H.H. & LAKE, L.W. (1984) A new approach to shale management in field-scale models. *SPE Journal*, (August), 442–452.

HALVORSEN, C. & HURST, A. (1990) Principles, practice and applications of laboratory minipermeametry. In: *Advances in Core Evaluation: Accuracy and Precision in Reserves Estimation (EUROCAS I)* (Ed. by P.F. Worthington), pp. 521–549. Gordon & Breach, London.

HAZEU, G.J.A., KRAKSTAD, O.S., RIAN, D.T. & SKAUG, M. (1988) The application of new approaches for shale management in a three-dimensional simulation study of the Frigg Field. *SPE Formation Evaluation*, **3**, 493–502.

HEARN, C.J., EBANKS, W.J.Jr., TYE, R.S. & RANGANATHAN, V. (1984) Geological factors influencing reservoir performance of the Hartzog Draw Field, Wyoming. *J. Petrol. Technol.*, **36**, 1335–1344.

HURST, A. (1987) Problems of reservoir characterisation in some North Sea sandstone reservoirs solved by the application of microscale geologic data. In: *North Sea Oil & Gas Reservoirs* (Ed. by J. Kleppe, E.W. Berg, A.T. Buller, O. Hjelmeland & O. Torsæter), pp. 153–167. Graham and Trotman, London.

HURST, A. & ROSVOLL, K.J. (1991) Permeability variations in sandstones and their relationship to sedimentary structures. In: *Reservoir Characterisation—II* (Ed. by L.W. Lake, H.B. Carroll Jr & T.C. Wesson), pp. 166–196, Academic Press, New York.

JENSEN, J.L. (1990) A model for small-scale permeability measurements with applications to reservoir description. *SPE Paper* **20265**.

JENSEN, J.L. (1991) Use of the geometric average for effective permeability estimation. *Math. Geol.*, **23**, 833–840.

JOHNSON, H.D. & STEWART, D.J. (1985) Role of clastic sedimentology in the exploration and production of oil and gas in the North Sea. In: *Sedimentology: Recent Developments and Applied Aspects* (Ed. by P.J. Brenchley & B.P.J. Williams), Spec. Publ. geol. Soc. London, Blackwell Scientific Publications, Oxford, **18**, pp. 249–310.

JOURNEL, A.J. & ALABERT, F.G. (1988) Focusing on spatial connectivity of extreme-valued attributes: stochastic indicator models of reservoir heterogeneities. *SPE Paper* **18324**.

KELKAR, M.G. & GUPTA, S.P. (1988) The effects of small-scale heterogeneities on the effective dispersivity of a porous medium. *SPE/DOE Paper* **17339**.

KORTEKAAS, T.F.M. (1985) Water/oil displacement characteristics in crossbedded reservoir zones. *SPE Journal*, 917–926.

KYTE, J.R. & BERRY, D.W. (1975) New pseudo functions to control numerical dispersion. *SPE Journal*, August, 269–276.

LAKE, L.W., KASAP, E. & SHOOK, M. (1990) Pseudofunctions—the key to practical use of reservoir description. In: *North Sea Oil and Gas Reservoirs—II* (Ed. by A.T. Buller, E. Berg, O. Hjelmeland, J. Kleppe, O. Torsæter & J.O. Aasen), pp. 297–308, Graham and Trotman, London.

LUTHI, S.M. & BANAVAR, J.R. (1988) Application of borehole images to the 3-D modelling of eolian sandstone reservoirs, Permian Rotliegende, North Sea. *Bull. Am. Assoc. Petrol. Geol.*, **72**, 1074–1089.

MARTIN, J.H. & COOPER, J.A. (1984) An integrated approach to the modelling of permeability barrier distribution in a sedimentogically complex reservoir. *SPE Paper* **13051**.

MATTAX, C.C. & DALTON, R.L. (1990) Reservoir simulation. *J. Petrol. Technol.*, **42**, 692–695.

MUGGERIDGE, A.H. (1991) Generation of pseudo relative permeabilities from detailed simulation of flow in heterogeneous porous media. In: *Reservoir Characterization—II* (Ed. by L.W. Lake, H.B. Carroll Jr & T.C. Wesson), pp. 197–225. Academic Press, New York.

NYBRÅTEN, G., SKOLEM, E. & ØSTBY, K. (1990) Reservoir simulation of the Snorre Field. In: *North Sea Oil and Gas Reservoirs—II* (Ed. by A.T. Buller, E. Berg, O. Hjelmeland, J. Kleppe, O. Torsæter & J.O. Aasen), pp. 91–114, Graham and Trotman, London.

ODEH, A.S. & HEINEMANN, R.F. (1988) Reservoir simulation—the present and the future. *In Situ*, **12**, 153–180.

PETTERSON, O., STORLI, A., LJOSLAND, E. & MASSIE, I. (1990) The Gullfaks Field: geology and reservoir development. In: *North Sea Oil and Gas Reservoirs—II* (Ed. by A.T. Buller, E. Berg, O. Hjelmeland, J. Kleppe, O. Torsæter & J.O. Aasen), pp. 67–90, Graham and Trotman, London.

RIDER, M.H. (1986) *The Geological Interpretation of Well Logs*. Blackie Halsted Press, Glasgow, 175 p.

VAN DER HELM, A.A., GRAY, D.I., COOK, M.A. & SCHULTE, A.M. (1990) Fulmar: the development of a large North Sea field. In: *North Sea Oil and Gas Reservoirs—II* (Ed. by A.T. Buller, E. Berg, O. Hjelmeland, J. Kleppe, O. Torsæter & J.O. Aasen), pp. 25–45, Graham and Trotman, London.

VAN OORT, B. (1988) Lessons learned in North Sea oil field developments. *J. Can. Petrol. Technol.*, **27**, 123–132.

WEBER, K.J. (1986) How heterogeneity affects oil recovery. In: *Reservoir Characterisation* (Ed. by L.W. Lake & H.B. Carroll Jr.), pp. 487–544, Academic Press, New York.

WEBER, K.J. (1987) Computation of initial well productivities in aeolian sandstone on the basis of a geological model, Leman Gas Field, U.K. In: *Reservoir Sedimentology* (Ed. by R.W. Tillman & K.J. Weber), pp. 333–354, Spec. Publ. Soc. econ. Paleon. Miner., Tulsa, **40**.

WEBER, K.J. & VAN GEUNS, L.C. (1990) Framework for constructing clastic reservoir simulation models. *J. Petrol. Technol.*, **42**, 1248–1297.

WILLS, J.M. & PEATTIE, D.K. (1990) The Forties Field and the evolution of a reservoir management strategy. In: *North Sea Oil and Gas Reservoirs—II* (Ed. by A.T. Buller, E. Berg, O. Hjelmeland, J. Kleppe, O. Torsæter & J.O. Aasen), pp. 1–23, Graham and Trotman, London.

WORTHINGTON, P.F. (1991) Reservoir characterisation at the mesoscopic scale. In: *Reservoir Characterization—II* (Ed. by L.W. Lake, H.B. Carroll Jr & T.C. Wesson), pp. 123–165. Academic Press, New York.

YUAN, S.W. (1967) *Foundations of Fluid Mechanics*. Prentice-Hall, London.

Spec. Publs int. Ass. Sediment. (1993) **15**, 205–212

The use of length distributions in geological modelling

G. GEEHAN *and* J. UNDERWOOD*

BP Research International Ltd, Chertsey Road, Sunbury upon Thames, UK

ABSTRACT

Geological data are often evaluated via statistics describing the frequency of occurrence of various attributes. Distributions of the lengths (or widths) of geological features, such as sandbodies or shales, are of prime importance in reservoir simulation; however, these distributions can easily be misapplied. This paper addresses three issues regarding the use of length distributions: (i) dimensionality, (ii) partial lengths, and (iii) shape/orientation.

1 Apparent length distributions vary according to the number of *dimensions* within which they are sampled: 1-D, 2-D or 3-D. Thus, a distribution measured from outcrop (2-D) must be modified if it is to be modelled in a 3-D volume. In going from a 2-D distribution to 3-D (or 1-D to 2-D) the relative frequency of a particular length class will decrease in direct proportion to its length, assuming an isotropic shape.

2 Length data derived from outcrop studies usually include complete lengths and *partial lengths*—those limited by the edge of the outcrop. The expected proportion of partial occurrences is a function of the relative lengths of the feature and the outcrop. This relationship can be used to estimate full-length distributions.

3 It has previously been shown that length distributions observed in a 2-D slice (outcrop) are greatly dependent on the *shapes* and *orientations* of the features. Thus, outcrop-derived distributions are ambiguous and can only be applied to modelling with appropriate shape/orientation assumptions.

The effects of these considerations are shown by hypothetical and real examples. They can have a critical influence on model results, particularly if length distributions are spread over a wide band.

INTRODUCTION

Geological data are often evaluated via statistics describing their frequency of occurrence. Figure 1 (from Weber, 1982) shows an example of shale length frequencies for different depositional environments. Distributions of the lengths (or widths) of geological features, such as sandbodies or shales, are of prime importance in reservoir simulation (see, for example, Haldorsen & Chang, 1986); however, these distributions can easily be misapplied. This paper addresses three issues regarding the use of length distributions: (i) apparent length distribution vs. sample dimensions; (ii) using incomplete data; and (iii) shape and orientation considerations.

* Present address: Hamilton Brothers Oil and Gas Ltd, Devonshire House, Piccadilly, London W1X 6AO, UK

APPARENT LENGTH DISTRIBUTION VS. SAMPLE DIMENSIONS

Length distributions vary with sample dimensionality. For example Fig. 2 shows three sandbodies, two short (narrow) and one long (wide)—a relative length-frequency distribution of 2 to 1. What would the length-frequency distribution be based on well observations? Note that any vertical well will encounter only 1 short to 1 long. The difference is that the diagram is a two-dimensional (2-D) sample of the distribution, and the wells are taking one-dimensional (1-D) samples.

A volume or three-dimensional (3-D) sample would also have a different length-frequency distribution. The sandbodies in Fig. 2 were drawn in channel form, suggesting much greater lengths in the orthogonal direction; however, if we assume

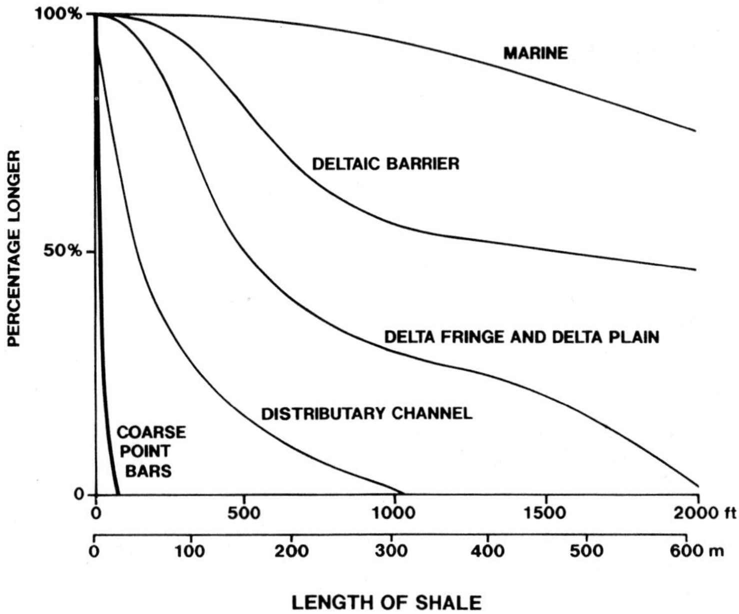

Fig. 1. Shale length distribution as a function of depositional environment. From Weber (1982).

them to be isotropic features, then the 3-D volume frequency distribution is four shorts to one long.

In going from a 2-D distribution to 3-D (or 1-D to 2-D) the relative frequency of a particular length class will decrease in proportion to its length, assuming an isotropic shape. Or:

$$F_{3\text{-D}}(L) = \frac{F_{2\text{-D}}(L) \times 1/L}{\int_0^\infty [F_{2\text{-D}}(L) \times 1/L]\,\mathrm{d}L} = \frac{F_{1\text{-D}}(L) \times 1/L^2}{\int_0^\infty [F_{1\text{-D}}(L) \times 1/L^2]\,\mathrm{d}L} \quad (1)$$

where F is the frequency of a particular length as a function of L, and L is the length. If the features are not isotropic, the (length)^2 term is replaced by the (length $\times$ the width)^2. These are standard formulations of stereology; however, the concept is easily overlooked, which can lead to erroneous results.

Selection of the proper version of a length distribution depends on the application. Use in 2-D or 3-D reservoir simulation models would require distributions of the corresponding dimensionality. The appropriate version to use with analytical techniques is not always clear; see the example data set (below) where the 1-D version is selected to use with the 'streamline method' of Begg & King (1985).

USING INCOMPLETE DATA FROM OUTCROP

Much of the existing length data have come from outcrop studies, yet outcrops commonly give an incomplete sample of length distributions (Fig. 3). Many lengths are limited by the outcrop rather than natural terminations (partial shales), some at both ends (unlimited shales). Using only the 'complete' lengths biases the distribution toward the short end, because a disproportionate share of the longer values will be outcrop-limited. All the data must be considered to provide the best estimate of the actual length distribution.

A method we have used involves calculation of the probable position of a feature of given length relative to the boundaries of an outcrop of given length.

Features *shorter* than the outcrop

Figure 4 illustrates the situation in which the features are shorter than the outcrop. We assume that

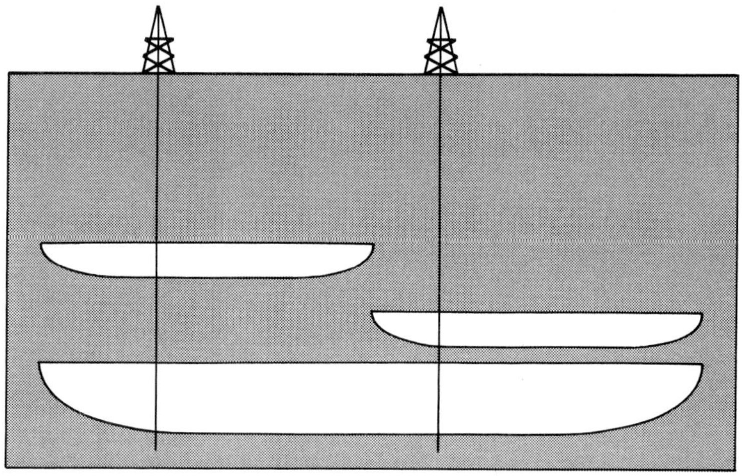

Fig. 2. Diagram illustrating the variation of apparent length distribution with sample dimensionality. This 2-D diagram shows a simple length distribution of 2 : 1, two short sandbodies to one long. Note that each well, sampling in 1-D, encounters a 1 : 1 distribution, one short sandbody to one long.

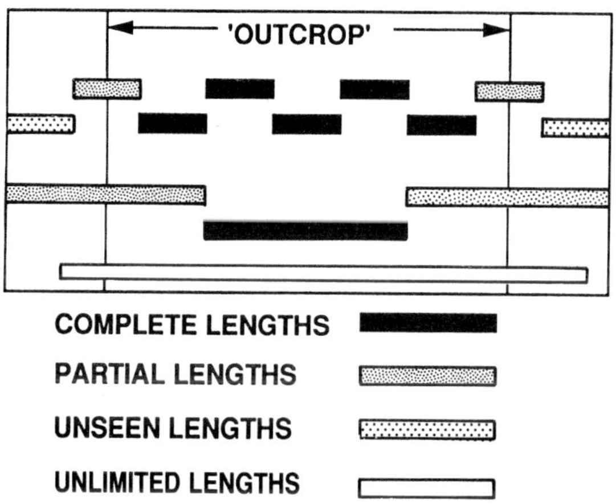

Fig. 3. Categories of length occurrence for features exposed at outcrop.

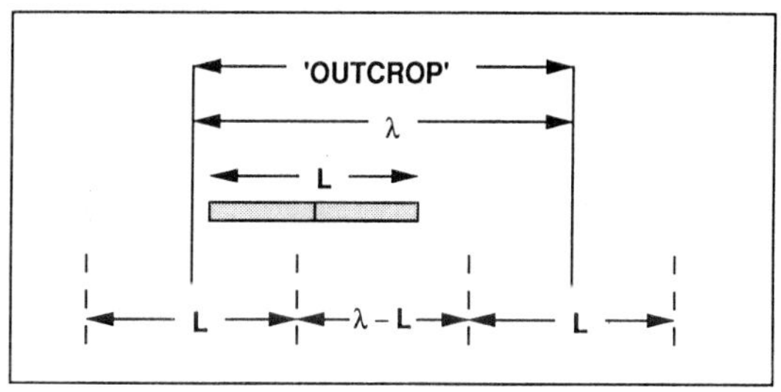

Fig. 4. Diagram of the relationship between the length (L) of a geologic feature and the length (λ) of the outcrop section on which it is observed when the feature is shorter than the outcrop. The length will be partial if the centre of the feature falls within either of the areas marked 'L' centred on the outcrop edges. The length will be complete if the centre falls within the area marked '$\lambda - L$'.

all possible positions are equally likely. For the probability that a feature will be 'partial', we take as the numerator the total length of the diagram in which a feature would overlie *either* of the boundaries: $2L$, where L is the feature length. The denominator becomes the total length within which the feature is either 'partial' or 'complete': $2L + (\lambda - L)$, where λ is the outcrop length. Thus, the probability of partial occurrence (P_p) is:

$$P_p = \frac{2L}{2L + (\lambda - L)} \tag{2}$$

or

$$P_p = \frac{2L}{L + \lambda} \tag{3}$$

In practice, we use this equation to calculate the expected number of partial lengths for each observed complete length class. The total of the expected numbers of partial lengths are then compared to the observed number of partial lengths. If there is a significant discrepancy, then the base distribution (complete lengths only) is altered iteratively until a good match is obtained. Finally, the partial length occurrences are given their expected actual lengths to provide the estimated total distribution including partial lengths.

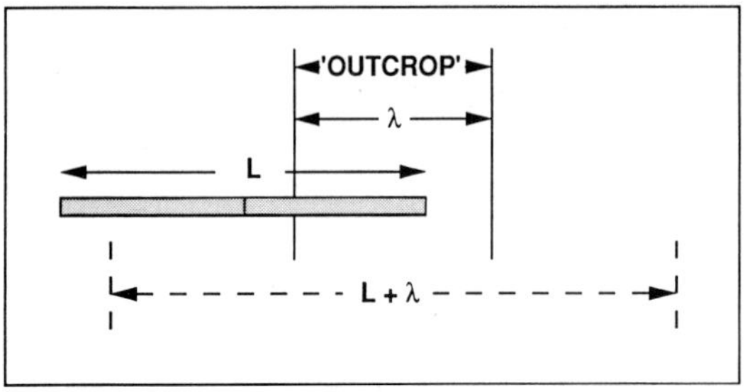

Fig. 5. Diagram of the relationship between the length (L) of a geologic feature and the length (λ) of the outcrop section on which it is observed, when the feature is longer than the outcrop. The length will be partial if either edge of the feature falls on the outcrop section. The length will be 'unlimited' it the centre falls anywhere within the area marked '$L + \lambda$' and neither edge falls on the outcrop.

Normalizing the distribution to a uniform area

In adjusting a distribution for the partial lengths, we have substituted an unwanted bias for the bias that was corrected. Figure 3 shows that there are also 'unseen' shorter length occurrences that lie within the same area as some of the longer partial lengths. To correct for this new bias, we must normalize the distribution; that is adjust the frequency for each length class so that it is representative of the same uniform area. This is done as follows:

$$F_n = F \frac{\lambda}{L + \lambda} \tag{4}$$

where F_n is the normalized frequency of a particular length class and F is the initial frequency adjusted for partial occurrences.

Features *longer* than the outcrop

Length estimations can also be made even without complete length occurrences, that is when the features are longer than the outcrop (see Fig. 5). Similarly to the situation described above, the probability that a shale will be partial is a function of its length relative to the outcrop length; however, in this case, the probability of partial occurrence (P_p) is:

$$P_p = \frac{2\lambda}{L + \lambda} \tag{5}$$

Therefore, using an observed relative proportion of 'partials' as an indication of their probability of occurrence, we can use the above equation to calculate a most likely single length:

$$L = \frac{2\lambda}{P_p} - \lambda$$

Here we have estimated what is probably a distribu-

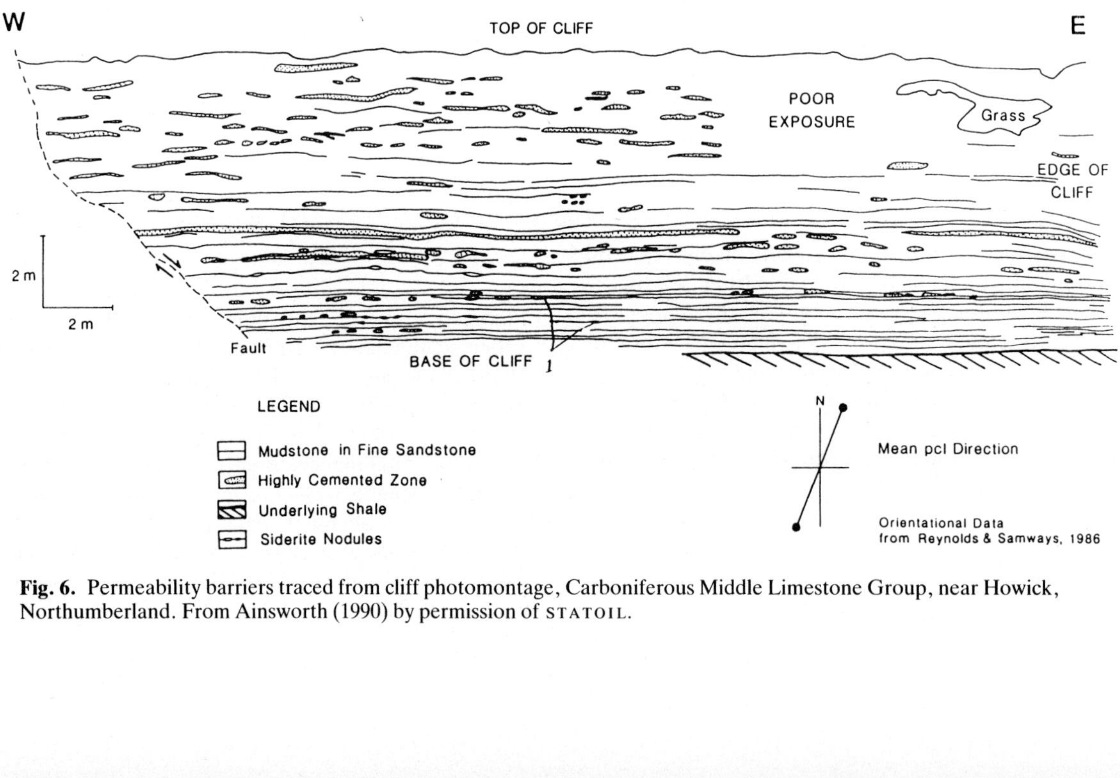

Fig. 6. Permeability barriers traced from cliff photomontage, Carboniferous Middle Limestone Group, near Howick, Northumberland. From Ainsworth (1990) by permission of STATOIL.

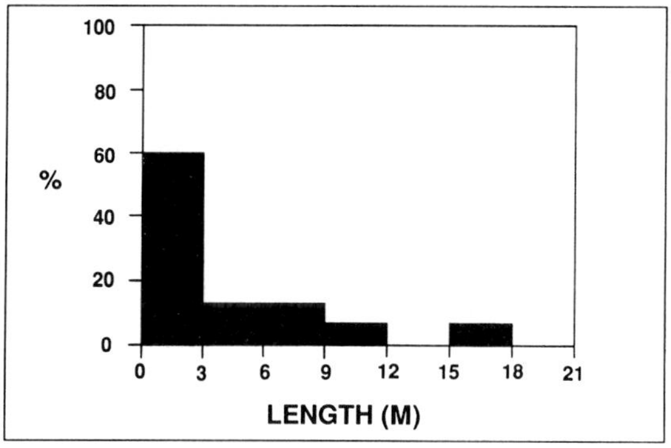

Fig. 7. Length distribution for 'complete' shales from the middle portion of the Howick outcrop.

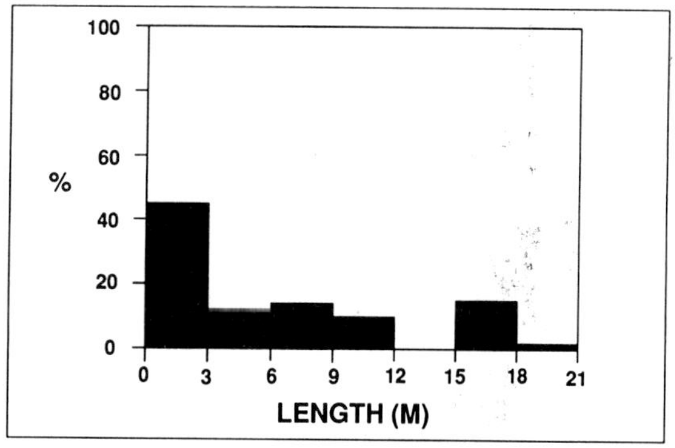

Fig. 8. Howick outcrop (2-D) shale length distribution adjusted for the estimated full lengths of partial length occurrences.

tion of lengths with a single characteristic length. Note that this characteristic length is strictly 2-D. In 1-D and 3-D, characteristic lengths probably will be different, the amount depending on the spread of the actual distribution which we have not estimated.

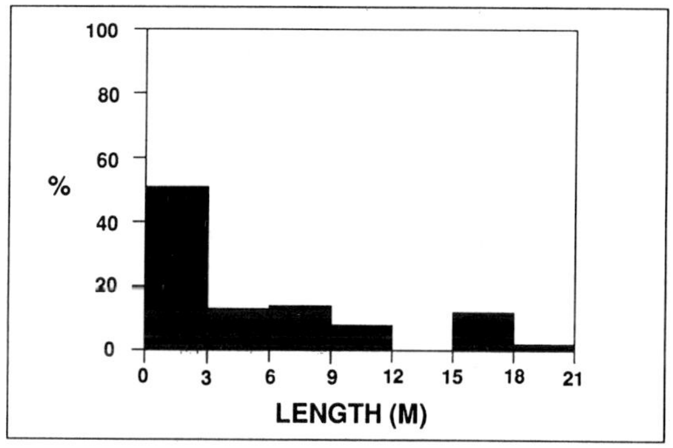

Fig. 9. Howick outcrop (2-D) shale length distributions adjusted for partial lengths and normalized to equal area.

EXAMPLE DATA SET

Figure 6 shows the distribution of thin shales on an outcrop face, interpreted and mapped by Ainsworth (1990, personal communication). The sandstone is from the Middle Limestone Group of the Lower Carboniferous of Northumberland, England. Outcropping on the beach near the coastal village of Howick, it coarsens upwards and is interpreted as storm/wave influenced, shallow marine in origin (Brenchley & Gowland, 1985; Reynolds, 1992; Reynolds & Samways, 1986).

As an example data set, we have taken 43 shales from the middle portion of the outcrop, which is characterized by a reasonably consistent sedimentary facies (grain size and structures). In the following, we (i) adjust the shale length frequency distribution, incorporating the incomplete data in a step-by-step process; (ii) consider the variability of the distribution according to sample dimensionality; and (iii) summarize the importance of these steps in terms of a prime reservoir performance parameter: K_v/K_h.

Adjustments for incomplete data

Figure 7 shows the distribution of 'complete' shale lengths, that is shales with natural terminations within the outcrop. In this case, when we calculated the *expected* partial lengths for the distribution of complete lengths in Fig. 7, the total number was very close to the *actual* observed number of partial lengths. Adjusting the distribution for the expected values of those partial lengths leads to the distribution shown in Fig. 8, with a much higher proportion of longer shales. Normalizing the distribution to an equal area results in the distribution shown in Fig. 9, somewhat in-between the two previous distributions.

Sample dimensionality

Figure 9 represents a best estimate of the actual 2-D distribution. If we assume that the shales are isotropic in shape, then the expected 3-D volume

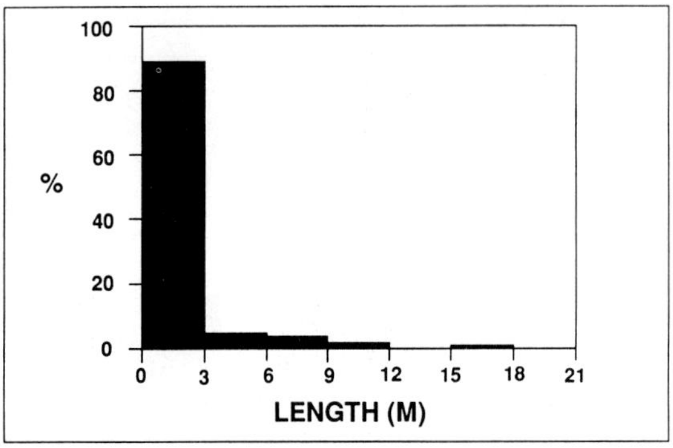

Fig. 10. Volume (3-D) shale length distribution inferred from the normalized outcrop distribution shown in Fig. 9, assuming shales are isotropic in plan shape.

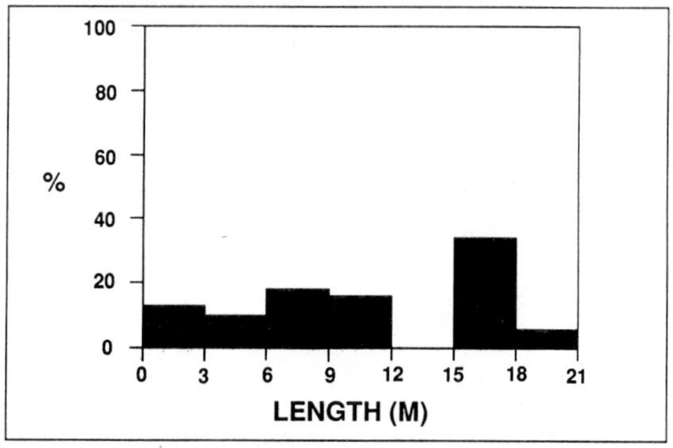

Fig. 11. 1-D (vertical) shale length distribution calculated from the normalized outcrop distribution shown in Fig. 9. This is the distribution that vertical wells would sample and that vertically moving fluid particles would encounter.

distribution is shown in Fig. 10. Volumetrically, the shorter shales must be much more numerous in the 3-D case.

The expected 1-D distribution is shown in Fig. 11. This is the frequency distribution that would be encountered by a vertical well or by a fluid particle moving vertically through the sandstone. Here the longer shales are encountered much more frequently simply because they cover much more area. Obviously a vertical well will not actually sample the lengths of the shales it encounters; however, lengths often can be inferred from other criteria such as length–thickness relationships or depositional facies interpretation. This distribution is important for some methods of effective permeability estimation, as in the following.

Effects on K_{ve}/K_{he} estimation

A distribution of thin discontinuous shales, such as in our example, will reduce the effective vertical permeability (K_{ve}) relative to the effective horizontal permeability (K_{he}). There are a variety of different methods for estimating this effect (Begg *et al.*, 1985; Begg & King, 1985). Table 1 shows K_{ve}/K_{he} values for various versions of the shale distribution calculated using the 'streamline method (2-D)' of Begg & King (1985, p. 3). It is assumed that the shales have 0 permeability.

Since the streamline method treats a distribution of shale lengths as encountered by fluid moving vertically, the proper version to use in this case is the 'inferred 1-D' (bottom line of Table 1). Calculated K_{ve}/K_{he} values for the other versions are given to show the relative effects of the various adjustments made to improve the initial distribution taken from only the 2-D sample of complete length measurements. Note that while inclusion of incomplete data does significantly affect K_{ve}/K_{he} estimates, dimensionality is critical. Improperly using the 3-D distribution results in an error of more than an order of magnitude.

Table 1. Effective K_v/K_h calculated by the streamline method of Begg & King (1985) using various versions of the example shale length distribution. Note that the *inferred 1-D* is the only correct distribution to use for *this calculation method* and, therefore, 0.7 is the proper estimate

Version	K_{ve}/K_{he} (%)
Complete lengths only	(4.7)
Including incomplete data	(2.1)
Normalized 2-D	(2.9)
Inferred 3-D	(22.8)
Inferred 1-D	*0.7*

ORIENTATION AND SHAPE CONSIDERATIONS

The points made in the preceding discussion relating to 2-D outcrop sampling and the inclusion of incomplete data are independent of the shapes or orientations of the features considered. However, knowledge of orientation and shape is very important

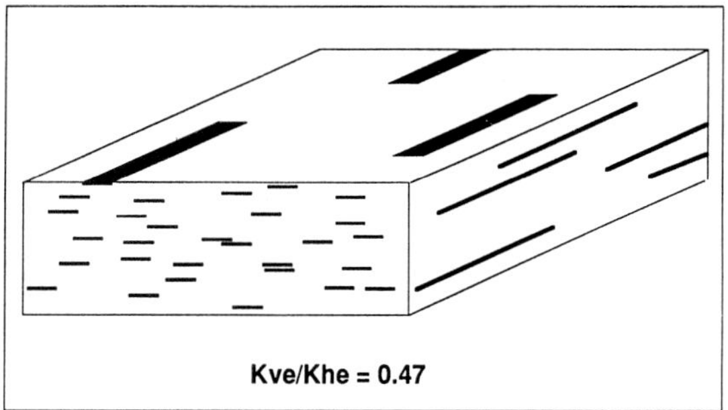

Fig. 12. Hypothetical shale distribution. The front face is a lateral cut of uniformly oriented shales with 5 : 1 aspect ratio. Effective K_v/K_h estimation using the streamline method (3-D) of Begg & King (1985).

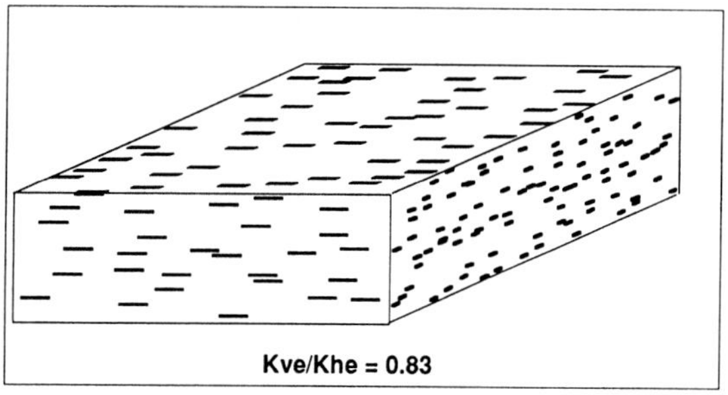

Fig. 13. Hypothetical shale distribution. The front face is identical to that in Fig. 12; however, here we assume that it is a longitudinal cut of uniformly oriented shales with 5 : 1 aspect ratio. Effective K_v/K_h estimation using the streamline method (3-D) of Begg & King (1985).

to a full 3-D characterization of any distribution of features.

Relationships between 2-D and 3-D length distributions

A key question is the relationship between primary length/width axes values and the distribution of lengths from a 2-D slice (Eller, 1988; Mijnssen, *et al.*, 1990). If the slice is orthogonal to the major axes of uniformly oriented features, such as in Figs 12 and 13, then the sampled lengths/widths will match the primary length/widths. However, in the more realistic situation of an oblique slice orientation and/or varying orientations of the features themselves, the distribution of measured lengths will vary even if the features themselves are all the same size and shape.

King (1988, personal communication) has derived analytical solutions to the relationship between true length statistics and outcrop samples for circular and elliptical features of known eccentricity and orientation. Mijnssen *et al.* (1990) showed that for the case of randomly sliced rectangles, the observed length-frequency distribution will be bimodal with the tallest peak corresponding to the width and the shorter peak corresponding to the length of the rectangles. Thus, the sample distributions shown in Fig. 9 could represent shales of roughly the same size but varying orientation. From the normalized case, we would interpret shales of uniform size but variable orientation about 1.5 m in width and 16.5 m in length.

Clearly, while there are methods of estimating actual 3-D geometries from 2-D data, they all rely on orientation and shape assumptions; thus, there will always be a degree of ambiguity in evaluating 2-D data. In the following section, we present two examples indicative of the general level of uncertainty in estimation of effective permeability due to orientation/shape ambiguity.

Effect on permeability

In the first example, we consider the effect of orientation only. Consider that the front face of the block diagrams in Figs 12 and 13 is a sample shale distribution; notice that the front face is identical in the two figures. If we hold the shape to be known and uniform—in this example we use a 5:1 aspect ratio—then extremes in the uncertainty of orientation are represented by Figs 12 and 13. The extremes in uncertainty of effective K_v/K_h are also shown in the two figures.

In the second example, we consider the effect of shape only (Fig. 14). Here we have held the overall size of the shales at a constant 500 m^2 and the *overall* vertical frequency at a constant one shale per 2 m. The shales are otherwise assumed to be distributed randomly. In this case, effective K_v/K_h is fairly constant until aspect ratio drops below 0.5.

CONCLUSIONS

1 Dimensionality is critical in the sampling and application of length-frequency distributions. We have presented a simple method to transform

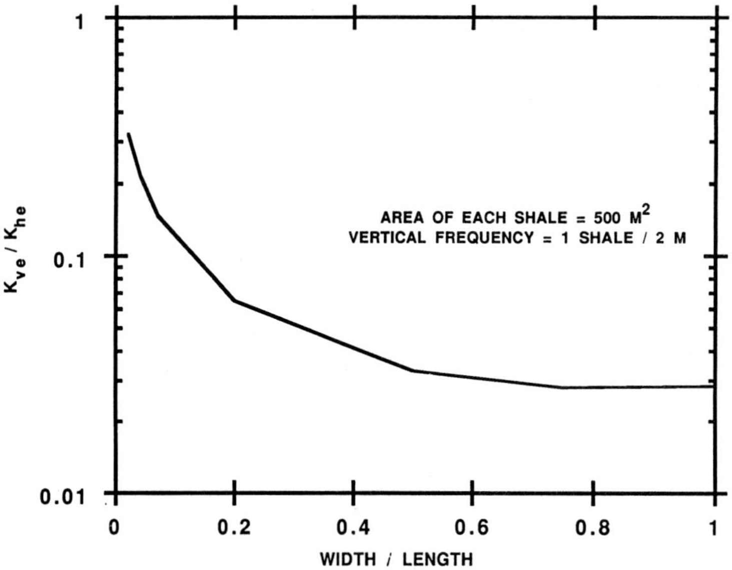

Fig. 14. The variation of the ratio of effective vertical to effective horizontal permeability with aspect ratio for a given shale size and frequency distribution.

distributions from one dimensionality situation (1-D, 2-D, and/or 3-D) to another.

2 Incomplete data (partial and unlimited length occurrences) can and should be used to improve the estimation of actual 2-D length distributions from outcrop. We have presented a two-step approach to handling partial lengths and normalizing a distribution, and have also presented a method to estimate a single characteristic length for a distribution of partial and unlimited lengths.

3 Methods are being developed to estimate 3-D geometries from 2-D sample distributions given assumptions of orientation and shape. Ambiguities in orientation and/or shape can significantly affect resulting reservoir performance estimation, and these parameters should be addressed during the evaluation and use of length distribution data.

ACKNOWLEDGEMENTS

We would like to thank Paul Kirwan (BP) for helping with the development of the ideas presented here. We are also grateful to STATOIL, in particular Charles Jordan and Andrew Hirst (now of UNOCAL) for allowing us to use the Howick shale data. We also thank Chris Mijnssen for his constructive comments after the initial presentation of this material.

REFERENCES

Begg, S.H., Chang, D.M. & Haldorsen, H.H. (1985) A simple statistical method for calculating the effective vertical permeability of a reservoir containing discontinuous shales. *SPE Paper* **14271**.

Begg, S.H. & King, P.R. (1985) Modelling the effects of shales on reservoir performance: calculation of effective vertical permeability. *SPE Paper* **113529**.

Brenchley, P.J. & Gowland, S. (1985) *The Sedimentology of Wave and Storm Influenced Shelf Facies: Carboniferous of Northumberland and Jurassic of Yorkshire*. University of Liverpool, Department of Earth Sciences Field Guide.

Eller, M.G. (1988) The limitations of shale statistics measured from outcrop. In: *Quantification of Sediment Geometries and their Internal Heterogeneities*, Geol. Soc. London. (Abstract.)

Haldorsen, H.H. & Chang, D.M. (1986) Notes on stochastic shales; from outcrop to simulation model. In: *Reservoir Characterization* (Ed. by L.W. Lake & H.B. Carroll, Jr.) Academic Press, Orlando, Florida, 659 pp.

Mijnssen, F.C.J., Weber, K.J., Rozendal, S., Kool, H. & Vonck, J.D.N. (1990) *The Effect of Horizontal Shales on Vertical Permeability*. EAPG (Abstract.)

Reynolds, A.D. (1992) *Storm, wave and tide dominated sedimentation in the Dinantian Middle Limestone Group, Northumbrian Basin*. Proc. Yorks. Geol. Soc., vol. 49, part 2.

Reynolds, A.D. & Samways, G.M. (1986) *The Middle Limestone Group, uppermost Visean, Lower Carboniferous, Northumberland*. University of Liverpool, Department of Earth Sciences Field Guide.

Weber, K.G. (1982) Influence of common sedimentary structures on fluid flow in reservoir models. *J. Petrol. Technol.* **34**, 665–672.

Spec. Publs int. Ass. Sediment. (1993) **15**, 213–236

A theoretical study of fluvial sandstone body dimensions

J.S. BRIDGE *and* S.D. MACKEY

Department of Geological Sciences, State University of New York, PO Box 6000, Binghamton, New York 13902-6000, USA

ABSTRACT

The geometry and distribution of channel-belt sandstone bodies in alluvial deposits have been simulated recently using stochastic models. These models are based on Monte Carlo sampling of empirical distributions of channel-belt widths and thicknesses for channels of varying planform geometry. However, such distributions are difficult to determine in practice and the location of channel-belt sandstone bodies in these models is essentially random. In order to overcome these difficulties, a process-based model of alluvial stratigraphy is used here to simulate the widths and thicknesses of *avulsion controlled* channel-belt sandstone bodies. Our results suggest that sandstone body width and thickness depend critically on the proportion of channel deposits (*CDP*) as this controls the degree of connectedness of individual channel belts. For low values of *CDP*, channel belts are unconnected, sandstone body width equals channel-belt width, and sandstone body thickness equals aggraded channel-belt thickness. As *CDP* increases and a few channel belts become connected, width and thickness frequency distributions become polymodal (and standard deviation increases) with the largest modes at the smallest width and thickness values. As *CDP* increases further, the distributions are still polymodal but larger modes occur at higher width and thickness values. For *CDP* values greater than 0.75, all channel belts are connected, and the single sandstone body has a width equal to floodplain width and a thickness equivalent to the whole section. Channel deposit proportion, connectedness, and sandstone body widths and thicknesses are controlled by channel-belt width/floodplain width, maximum channel depth, cross-floodplain variation in deposition rate, avulsion frequency, depth of burial (affecting compaction), and tectonic tilting of the floodplain.

INTRODUCTION

The geometry and spatial distribution of channel-belt sandstone bodies in alluvial successions are of major concern to geologists. Recently, reservoir geologists have employed a variety of stochastic models to simulate the architecture of fluvial reservoirs (e.g. Haldorsen & Macdonald, 1987; Begg & Williams, 1989; Budding *et al.*, 1988). A common approach is to specify sandstone body widths and thicknesses by sampling empirical distribution functions, and then randomly place the sandstone bodies in space until the required sandstone proportion is reached. A series of arbitrary overlap rules may be employed to produce 'realistic' spatial distributions of sandstone bodies. A variant of this approach (Budding *et al.*, 1988) is to place the first sandstone bodies at well locations such that their vertical positions and thicknesses agree with well data. The space between wells is then filled up appropriately.

The empirical distribution functions employed in the approaches described above are supposed to represent the widths and thicknesses of unconnected channel-belt sandstone bodies measured in a cross-section normal to their direction of elongation (i.e. regional palaeocurrent direction). They are derived from observations of modern and ancient deposits of specific types of channels. Therefore, in order to use this approach it is necessary to determine the regional palaeocurrent direction, type of palaeochannel, and appropriate distribution functions for channel-belt width and thickness. Herein lies a major problem as discussed below.

It is imperative to separate data on the dimensions

of ancient channel belts from those of modern channel belts, as many of the parameters used to describe modern channel-belt dimensions are difficult (or impossible) to determine in ancient deposits. For example, in the case of ancient channel belts, determination of channel type is a formidable task even with large exposures, and is virtually impossible from subsurface data (e.g. Bridge, 1985). In order to determine channel-belt width and thickness it is necessary to observe the full width of sandstone bodies in sections nearly normal to palaeoflow direction, and to ensure that these sandstone bodies are not composed of connected channel belts. The thicknesses of aggraded channel-belt sandstone bodies of a given width are likely to be quite variable, as they are controlled by maximum bankfull channel depth, aggradation rate, and avulsion frequency (Bridge & Leeder, 1979). It is important to note that maximum bankfull channel depth is not necessarily equivalent to channel-belt thickness as suggested by Collinson (1978). Fielding & Crane's (1987) data indicate that channel-belt thickness/maximum channel depth commonly ranges between 1 and 4. For these reasons, it is unlikely that limited data from one or two well-exposed formations (e.g. Collinson, 1978; Walderhaug & Mjøs, 1991; Lowry & Raheim, 1991) will be generally representative or even specifically representative of adjacent formations.

The width and channel pattern of modern channel belts are relatively easy to determine. However, determination of their thickness requires extensive coring and some projection of how much aggradation will occur before the channel belt is abandoned. There have been a number of attempts to relate the widths of modern channel belts to channel width and/or depth (e.g. Collinson, 1978; Lorenz *et al.*, 1985; Fielding & Crane, 1987) with the ultimate objective of predicting channel-belt widths from channel dimensions reconstructed from ancient deposits. The pitfalls of these approaches are discussed in a following section.

Another problem with the stochastic modelling approach is the random placement of channel-belt sandstone bodies. This problem is countered, to a certain extent, by 'sequence based' stochastic simulations (reviewed by Dubrule, 1989). Here, the lithological characteristic predicted to occur at any point in space is dependent on what is found at a neighbouring site. The conditional probabilities of occurrence are based on, for instance, a Markov transition matrix or an autocorrelation function. Unfortunately, such probabilities are very difficult to define in the horizontal direction (between wells), and these methods are unable to incorporate information about the geometry of channel-belt sandstone bodies.

Process-based alluvial architecture models are a viable alternative to these stochastic models (e.g. Allen, 1978; Leeder, 1978; Bridge & Leeder, 1979). These models have the potential for predicting realistic geometries and spatial distributions of channel-belt sandstone bodies, but have the disadvantage that they cannot simulate the exact sequence observed in a well. The purpose of this paper is to use a modified version of Bridge & Leeder's (1979) two-dimensional model of alluvial stratigraphy (see Bridge & Mackey, 1993; Mackey & Bridge, 1992) to

Table 1. List of symbols used

a	channel-belt deposition rate (m/yr)
b	'b' exponent
CDP	channel deposit proportion
CR	connectedness ratio
d	maximum bankfull depth of channel (m)
d_m	mean bankfull depth of channel (m)
ϕ	meander arc-angle (radians)
L	meander wavelength (m)
n	exponent in power function for $\overline{SBW}/W$ versus CDP
r	time-averaged overbank deposition rate (m/yr)
$\bar{r}$	mean overall floodplain deposition rate (m/yr)
R	meander radius of curvature (m)
SBW	width of sandstone body (m)
$\overline{SBW}$	mean width of sandstone bodies (total section) (m)
SBT	mean thickness of an individual sandstone body
$\overline{\mathrm{SBT}}$	mean thickness of sandstone bodies (total section) (m)
$\overline{SBW}/W$	dimensionless mean sandstone body width
$\overline{SBT}/\bar{T}$	dimensionless mean sandstone body thickness
$\overline{SBW}/\overline{SBT}$	mean sandstone body width over mean thickness
Sn	channel sinuosity
t	avulsion period (yrs)
$\bar{t}$	mean avulsion period (yrs)
t_{tot}	total time elapsed for section (yrs)
$\bar{T}$	$a\bar{t} + d$, mean thickness of individual channel belts (m)
w	width of channel belt (m)
w_c	width of channel (m)
W	width of floodplain (m)
z	distance from edge of channel belt (m)
z_m	$W-w$, width of floodplain minus width of channel belt (m)

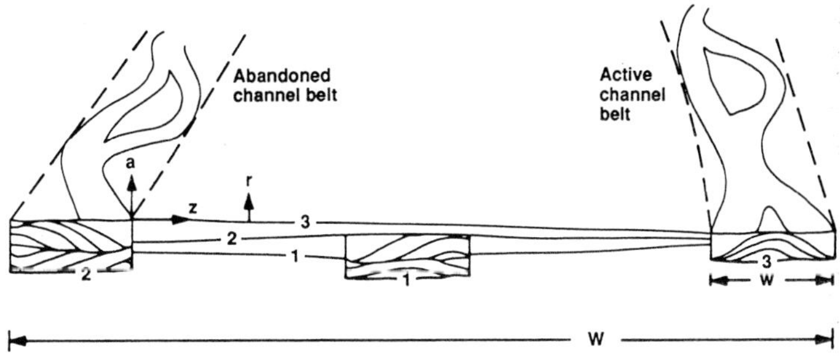

Fig. 1. Definition diagram for revised FPSM model. Floodplain surfaces are represented at a point in time immediately prior to an avulsion. Note that channel belts are formed by either meandering or braided channels.

explore the controls on sandstone body geometry and distribution in *avulsion controlled* alluvial sequences.

MODEL DESCRIPTION

In the revised model, a floodplain of finite width, W, is occupied by a single active channel belt of width, w, formed by *either* a meandering or braided channel with maximum bankfull depth, d, which is free to migrate laterally within the channel belt (see Fig. 1 and Table 1). Sediments within the channel belt are dominantly sands and/or gravels while the adjacent overbank deposits are dominantly silts and clays. Channel belts are aggraded at a constant rate during the time between avulsions. Adjacent floodplain surfaces are aggraded concurrently, with overbank deposition decreasing away from the edge of the channel belt. Immediately prior to an avulsion (sudden abandonment of a channel belt in favour of a new location on the floodplain surface), both fine-grained overbank deposits and coarse-grained channel-belt deposits are compacted. If selected, tectonic tilting of the floodplain surface is simulated by changing the relative elevations of the floodplain margins. The number of tectonic events and associated movements (direction and magnitude) are determined stochastically. The elevation of the floodplain surface is recalculated incorporating the effects of differential aggradation, compaction, and tectonic tilting. The channel belt is then relocated to the position of lowest elevation on the floodplain surface (avulsion). If there is more than one possible location, the point nearest to the currently active channel belt is chosen. It is assumed that the point of avulsion occurs far enough upstream to allow free migration of the channel to any point on the floodplain surface. When the new channel belt is initially established, the underlying floodplain deposits are eroded to a depth equal to the bankfull depth of the channel. Then the cycle of aggradation, compaction, tectonism and avulsion is repeated until the desired number of avulsions is reached. The time between avulsive events, t, is determined stochastically.

A two-dimensional stratigraphic cross-section (oriented normal to the valley axis) is plotted illustrating coarse-grained channel-belt deposits, fine-grained overbank deposits, and time lines representing the compacted floodplain surface at the time an avulsion occurs. Information on the thickness and lithology of each avulsion-bounded depositional increment for a series of reference verticals across the floodplain section is used to calculate the areal proportion of coarse-grained channel-belt deposits (channel deposit proportion) and the proportion of horizontal contact between channel belts (connectedness ratio). Channel-belt sandstone body dimensions and associated statistics are calculated for both individual sandstone bodies and the total section (see Bridge & Mackey, 1993; Mackey & Bridge, 1992).

INPUT PARAMETERS AND EXPERIMENTAL PROCEDURES

In this simulation model, and in other approaches discussed previously, it is necessary to make realistic estimates of channel-belt width, w, in relation to the average maximum bankfull depth of the channel, d, or bankfull channel width, w_c (Fig. 2). Given laterally extensive outcrops oriented almost normal to palaeoflow direction, it may be possible to directly measure channel-belt widths and thicknesses along with the dimensions of individual channels (i.e. width and depth). However, if suitable outcrops or subsurface data are unavailable for direct

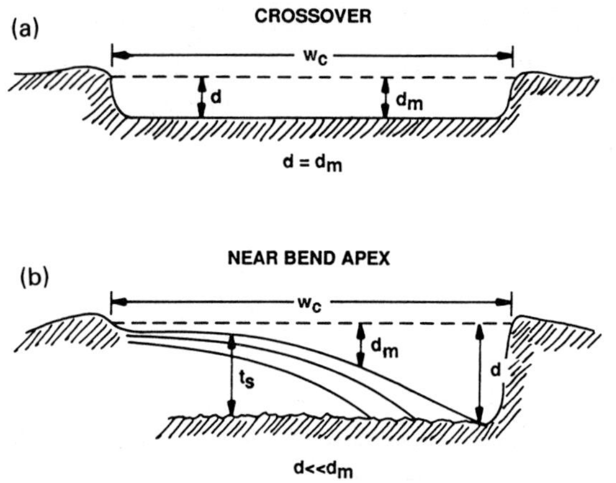

Fig. 2. Definition diagram for bankfull depth measurements. (a) At crossovers, mean bankfull depth, d_m, is equal to maximum bankfull depth, d. (b) Near bend apices, maximum bankfull depth may be up to three times mean bankfull depth. Note that storey thickness, t_s, is slightly less (10%) than the maximum bankfull depth. In ancient deposits, storey thicknesses must be corrected for compaction.

measurement, channel belt dimensions may have to be estimated using theoretical relationships based on planform geometry and/or empirical regression expressions derived from modern river data (Tables 2, 3 and 4).

Several attempts have been made to apply empirical expressions to predicting channel-belt widths in the subsurface. Collinson (1978) combined Carlston's (1965) empirical equations for meander-belt width and channel width against discharge and Leeder's (1973) empirical relationship between channel width and maximum bankfull depth to produce a relationship between meander-belt width and channel depth. The resulting equation (Collinson, 1978) is in error because the proportionality coefficients in Carlston's and Leeder's equations are dimensional but based on different measurement units. Lorenz *et al.* (1985) also used Leeder's (1973) equation for channel width versus maximum depth and combined the data of Leopold & Wolman (1960) and Carlston (1965) to derive an empirical relationship between meander-belt width and channel width. Lorenz *et al.*

Table 2. Regression relationships between channel-belt width, w, and bankfull channel width, w_c, mean bankfull channel depth, d_m, and maximum bankfull channel depth, d

Initial parameter	Channel width	R	SE	Channel depth	R	SE
Jefferson (1902)	(a) $w = 17.6\,w_c$	n/a	n/a	—	—	—
Inglis (1949)	(a) $w = 18.43\,w_c^{0.99}$	n/a	n/a	—	—	—
	$[w = 17.38\,w_c]$					
	(a) $w = 11.43\,w_c^{1.04}$					
	$[w = 14\,w_c]$					
Leopold & Wolman (1960)	(a) $w = 3.4\,w_c^{1.1}$	n/a	n/a	—	—	—
Zeller (1967)	(a) $w = 4.5\,w_c$	n/a	n/a	—	—	—
Collinson (1978)	—	—	—	(b) $w = 65.6\,d^{1.57}$	n/a	n/a
Lorenz *et al.* (1985)	(a,c) $w = 7.53\,w_c^{1.01}$	0.9600	n/a	—	—	—
Williams (1986)	$w = 4.3\,w_c^{1.12}$	0.9600	+74/−42%	$w = 148\,d_m^{1.52}$	0.9000	+115/−53%
Fielding & Crane (1987)	—	—	—	$w = 12.1\,d_m^{1.85}$	n/a	n/a
Regression (this study) (Fig. 4)	$w = 6.89\,w_c^{0.99}$	0.9436	0.233	$w = 192.01\,d_m^{1.37}$	0.8149	0.324
Combined equations (this study) (Fig. 3)						
dataset (this study)	—	—	—	(c) $w = 59.86\,d_m^{1.80}$	n/a	n/a
with Leeder (1973)	—	—	—	(c) $w = 45.76\,d^{1.52}$ or (c) $w = 107.53\,d_m^{1.52}$	n/a	n/a
with Crane (1982)	—	—	—	(c) $w = 86.10\,d_m^{1.57}$	n/a	n/a

Units are in metres.
SE in log units unless otherwise noted.
(a) Mean channel width.
(b) Derived from combined equations with different dimensional units. Corrected expression is listed here.
(c) Derived from combined equations.

Table 3. Relationships between channel-belt width, w, and channel planform geometry. Theoretical expressions of Bridge (1975) and Crane (1982) are also included here as they are based on geometrical models of river meanders

Source	Equation	R	SE	Comments
Schumm (1963)	$Sn = 3.5\,(w_c/d_m)^{-0.27}$	n/a	n/a	Small rivers from Midwest, US
Chitale (1970)	$Sn = 1.145\left(\frac{w}{w_c}\right)^{0.134}$	n/a	18.8%	for $Sn < 1.145$, $w/w_c < 1$ (*not possible*)
Crane (1982)	$w = w_c + 2 \cdot (R - R\cos\phi)$ or $w = w_c + (SnL/2\phi) \cdot (1 - \cos\phi)$	n/a	n/a	Circular arc method where: $\phi = 2.2 \cdot \sqrt{\frac{Sn^{1.3} - 1}{Sn}}$
Bridge (1975)	$w = w_c + L(1.186Sn - 0.247\,Sn^2 + 0.0252\,Sn^3 - 0.820)$	n/a	n/a	Sine generated curve

Units are in metres.
SE in log units unless otherwise noted.

Table 4. Regression relationships between bankfull channel width, w_c, and mean bankfull channel depth, d_m, and maximum bankfull channel depth, d

Source	Equation	R	SE	Comments
Leeder (1973)	$w_c = 6.8\,d^{1.54}$	0.9100	0.300	Sn > 1.7
Crane (1982)	$^{(a)}w_c = 12.82\,d_m^{1.59}$	n/a	n/a	Sn > 1.0
Williams (1986)	$w_c = 15.5\,d_m^{1.40}$	n/a	n/a	Sn > 1.7
	$w_c = 21.3\,d_m^{1.45}$	0.8100	+160/−62%	Sn > 1.3
	$w_c = 96\,d_m^{1.23}Sn^{-2.35}$	0.8700	+121/−55%	Sn dependent
Regression (this study)	$w_c = 8.88\,d_m^{1.82}$	0.6063	0.444	Sn > 1.0
Combined equations (this study)	$^{(a)}w_c = 15.85\,d_m^{1.58}$	n/a	n/a	Sn > 1.0

Units are in metres.
SE in log units unless otherwise noted.
(a) Derived from combined equations.

(1985) then reconstructed channel-belt width using these equations by taking maximum channel depth as the average thickness of untruncated storeys (channel bars) at the top of multistorey sandstone bodies (decompacted by an arbitrary 10%).

There are a number of problems with this approach.

1 Relationships between channel-belt width and channel width and channel width and depth are dependent on wavelength and sinuosity. For example, Leeder's (1973) equation relating bankfull width to maximum bankfull depth is derived from channel data with sinuosities greater than 1.7, and Williams' (1986) equation relating bankfull width and mean bankfull depth includes sinuosity in the regression. Table 2 clearly shows that channel-belt width is approximately proportional to channel width but with proportionality coefficients that vary from 4 to at least 17 as sinuosity varies. Therefore, it is necessary to reconstruct channel sinuosity before this method can be used.

2 Maximum bankfull depth and channel bar thickness in a given reach of a river are spatially variable, such that relationships between bankfull width and maximum bankfull depth will have much scatter (see Fig. 2). In general, channel bar thickness is not equivalent to maximum bankfull channel depth, except near bend apices.

3 Empirical data and associated regression equations exhibit considerable variation and are generally most useful when establishing statistical limits of observed channel and channel belt dimensions (Leeder, 1973). The use of empirical regression expressions (as in the above studies) without

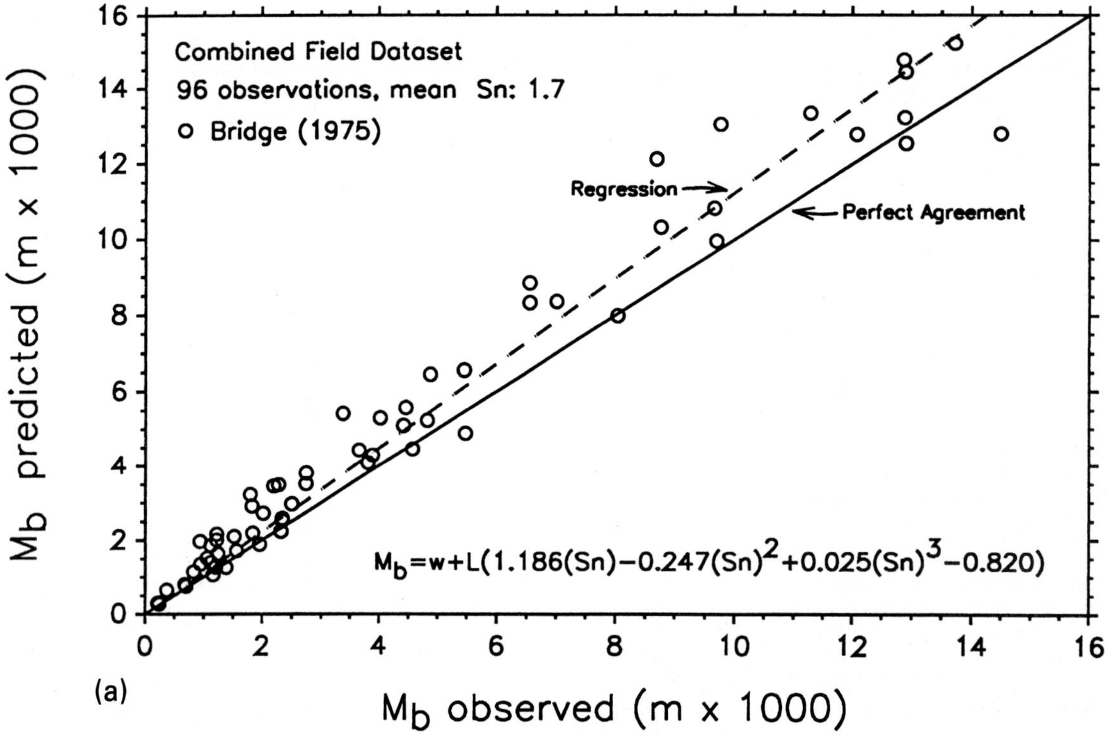

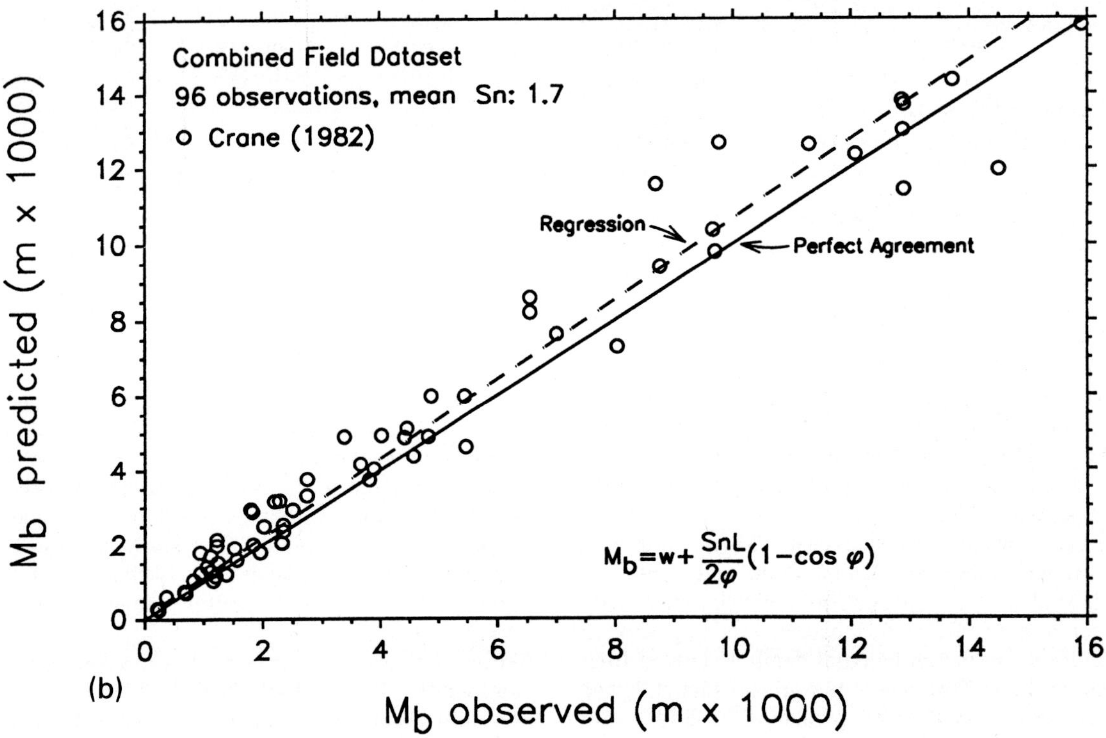

Fig. 3. Comparison of predicted versus observed channel-belt widths for theoretical expressions of: (a) Bridge (1975), and (b) Crane (1982) using published data from modern rivers. Regression lines are the 'best fit' for predicted values. For Bridge (1975), R = 0.9800 with an SE of 526 m. For Crane (1982), R = 0.9795 with an SE of 507 m.

consideration of the associated error is statistically unacceptable.

4 Channel geometry is a function of discharge, sediment supply, and valley slope. Bivariate empirical regression analyses do not include all of the relevant controlling variables, and are biased towards channels with relatively high sinuosities.

Accurate estimates of channel-belt width can be obtained by using the theoretical equations of Bridge (1975) and Crane (1982) that relate channel-belt width to channel width, wavelength, and sinuosity (Table 3). For modern rivers, where channel widths and patterns are relatively easy to determine, these theoretical expressions agree well with natural data (Fig. 3). However, to use these equations in ancient deposits, it is necessary to obtain values for channel width, sinuosity, and wavelength (or radius of curvature) in terms of maximum bankfull channel depth, d, which are difficult to estimate from outcrops or subsurface data (Bridge, 1985). In practice, these parameters may be estimated using high-resolution seismic data, theoretical models (e.g. Bridge 1978, 1984), or direct regression expressions relating channel dimensions to planform geometry (Table 2). Although relationships between bankfull width and depth are based on limited data and depend on at least sinuosity and discharge (see above, and Table 4), channel width can be estimated once depth is specified. Then wavelength, L, can be estimated from $L \approx 10w_c$ (Bridge, 1978). The remaining variable required to define channel-belt width is sinuosity, Sn, which can be assigned a value appropriate for the channel width and depth used (Table 4). Accurate specification of sinuosity is not necessary because the relationship between channel-belt width and bankfull depth is not likely to vary much for channels with different planform geometry. For instance, for low sinuosity and braided channels, $w_c = O(10^2)d_m$ and $w_c \approx w$, so that $w = O(10^2)d_m$, where O denotes 'on the order of'. For high-sinuosity meandering channels, $w_c = O(10)d_m$ and $w = O(10)w_c$, so that $w = O(10^2)d_m$. Indeed, an alternative way of estimating channel-belt width is to use direct regression expressions between channel-belt width and bankfull width and depth based on data from modern rivers (Table 2 and Fig. 4).

In ancient deposits *maximum* bankfull depths can be estimated from the vertical extent of sandstone or conglomerate storeys (channel bars) from either outcrops or subsurface data (e.g. Allen, 1965; Moody-Stuart, 1966; Elliott, 1976; Stewart, 1981; Gardner, 1983; Lorenz *et al.*, 1985) (but see Fig. 2). With the exception of data from Leeder (1973), channel depths reported in the literature for modern rivers are either mean or mean bankfull values that are generally *not* equivalent to the thickness of channel bar deposits in an alluvial channel (see Fig. 2). Equations relating mean bankfull depth to channel width can be combined with Leeder's equation for maximum bankfull depth and channel width to develop an approximate relationship between mean bankfull depth, d_m, and maximum bankfull depth, d, i.e.

$$d_m = 0.57d \tag{1a}$$

which compares favourably with a similar expression reported by Ethridge & Schumm (1978):

$$d_m = 0.585d \tag{1b}$$

For ancient deposits, the thickness of untruncated storeys must be decompacted before these expressions are used. Cotter (1971), Ethridge & Schumm (1978), and Lorenz *et al.* (1985) decompact channel bar thicknesses by an arbitrary 10%. However, if burial depth is known, porosity-depth curves (Sclater & Christie, 1980; Baldwin & Butler, 1985; Scherer, 1987) can also be used to decompact these thicknesses.

Realistic values of the ratio of channel-belt width, w, to floodplain width, W, were obtained from the data of Carlston (1965), Bridge & Leeder (1979), and Crane (1982) and range from less than 0.1 to 0.9. Mean avulsion periods are expected to be of the order of 10^2 to 10^3 years (Bridge & Leeder, 1979). In the floodplain deposition rate equation

$$r = \frac{a}{(1 + z/z_m)^b} \tag{2}$$

the exponent, b, is expected to range from approximately 0.5 to 10, based on comparison with natural data (Bridge & Mackey, 1993). Here, r is overbank deposition rate, a is channel-belt aggradation rate, z is distance from edge of the channel belt, and z_m is the maximum value of z (see Bridge & Mackey, 1993). Estimates of channel-belt aggradation rates were obtained from Bridge & Leeder (1979) and Sadler (1981).

Sandstone body widths, thicknesses, and width/thickness ratios were determined for a series of runs of the simulation model in which the main input variables, channel-belt width, w, floodplain sedimentation (b exponent), channel-belt aggradation

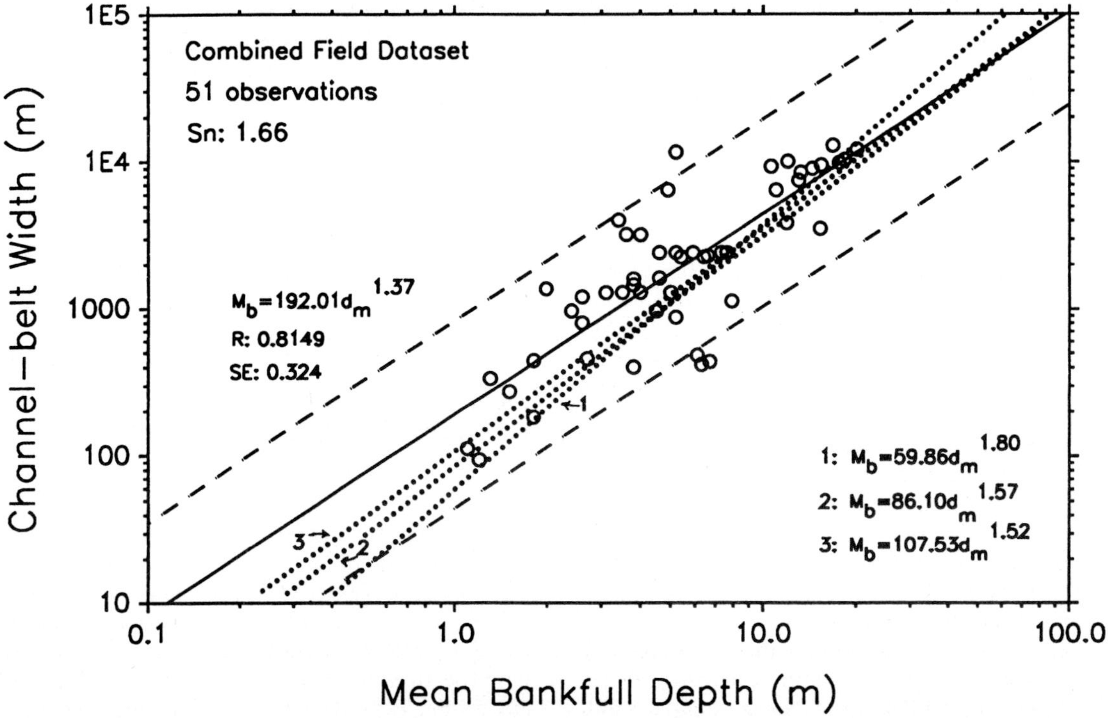

Fig. 4. Regression plot of channel-belt width versus mean bankfull depth for published data from modern rivers (solid line). The 95% confidence limits (dashed lines) are plotted along with three combined expressions (dotted lines) based on regressions from different datasets (Table 2). Regression expressions and associated confidence limits can be used to constrain estimates of channel-belt width in the model.

rate, a, maximum bankfull depth, d, and mean avulsion period, $\bar{t}$,were varied systematically. A detailed listing of all input parameters used in these experiments is given in a companion paper (Bridge & Mackey, 1993). In each experiment, a stratigraphic section was produced by running the model for 50 avulsions (Fig. 5). The average channel deposit proportion, *CDP*, and connectedness ratio, *CR*, for the section were calculated and individual sandstone bodies were identified. Channel deposit proportion, *CDP*, is defined as the areal proportion of channel-belt deposits relative to total area of the generated section. *CDP* values range from 0.0 (all overbank deposits) to 1.0 (all channel-belt deposits). The connectedness ratio, *CR*, is defined as the width of horizontal contact between connected channel belts over the total width of all channel belts in the section. *CR* values range from 0.0 (isolated channel belts) to 1.0 (all channel belts are connected along their entire width). A sandstone body is defined as any number of channel belts that are connected to each other but not connected to any other sandstone body (see Fig. 6). Channel-belt sandstone bodies may contain fine-grained channel-fill deposits and/or overbank deposits, but can probably be considered to be a single reservoir sandstone body for large-scale simulations. The width of an individual sandstone body, *SBW*, is defined as the maximum horizontal distance between edges of the sandstone body. Frequency distributions of sandstone body widths were constructed, and overall mean width, $\overline{SBW}$, and standard deviation were calculated for each simulated section. As the thickness of an individual sandstone body may vary across its width, the mean thickness of each sandstone body, *SBT*, was also calculated. Then the overall mean thickness, $\overline{SBT}$, standard deviation, and frequency distribution of mean sandstone body thickness were calculated for each simulated section. Section averaged sandstone body widths and thicknesses are expressed in

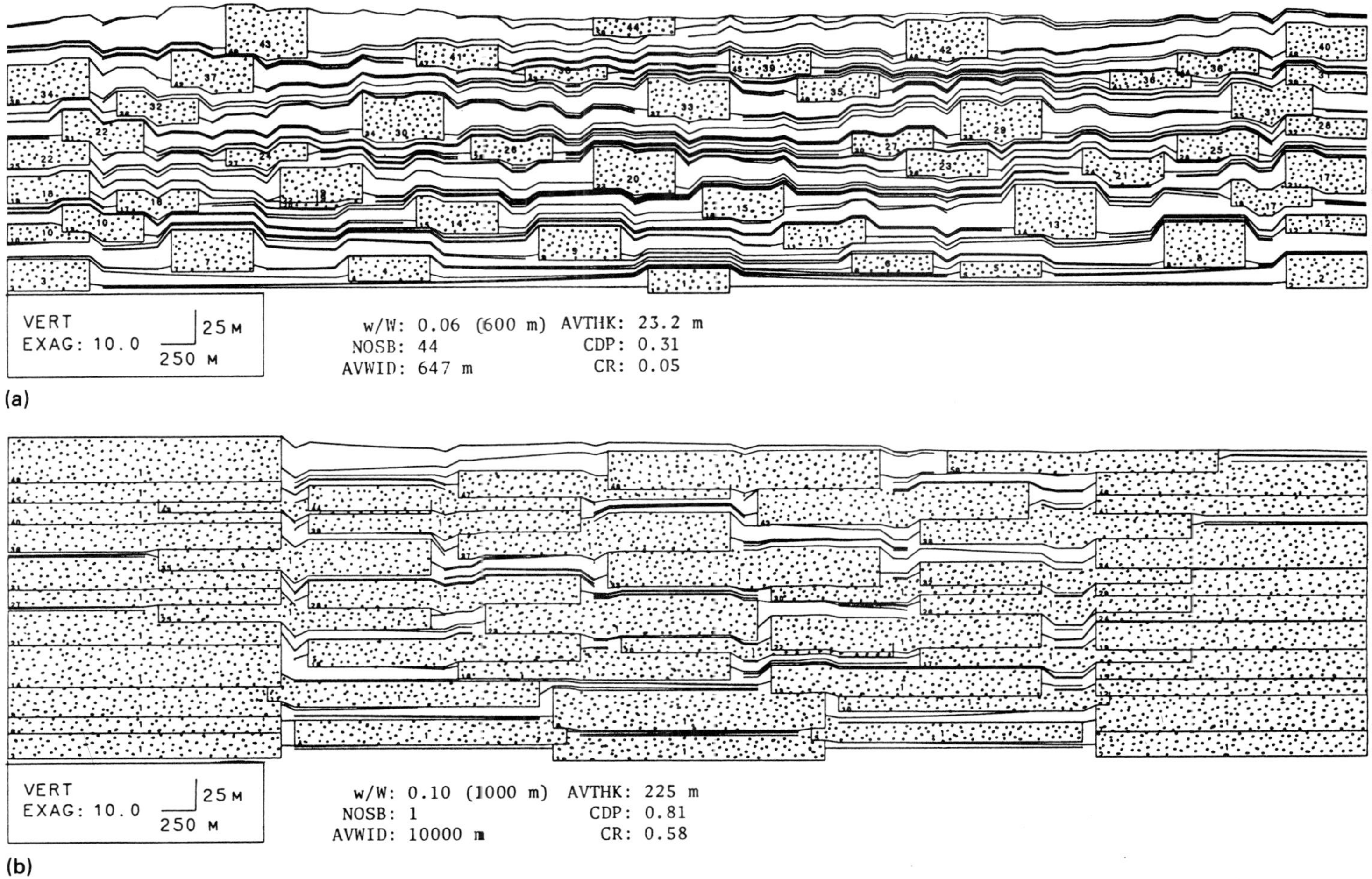

Fig. 5. Examples of simulated cross-sections generated by the model. (a) Isolated channel belts where mean sandstone body width is equal to the mean width of individual channel belts. (b) Case where channel belts are connected and sandstone body width is equal to floodplain width. Note that assigned sandstone body numbers are plotted in the centre of each channel belt and each individual channel belt is identified by an avulsion increment number in the lower left-hand corner of the channel belt.

dimensionless form. i.e. sandstone body width over floodplain width, $\overline{SBW/W}$, sandstone body width over thickness, $\overline{SBW/SBT}$, and sandstone body thickness over channel-belt thickness, $\overline{SBT}$, where $\bar{T}$ is the mean thickness of individual channel belts in the section.

MEAN SANDSTONE BODY WIDTH

In an earlier paper, Bridge & Mackey (1993) demonstrated the effects of varying channel belt and overbank aggradation rates, bankfull depth, avulsion period, and normalized channel-belt width, w/W, on CDP and connectedness. These parameters can also have a profound influence on the width of channel-belt sandstone bodies (Fig. 7). For channel belt aggradation rates greater than 0.02 m/yr (or avulsion periods greater than 1000 years), there is little effect on $\overline{SBW/W}$. However, as a decreases below 0.02 m/yr (or for $\bar{t}$ less than 1000 years), $\overline{SBW/W}$ increases rapidly (Fig. 7a). For w/W values less than 0.2, variation in overbank sedimentation rate or bankfull channel depth has little effect on $\overline{SBW/W}$ (Figs 7b,c). For w/W values of 0.2 or greater, $\overline{SBW/W}$ rapidly approaches unity as the overbank sedimentation rate decreases (increasing b) or bankfull depth increases. In general, for realistic values of a, b, d, and $\bar{t}$, sandstone body widths increase as bankfull depth or channel belt width *increase*, or overbank aggradation rate, channel belt aggradation rate, or avulsion period *decrease*.

Mean sandstone body width is expected to increase with increasing CDP and connectedness, that is, as more channel belts become connected. When there are no connected channel belts, mean sandstone body width is equal to mean channel-belt width and $\overline{SBW}$ equals w/W. When all channel belts are connected, a single sandstone body extends across the entire section and $\overline{SBW/W}$ equals unity. The standard deviation of SBW/W will be zero for both end member cases. $\overline{SBW/W}$ varies with channel deposit proportion as a power function, i.e.

$$CDP \leq 0.75$$

$$\frac{\overline{SBW}}{W} = \left(1 - \frac{w}{W}\right)\left(\frac{CDP}{0.75}\right)^{n} + \frac{w}{W} \qquad (3)$$

where n depends on w/W (Fig. 8c). $\overline{SBW/W}$ and its standard deviation increase with increasing CDP and connectedness until a critical threshold proportion is reached. At this limit, $\overline{SBW}$ goes to W as

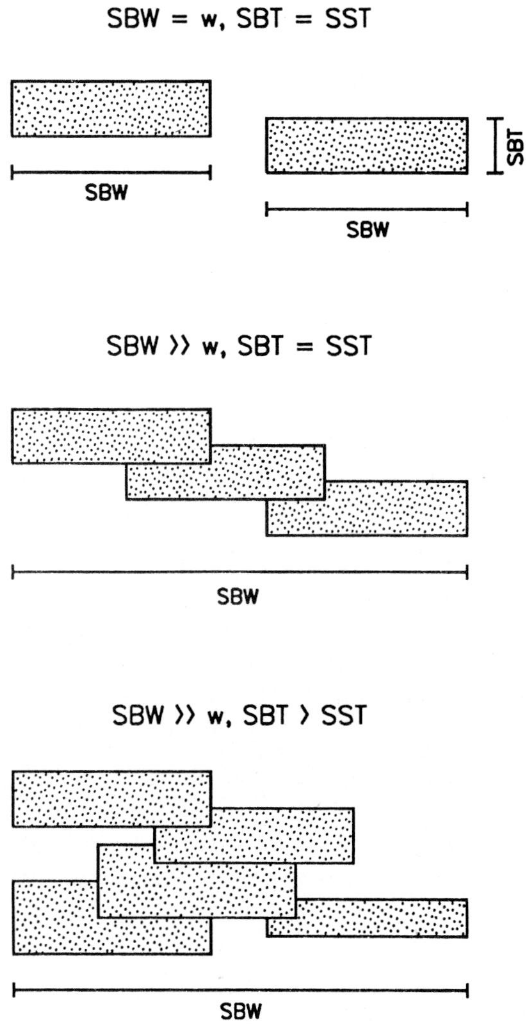

Fig. 6. Definition diagram for channel-belt sandstone body dimensions calculated by the model. Note that fine-grained overbank and channel-fill deposits may be included *within* a single sandstone body. SBT is the mean total thickness of the sandstone body and SST is the mean thickness of coarse-grained deposits within the sandstone body.

CDP increases beyond 0.75 and connectedness becomes greater than 0.54. A similar effect is described in percolation theory where clustering behaviour is controlled by change in areal proportion (Stauffer, 1985). For w/W values greater than 0.2, equation (3) does not apply as CDP values are above the critical threshold value for reasonable values of a, b, d, or $\bar{t}$. For a given w/W, the

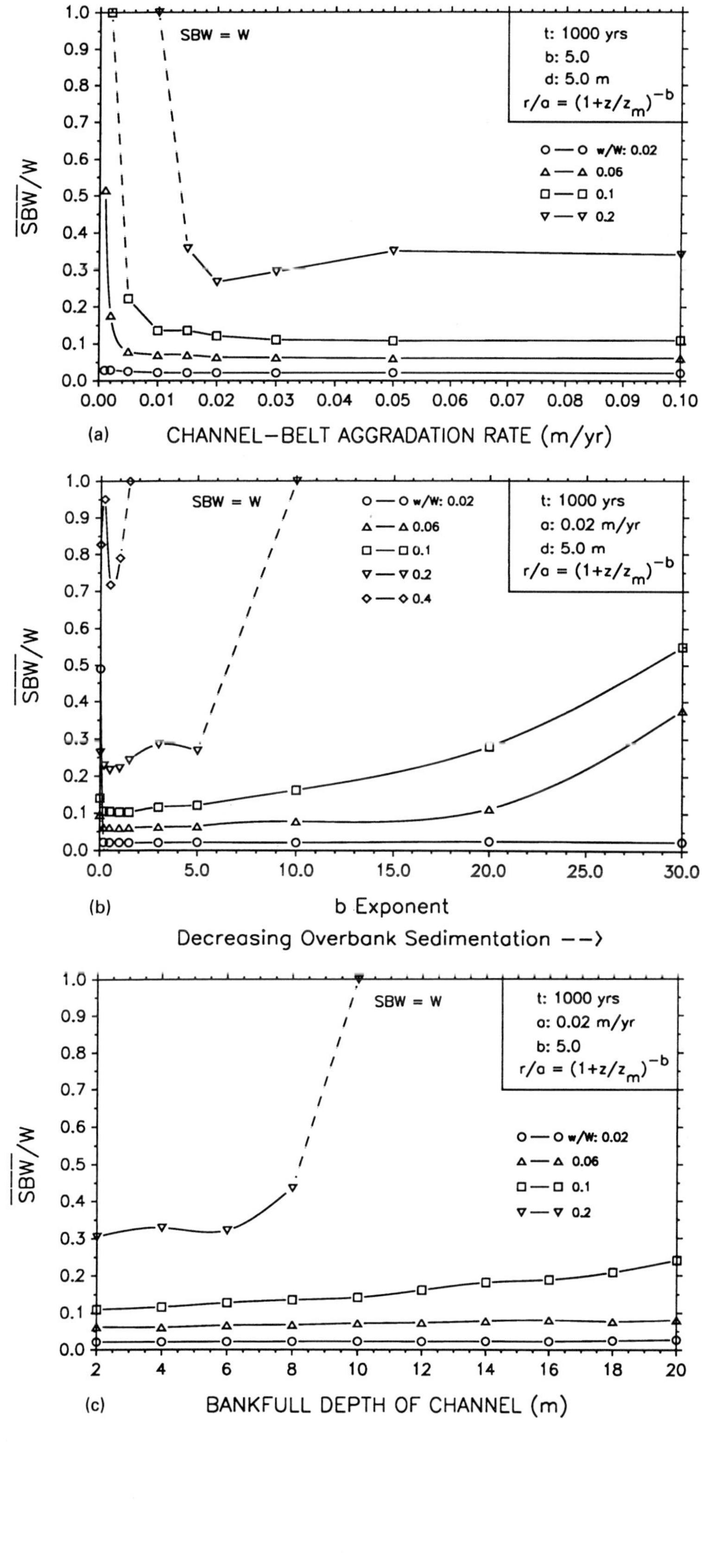

Fig. 7. (a–c) Effect of varying channel-belt aggradation rate, overbank sedimentation rate, and bankfull channel depth on mean sandstone body width, $\overline{SBW}/W$, for different values of w/W. Dashed lines denote conditions where $\overline{SBW}/W$ approaches, or is equal to unity. Experimental results for mean avulsion period are identical to those for channel-belt aggradation rate and are not presented here.

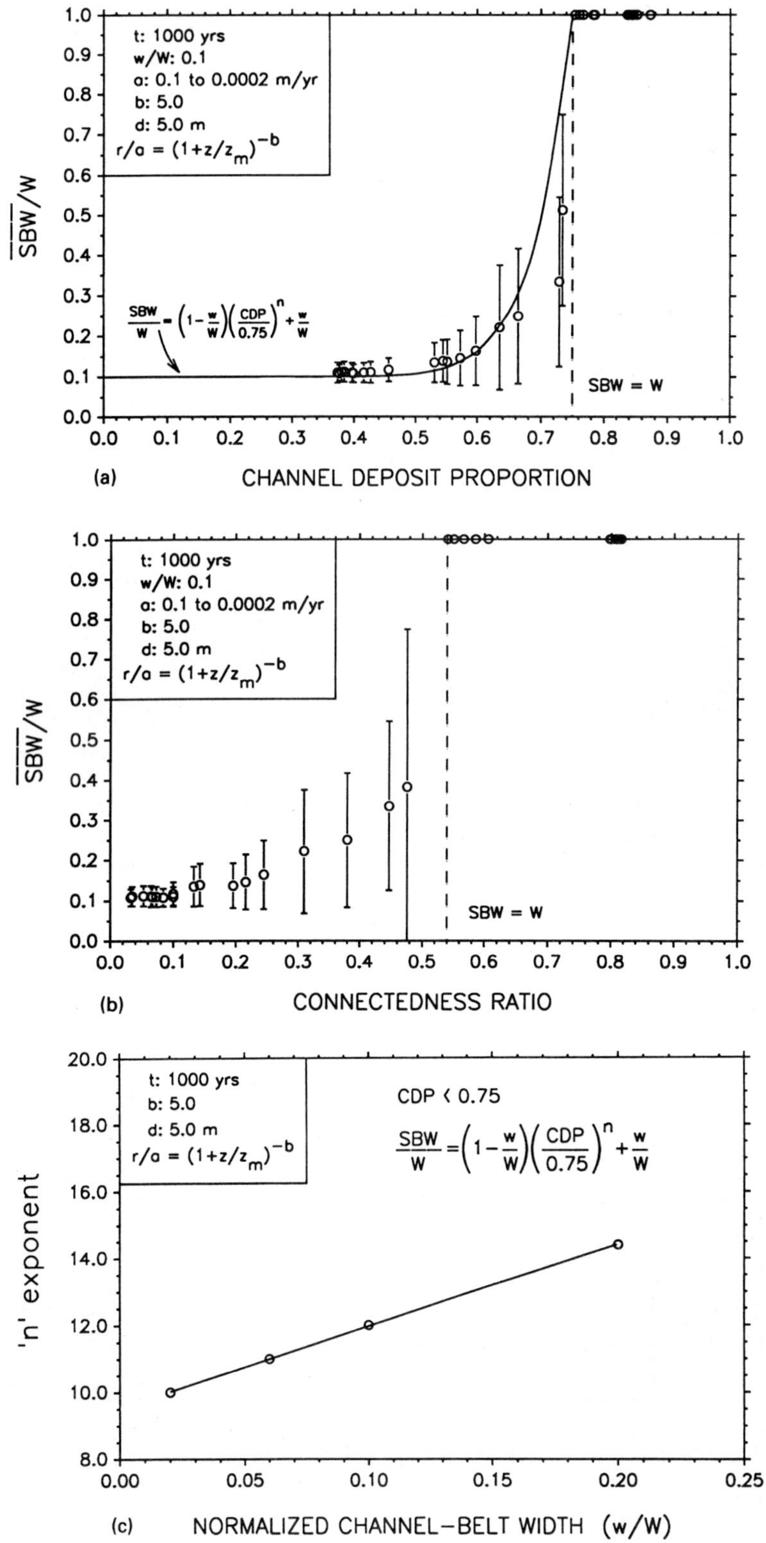

Fig. 8. (a,b) Variation of mean sandstone body width ($\overline{SBW/W}$) and its standard deviation with *CDP* and connectedness for a w/W value of 0.1. Increasing *CDP* and connectedness are produced by decreasing a. Error bars are one standard deviation from the mean. 'Best fit' curve is from equation (3) with $n = 12$. Vertical line denotes the threshold proportion where $\overline{SBW/W}$ equals W. Similar results are obtained if *CDP* and connectedness are changed by varying exponent b, bankfull channel depth, d, and mean avulsion period, $\bar{t}$. (c) For *CDP* values less than 0.75, relationship of n with increasing w/W values for equation (3) relating $\overline{SBW/W}$ to *CDP*. For w/W values greater than 0.2, equation (3) does not apply as *CDP* values are greater than 0.75 for realistic values of a, b, d, or $\bar{t}$.

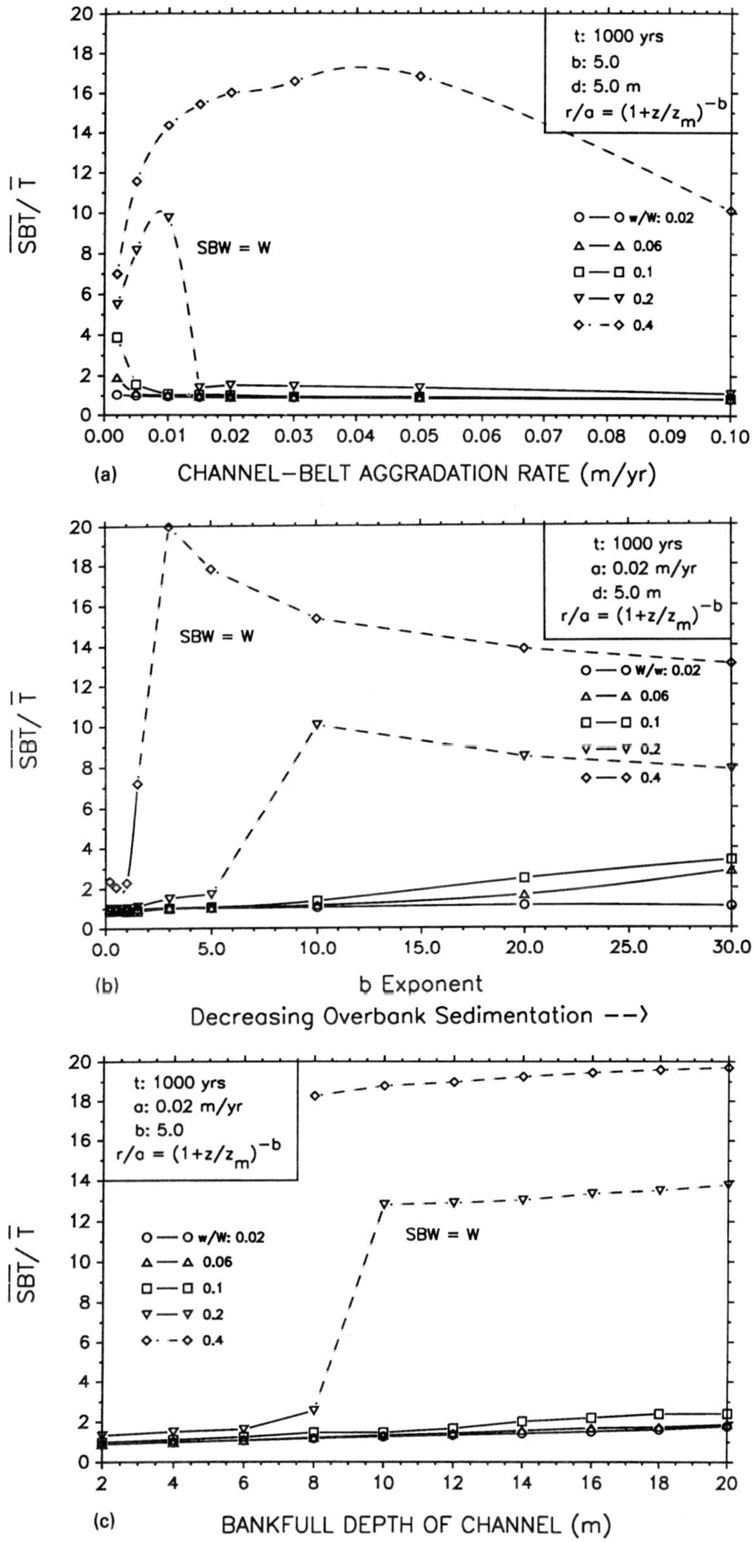

Fig. 9. (a–c) Variation of mean sandstone body thickness ($\overline{SBT}/\overline{T}$) with channel-belt aggradation rate, *a*, exponent *b*, and bankfull channel depth, *d*. Dashed lines denote conditions where $\overline{SBW}/W$ approaches, or is equal to unity. Experimental results for mean avulsion period are identical to those for channel-belt aggradation rate and are not presented here.

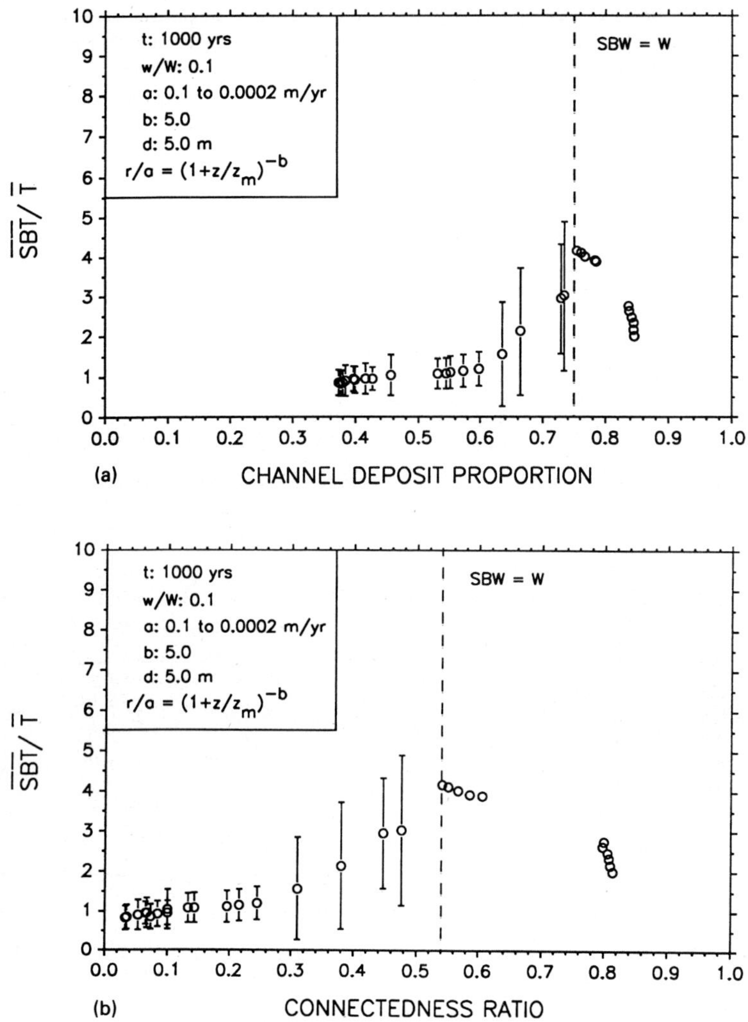

Fig. 10. (a,b) Variation of mean sandstone body thickness ($\overline{SBT/T}$) and its standard deviation with *CDP* and connectedness for a *w/W* value of 0.1. Increasing *CDP* and connectedness are produced by decreasing *a*. Error bars are one standard deviation from the mean. Vertical line is at the critical threshold proportion. Similar results are obtained if *CDP* and connectedness are changed by varying mean avulsion period, $\bar{t}$, or exponent *b*.

relationship of $\overline{SBW}/W$ to *CDP* and connectedness is essentially the same irrespective of whether changes in *CDP* and connectedness are produced by variation in *a*, *b*, *d*, or $\bar{t}$.

MEAN SANDSTONE BODY THICKNESS

When all channel belts are unconnected, mean sandstone body thickness averaged over the whole section, $\overline{SBT}$, is given by:

$$\frac{\overline{SBT}}{\bar{T}} = \frac{\overline{(at + d)}}{(a\bar{t} + d)} = 1 \qquad (4)$$

where $\bar{T}$ is the mean thickness of individual channel belts ($a\bar{t} + d$). For *w/W* values less than 0.2 and realistic values of *a*, *b*, *d* and $\bar{t}$, equation (4) is a reasonable approximation as simulated $\overline{SBT}/\bar{T}$ values are approximately equal to unity for low values of *CDP* and connectedness (Figs 9, 10). As *a*, *d*, and $\bar{t}$ are held constant for each simulation, variation in $at + d$ about the mean is due to Monte Carlo sampling of the Weibull distribution used to simulate the time between avulsions (see Bridge & Leeder, 1979).

As more channel belts become connected, the mean and the standard deviation of $\overline{SBT}/\bar{T}$ gradually increases up to a point where all channel belts become connected (*CDP* equals 0.75) and mean sandstone body width equals the width of the section ($\overline{SBW}$ equals W). At this limit, maximum values of $\overline{SBT}/\overline{T}$ are reached (Fig. 10). When $\overline{SBW}$ equals W.

$$\frac{\overline{SBT}}{\overline{T}} = \frac{\bar{r}t_{tot}}{(a\bar{t} + d)} \tag{5}$$

and the standard deviation of $\overline{SBT}/\overline{T}$ goes to zero as there is only one sandstone body in the section. In equation (5), $\bar{r}$ is the overall mean floodplain aggradation rate and t_{tot} is total elapsed time represented by the section. When $\overline{SBW}$ equals W, variation in $\overline{SBT}/\overline{T}$ depends on t_{tot} and $\bar{r}$ as $\bar{t}$, a, and d are held constant during each experiment. Decreases in $\overline{SBT}/\overline{T}$ for high *CDP* and connectedness values are the result of a reduction in $\bar{r}$, either due to a decrease in overbank sedimentation rate (causing channel belts to stack together more closely) or a decrease in channel-belt width. In this example, $\bar{r}$ is decreasing with decreasing a, which controls in part the overbank sedimentation rate.

When channel belts are connected, overbank deposits may be enclosed within sandstone bodies. The calculated thickness, *SBT*, includes these overbank deposits. The ratio of sandstone thickness, *SST*, to total thickness, *SBT*, ranges from 0.86 to 1.0 for this suite of experiments.

SANDSTONE BODY WIDTH/THICKNESS

When all channel belts are unconnected, the mean sandstone body width/thickness ratio is at a minimum, i.e.

$$\frac{\overline{SBW}}{\overline{SBT}} = \frac{w}{(d + at)} \tag{6a}$$

When all channel belts are connected and mean sandstone body width is equal to the floodplain width, the mean sandstone body width/thickness ratio is at a maximum, i.e.

$$\frac{\overline{SBW}}{\overline{SBT}} = \frac{W}{\bar{r}t_{tot}} \tag{6b}$$

For w/W less than 0.2, $\overline{SBW/SBT}$ gradually increases as channel-belt aggradation rate (or avulsion period) decreases (Fig. 11a). In this case, *CDP* and connectedness are low and equation (6a) generally applies. However, once a drops below 0.02 m/yr, *CDP* and connectedness increase rapidly as $\overline{SBW}$ approaches W and equation (6b) starts to apply. The effect of varying overbank sedimentation rates within realistic limits (say, b values from 1 to 10) is minimal as long as w/W is 0.1 or less (Fig. 11b). For w/W values greater than 0.1, $\overline{SBW/SBT}$ decreases as b (hence *CDP*) increases, up to a point where $\overline{SBW}$ equals W (all channel belts are connected). This is due to a very rapid increase in $\overline{SBT}$ relative to the smaller increase in $\overline{SBW}$ (which is already quite large). Similarly, varying bankfull depth, d, within realistic limits has little effect on $\overline{SBW/SBT}$ for w/W values less than 0.2 (Fig. 11c). In this case, $\overline{SBW/SBT}$ gradually decreases with increasing bankfull depth and equation (6a) generally applies. For w/W values greater than 0.1, decreases in $\overline{SBW/SBT}$ are due to $\overline{SBW}$ increasing at a slower rate than $\overline{SBT}$. With increasing bankfull depth, more section is removed from under each channel belt (during avulsion) which increases both channel-belt thickness and connectedness. When $\overline{SBW}$ equals W (i.e. $\overline{SBW}$ is constant). $\overline{SBT}$ continues to increase further reducing values of $\overline{SBW/SBT}$.

Figure 12 shows that as *CDP* and connectedness become larger, $\overline{SBW/SBT}$ and its standard deviation also increase until a critical channel deposit proportion is reached. At this point, $\overline{SBW}$ equals W, equation (6b) applies, and the standard deviation of $\overline{SBW/SBT}$ goes to zero as there is only one sandstone body in the section. When all channel belts are connected (*CDP* greater than 0.75), $\overline{SBW/SBT}$ increases rapidly with decreasing a. This is caused by a decrease in $\bar{r}$ which reduces the thickness of both channel belt and overbank deposits in the section (thus affecting $\overline{SBT}$, see equation (6b)). Unlike mean sandstone body width, however, $\overline{SBW/SBT}$ and its standard deviation are *not* influenced by overbank sedimentation rate and bankfull depth as both $\overline{SBW}$ and $\overline{SBT}$ vary with b and d at approximately the same rate when *CDP* and connectedness values are less than 0.7 and 0.3, respectively.

FREQUENCY DISTRIBUTIONS OF *SBW*, *SBT* AND *SBW/SBT* IN A SECTION

At low *CDP* and zero connectedness (no connected channel belts), *SBW/W* is constant and equal to w/W for all sandstone bodies. As *CDP* and connectedness increase, the distribution of *SBW/W* of individual

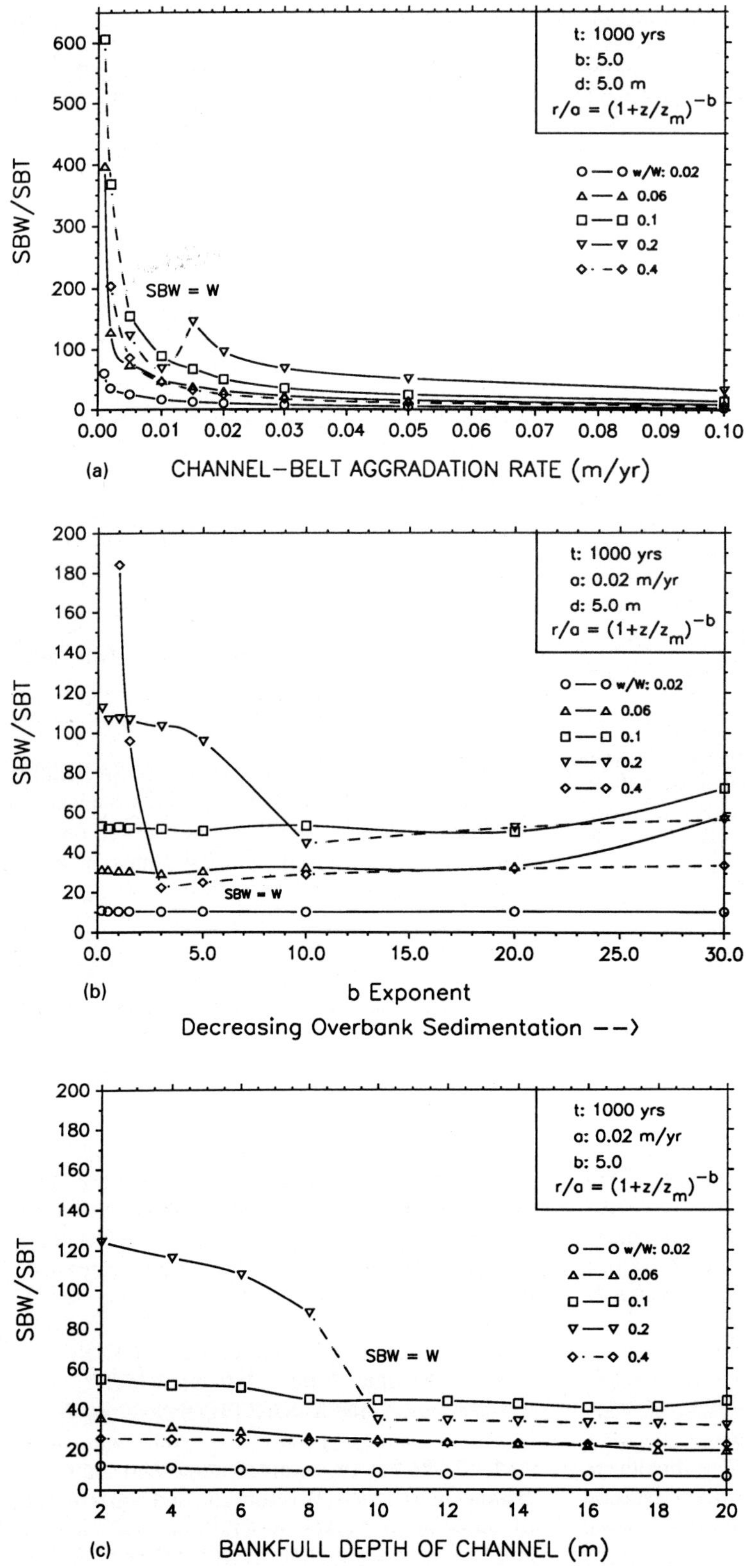

Fig. 11. (a–c) Variation of mean sandstone body width/thickness ($\overline{SBW/SBT}$) with channel-belt aggradation rate, *a*, exponent, *b*, and bankfull channel depth, *d*. Dashed lines denote conditions where $\overline{SBW}/W$ approaches, or is equal to unity. Experimental results for mean avulsion period are identical to those for channel-belt aggradation rate and are not presented here.

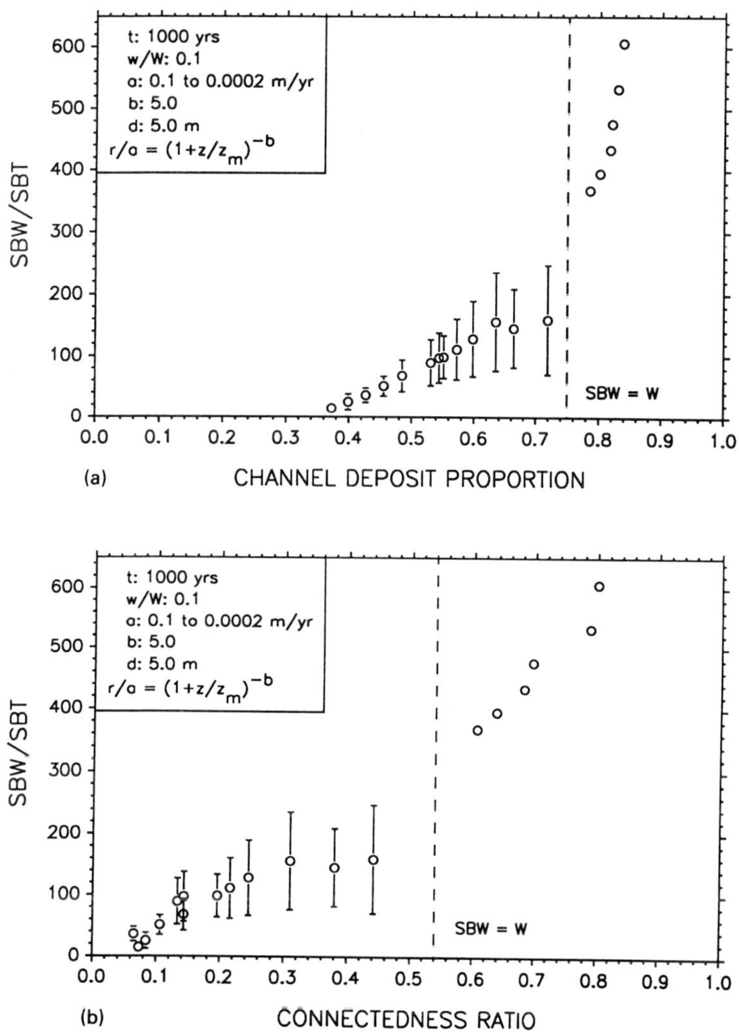

Fig. 12. (a,b) Variation of mean sandstone body width/thickness ($\overline{SBW}/\overline{SBT}$) and its standard deviation with *CDP* and connectedness for a *w*/*W* value of 0.1. Increasing *CDP* and connectedness are produced by decreasing *a*. Error bars are one standard deviation from the mean. The vertical line is at the critical threshold proportion. Similar results are obtained if *CDP* and connectedness are changed by varying mean avulsion period, $\bar{t}$.

sandstone bodies becomes progressively more polymodal, with the proportions of each mode decreasing for higher values of *SBW*/*W* (Fig. 13). Thus the dominant mode occurs where *SBW*/*W* equals *w*/*W*, the next highest mode is for two connected channel belts, and so on. As the number of modes increases, so does the overall standard deviation of the distribution. At this stage, the overall distribution has a 'coarse' tail. As *CDP* and connectedness increase still further, the distribution becomes dominated by sandstone bodies composed of many connected channel belts, and unconnected channel belts are rare. At this stage the distribution remains polymodal with a large standard deviation, but has a 'fine' tail. For *CDP* and connectedness values greater than the critical proportion, all channel belts are connected and there is just one sandstone body where the mean sandstone body width equals the floodplain width.

The mean thickness of individual sandstone bodies varies in a similar way for the same reasons (Fig. 14). However, the individual modes in the distribution of *SBT*/*T* are broader than for *SBW*, because *SBT* of individual unconnected channel belts varies with avulsion period (see Fig. 10). Frequency distributions of thickness variations within *individual* sandstone bodies composed of more than one channel belt are polymodal, with modes corresponding to connected and unconnected parts of the component channel belts. The number of modes

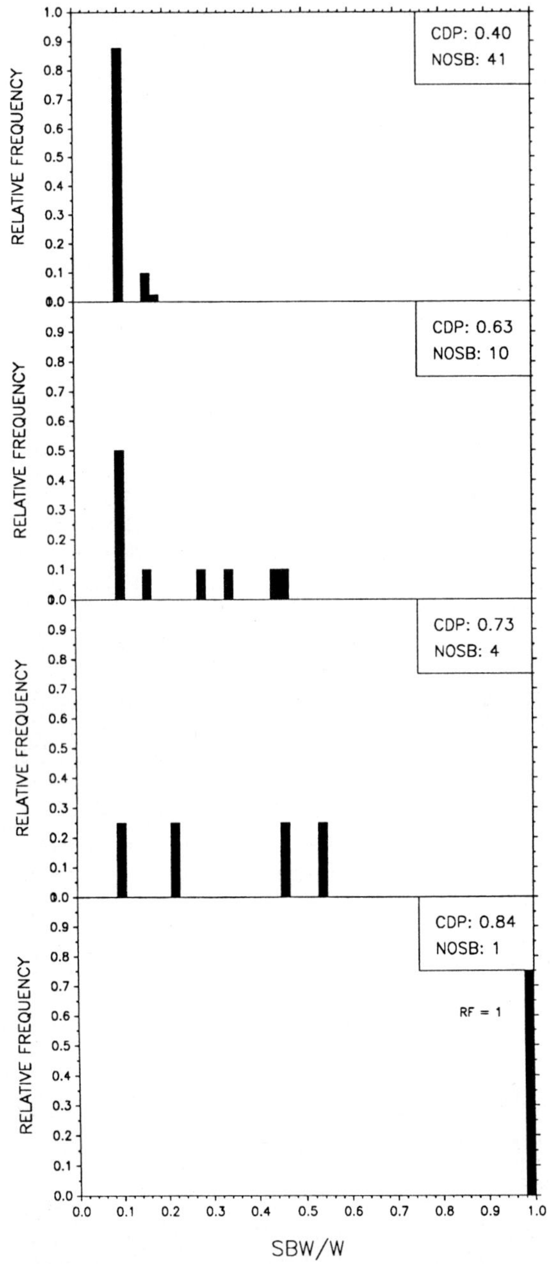

Fig. 13. Frequency distributions of sandstone body width ($\overline{SBW/W}$) for simulated cross-sections with increasing *CDP* and connectedness, produced by decreasing channel-belt aggradation rate, *a*, from 0.05 to 0.001 m/yr. Similar patterns are produced if *CDP* and connectedness are caused to vary by changes in w/W, $\bar{t}$, *d*, or *b*.

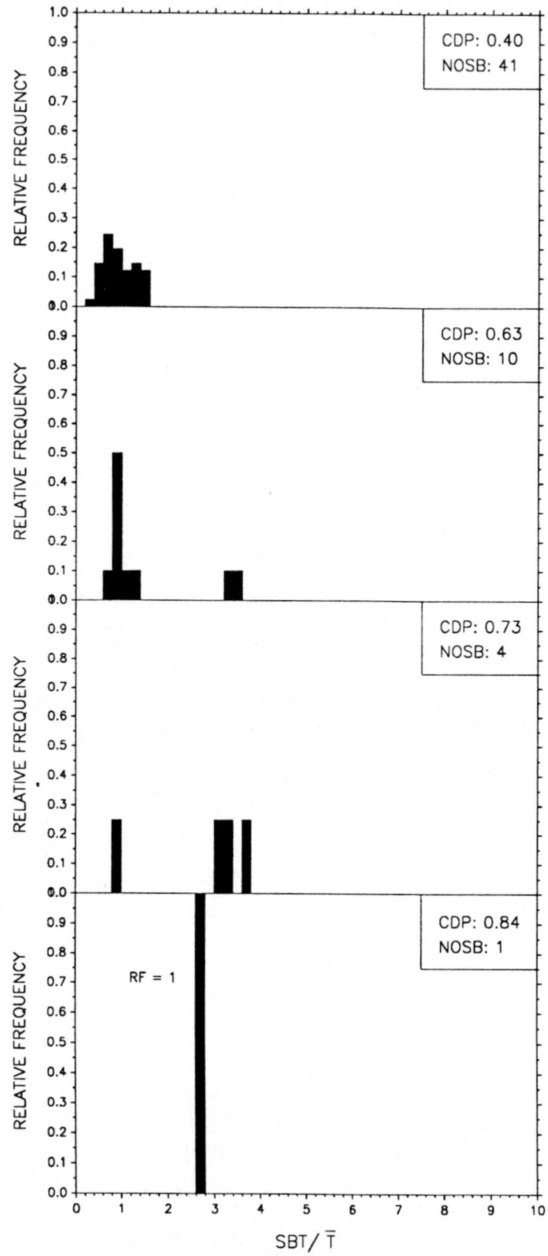

Fig. 14. Frequency distributions of sandstone body thickness ($\overline{SBT/T}$) for simulated cross-sections with increasing *CDP* and connectedness, produced by decreasing channel-belt aggradation rate, *a*, from 0.05 to 0.001 m/yr. Similar patterns are produced if *CDP* and connectedness are caused to vary by changes in w/W, $\bar{t}$, *d*, or *b*.

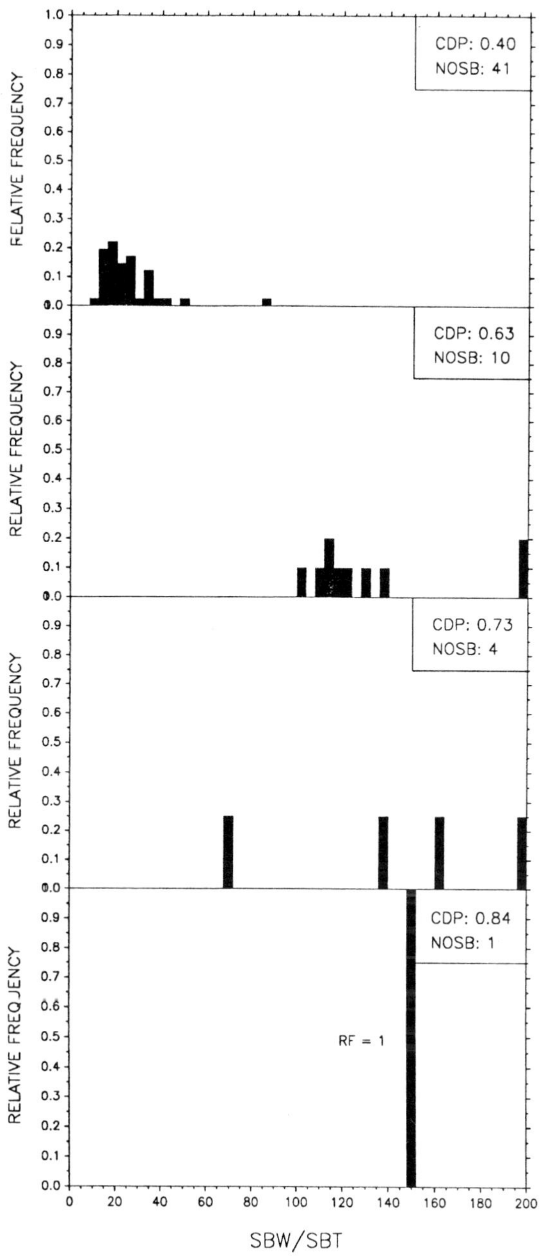

Fig. 15. Frequency distributions of sandstone body width/thickness ($\overline{SBW/SBT}$) for simulated cross-sections with increasing *CDP* and connectedness, produced by decreasing channel-belt aggradation rate, *a*, from 0.05 to 0.001 m/yr. Similar patterns are produced if *CDP* and connectedness are caused to vary by changes in w/W, $\bar{t}$, *d*, or *b*.

increases as the number of channel belts in a sandstone body increases.

The frequency distributions of *SBW/SBT* also vary in a similar way to those for *SBW* and *SBT* (Fig. 15), but the modes in this distribution are still broader that those for *SBW* and *SBT*. This is because a mode is not, in general, a simple combination of the *SBW* and *SBT* modes.

EFFECTS OF COMPACTION AND FLOODPLAIN TILTING

The results presented in this paper are applicable to sections that have *not* undergone appreciable compaction and/or preferred tectonic tilting of the floodplain. Sandstone body dimensions will undoubtedly change in response to increasing depth of burial (compaction) or tectonic tilting of the floodplain. Simulation experiments indicate that channel deposit proportion initially increases with burial depth as overbank fines are preferentially compacted (Bridge & Mackey, 1993). At greater depths, channel deposit proportion gradually decreases in response to the onset of appreciable sandstone compaction. As is expected, with increasing depth of burial, the mean thickness of sandstone bodies decreases and $\overline{SBW/SBT}$ increases as mean sandstone body width remains constant (Fig. 16). Width/thickness ratios calculated from ancient deposits must therefore be decompacted before comparison with model results.

Experiments varying the degree of tectonic tilting suggest that there is little effect on *cross-section averaged* values of channel deposit proportion, connectedness, mean sandstone body width, and mean sandstone body thickness. However, the model does predict lateral variation in vertical channel deposit proportion and it is clear that the number and proportion of channel belts increases on the downthrown side and decreases on the upthrown side of the section. This suggests that sandstone body widths and thicknesses will also be greater on the downthrown side. However, due to limitations in the current model, we cannot quantitatively assess the lateral variation of sandstone body dimensions within a generated section.

DISCUSSION

To use this model, it is first necessary to constrain channel-belt widths using the theoretical expressions

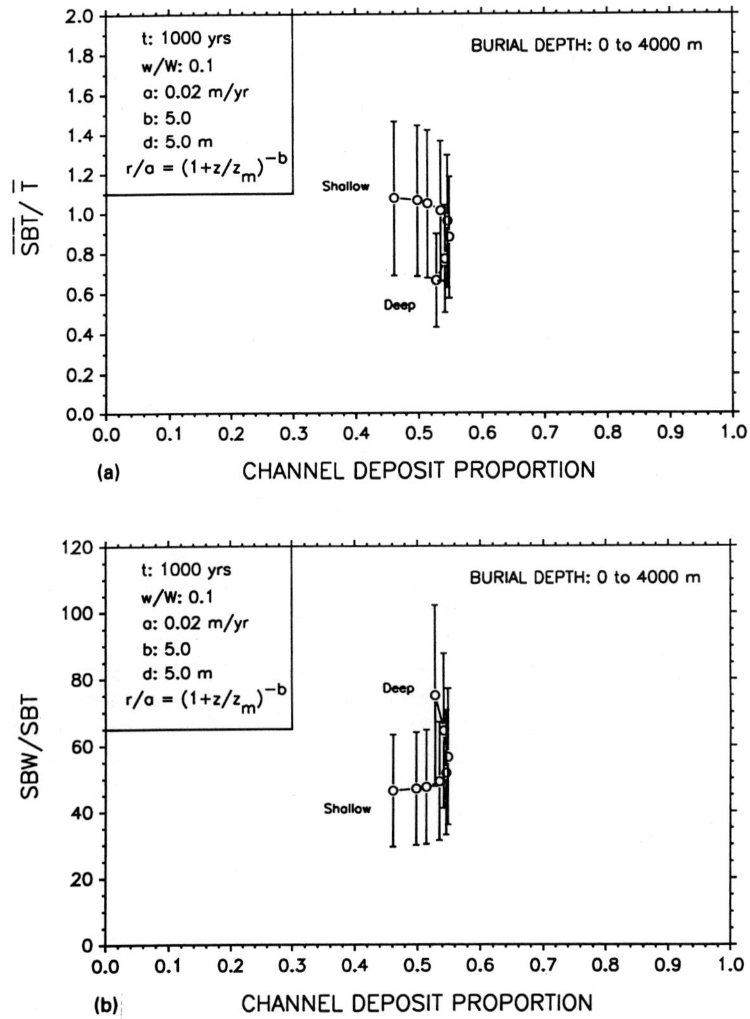

Fig. 16. (a,b) Effect of burial depth (compaction) on $\overline{SBT}/\overline{T}$ and $\overline{SBW}/\overline{SBT}$ for a representative simulation. Sandstone body width ($\overline{SBW}$) is unaffected by compaction.

of Bridge (1975) or Crane (1982) or empirical regression expressions derived from data from modern rivers (previous discussion and Tables 2, 3, 4). Well data may be used to constrain input data such as maximum bankfull channel depth, d. This may be approximated by the maximum thickness of untruncated sandstone storeys (bars) encountered in the wells (Fig. 2). However, it is very difficult to estimate channel-belt widths from a series of wells, as a single sandstone body penetrated by the wells may consist of more than one channel belt. Seismic data may be used to estimate floodplain widths and to detect the presence of marginal faults. Absolute age dating is required to estimate compacted deposition rates. In natural sequences, estimates of channel deposit proportion can be obtained from outcrops or subsurface data. Sandstone body dimensions can then be simulated by varying selected input parameters (within reasonable limits) so that the simulated sequence has an appropriate channel deposit proportion and other calculated architectural parameters are consistent with available sedimentological data. Using the dimensional data generated by this model, it is then possible to calculate the probability of penetration for a given number of wells with a fixed spacing. Examples of this technique are given by Lorenz *et al.* (1985) and Fielding & Crane (1987).

Table 5. Summary table of effects on sandstone body width and thicknesses of increasing channel deposit proportion and connectedness

CDP	Low	Increases	High
CR	Unconnected channel belts	Increases	All channel belts connected (one sandstone body)
SBW	Channel-belt width	Increases	Floodplain width
std dev	Zero	Increases	Zero
SBT	Channel-belt thickness	Increases	Total section thickness
std dev	Low	Increases	Zero
frequency histogram	Unimodal to polymodal (coarse tail)	to	Polymodal (fine tail) to unimodal

Our results suggest that channel-belt sandstone body dimensions are controlled, in part, by changes in channel deposit proportion and connectedness (Table 5). As the proportion and connectedness of channel-belt deposits increase, the mean sandstone body width and thickness and associated standard deviations increase rapidly as a power function until a critical threshold proportion is reached. At this point, all channel belts become connected and mean sandstone body width equals the width of the floodplain. For a reasonable range of *w/W* values, the critical channel deposit proportion is approximately 0.75. In sequences with channel deposit proportions greater than 0.75, most channel-belt sandstone bodies will be connected and may extend laterally across the entire width of the floodplain. The thickness of the sandstone body will approximate the thickness of the total section. For sequences with channel deposit proportions somewhat less than 0.75 (say 0.40–0.75), sandstone body widths and thicknesses and associated standard deviations will increase with increasing channel deposit proportion. For sequences with channel deposit proportions less than 0.4, sandstone body widths and thicknesses will generally approximate the dimensions of individual channel belts.

These results complement those reported by Bridge & Leeder (1979, p. 632) where they concluded 'as channel deposit density (proportion) falls below 0.4, interconnectedness ratio becomes very small, less than 0.05'. In the simulations presented here, sandstone body widths and thicknesses are equivalent to the dimensions of individual channel belts when connectedness is less than approximately 0.1 (Fig. 8b). Moreover, this 'critical proportion for connectedness' is clearly dependent on values of *w/W* (Leeder, 1978; Bridge & Leeder, 1979). Figure 17 demonstrates the effect of varying *w/W* on channel deposit proportion and connectedness. Note that the critical channel deposit proportion for a connectedness value of 0.1 ranges from approximately 0.38 to 0.65 and increases with increasing *w/W*. In simulations similar to those run by Bridge & Leeder (1979), connectedness drops below 0.1 when the channel deposit proportion is approximately 0.45 (*w/W* equal to 0.1). For *w/W* values less than 0.4, the critical channel deposit proportion for which $\overline{SBW}$ equals *W* (*CDP* approximately 0.75) occurs at progressively larger connectedness ratios as *w/W* increases. For *w/W* values of 0.4 or greater, *CDP* values exceed 0.75, $\overline{SBW}$ equals *W* as all sandstone bodies are connected, and connectedness increases rapidly with increasing proportion.

It is important to recognize that changes in sandstone body geometry can result from changes in *several* controlling variables. These simulations demonstrate that normalized channel belt width (*w/W*), channel-belt aggradation rate, mean avulsion period, and overbank sedimentation rate are critical parameters that control the width and thickness of alluvial channel-belt sandstone bodies. Changes in bankfull depth are somewhat less critical, even though bankfull depth may significantly influence mean sandstone body thickness under certain conditions.

Based on these simulations, theoretical expressions can be derived to estimate selected sandstone body dimensions from field data. For example, working curves can be constructed relating mean sandstone body width to normalized channel-belt width, *w/W*, and channel deposit proportion, *CDP*, using equation (3). Estimates of mean sandstone body width can then be obtained if the channel deposit proportion, channel-belt width, and floodplain width are known.

The simulations presented here are based on the

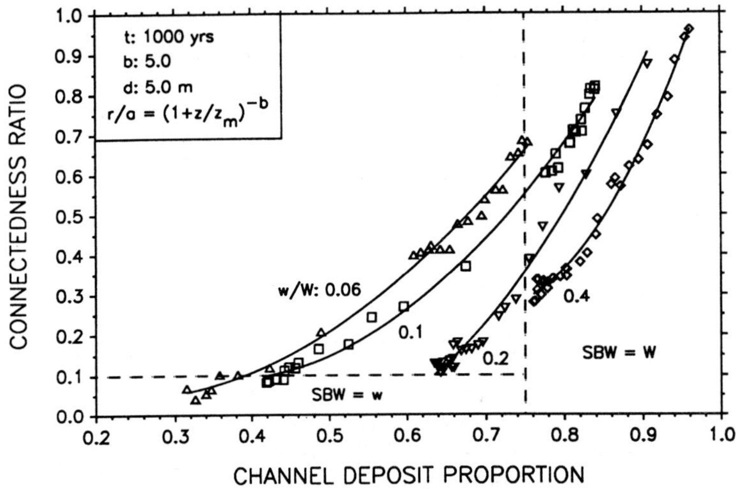

Fig. 17. Variation of connectedness with *CDP* for different values of *w*/*W*. Increasing *CDP* and connectedness are produced by decreasing *a*. Vertical line denotes critical channel deposit proportion for maximum width. Horizontal line denotes critical proportion for connectedness. Note that proportion varies from 0.38 to 0.65 for connectedness values of 0.1 with increasing *w*/*W*. For *w*/*W* values greater than 0.2, *CDP* values are greater than 0.75, $\overline{SBW}$ equals *W* as all sandstone bodies are connected, and connectedness increases rapidly with increasing proportion.

least complicated set of input parameters, i.e. fixed channel-belt width, constant aggradation rates through time, and architectural-dimensional parameters calculated for a two-dimensional section oriented normal to all channel-belt directions. As such, these results represent the *minimum* amount of variability in a cross-section. Nevertheless, these results can be reasonably extrapolated for different cross-section orientations and for time-varying input parameters, as long as there is no variation in alluvial architecture in the down-valley direction. However, in reality, down-valley variation in the geometry, orientation, and location of individual channel belts can be expected. Channel belts may *not* always be located at the lowest point on the floodplain surface as their position and location are constrained by the point of upstream avulsion and topographic relief of the floodplain surface. In three dimensions, channel deposit proportion is a volume ratio and connectedness is an area ratio. The two-dimensional channel deposit proportion and connectedness may differ significantly from the three-dimensional case and may vary appreciably with cross-section orientation. These problems can only be resolved with the development of a three-dimensional model of alluvial architecture, which is in progress.

CONCLUSIONS

1 Process-based models are a viable alternative to stochastic-based models when simulating the geometry and distribution of avulsion-controlled, alluvial channel-belt sandstone bodies in the subsurface. The process-based model used here can simulate changes in the width and thicknesses of channel-belt sandstone bodies in response to varying input conditions, such as channel-belt aggradation rate, avulsion period, overbank sedimentation rate, bankfull depth, channel-belt width, depth of burial (compaction), and tectonic tilting of the floodplain.

2 These simulations demonstrate that for realistic values of channel-belt width/floodplain width, channel-belt aggradation rate, mean avulsion period, and overbank sedimentation rate, sandstone body widths increase as bankfull depth or channel-belt width/floodplain width increase, or overbank aggradation rate, channel-belt aggradation rate, or avulsion period decrease. Sandstone body thicknesses increase with increasing channel-belt width/floodplain width, and increase slightly as bankfull depth increases or overbank aggradation rate decreases.

3 In sequences with channel deposit proportions greater than 0.75, all channel belts will be connected and the resulting single sandstone body may extend laterally across the entire floodplain. For sequences with channel deposit proportions somewhat less than 0.75 (say 0.4 to 0.75), sandstone body widths and thicknesses increase and become more variable with increasing channel deposit proportion and connectedness. For sequences with channel deposit proportions less than 0.4, sandstone body widths and thicknesses will generally approximate the dimensions of individual channel belts.

4 These results represent the *minimum* amount of

variability in an alluvial sequence and are limited by the two-dimensional nature of the current model. More realistic results will be obtained with the development of a three-dimensional model of alluvial architecture (in progress).

ACKNOWLEDGEMENTS

This research has been supported by grants from Conoco, Inc. and ARCO Oil and Gas Company. We are grateful for their encouragement and support of this research. This manuscript benefited from the constructive comments of Ian Bryant and Alf Ryseth.

REFERENCES

ALLEN, J.R.L. (1965) A review of the origin and characteristics of recent alluvial sediments. *Sedimentology*, **5**, 89–191.

ALLEN, J.R.L. (1978) Studies in fluviatile sedimentation: an exploratory quantitative model for the architecture of avulsion-controlled alluvial suites. *Sedim. Geol.*, **21**, 129–147.

BALDWIN, B. & BUTLER, C.O. (1985) Compaction curves. *Bull. Am. Assoc. Petrol. Geol.*, **69**, 622–626.

BEGG, S.H. & WILLIAMS, J.K. (1991) Algorithms for generating and analysing sand-body distributions. In: *Reservoir Characterization—II* (Ed. by L.W. Lake, H.B. Carroll Jr & T.C. Wesson), pp. 613–642. Academic Press, London.

BRIDGE, J.S. (1975) Computer simulation of sedimentation in meandering streams. *Sedimentology*, **22**, 3–43.

BRIDGE, J.S. (1978) Palaeohydraulic interpretation using mathematical models of contemporary flow and sedimentation in meandering channels. In: *Fluvial Sedimentology* (Ed. by A.D. Miall), Canadian Society Petroleum Geologists Memoir **5**, pp. 723–742.

BRIDGE, J.S. (1984) Flow and sedimentary processes in river bends: comparison of field observations and theory. In: *River Meandering* (Ed. by C.M. Elliot), Proc. Rivers '83, American Society Civil Engineers, pp. 857–872.

BRIDGE, J.S. (1985) Paleochannel patterns inferred from alluvial deposits: a critical evaluation. *J. sedim. Petrol.*, **55**, 579–589.

BRIDGE, J.S. & LEEDER, M.R. (1979) A simulation model of alluvial stratigraphy. *Sedimentology*, **26**, 617–644.

BRIDGE, J.S. & MACKEY, S.D. (1993) A revised alluvial stratigraphy model. In: *Alluvial Sedimentation* (Ed. by M. Marzo & C. Puigdefabregas). Spec. Publs int. Ass. Sediment, Blackwell Scientific Publications, Oxford, **17** (in press).

BUDDING, M.C., EASTWOOD, K.M., HERWEIJER, J.C., LIVERA, S.E., PAARDEKAM, A.H.M. & REGTIEN, J.M.M. (1988) Probabilistic modelling of discontinuous reservoirs: IPA 88-21.04. *Proceedings Indonesian Petroleum Association*, 17th Annual Convention, October, 1988.

CARLSTON, C.W. (1965) The relation of free meander geometry to stream discharge and its geomorphic implications. *Am. J. Sci.*, **263**, 864–885.

CHITALE, S.V. (1970) River channel patterns. *Journal Hydraulics Division, Proceedings American Society Civil Engineers*, vol. 96, HY1, pp. 201–221.

COLLINSON, J.D. (1978) Vertical sequence and sand body shape in alluvial sequences. In: *Fluvial Sedimentology* (Ed. by A.D. Miall), Canadian Society Petroleum Geologists Memoir **5**, pp. 577–586.

COTTER, E. (1971) Paleoflow characteristics of a Late Cretaceous river in Utah from analysis of sedimentary structures in the Ferron Sandstone. *J. sedim. Petrol.*, **41**, 129–138.

CRANE, R.C. (1982) A computer model for the architecture of avulsion controlled alluvial suites. Unpublished PhD Dissertation, University of Reading, 534 pp.

DUBRULE, O. (1989) An overview of stochastic models for petroleum reservoirs. In: *Geostatistics* (Ed. by M. Armstrong). Proceedings 3rd International Geostatistics Congress, September 5–9, Avignon, France, 1038 pp.

ELLIOTT, T. (1976) The morphology, magnitude, and regime of a Carboniferous fluvial-distributary channel. *J. sedim. Petrol.*, **46**, 70–76.

ETHRIDGE, F.G. & SCHUMM, S.A. (1978) Reconstructing paleochannel morphologic and flow characteristics: methodology, limitations, and assessment. In: *Fluvial Sedimentology* (Ed. by A.D. Miall), Canadian Society Petroleum Geologists Memoir **5**, pp. 703–721.

FIELDING, C.R. & CRANE, R.C. (1987) An application of statistical modelling to the prediction of hydrocarbon recovery factors in fluvial reservoir sequences. In: *Recent Developments in Fluvial Sedimentology* (Ed. by F.G. Ethridge, R.M. Flores & M.D. Harvey), Society Econonomic Paleontologists Mineralogists Special Publication No. **39**, pp. 321–327.

GARDNER, T.W. (1983) Paleohydrology and paleomorphology of a Carboniferous, meandering, fluvial sandstone. *J. sedim. Petrol.*, **53**, pp. 991–1005.

HALDORSEN, H.H. & MACDONALD, C.J. (1987) Stochastic modelling of underground reservoir facies (SMURF). *SPE Paper* **16751**, 7 pp.

INGLIS, C.C. (1949) *The Behavior and Control of Rivers and Canals (with the Aid of Models)*. Research Publication 13, pt. 1, Central Waterpower Irrigation Navigation Research Station, Poona, India, 298 pp.

JEFFERSON, M.S.W. (1902) Limiting width of meander belts. *National Geographic Magazine*, **13**, 373–384.

LEEDER, M.R. (1973) Fluviatile fining-upwards cycles and the magnitude of palaeochannels. *Geol. Mag.*, **110**, 265–276.

LEEDER, M.R. (1978) A quantitative stratigraphic model for alluvium, with special reference to channel deposit density and interconnectedness. In: *Fluvial Sedimentology* (Ed. by A.D. Miall), Canadian Society Petroleum Geologists Memoir **5**, pp. 587–596.

LEOPOLD, L.B. & WOLMAN, M.G. (1960) River meanders. *Bull. geol. Soc. Am.*, **71**, 769–794.

LORENZ, J.C., HEINZE, D.M., CLARK, J.A. & SEARLS, C.A. (1985) Determination of widths of meander-belt sandstone reservoirs from vertical downhole data, Mesaverde Group, Piceance Creek Basin, Colorado. *Bull. Am. Assoc. Petrol. Geol.*, **69**, 710–721.

LOWRY, P. & RAHEIM, A. (1991) Characterization of delta front sandstones from a fluvial-dominated delta system. In: *Reservoir Characterization—II* (Ed. by L.W. Lake, H.B. Carroll Jr & T.C. Wesson), pp. 665–676. Academic Press, London.

MACKEY, S.D. & BRIDGE, J.S. (1992) A revised FORTRAN program to simulate alluvial stratigraphy. *Computers and Geosciences*, **18**, 119–181.

MOODY-STUART, M. (1966) High and low sinuosity stream deposits, with examples from the Devonian of Spitsbergen. *J. sedim. Petrol.*, **36**, 1102–1117.

SADLER, P.M. (1981) Sediment accumulation rates and the completeness of stratigraphic sections: *J. Geol.*, **89**, 569–584.

SCHERER, M. (1987) Parameters influencing porosity in sandstones: a model for sandstone porosity prediction. *Bull. Am. Assoc. Petrol. Geol.*, **71**, 485–491.

SCHUMM, S.A. (1963) Sinuosity of alluvial rivers on the Great Plains. *Bull. geol. Soc. Am.*, **74**, 1089–1100.

SCLATER, J.G. & CHRISTIE, P.A.F. (1980) Continental stretching: an explanation of the post-mid Cretaceous subsidence of the central North Sea Basin. *J. Geophys. Res.*, **85**, 3711–3739.

STAUFFER, D. (1985) *Introduction to Percolation Theory*. Taylor and Francis, London, 124 pp.

STEWART, D.J. (1981) A meander-belt sandstone of the Lower Cretaceous of Southern England. *Sedimentology*, **28**, 1–20.

WALDERHAUG, O. & MJØS, R. (1991) Sandbody geometry in fluvial systems—an example from the Jurassic of Yorkshire, England. In: *Reservoir Characterization—II* (Ed. by L.W. Lake, H.B. Carroll Jr & T.C. Wesson), pp. 701–703. Academic Press, London.

WILLIAMS, G.P. (1986) River meanders and channel size. *J. Hydrol.*, **88**, 147–164.

ZELLER, J. (1967) *Meandering channels in Switzerland*. International Association Scientific Hydrologists, Publication No. **75**, pp. 174–186.

Spec. Publs int. Ass. Sediment. (1993) **15**, 237–252

Stochastic modelling of fluvial sandstone bodies

J.P.P. HIRST*, C.R. BLACKSTOCK† *and* S. TYSON**

* *BP Exploration Company Limited, 301 St Vincents Street, Glasgow G2 5DD, UK;*
† *BP Exploration Company Limited, Britannic House, Moor Lane, London EC2Y 9BU, UK;*
** *Pipers Ash Limited, 42 High Street, Girton, Cambridge CB3 OPU, UK*

ABSTRACT

This study concerns the development of a stochastic routine for modelling fluvial sequences in which the channel belts are variably enclosed within floodplain fines. The method has application in appraising hydrocarbon discoveries and development planning.

In mixed load fluvial systems, coarse, bedload sediment is mainly deposited within or immediately adjacent to the active channel (coarse member) while finer, suspended sediment dominates the floodplain deposits (fine member). The resulting sequence will be heterogeneous, comprising interconnected channel, levee and crevasse splay sandstones variably enclosed within floodplain fines. The volume of interconnected sandstone contacted by a well penetrating such a sequence is uncertain but may be constrained by stochastic modelling.

The aim is to produce geologically realistic three-dimensional models of sandstone body geometries. Both modern channel belts and ancient fluvial systems have been studied to determine the range of variables necessary to define the coarse members. These variables may be divided into three categories: cross-sectional geometry (e.g. mean sandbody thickness, width/thickness ratio), plan geometry (e.g. reach length, relative orientation of connected reaches) and three-dimensional arrangement (e.g. coarse/fine member ratio, palaeoflow).

The interactive software developed in this study to stochastically reproduce channel scale heterogeneity is termed SIRCH (System for Integrated Reservoir Characterization). Within SIRCH, the geostatistical method is object based; the objects are specified geometric shapes. Each generated shape has a set of attributes which define its dimensions, position and relationship to other shapes. The attributes are based on the geological variables. Values are derived from distribution functions appropriate to the system being modelled. Sandstone units are constructed as a series of linked shapes and thus a series of channel belts may be built-up using well data and empirical relationships. Multiple realizations may be output and evaluated to determine the volume of sandstone connected by one or more wells.

INTRODUCTION

Fluvial channels typically range in depth from a metre up to a few tens of metres. At this channel depth scale, fluvio-deltaic sequences are invariably heterogeneous, reflecting the range of hydrodynamic processes and sediment grade that may be encountered within a single system. These heterogeneities fall into two broad categories:

1 Bedload dominated rivers deposit predominantly coarse-grained sequences. In hydrocarbon reservoir terms, the coarse material is potentially net reservoir and these sequences may be termed high net/gross deposits (high proportion of coarse member within the total interval). Heterogeneities occur as thin, low permeability units of variable extent and continuity. Examples of these potential transmissibility barriers include shale/silt drapes on barforms, pedogenic horizons and intraclast lags (Allen, 1983; Miall, 1988). Typically these units are subhorizontal and, from a reservoir performance perspective, will reduce vertical permeability (K_v).

2 In mixed-load fluvial systems, coarse, bedload sediment is mainly deposited within the active

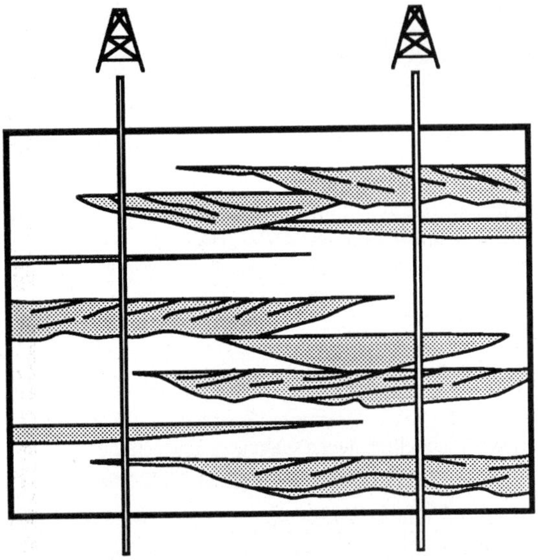

Fig. 1. Schematic representation of a low net/gross fluvial sequence in which the coarse component constitutes less than 50% of the total. Stipple = coarse component. The problem is determining the volume of sandstone connected to a single well and the connectivity between wells.

channel or immediately adjacent to the channel in the form of levees and crevasse splays (coarse member). Finer, suspended sediment dominates the floodplain deposits away from the main channels (fine member). The resulting sequence will likely have a low proportion of the coarse member within the total sequence (low net/gross sequence). The sequence will be heterogeneous, comprising interconnected, geometrically distinct sandstone bodies (channel, levee and crevasse splay) variably enclosed within floodplain fines (Fig. 1; Elliott, 1965; Friend *et al.*, 1979; Fielding, 1986; Hirst, 1992). A major reservoir appraisal problem is determining the connectivity of the sandstone units.

In this study, low (to medium) net/gross sequences are modelled. Where such heterogeneous fluvio-deltaic sequences form hydrocarbon reservoirs (Livera, 1989), major problems are encountered in estimating important aspects including (i) net reservoir volume, (ii) the volume of sandstone connected to a single well and (iii) the recoverable hydrocarbons from the whole field and from single wells. In attempting to provide answers to these problems, geoscientists use a variety of methods ranging from well-to-well correlations, through simple models (Knutson, 1976; Lorenz *et al.*, 1985) to complex, multiple realization, computer simulations (e.g. papers in Buller *et al.*, 1990). Stochastic routines have been widely used in computer models (Haldorsen & MacDonald, 1987; Haldorsen & Damsleth, 1990; Rudkiewicz *et al.*, 1990; Aasen *et al.*, 1990). 'Stochastic' implies that a given set of circumstances does not always produce the same outcome but gives rise to different results such that there is a statistical regularity (Haldorsen & Damsleth, 1990). The interactive software developed in this study to stochastically reproduce channel scale heterogeneity is termed SIRCH (System for Integrated Reservoir CHaracterization; see Appendix). It is not a deterministic process and neither is it an attempt to model the dynamic processes of the environment (Alexander, 1986; Strobel *et al.*, 1989; Lee *et al.*, 1990). The aim of the SIRCH software is to stochastically generate realistic reservoir heterogeneities and provide tools for the analysis of these models.

THE SIRCH SOFTWARE

Geostatistical method

A variety of geostatistical methods may be used in stochastic modelling routines (Haldorsen & Damsleth, 1990). Within SIRCH, the geostatistical method is object based; the objects are specified geometric shapes. Each generated shape has a set of attributes that are derived from user-defined distribution functions appropriate to the system being modelled. The attributes of each shape determine its size, position and its relationship to other shapes.

Object-based modelling has been applied elsewhere (Fielding & Crane, 1987; Henriquez *et al.*, 1988, Gundesø & Egeland, 1990; Clemetsen *et al.*, 1990). Simplifying assumptions, particularly in earlier routines, have resulted in potentially important features observed in fluvial sequences not being fully addressed; these include the variability in channel-belt orientation and geometry, the contribution of coarse overbank deposits to net reservoir, channel bifurcations and variable net/gross. The guiding philosophy behind the development of SIRCH is that realistic generations can only be achieved by taking full account of the geological variability. The design has therefore been driven by geological observations of depositional systems. The basis for geometric construction by the modelling routine is a range of sedimentologically defined parameters which describe the geometry of channel belts and their coarse overbank deposits (Table 1). These parameters are discussed in detail below.

Other requirements of the routine are the abilities to produce multiple realizations and test sensitivities quickly and easily. A further important rationale for developing the software concerns the requirement for real input data. Field studies collecting proprietary data from potential analogues are an important part of many modern reservoir modelling studies (e.g. Ravenne *et al.*, 1987; Dreyer, 1990). Before embarking upon such expensive activities, it is important to know what numerical data are important. Sensitivity testing is important for highlighting the critical variables within a given system and thus providing an indication of what data are required and which aspects are trivial.

Table 1. Parameters which describe the geometry of fluvial sandstone bodies in low net/gross systems

Cross-section geometry
T_b thickness of sandstone body (channel belt)
W_b width of sandstone body (channel belt)
X 2-D cross-sectional shape

Plan geometry
δW variation in sandstone body width along reach
δT_b variation in sandstone body thickness along reach
L_r length of reach
α angle between connected reaches
Y frequency of bifurcation/unification of sandstone bodies
β orientation of system in space (i.e. palaeoflow range)—the polarity of the total system

Three-dimensional relationships
F frequency of sandstone bodies in the succession, equivalent to the volume of coarse component (net/gross)
S stacking patterns—random or preferred

In addition, by using the more complex shapes, levees and crevasse channels can be included:

Levees/overbanks
W_o width of (effective) overbanks

Crevasse channels
L_c length of crevasse channels
W_c width of crevasse channels
T_c thickness of crevasse channels
F_c frequency of crevasse channels

Sandstone body generation

SIRCH is a shape generator. The process of generating sandstone bodies begins by accessing a library of shapes. These are specified geometrical descriptions, the simplest shape in the library being a hexahedra (=box, Fig. 2). Each generated box has a set of attributes which define the:

1 position in space;
2 dimensions of each face;
3 orientation;
4 connections with other shapes and geometry of these attached shapes;
5 assigned properties (porosity, permeability).

Shapes are joined together as specified to create a larger, more complex shape. SIRCH geometrically represents channel-belt sandstones as continuous links of the specified shape (Fig. 2). The attributes of each shape reflect the geological variables determined for fluvial sandstone bodies (Table 1). The values for the attributes are repeatedly sampled from distribution functions of appropriate geological data as a channel belt is constructed.

The model volume in which shapes are to be fitted is a cuboid defined by specifying a range of coordinates for each of three directions (X, Y and Z; Fig. 3). All sandstone bodies start from the Y–Z plane (the X = 0 plane) and may exit the model at the opposite plane or along one of the sides. To allow for channel belts which may enter the volume of interest from adjacent areas, a larger volume than required is specified (Fig. 3); upon completion of a generation, the program trims back the final model to the smaller, inner area of interest.

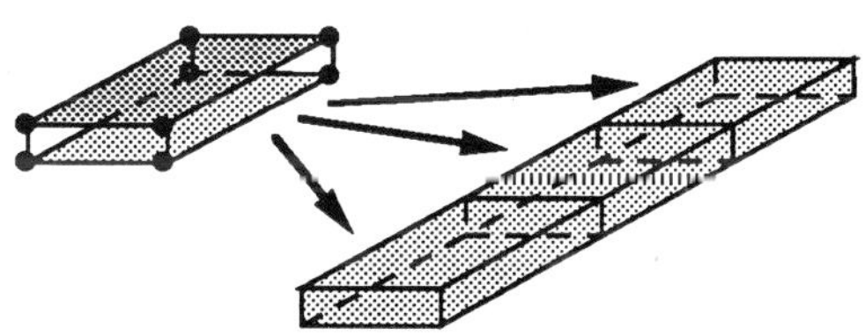

Fig. 2. The corner points define the shape of a single box (hexahedra). Fluvial sandstone bodies are constructed as a series of linked boxes. A channel belt may be constructed of many such boxes, the dimensions and the internal angles being defined by actual or empirical data.

When generating a model, the shape geometry is read from the data library and the program takes the following steps:

1 A random location is chosen on the vertical Y–Z plane as the start point (at the minimum X coordinate of the outer volume).
2 Linked hexahedra are generated as described by

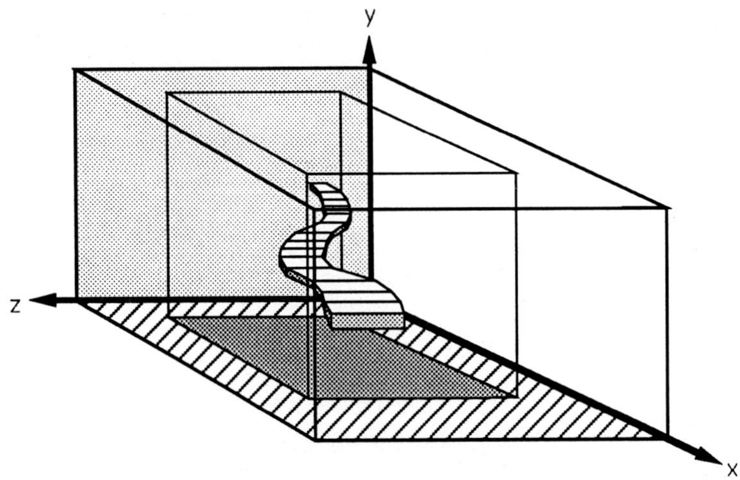

Fig. 3. Model volume defined by X, Y and Z axes. Shape generation begins on the X = 0 plane (light stipple) and is built up until another side of the model volume is reached. When all the units have been generated, the model is trimmed back to the smaller volume (darker stipple). This is to eliminate edge effects (e.g. enabling units to enter the model from the side as well as X = 0 plane).

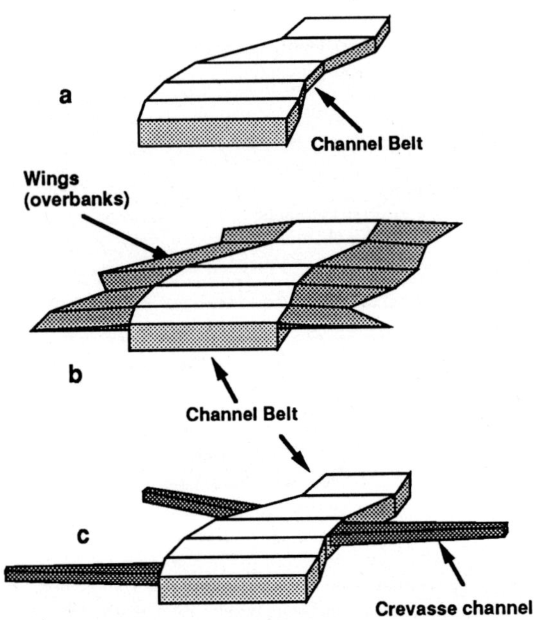

Fig. 4. Geometric shapes representing fluvial sandstone bodies: (a) channel belts, (b) with associated overbank sandstones, and (c) crevasse channels. Extra-channel sandstones thin, split and diminish in quality away from the depositing channel and thus a tapered profile is used in the models.

the geometrical description. If porosity and permeability data are input then these are assigned to each hexahedra at this stage. The belt is complete when another side of the model volume is reached (Fig. 3).

3 If well conditioning is required (i.e. actual positions of sandstones in wells are to be honoured), a completed channel belt is scanned to determine if the thickness matches that of a sandbody in one of the wells. As exact fits would require considerable CPU time, an acceptable error for the thickness of the channel belt at the well locations can be specified in the geometrical shape description; for example, the fit should be within 10% of the thickness of the sand unit. If a suitable match is found, the channel belt is moved within the model so that it ties to the well at the appropriate location. To be accepted, the channel belt must correlate with or miss the other wells, as specified by the input. Direct channel sand correlations between multiple wells may be specified.
4 Further sandstone bodies are generated sequentially, i.e. one is completed before the next is started. When all the well-conditioned channel belts have been generated then the program continues by placing more channel belts in the model volume. These additional sandstones are prevented from intersecting any well where a channel belt is not expected. The generation stops when the required net/gross for the system (the coarse-member proportion determined from wells) within the volume of interest is attained.

Geometric shapes

The SIRCH shapes library currently contains:
1 Boxes. These may be linked together to model trunk channel belts (Figs 2, 4).
2 Boxes with wings (Fig. 4). Linking these units

enables trunk channel belts to be modelled with overbank/levee/minor splay (Fig. 4).
3 Boxes to which tapered boxes may be attached enabling trunk channel belts with crevasse channels to be modelled (Fig. 4).

A user-defined shape may be added to this library where the above are deemed inappropriate for the system to be modelled.

The internal angles and the edges of these shapes may be varied; the dimensions of each box in a link and its relative orientation need to be defined by geologically realistic data.

GEOLOGICALLY DEFINED PARAMETERS DESCRIBING FLUVIAL SANDSTONE BODIES

Sandstone bodies

The SIRCH routine described here is targeted at modelling channel sandstone bodies. In depositional terms, sandstone bodies represent the coarse sediments of a river deposited during the period between its scour and subsequent abandonment. It is important to distinguish between *channels* (i.e. the depositing rivers) and *channel belts* (the 'sandstone' bodies). Where a channel is simply scoured and plugged by sediment prior to any lateral migration of that channel, the sandstone body may preserve the channel geometry (Friend *et al.*, 1986). However, most channel systems, whether meandering, braiding or anastomosing, are laterally unstable. The channel locations move by lateral migration, neck cut-off (meandering systems) or shifting (braided systems) prior to avulsion to another part of the floodplain. The width of the resulting sandstone body reflects the lateral instability of the channel; the channel belt is generally much wider than the depositing channel. In terms of hierarchical bedding surfaces, the third order elements of Allen (1983) and the fifth order elements of Miall (1988) define sandstone bodies/channel belts.

Examples are evident in the form of abandoned meanderbelts/braidbelts at the surface on modern floodplains such as the Mississippi River (Saucier, 1974). Ancient meanderbelts have been defined in the subsurface where boreholes are closely spaced (Curry & Curry, 1972; Cornish, 1984; Wingate, 1988) and in vertical outcrop (Elliott, 1976; Hirst, 1992).

In addition to the channel belts, crevasse splays and crevasse channels are likely to be components of a reservoir system but are generally less well defined. Studies have documented the shapes and dimensions of these deposits in both modern (Farrell, 1987) and ancient systems (Galloway, 1981; Flores, 1984).

Although the general term 'sandstone body' is used in this study, channel belts are heterogeneous units with sediment grades ranging from conglomerate to mud. Coarse sediment (potential reservoir) predominates in most channel belts and connectivity is maintained throughout the channel belt, although with internal baffles. Reservoir quality is accounted for in the porosity values applied to the sandstone unit which should reflect both the reservoir and non-reservoir portions of the unit. However, some channel belts may be effectively compartmentalized by channel plugs of mudstone and siltstone (Swanson, 1979; Ebanks & Weber, 1982; Cornish, 1984). These clearly cannot be modelled as simple contiguous channel belts. If sandstone bodies of this nature are interpreted, a different geometric model is called for.

Geometric variables

The parameters which describe the variations in the geometry of fluvial sandstone bodies have been determined from observation of both ancient and modern fluvial deposits (Table 1). For trunk channel belts, these can be divided into three main groups:
1 cross-section geometry,
2 plan geometry,
3 three-dimensional relationships,
with additional parameters to describe the levees/overbanks and the crevasse channels. Each of the variables listed in Table 1 is discussed below. Included in this discussion is a basic description of the parameter and the problems in the derivation and/or application of the variable.

Cross-sectional parameters

Thickness

The thickness distribution of sandstone bodies may be obtained from well results. Within a sequence, more than one type of geometrically distinct sandstone unit may be present; for example, overbank deposits (crevasse splays, etc.) and channel-belt sandstones. These geometrically distinct components will have contrasting empirical relationships and rock properties and therefore separate thickness distributions need to be collated. Several problems are encountered when attempting to determine sandstone body thickness in fluvial sequences:

1 Are definitions of sandstone bodies used at surface suitable to enable recognition in a borehole sequence?

Resolving sandstone bodies encountered in a well into individual channel belts as defined above is essential. Single storey, broadly fining-upward sandstones are easily identified, particularly if they can be correlated between wells. Stacked sandstones present a problem; are they multistorey channel belts or two separate channel belts? Interpretation will sometimes be equivocal, but, provided core is available and particularly if several wells have been drilled, resolving individual sandstone bodies is possible (Eaves, 1976). In a single well situation, or where core is unavailable, interpretation is limited to fining/coarsening-up trends determined from electric logs. A unique answer is unlikely. Hence the power of a multi-realization, stochastic model in which different interpretations/sensitivities can be tested.

2 In order to use empirical relationships, the depositional thickness of a sandstone body has to be determined. Where partly eroded by a subsequent deposit, the depositional thickness has to be estimated by extrapolating any trends (fining-upwards, gamma log, etc.). Mudstone plugs are also indicative of an 'incomplete' sandstone body and their thickness should thus be included in the thickness.

3 How representative/complete is the distribution determined from the well? Where only a few sandstones are encountered in the well (or wells), the thickness distribution obtained is unlikely to represent the full range of variability in the sequence but merely provide an indication of the sandstone dimensions. In such cases, an analogue distribution should be compared to assess the degree of this problem—the analogue being used to 'fill in' the distribution if necessary.

The input data required by the model is a sandstone body thickness distribution for each sandstone body type in the model.

Width

Compilations of field data indicate that sandstone body (channel belt) width is related empirically to the palaeochannel depth (Fielding, 1986). Relationships of this form have been used to estimate sandstone body width (Fielding & Crane, 1987). The weakness of this method is that palaeochannel depth has to be derived for each sandstone body. This is achieved by using poorly constrained relationships between sandstone body thickness and channel depth (Fielding & Crane, 1987) or by equating palaeochannel depth to the sandstone body thickness (Collinson, 1978; Lorenz *et al.*, 1985). Alternatively, it is observed that thicker sandstone bodies tend to be wider and empirical compilations support the use of a thickness–width relationship (Hirst, 1992). This avoids the extra step of estimating palaeochannel depth, which would be a source of error.

The field data compiled in empirical relationships invariably relate the maximum thickness of a sandstone body to its width. However, it is improbable that a bore hole will penetrate the maximum thickness of a unit—the measured value from a well will be a *fraction* of the maximum. By measuring vertical sections through exposed channel belts, the ratio of the average thickness to the maximum thickness can be estimated. For fluvial sandstone bodies, this can range from 0.6 to about 0.8, i.e. on average the thickness encountered in a well is likely to be 0.6 to about 0.8 times the maximum sandstone body thickness. Borehole thicknesses should be appropriately corrected when using empirical relationships of maximum thickness against width. Clearly, the greatest inaccuracies will occur where only a few sandstone bodies are encountered but the errors will average out where many units are measured.

The data requirement is a plot of width against thickness from which a linear equation can be obtained.

Shape of cross-section

The thickness (measured distribution) and the width (determined empirically) constrain the dimensions but not the shape of the sandstone body cross-section (Fig. 5). The current model assumes an approximate tabular cross-section to the sandstone body.

Plan geometry

The width and thickness attributes enable the cross-sections of the channel belts to be constructed; these are effectively the common faces of the linked boxes (Fig. 2). The plan geometry of channel belts determines how these faces link up.

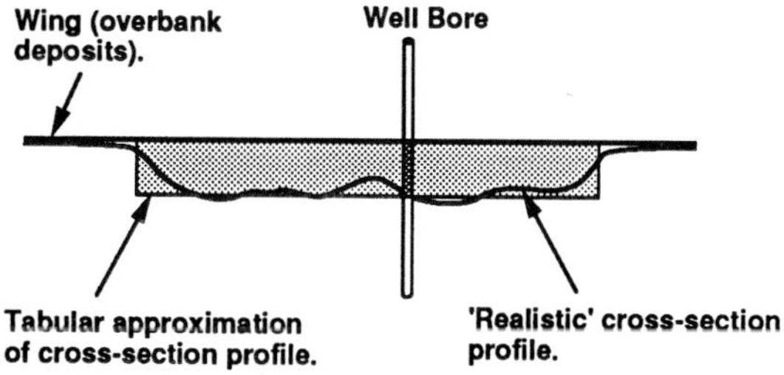

Fig. 5. Section across a channel sandstone body perpendicular to palaeoflow.

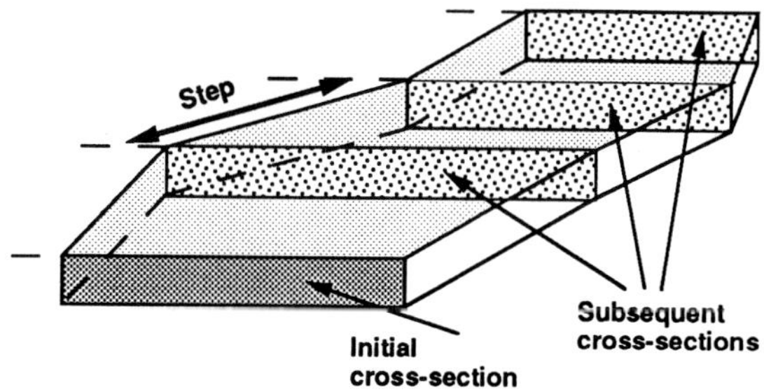

Fig. 6. Successive tabular cross-sections (determined using appropriate distributions of real data) are positioned at specified steps along the channel belt enabling the sandstone body to be built away from the starting plane as a series of interconnected boxes.

Step length

The length of each individual box (i.e. down mean palaeocurrent) forming a channel belt has to be defined (Figs 2, 6); this is termed the step length. Values of thickness, width, porosity, permeability, etc. will be assigned at each step; the step length represents the smallest scale of variability along a channel belt and its value should be small enough to accommodate the perceived variation without being so small that CPU time is excessive. Porosities and permeabilities will typically be obtained from core plug data and should be appropriately averaged (arithmetic and geometric respectively). A step length similar to mean channel-belt width is a guide but may be varied as required.

Variation in sandstone body thickness

Sandstone bodies vary in thickness along the length of channel belts (e.g. Wingate, 1988). This is incorporated by sampling from the thickness distribution (as used for the initial thickness) each time a box is generated. The problem encountered is that two consecutive thicknesses could be sampled from extreme ends of the input distribution resulting in a channel belt that suddenly 'necks' or expands. SIRCH prevents this by including a degree of correlation between adjacent thicknesses constraining them to being within 25% of the total range of the distribution.

Variation in sandstone body width

Channel belts, as seen in planform in recent systems (Saucier, 1974) and in ancient deposits (Busch, 1974; Siemers, 1976; Swanson, 1979) vary in width along their length. This width variation is partly the reason for the scatter around the median line in width–thickness plots. Thus, as the channel belt is constructed (Fig. 6), different widths are selected using the same equation specified, but to incorporate the element of width variation, the selected width can fall anywhere within defined scatter limits.

Length of reach

Channel belts are not infinitely straight across their floodplain but periodically change direction in response to floodplain inhomogeneities, tectonic deformation, avulsion, etc. Channel belts may be divided into approximately straight lengths (Siemers, 1976; Fig. 7a). There will be a degree of subjectivity associated with these divisions which will be of little consequence provided that reach lengths measured from an analogue system are used in combination with the angles between the reaches from the same system (see next geometric variable).

The input to the model will be a distribution of reach lengths appropriate to the system being modelled. Possible sources of data are: (i) active modern systems, particularly where abandoned channel belts can be defined. Examples include the modern Mississippi floodplain (Saucier, 1974), the Rio Grande (Nanz, 1954) and other systems accessible from maps and Landsat images; (ii) ancient systems which are constrained by mine and/or borehole information (Elliott, 1965; Curry & Curry, 1972; Eaves, 1976); (iii) exhumed channel belts. These tend to be of limited application due to the short lengths of reaches exposed (Siemers, 1976; Smith, 1987; Friend *et al.*, 1986).

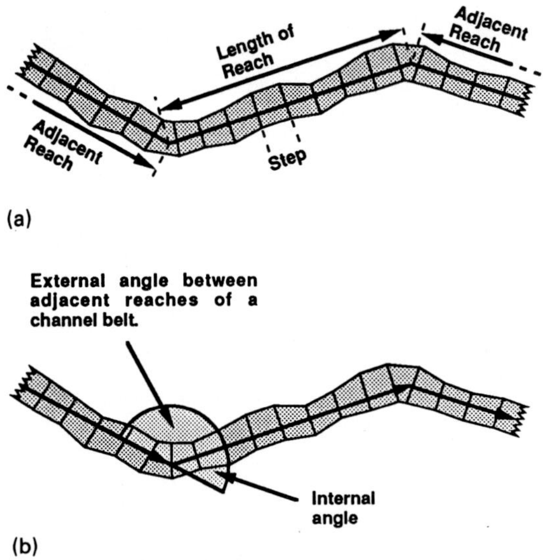

Fig. 7. (a) Plan view of a channel belt composed of linked boxes with equal steps between adjoining faces. Periodic changes in the mean orientation of the channel belt are the basis of its division into reaches. (b) Internal and external angles between adjacent reaches of a channel belt.

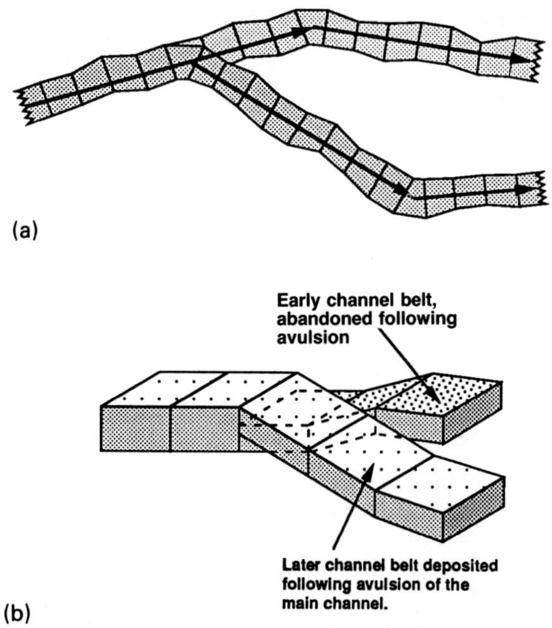

Fig. 8. (a) Plan view of a bifurcating channel belt; the arrows indicate palaeoflow. (b) Modelled geometry at a channel-belt bifurcation.

Angle between successive reaches

The division of a channel belt into reaches implies an angular variation between successive reaches. The internal angle is used (Fig. 7b) and the distribution used in the model should be derived from the same source as the reach length data. The distribution will be about 0° and will determine the size of the angle and whether it is to the right or the left. The input format has to be a PDF (probability distribution function) in which 0° and a small range each side of 0° (5°) is excluded from the range to prevent two reaches forming a single straight reach.

Bifurcation/Unification of channel belts

When a channel avulses, the new channel belt which evolves will be 'attached' at the point of avulsion to the now partly abandoned channel belt. A bifurcating sandstone body will result (Fig. 8; Nanz, 1954; Eaves, 1976). At some point downstream, the old and new channel belts may reunite by crossing.

The frequency of bifurcation is input as the probability of a box within a channel belt bifurcating. For example, if a bifurcation is observed on average every 20 km and the step length of the boxes is 800 m, then the input is 1 bifurcation per 25 boxes.

Systematic variations in channel belt size

The dimensions of channels, and thus the sandstone bodies they deposit, vary considerably between source and downstream termination. Typically these variations take place over 10 s–100 s of kilometres. In terms of reservoir scale, the steady enlargement (or decrease in size in distributive networks; Hirst & Nichols, 1986) of channel belts downsystem will be negligible. Thus over the order of 10 km, this aspect is not considered important.

Orientation

Absolute orientation is essential when conditioning to wells. Information required is the maximum range of orientation of all the channel belts in the sequence. Palaeocurrents determined from cross-beds in boreholes, whether by dipmeter or orientated core goniometry, will give a wider azimuth spread than the channel-belt orientations (Allen, 1966). However, it is not possible to measure channel-belt orientations in boreholes and the mean derived from cross-beds should be used in conjunction with a channel-belt distribution from an outcropping analogue. In some circumstances, direct imaging of

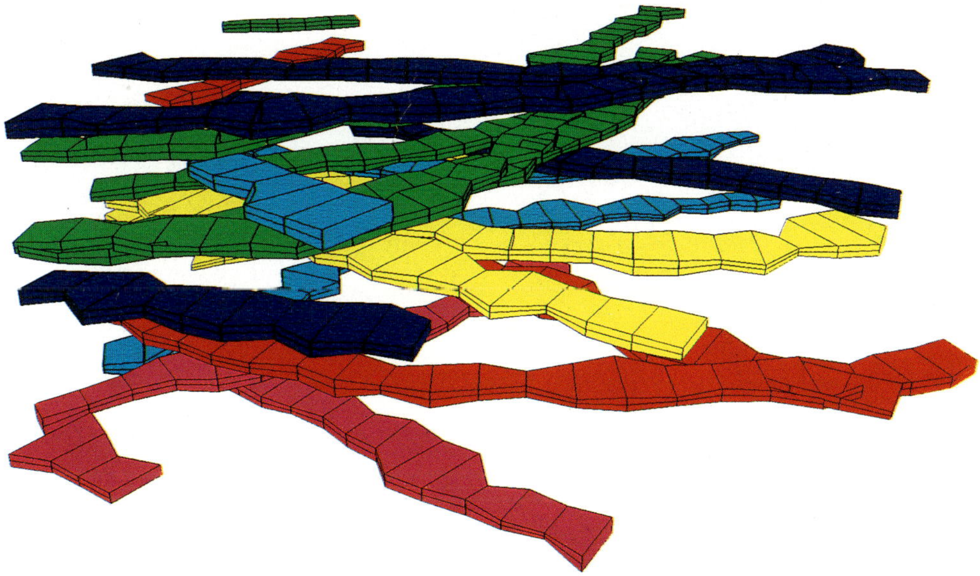

Plate 1. Example of output from the SIRCH modelling routine. The net/gross of the model is approximately 10% and channel belts have been constructed without bifurcations, crevasse splays or crevasse channels. Variations in channel-belt orientation, thickness and width are evident. In this oblique view, it is not obvious which channel belts are connected. The colours indicate clustered (i.e. connected) channel belts although only six colours are possible and thus some colours are repeated. The volume of each cluster may be computed, either for the total cluster within the model or within a specified volume such as a well radius.

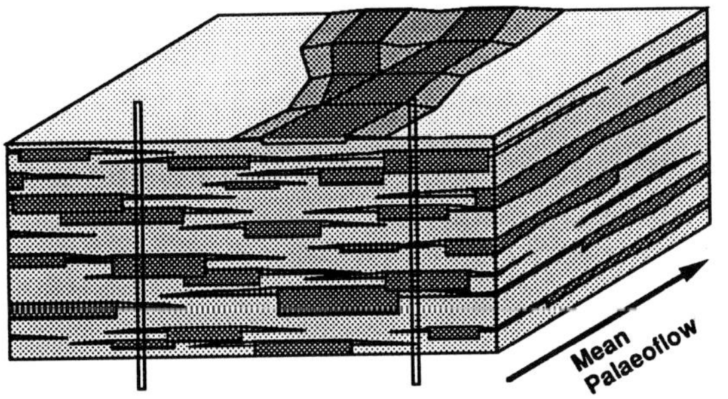

Fig. 9. Schematic representation of a fluvial sequence incorporating many of the variables described in the text.

channel-belt orientation from seismic may be possible (e.g. Brown, 1988). The input range will limit the orientations of channel belts and prevent them turning back on themselves.

Three-dimensional parameters

The relative arrangement and number of channel belts generated is determined by the three-dimensional attributes of the system (Fig. 9).

Frequency of sandbodies in the succession

The total thickness of all the sandstone bodies encountered in a well (net/gross) will give an indication of their volume in the system. SIRCH fills the model volume with sandstone bodies until this value is attained although at any veritical line within the model, the net/gross may vary slightly around the input value.

The more wells available, the better the control on the proportions of each sandstone body type. Combined with regional data, it may be possible to infer lateral variation in the coarse member proportion over the area concerned. This may be incorporated into the model by biasing the starting point of sandstone bodies on the Y–Z plane (of the total model) to satisfy the inferred variation.

Stacking patterns

Abandoned channel belts create a floodplain topography which will decrease the potential for subsequent channels to avulse to that area until the topography has been subdued. In semi-arid sequences, differential compaction between floodplain fines and channel-belt sandstones is observed to be minimal (Hirst, 1983). Any topography is short lived and systematic stacking patterns are not evident. In other systems, particularly where there is strong differential compaction between the floodplain and the channel belt (e.g. coal-bearing sequences), there may be a tendency for channel-belt location to be influenced by the underlying deposits. Data on this aspect are not yet quantified (e.g. repulsion function of Clemetsen *et al.*, 1990) and preferred stacking patterns cannot be directly incorporated into the model. If, however, this aspect is considered significant, the orientation range can be reduced to decrease the number of possible belt crossings.

Multiple channel facies

In cetain depositional systems, more than one 'facies' of channel deposit is developed (Friend, 1989). SIRCH can incorporate more than one type of sandstone body within a single model, for example large and small channel belts—with and without levees. Each facies has to be separately defined with its own range of geological data.

Levees/overbank deposits

Coarse-grained flood deposits, contiguous with the upper parts of channel-belt sandstone, may be of reservoir quality. Where this is the case, they need to be included in the model. These deposits, often referred to as wings in red bed sequences (Friend *et al.*, 1979), typically thin and split away from their depositing channel. To reflect this, a wedge-shaped vertical profile, thickest at the end attached to the channel belt, is incorporated into the shapes library (Fig. 4).

The data requirement is a distribution of levee widths. In the model, the thickness of the levee where it is attached to the channel belt is related to channel-belt thickness. At the point of connection to the channel belt, the default thickness for the levees is set at 30% of the channel-belt thickness (but can be user defined). Away from the channel belt the levees thin uniformly to zero. In systems where levees are absent or not of reservoir quality, this shape can be ignored.

Crevasse channels

Typical of many fluvial systems are minor channels which develop during flood stage as 'relief channels' carrying water and sediment away from the main channel onto the floodplain. These crevasse channels (Fig. 4) cut through the levees and tend to feed into crevasse splays. They are often plugged by reservoir quality sandstone and, being at a high angle to the main channel axis, they may represent important components in the connectivity of a reservoir. In the geometric shape constructed for crevasse channels (Fig. 4), these channel fills are modelled as straight units orthogonal to the main channel belt axis. They narrow distally to half their initial width in the horizontal plane.

The data input is: (i) a distribution of initial widths, thicknesses and lengths from an appropriate outcrop analogue; (ii) a frequency of crevasse channels along the length of a channel belt (e.g. one per 2.5 step lengths).

Data on the areal extent and thickness variations of levees, crevasse splays and crevasse channels is difficult to obtain and there are few mapped examples (e.g. Galloway, 1981; Flores, 1984). The frequency of crevasses may be roughly estimated from a few studies (e.g. Swanson, 1979) or from well observations. The frequency of splays multiplied by the splay volumes (dimension input to the model) should be balanced so that the volume ratio of channel belts to crevasses observed in the well(s) is reproduced.

APPLICATIONS OF THE SIRCH SOFTWARE

The product is a rock volume in which stochastically generated sandstones are constructed based on a specific sedimentary model, well data and analogue data. The individual sand connections and their groupings (clusters) are determined and the model is saved on computer disk. A hard copy of the model may be obtained (see Plate 1, facing p. 244) in which the separate clusters (connected sandstones) are differently coloured. The realizations are probabilistic rather than actual representations, and several model runs (at least 10) are necessary to produce a distribution of results. However, the cuboid model volume bears little relationship to the volume of economic interest which may range from the whole field down to an individual well radius of a few 100 meters. The SIRCH package therefore has further features added.

1 The models can be 'faulted' to adhere to mapped faults and structured to account for regional dip. This is done by 'distorting' the model so that it conforms to the appropriate structure map.

2 The volume generated can be 'clipped back' to the volume within closure using the mapped top and bottom surfaces of the reservoir. These 'clipping' surfaces may be the gas or oil/water contact (GWC/OWC) and the structural map of the cap rock respectively.

Estimates of the following may be output:

1 Volume of reservoir sandstone within the specified interval.

2 Connectivity of the reservoir, or the size of each connected portion (cluster) of the reservoir.

3 Sand volume, connectivity and hydrocarbons in place for a specified radius around a well location.

The role of the model is to determine realistic reservoir geometry and to indicate how sensitive this model is to variations in the input parameters. This should assist in determining the economics of marginal fields and help in the design of production well patterns. It is not for use in determining individual well placements because of the probabilistic nature of the output.

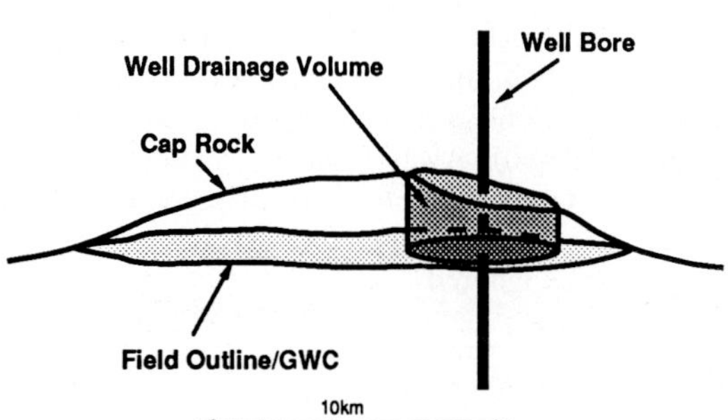

Fig. 10. Oblique view of a hypothetical field illustrating the volume drained by a single randomly positioned well.

Example

The application of SIRCH is illustrated by the following hypothetical worked example. The problem presented to an appraisal team consists of a sequence in which gas is encountered in several similar but separate sandstone bodies. The net/gross is 0.4 and sedimentological studies indicate the sandstones are fluvial channel belts plus crevasse channels (0.34 and 0.06 of the sequence, respectively) and that both components are of reservoir quality. The hypothetical structure is a four-way dip closure (Figs 10, 11) while the bottom of the reservoir interval is a sharp gas/water contact. The reservoir drive mech-

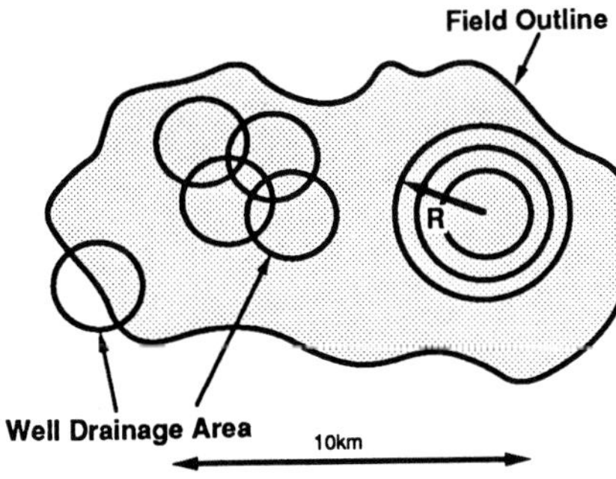

Fig. 11. Plan view of the hypothetical field illustrating aspects of the randomly placed well drainage areas—their overlapping nature, variable radius and probable overlapping of the edge of the field.

anism is predicted to be gas expansion (natural depletion). The critical uncertainty is considered to be the connected gas initially in place (GIIP) per well and it is estimated that 30 billion cubic feet (BCF) per well is the economic threshold for the field. The aim is to constrain the uncertainty on GIIP per well and to assess whether this economic threshold is likely.

The first step is to decide which of the geometric shapes (Fig. 4) is appropriate. The sequence consists of trunk channel belts with crevasse channels which dipmeter data indicate are typically at a high angle to the trunk channels. The SIRCH geometric shape to be used is therefore one of channel belts with crevasse channels (Fig. 4c). The input data is then compiled from the wells and analogue examples appropriate to a humid environment, coal bearing, fluvial sequence (Table 2).

The software is then accessed, the model volume defined and the data input. The model can be run to generate multiple realizations which are then saved to computer disk. The modelled volume defined was larger than the field volume so that the mapped top unconformity (seal) and the GWC can be used to 'trim' the realizations back to the appropriate size. The resulting realization may be used to indicate the net reservoir volume (NRV) within closure. Also, using appropriate gas saturation/depth relationships, the GIIP of the field can be estimated. Being a stochastic process, no two realizations will be the same and multiple realizations must be used to obtain an estimate of the distribution and mean NRV and GIIP results.

These outputs are not particularly useful to estimate potential GIIP per well. Results from the complete modelled volume indicated that all the sandstones generated were connected via variably circuitous routes. What is required is an indication of the gas connected to a well within the field. This is achieved by positioning multiple random wells within a realization; effectively determining the NRV

Table 2. Input values to hypothetical field model discussed in text

Model volume				
Generated volume: 25 km × 19 km × 195 m				
Inner volume of interest: 17 km × 15 km × 195 m				
Trunk channels: cross-section geometry				
T_b	thickness of sandstone body (channel belt), normal distribution, units in metres:			
	5%	Mode	95%	
	3.0	6.5	14	
W_b	median width of channel belt = $(9.5 \times \text{thickness})^{2.2}$			
	Rectangular distribution ± 20% each side of median			
Trunk channels: plan geometry				
Step length 1500 m				
δT_b	variation in sandstone-body thickness along reach achieved by repeated selection from initial distribution			
δW	variation in sandstone-body width along reach by repeated selection using empirical relationship			
L_r	reach lengths input as a triangular distribution, units in metres:			
	Min 5000	ML 10 000	Max 15 000	
α	angle between reaches input as a PDF, units degrees from north			
	50% between +30° and +5°			
	50% between −30° and −5°			
Y	no bifurcations of sandstone bodies			
β	orientation of system (palaeoflow range) input as a triangular distribution, units in degrees from north			
	Min 55°,	ML 85°,	Max 115°	
Three-dimensional relationships				
F	single values for net/gross; 0.34 for trunk channels, 0.06 for crevasse channels			
Crevasse channels (perpendicular to trunk channels)				
L_c	length of crevasse channels, rectangular distribution, units metres			
	Min 800	Max 4000		
W_c and T_c	width and thickness of crevasse channels, normal distribution, units metres			
		5%	Mode	95%
	width	60	200	500
	thickness	0.25	2.5	5.0
F_c	frequency of crevasse channels = one channel per 2400 m			

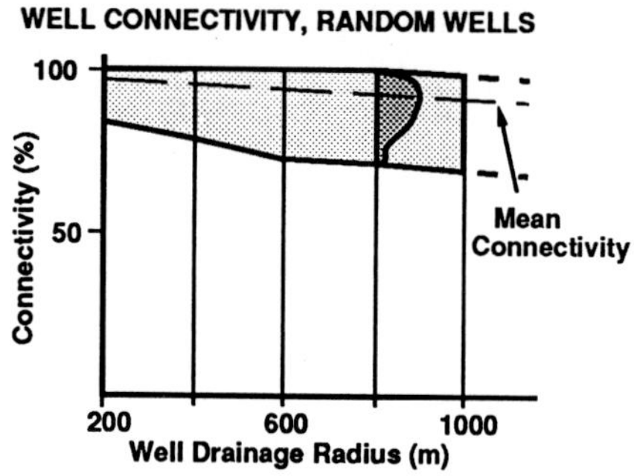

Fig. 12. Drainage radius of 50 randomly positioned wells plotted against the % of sandstone within the radius which is directly connected to the well bore. The spread of results lies within the light stipple while the distribution of results for an 800 m radius is indicated by the dark stipple.

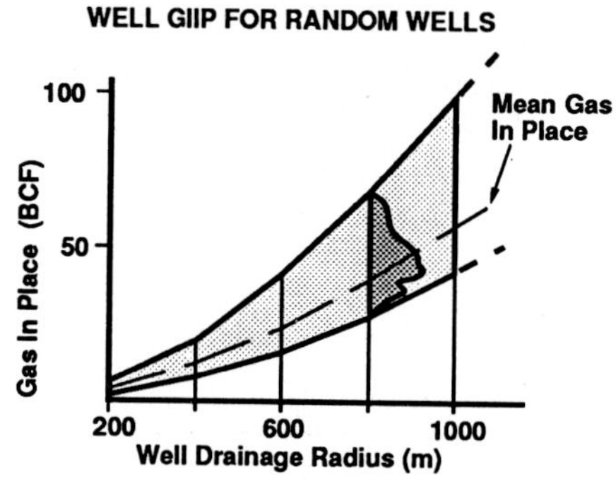

Fig. 13. Drainage radius of 50 randomly positioned wells plotted against gas initially in place connected to each well bore. The spread of results lies within the light stipple while the distribution of results for an 800 m radius is indicated by the dark stipple.

and GIIP of a cylinder rather than the GIIP of the whole field (Fig. 10). The radius of the cylinders may be varied, and, with numerous random wells, they will inevitably overlap each other and some will extend partially beyond the 'trimmed' edge of the field (Fig. 11). The NRV and GIIP determined for overlapping wells assumes each well is the only well draining the area, i.e. the reserves are not shared between the wells. The overlap is generated stochastically and would be avoided in a production grid.

The NRV and GIIP for 50 such wells randomly positioned within one realization using the above input parameters are plotted against drainage radius in Figs 12, 13. Only wells wholly within the field are plotted. The results of both parameters display a distribution. If the reservoir engineer determines from his test results that an 800 m drainage radius is most likely, then, using these graphs, the range of likely gas volumes from wells can be estimated for this specified radius. For 800 m, the distribution of results is indicated to range from 25 to 70 BCF, with a mean GIIP of 42 BCF.

The mean value is above the inferred economic threshold of 30 BCF per well. However, critical uncertainties need to be determined to see how they affect the results. The stochastic routine now becomes invaluable for testing sensitivities and the input data can be altered to reflect the uncertainties in the sedimentological model and empirical relationships. The above model was rerun on the assumption that the crevasse channels, which are important components connecting separate channel belts, had been misinterpreted. The GIIP determined for the revised model which excludes the crevasse channels were approximately 25% lower, enough to convince any development manager not to proceed until the uncertainties were better constrained, perhaps via outcrop studies of analogue sequences.

SUMMARY

Sequences deposited by mixed-load fluvial systems are heterogeneous, comprising interconnected channel, levee and crevasse splay sandstones variably enclosed within floodplain fines. The variable connectivity of the coarser components leads to problems in determining net reservoir volumes and recoverable hydrocarbons from both the whole field and from single wells.

The interactive software developed in this study to stochastically reproduce channel-scale heterogeneity is termed SIRCH (System for Integrated Reservoir CHaracterization). The aim of the SIRCH software is to generate realistic reservoir heterogeneities and provide tools for the analysis of these models. The design has been driven by geological observations of depositional systems. A range of sedimentologically defined variables has been compiled to describe the geometry of the coarse components. Included are parameters defining the variability in channel-belt orientation and geometry, the contribution of coarse overbank deposits, channel bifurcations and systematic variations in net/gross.

A variety of geostatistical methods may be used in

stochastic modelling routines. Within SIRCH, the geostatistical method is object based; the objects are specified geometric shapes stored in a library. Each generated shape has a set of attributes which are based on the range of sedimentologically defined variables. The attributes of each shape determine its size, position and its relationship to other shapes. Channel belts are constructed as continuous links of a specified shape. As the belt is constructed, input values for the attributes are repeatedly sampled from user-defined distribution functions of appropriate geological data.

The product is a sequence in which stochastically generated sandstones have been constructed based on the sedimentary model, well data and analogue data. The generated channel belts may be conditioned to wells. The model volume may be 'faulted' and 'clipped back' to the volume within closure as defined by structure maps. Estimates of the following are possible:

1 Volume of reservoir sandstone within the specified interval.

2 Connectivity of the reservoir, or the size of each connected portion (cluster) of the reservoir.

3 Sand volume, connectivity and hydrocarbons in place for a specified radius around a well location.

The realizations will be a probabilistic rather than an actual representation and several model runs will be necessary to produce a distribution of results. The routine is of most value in indicating how sensitive results are to the various input parameters. The more sensitive aspects can then be actively constrained by appropriate work (refining the sedimentary model; analogue data compilation). This should assist in determining the economics of marginal fields.

SIRCH is a flexible modelling routine allowing multiple realizations and sensitivities to be easily produced. The shapes library may be augmented for more specific applications. Also, by determining the range of geological variables and designing suitable basic shapes, the routine may be extended to other depositional environments such as turbidites and offshore bars.

APPENDIX

SIRCH (System for Integrated Reservoir CHaracterization) has been developed as proprietary software by BP Exploration Company since August 1987 and is at release 2.1. It runs on a VAX system under VMS version 5.0 and above. It requires Ingres version 5 for its data storage, DI3000 for graphics and SMARTSTAR for the forms interface.

In the example described, approximately 30 sandstone bodies were generated, the number of units varying between runs. These sandstone bodies were initially conditioned to three wells. Between 30 and 60 minutes CPU time was required for each realization on the VAX 8800, although without well conditioning, CPU time would be considerably less.

ACKNOWLEDGEMENTS

The authors wish to acknowledge BP Exploration Company Ltd, London, for permission to publish this paper. We thank the reviewers, Daniel H. Horowitz, Carlton R. Johnson and Steve Flint for their comments on the manuscript.

REFERENCES

AASEN, J.O., SILSETH, J.K., HOLDEN, L., OMRE, H., HALVORSEN, K.B. & HOIBERG, J. (1990) A stochastic reservoir model and its use in evaluations of uncertainties in the results of recovery processes. In: *North Sea Oil and Gas Reservoirs—II* (Ed. by A.T. Buller, E. Berg, O. Hjelmeland, J. Kleppe, O. Torsaeter & J.O. Aasen), pp. 425–436. Graham and Trotman, London.

ALEXANDER, J. (1986) Idealised flow models to predict alluvial sandstone body distribution in the Middle Jurassic Yorkshire Basin. *Marine Petrol. Geol.*, **3**, 298–305.

ALLEN, J.R.L. (1966) On bedforms and palaeocurrents. *Sedimentology*, **6**, 153–190.

ALLEN, J.R.L. (1983) Studies in fluviatile sedimentation: bar-complexes and sandstone sheets (low sinuosity braided streams) in the Brownstones (L. Devonian), Welsh Borders. *Sedimen. Geol.*, **33**, 237–293.

BROWN, A.R. (1988) Interpretation of three-dimensional seismic data (second edition). *Mem. Am. Ass. petrol. Geol.*, **42**, 253 pp.

BULLER, A.T., BERG, E., HJELMELAND, O., KLEPPE, J., TORSAETER, O. & AASEN, J.O. (1990) *North Sea Oil and Gas Reservoirs—II*. Graham and Trotman, London.

BUSCH, D.A. (1974) Stratigraphic traps in sandstones—exploration techniques. *Mem. Am. Ass. petrol. Geol.*, **21**, 174 pp.

CLEMETSEN. R., HURST, A.R., KNARUD, R. & OMRE, H. (1990) A computer program for evaluation of fluvial reservoirs. In: *North Sea Oil and Gas Reservoirs—II* (Ed. by A.T. Buller, E. Berg, O, Hjelmeland, J. Kleppe, O, Torsaeter, & J.O. Aasen), pp. 373–385. Graham and Trotman, London.

COLLINSON, J.D. (1978) Vertical sequence and sand body shape in alluvial sequences. In: *Fluvial Sedimentology* (Ed. by A.D. Miall), Mem. Can. Soc. petrol. Geol., **5**, 577–586.

Cornish, F.G. (1984) Fluvial environments and palaeohydrology of the Upper Morrow 'A' (Pennsylvanian) meander belt sandstone, Beaver County, Oklahoma. *Shale Shaker*, **34.6**, 70–80.

Curry, W.H. & Curry, W.H. III (1972) South Glenrock oil field, Wyoming: prediscovery thinking and postdiscovery description. In: *Stratigraphic Oil and Gas Fields* (Ed. by R.E. King), Mem. Am. Ass. petrol Geol., **16**, 415–427.

Dreyer, (1990) Sand body dimensions and infill sequences of stable, humid-climate delta plain channels. In: *North Sea Oil and Gas Reservoirs—II* (Ed. by a.T. Buller, E. Berg, O, Hjelmeland, J. Kleppe, O. Torsæter, & J.O. Aasen), pp. 337–351. Graham and Trotman, London.

Eaves, E. (1976) Citronelle oil field, Mobile County, Alabama. In: *North American Oil and Gas Fields* (Ed. by J. Braunstein), Mem. Am. Ass. petrol. Geol., **24**, 259–275.

Ebanks, W.J. & Weber, J.F. (1982) Development of a shallow heavy-oil deposit in a Pennsylvanian fluvial sandstone reservoir, Eastburn Field, Missouri. *Oil and Gas Journal* (Sept 1982), 222–234.

Elliott, R.E. (1965) Swilleys in the coal measures of Nottinghamshire interpreted as palaeoriver courses. *Mercian Geologist*, **1**, 133–142.

Elliott, T. (1976) The morphology, magnitude and regime of a Carboniferous fluvial-distributary channel. *J. sedim. Petrol.*, **46**, 70–76.

Farrell, K.M. (1987) *Sedimentology and Facies Architecture of Overbank Deposits of the Mississippi River, False River Region, Louisiana*. SEPM Special Publication, **39**, 111–120.

Fielding, C.R. (1986) Fluvial channel and overbank deposits from the Westphalian of the Durham coalfield, NE England. *Sedimentology*, **33**, 119–140.

Fielding, C.R. & Crane, R.C. (1987) An application of statistical modelling to the prediction of hydrocarbon recovery factors in fluvial reservoir sequences. In: *Recent Developments in Fluvial Sedimentology* (Ed. F.D. Etheridge, R.M. Flores & M.D. Harvey), Spec. Publs. Soc. econ. Paleont. Miner., Tulsa, **39**, 321–327.

Flores, R.M. (1984) Comparative analysis of coal accumulation in Cretaceous alluvial deposits, southern United States Rocky Mountain basins. In: *The Mesozoic of Middle North America* (Ed. D.F. Stott & D.J. Glass), Mem. Can. Soc. petrol. Geol., **9**, 373–385.

Friend, P.F. (1989) Hydrological controls of alluvial geometry in the Indo-Gangetic sub-basins. *4th International Conference on Fluvial Sedimentology*, October 1989, Barcelona, Spain, 133. (Abstract.)

Friend, P.F., Slater, M.W. & Williams, R.C. (1979) Vertical and lateral building of river sandstone bodies, Ebro Basin, Spain. *J. geol. Soc. London*, **136**, 39–46.

Friend, P.F., Hirst, J.P.P. & Nichols, G.J. (1986) Sandstone-body structure and river process in the Ebro Basin of Argon, Spain. *Cuadernos Geologia Iberica*, **10**, 9–30.

Galloway, W.E. (1981) Depositional architecture of Cenozoic Gulf coastal plain fluvial systems. *Spec. Publs. Soc. econ. Paleont. Miner.*, **31**, 127–155.

Gundesø, R. & Engelund, O. (1990) *SESIMIRA*—a new geological tool for 3D modelling of heterogeneous reservoirs. In: *North Sea Oil and Gas Reservoirs—II* (Ed. by A.T. Buller, E. Berg, O. Hjelmeland, J. Kleppe, O. Torsaeter, & J.O. Aasen), pp. 363–371. Graham and Trotman, London.

Haldorsen, H.H. & Macdonald, C.J. (1987) Stochastic modeling of underground reservoir facies (SMURF). *SPE Paper* **16751**.

Haldorsen, H.H. & Damsleth, E. (1990) Stochastic modeling. *J. Petrol. Technol.* **42.4**, 404–412.

Henriquez, A., Tyler, K.J. & Hurst. A. (1988) Characterization of fluvial sedimentology for reservoir simulation modeling. *SPE Paper* **18323**.

Hirst, J.P.P. (1983) Oligo-Miocene alluvial systems in the northern Ebro Basin, Huesca Province, Spain. PhD. thesis, University of Cambridge, 247 pp.

Hirst, J.P.P. (1992) Variations in alluvial architecture across the Oligo-Miocene Huesca fluvial system, Ebro Basin, Spain. In: *The Three-dimensional Facies Architecture of Terrigenous Clastic Sediments, and its Implications for Hydrocarbon Discovery and Recovery* (Ed. by A.D. Miall & N. Tyler), SEPM 'Concepts and Models in Sedimentology and Paleontology', vol. 3, pp. 111–121.

Hirst, J.P.P. & Nichols, G.J. (1986) Thrust tectonic controls on Miocene alluvial distribution patterns, southern Pyrenees. In: *Foreland Basins* (Ed. P.A. Allen & P. Homewood). Spec. publs. int. Ass. Sediment., **8**, 247–258.

Knutson, C.F. (1976) Modeling of noncontinuous Fort Union and Mesaverde sandstone reservoirs, Piceance Basin, Northwestern Colorado. *SPE Paper* **5024**.

Lee, Y-H., Martinez, P.A. & Harbaugh, J.W. (1990) Dynamic 3-D graphics critical element in Stanford's SEDSIM project. *Geobyte*, **5**, 37–38.

Livera, S.E. (1989) Facies associations and sand-body geometries in the Ness Formation of the Brent Group, Brent Field. In: *Deltas: Sites and Traps for Fossil Fuels* (Ed. by M.K.G. Whateley & K.T. Pickering), Spec. Publs. Geol. Soc. London, **41**, 269–286.

Lorenz, J.C., Heinze. D.M., Clark, J.A. & Searls, C.A. (1985) Determination of widths of meander-belt sandstone reservoirs from vertical downhole data, Mesaverde Group, Piceance Creek Basin, Colorado. *Bull. Am. Assoc. Petrol. Geol.*, **69**, 710–721.

Miall, A.D. (1988) Architectural elements and bounding surfaces in fluvial deposits: anatomy of the Kayenta Formation (Lower Jurassic), southwest Colorado. *Sedim. Geol.* **55**, 233–262.

Nanz, R.H. (1954) Genesis of Oligocene sandstone reservoir, Seeligson Field, Jim Wells and Kleberg Counties, Texas. *Bull. Am. Assoc. Petrol. Geol.*, **38**, 96–117.

Ravenne, C., Eschard, R., Galli, A., Mathieu, Y., Montadert, L. & Rudkiewicz, J.L. (1987) Heterogeneities and geometry of sedimentary bodies in a fluvio-deltaic reservoir. *SPE Paper* **16752**.

Rudkiewicz, J.L., Guerillot, D., Galli, A. & Group Heresi (1990) An integrated software for modelling of reservoir lithology and property with an example from the Yorkshire Middle Jurassic. In: *North Sea Oil and Gas Reservoirs—II* (Ed. by A.T. Buller, E. Berg, O. Hjelmeland, J. Kleppe, O. Torsæter, & J.O. Aasen), pp. 399–406. Graham and Trotman, London.

SAUCIER, R.T. (1974) Quaternary geology of the lower Mississippi valley. *Research Series Arkansas Archaeological Survey*, **6**, 26 pp.

SIEMERS, C.T. (1976) Sedimentology of the Rocktown Channel Sandstone, upper part of the Dakota Formation (Cretaceous). Central Kansas. *J. sedim. Petrol.*, **46**, 97–123.

STROBEL, J., CANNON, R., KENDALL, C.G. St. C., BISWAS, G. & BEZDEK, J. (1989) Interactive (SEDPAK) simulation of clastic and carbonate sediments in shelf to basin settings. *Computers & Geosciences*, **15**, 1279–1290.

SMITH, R.M.H. (1987) Morphology and depositional history of exhumed Permian point bars in the southwestern Karoo, South Africa. *J. sedim. Petrol.*, **57**, 19–29.

SWANSON, D.C. (1979) Deltaic deposits in the Pennsylvanian Upper Morrow Formation of the Anadarko Basin. In: *Pennsylvanian Sandstones of the Mid-Continent* (Ed. N.J. Hyne), Spec. Publ. Tulsa Geol. Soc. **1**, 115–162.

WINGATE, T.P. (1988) Paleohydraulic parameters of a Morrowan point-bar complex, Salt Lake and Haswell fields, Kiowa County, Colorado. *Compass*, **65**, 58–69.

Spec. Publs int. Ass. Sediment. (1993) **15**, 253–258

A 3-D modelling approach for providing a complex reservoir description for reservoir simulations

G.L. KRUM *and* C.R. JOHNSON

Exxon Production Research Company, Houston, Texas, USA

ABSTRACT

The process of building a 3-D reservoir model of a geologically complex field involves first defining the stratigraphic framework and the field-wide reservoir parameters, and second, interpolating between well control points to develop the heterogeneity required to simulate the depositional environment. Heterogeneity is developed in the model by the choice of interpolation algorithm and by the addition of pseudowells. Interpolation algorithms can be discrete, continuous, or distance-weighted stochastic. When the model is complete, it is scaled up for the reservoir simulator. Advantages of 3-D modelling with this approach are validation of the interpretation and ready input for the reservoir simulator. Limitations include the lack of trained model builders and need for good quality, interpreted, well data.

INTRODUCTION

This paper discusses, in a step-by-step manner, our approach to building a 3-D reservoir description of a geologically complex field. There are two parts to the paper: (i) how we build models, and (ii) a discussion of advantages and limitations of our approach to 3-D modelling. Some of the figures used in this paper are from the Brage Fenfjord and Statfjord fields (Norwegian North Sea). Other figures are from our synthetic school data set.

The tools we use in building a 3-D reservoir description (model) are RESMAP, GPMAP, and GEOSET. GPMAP is a general-purpose mapping system and GEOSET is a 3-D modelling system. These mainframe programs have been developed by Exxon Production Research Company over the past two decades. RESMAP is a panel-driven system which builds batch jobs and invokes GPMAP for gridding and contouring and GEOSET for 3-D modelling.

The results of our modelling efforts are used first to validate the geologic interpretation, and then they are commonly used to build input arrays for a reservoir production simulator. The modelling process provides benefits to Exxon beyond just feeding the simulator—it serves as a communication tool between engineers and geologists, it acts as an archive for interpreted information, and it moves reservoir geology from a fuzzy concept, supported by sparse data, to a quantitative model.

BUILDING 3-D RESERVOIR MODELS

What we are building, at the most fundamental level, is a 3-D array of codes which represent reservoir attributes. These arrays comprise a GEOSET model in which each code represents the attribute value for a discrete, uniform volume: a GEOSET cell. Spatially, a GEOSET cell is tabular, with a typical cell being 100 m square by 3 dm thick. A model may consist of any number of cells, with 1–10 million cells being common.

A two-step approach is used when building a model for a complex field with GEOSET: (i) definition of the stratigraphic framework (architecture) and field-wide parameters; and (ii) building initial models from the definitions in step 1 and developing internal geologic heterogeneity within the layers. Step 2 is accomplished by the choice of interpolation algorithm and by the addition of pseudowells (dummy wells). The dummy wells add both heterogeneity and geologic interpretation. Step 2 is followed by a scale-up process to interface the geologic model to the lower resolution engineering reservoir simulator.

An important step precedes model building of any kind, and it is a step the authors normally are only peripherally involved in—data collection and analysis. The downhole information for each well in the area must be assembled because it is the starting point for attribute extrapolation into the framework. Seismic surfaces must be interpreted and depth converted; they usually define the reservoir structural framework.

Step 1—Framework building

The authors use standard gridding algorithms to build a 2-D grid for each structural surface and to assemble these surfaces into a framework which honours the stratigraphic relationships of the seismic and geologic interpretations. Those relationships are: (i) conformable, (ii) baselap, or (iii) truncation (Jones *et al.*, 1986).

Input surfaces are typically depth converted from seismic, but we may also bring in digitized, hand-contoured surfaces, or build surfaces directly from horizon tops in the well data. Each surface is flagged as either a sequence boundary or a correlation horizon. A sequence boundary subdivides the model vertically into zones. A correlation horizon controls the extension of foot-by-foot well attribute data into 3-space (a sequence boundary normally does not). The example framework on Plate 1 (facing p. 256) illustrates these concepts.

The dark blue layer on Plate 1 is interpreted as a lag deposit at the base of a fluvial sequence, a zone of approximately contemporaneous deposition, or a chronostratum (Jones, 1988). The beds above this layer are interpreted to be conformable to this surface. Plate 2 (facing p. 256) shows the porosity model for the same section; note the conformable bedding above the sequence boundary. Lateral facies changes are represented here as changes in porosity and are conformable to the interpreted time lines.

Additional surfaces may be added to the framework to provide correlation control for a zone. We classify these additional surfaces as supplementary or phantom horizons. A supplementary surface is within a zone and aids in controlling extrapolation in an area of limited stratigraphic control, i.e. an area with sparse well control in a complex framework or an area of truncation. It is an additional time line to help the model match the geologist's depositional interpretation. A phantom surface is usually at the top or base of the model and may simply control the volume above or below the last interpreted surface. A phantom surface can also be used to control correlation within a deeper zone, for example, when a zone does not have the minimum pair of bounding correlation horizons required by our software.

Field-wide parameters

At project inception, we define a suite of parameters which will control the gridding and modelling algorithms, define the geologic environment, and configure the display environment. The control parameters are the grid interval, the cell size, the coordinate limits, and the interpolation algorithms. The definition parameters for the geologic environment include names and numbers of the stratigraphic surfaces and their interrelationships, fault traces, fluid surfaces, the nature and amount of heterogeneity, and the type of data being modelled. The data typically modelled are porosity, permeability, and facies, but any qualitative or quantitative reservoir parameter could be modelled. The display environment includes such information as the plotter type, map scales, and titles.

Step 3—Initial models

GEOSET modelling begins with the attribute and the 2-D structural surfaces. These data are used to build the initial well models and horizon models. Well models and horizon models are internal data structures for use in GEOSET. The well model contains attribute data, automatically blocked into discrete units, and the spatial location of the wellbore. One well model is required for each attribute to be modelled.

In the case of deviated wells, GEOSET will honour the deviation survey for the well to properly position interval data and the top picks within the framework grids. Any top picks which are missing in the well data are automatically interpolated from the grids of the structural framework.

The GEOSET horizons models are built from the framework grids. RESMAP automatically follows the procedures described by Jones *et al.* (1986) to onlap and truncate surfaces and tie them to the stratigraphy in the well model. Three horizon models of the structural framework are maintained by RESMAP: a primary one for model building, and two additional models to store reference structures for modelling (fluid horizons and restored stratigraphic surfaces).

Once the parameters have been defined, the grids built for the stratigraphic framework and the well data read in, the difficult part of building a model with RESMAP/GEOSET is complete. RESMAP is used to build complex batch jobs which will construct the initial models, but this process requires active participation by the model geologist in entering and validating data and interpretation. The next steps, model building and display, are relatively automatic and require only minimal input from the model builder. The first model built is generally used to validate that the parameter file, the well data and the stratigraphic architecture all fit together. As such, it is generally followed by at least one repeat step. Data are seldom perfect.

Once past this validation point, there may be one more interpretive model-building step: adding geologic heterogeneity to the model. The heterogeneity of the reservoir strata that we model is generally insufficiently characterized by the available well information. Therefore, we need to add information (interpretation) to complete the model. Typically, we do this with a combination of pseudowells and our Random Interval Selection Algorithm (RISA).

Adding heterogeneity (pseudowells and RISA)

To build geologic features into a model and to add heterogeneity, we use pseudowells. We chose this method because the effect is easy to quantify and because it is relatively easy to get information from the field geologist with the question, 'What would you expect if you drilled a well here (and here . . .)?' Pseudowells can be used to add texture to an otherwise homogeneous model. Clearly, however, this procedure is highly influenced by the subjective views of the individual geologist.

Adding heterogeneity may require a large number of wells as in the case of the Brage Fensfjord field where 3000 pseudowells were required. Such wells are produced with an automatic well generator that matches a facies pattern and for each facies stochastically selects thickness and attribute values from a user-supplied facies description.

Specific features can be constructed with relative ease using pseudowells. For example, if a channel sandbody is interpreted, its geometry can be modelled with a line of triplet wells. The centre well of a triplet would represent the channel at its best, thickest section or best reservoir quality material, and the flank wells the shale-prone channel margin. Our RISA interpolation algorithm, described later, would handle the feathering of the channel margin to the channel centre, and a 2-D grid would form the sharp erosional base of the channel. Other interpolation algorithms could be used with the selection dependent on the interpretation. For example, a gradational change could be modelled with a continuously varying algorithm such as those routinely available in commercial mapping packages; least squares and moving weighted average to name two possible algorithms.

Small, random, geologic features can be built into a model by scattering pseudowells densely in an area. The size of a feature is inversely proportional to the density of the pseudowell; widely spaced pseudowells produce larger features. The concretionary lenses in the Brage Fensfjord reservoir were added with this type of pseudowell. The result is shown in Plate 3 (facing p. 256). Note the scattered concretionary lenses and also the erosional top of the reservoir, down-cutting to the viewer's left along the line of section.

In Plate 4 (facing p. 256) the top layer illustrates discrete interpolation which is sometimes referred to as polygonal, or nearest neighbour, interpolation. This interpolation provides a quick and simplistic first-pass evaluation of a reservoir. Its character is very blocky, and generally not as a geologist would draw it, but this technique has the advantage of providing a direct comparison with adjacent wells. The next lower layer in Plate 4 is an example of continuous interpolation (see Jones, 1988, for a more complete discussion). Notice the transitional facies created between wells by the continuous interpolation. They are not stratigraphically present in either well but are interpreted between wells by the mathematical progression of the gridding algorithm.

These new facies are not always indicative of a bad interpolation; that judgment is dependent on the attribute being modelled. If the attribute is temperature, then there is most certainly a progression of new values between wells. However, for sand–shale interfaces we do not usually see gradual transitions at reservoir scale. Rather, we see abrupt changes in rock type ('either/or' style changes). RISA provides an interpolation that is closer to what we would expect. It may not be as visually pleasing as a gradational change, but subjectively it has a more geologic character. Objectively, our experience indicates that it is easier (faster) to history-match a RISA model in a reservoir simulator than one built with other techniques.

Pseudowells and RISA interpolation are the tools we use to build heterogeneity and interpretation into a 3-D model. RISA interpolates well information into the space between wells, honouring facies from the surrounding wells without inventing new attributes such as a gridding type of algorithm might. Although RISA is our usual choice for interpolation, other algorithms are available and are illustrated in Plate 4. Two RISA examples are shown on this figure, at power 2 and power 4.

Power, with respect of the RISA algorithm, is the exponent of the distance in the weighting algorithm. The two RISA interpolations shown in Plate 4 represent a narrow (power 4) and a wider (power 2) transition zone between wells. With RISA, well attribute data are randomly selected from wells surrounding the cell to be interpolated with the probability of a given well supplying the cell value weighted by inverse-distance-to-a-power. The final selection is made stochastically with the highest weight being the most likely choice. Thus the nearer well is a more likely source for interpolation but a more distant well always has a chance of sourcing the cell. By increasing the exponent of the distance-weighting step (the power), the nearer well is more likely to be selected; at a power of 8, the result of a RISA interpolation is substantially similar to a discrete interpolation.

In the case of Brage Statfjord, the geologic interpretation called for discrete clay plugs formed as a result of channel abandonment. The plugs were important to the model because they represented permeability barriers within reservoir of unquantified economic effect (see Haldorsen *et al.*, 1987). The actual size of the clay plugs was estimated from analogous outcrop/experience because the plugs are not visible on seismic data and well control could not provide enough information to define the areal extent and continuity of the clay. For this model, we used RISA interpolation and 109 pseudowells to define the clay plugs. A section of the model is shown in Plate 5 (facing p. 256).

At this point the geologic model is essentially complete. What is left is to make displays in plan and section for final validation, for use in field management/problem solving, and for presentation to management. Given that the interpretation is approved, we are ready to pass the geologic model to the engineering reservoir simulator program. To do this, we use a simulator interface to scan the geologic model and build reservoir description arrays for the reservoir simulator.

Interfacing with a reservoir simulator

The geologic model is frequently of a higher resolution than the engineering reservoir simulator can accommodate, i.e. a simulator block may contain hundreds of geologic model cells. The task of manually averaging and assigning porosity and permeability values to individual reservoir blocks can be an onerous and lengthy task. The following discussion describes the major steps in moving information from the geologic model to the engineering simulator, the geology-to-simulator interface.

The cells of our 3-D geologic model are oriented parallel to the model edges and are square in plan view. Simulator blocks may be defined in either a Cartesian, cylindrical, or curvilinear coordinate system and are usually much larger than the GEOSET cells. Simulator block edges need not match the geologic cells. The top and bases of the reservoir blocks—the 'layers' of the reservoir-block model—are defined by a series of computer grids. In the same fashion as framework grids define the geologic layering, these grids define the flow layering.

The output of the simulator interface is a series of arrays containing block definitions and geometry, average block porosity, and permeability of the block faces. A typical array contains the X, Y, Z coordinates and depths of block corners calculated from the input block coordinates and the grids defining the layer boundaries. Porosity for each block is obtained by averaging the GEOSET cells or parts thereof that fall within each block, and is normally calculated for the block centres.

Permeability, unlike porosity, is usually computed across simulator block faces, and three directional arrays (kx, ky, kz) are compiled. A complete suite of permeability data is seldom available for modelling, so a porosity–permeability transform may be used to obtain a permeability model from the geologic porosity model. Different porosity–permeability relationships for various rock types can be defined, or separate GEOSET models may be built for permeability.

To allow feedback from the simulator to the geologic model, our interface is bidirectional. Problems encountered in history-matching the reservoir are resolved during simulation by altering block attributes. Logically, these problems reflect geologic information available to the production engineer

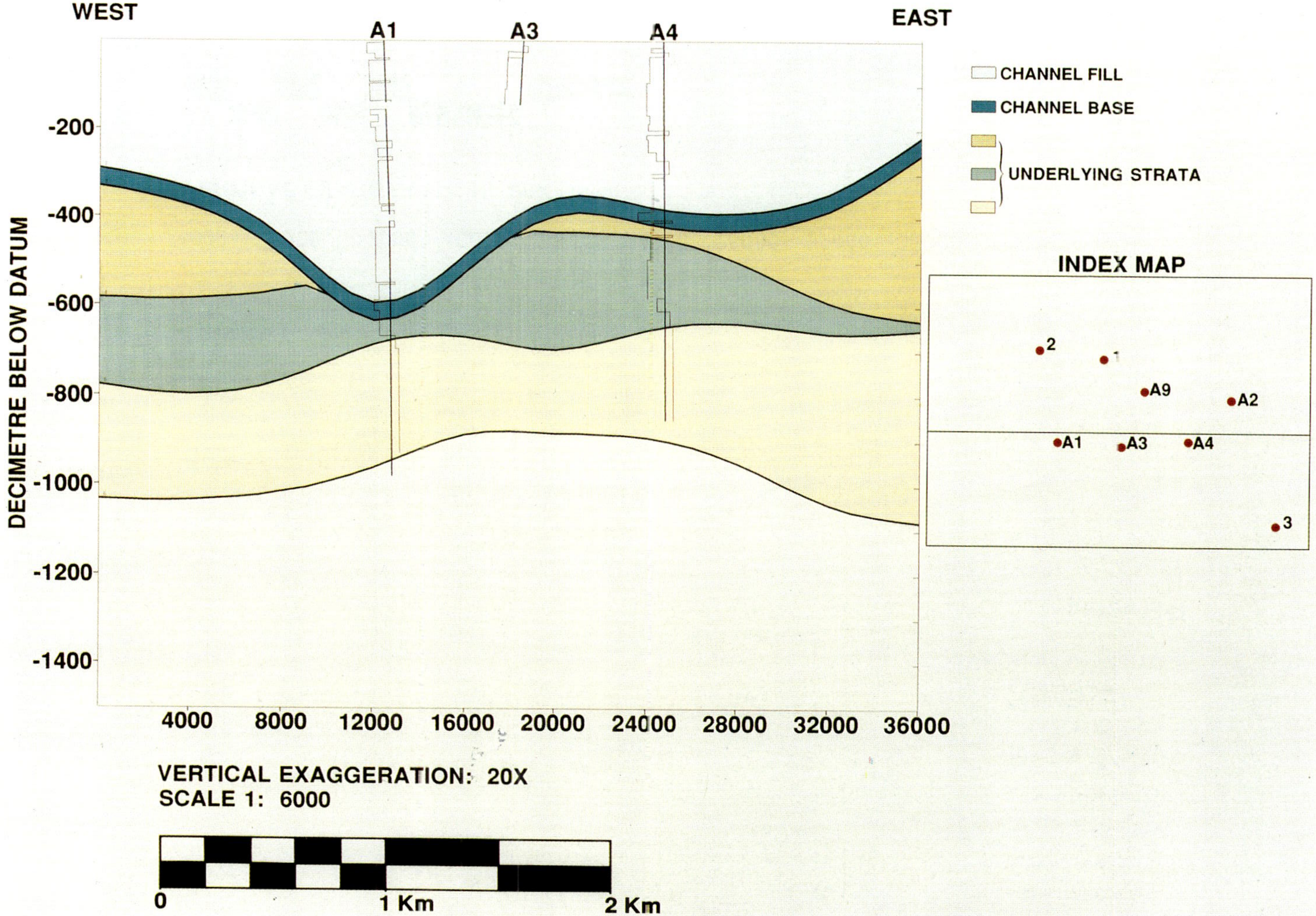

Plate 1. Example cross-section showing a stratigraphic framework.

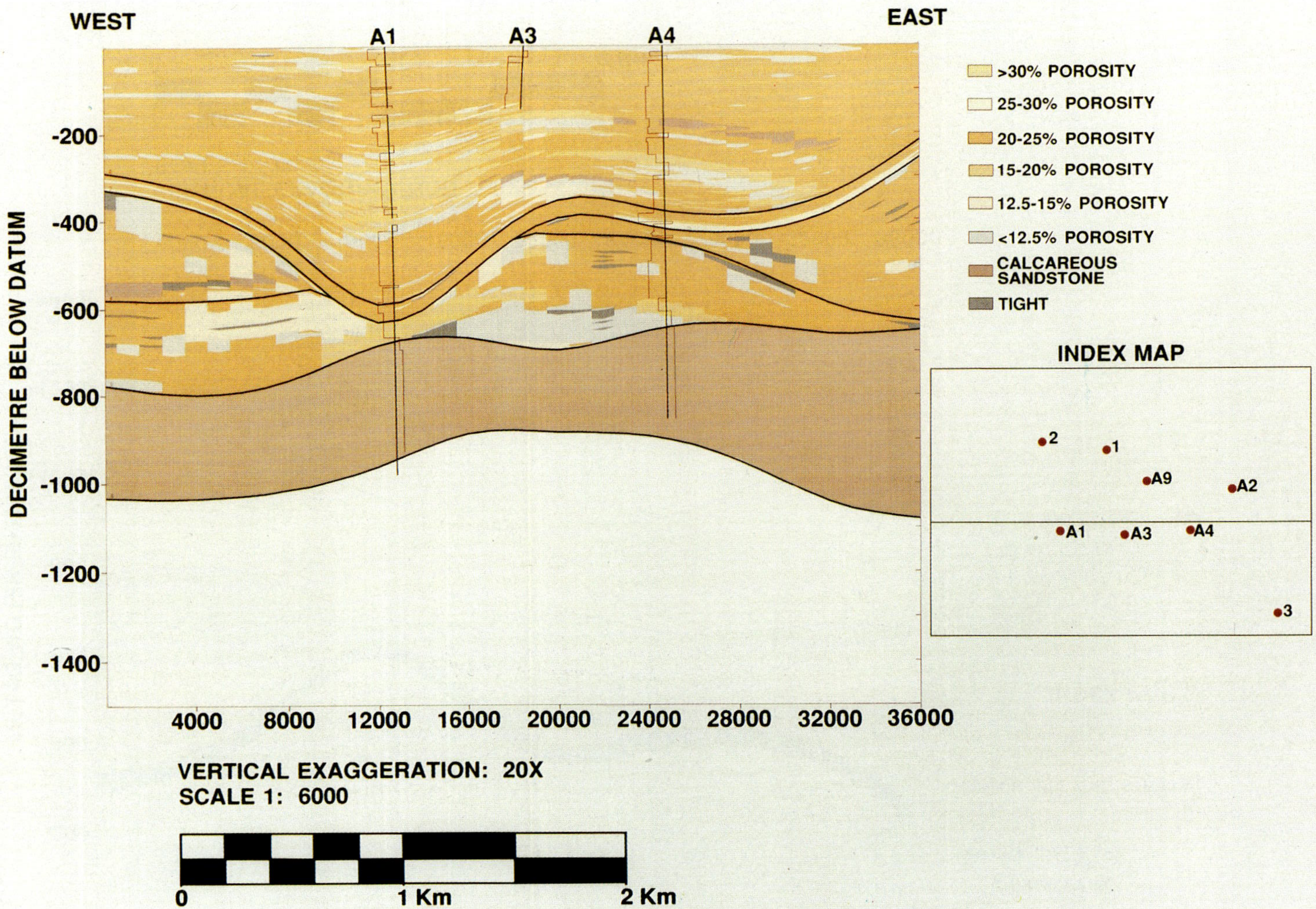

Plate 2. Example cross-section showing facies within a channel.

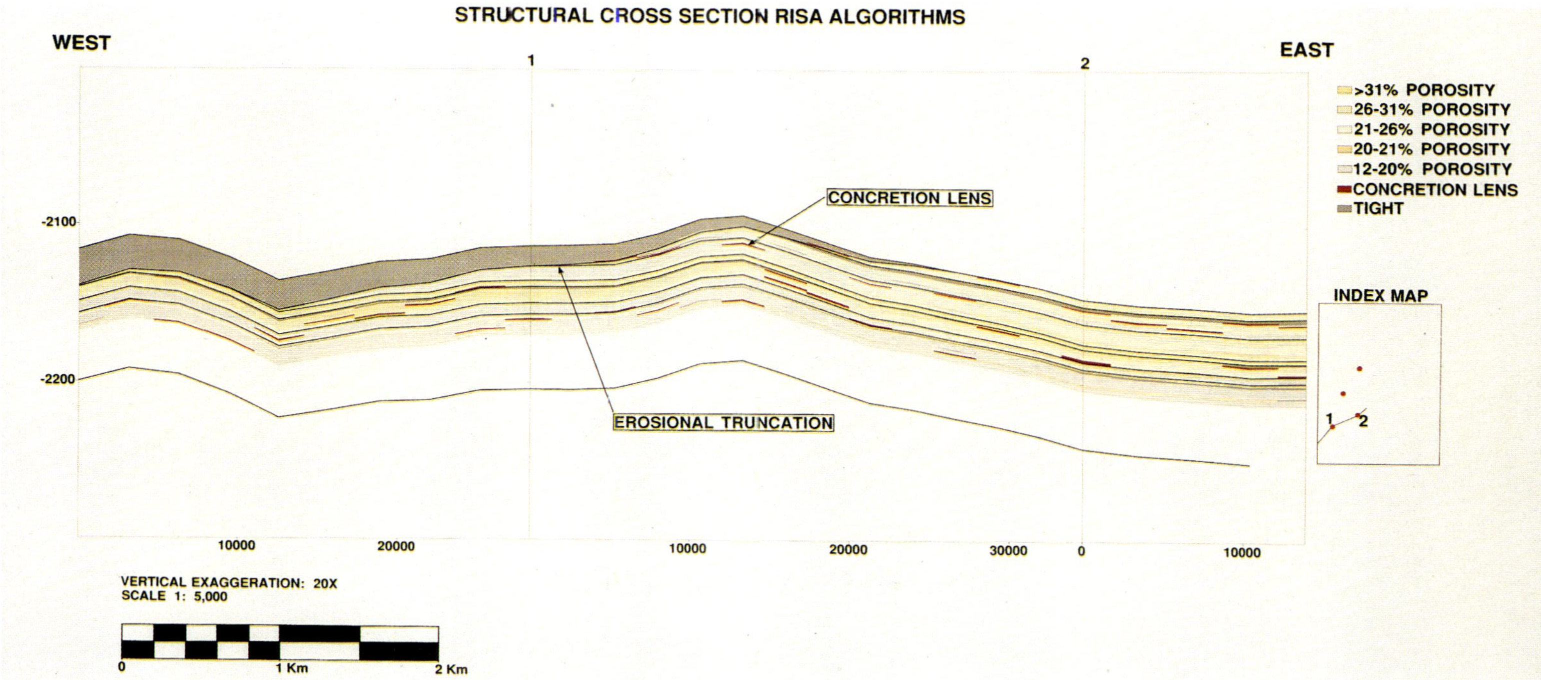

Plate 3. Brage Fensfjord cross-section.

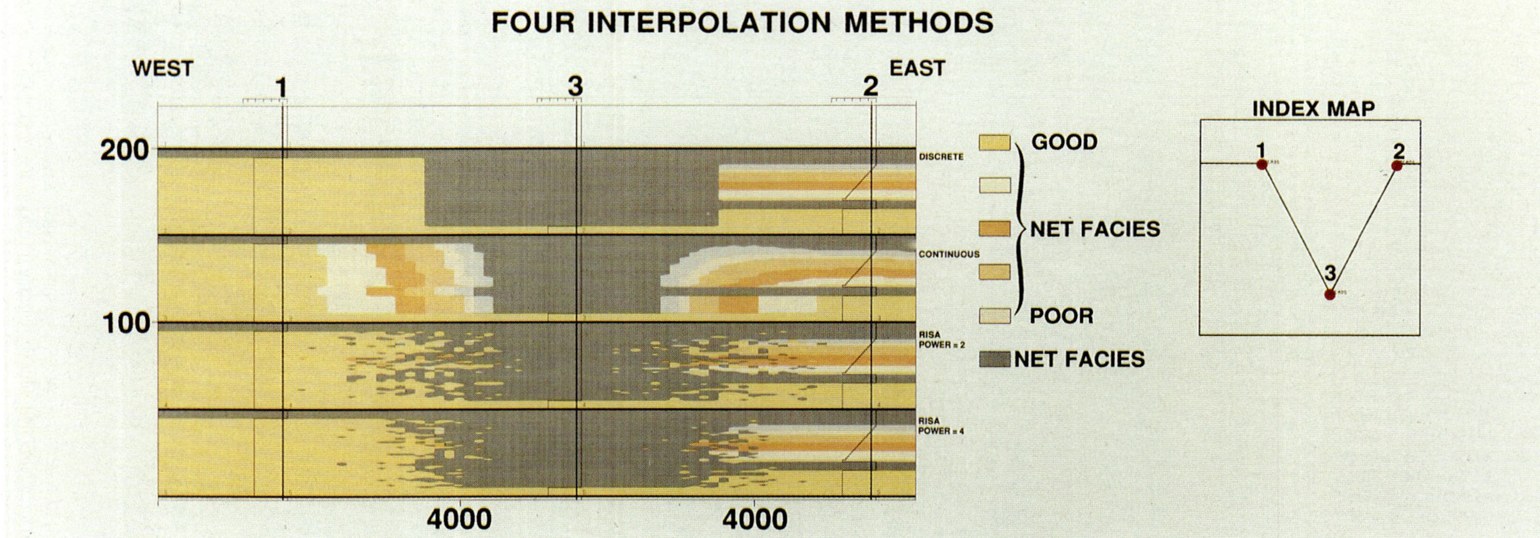

Plate 4. Interpolation methods.

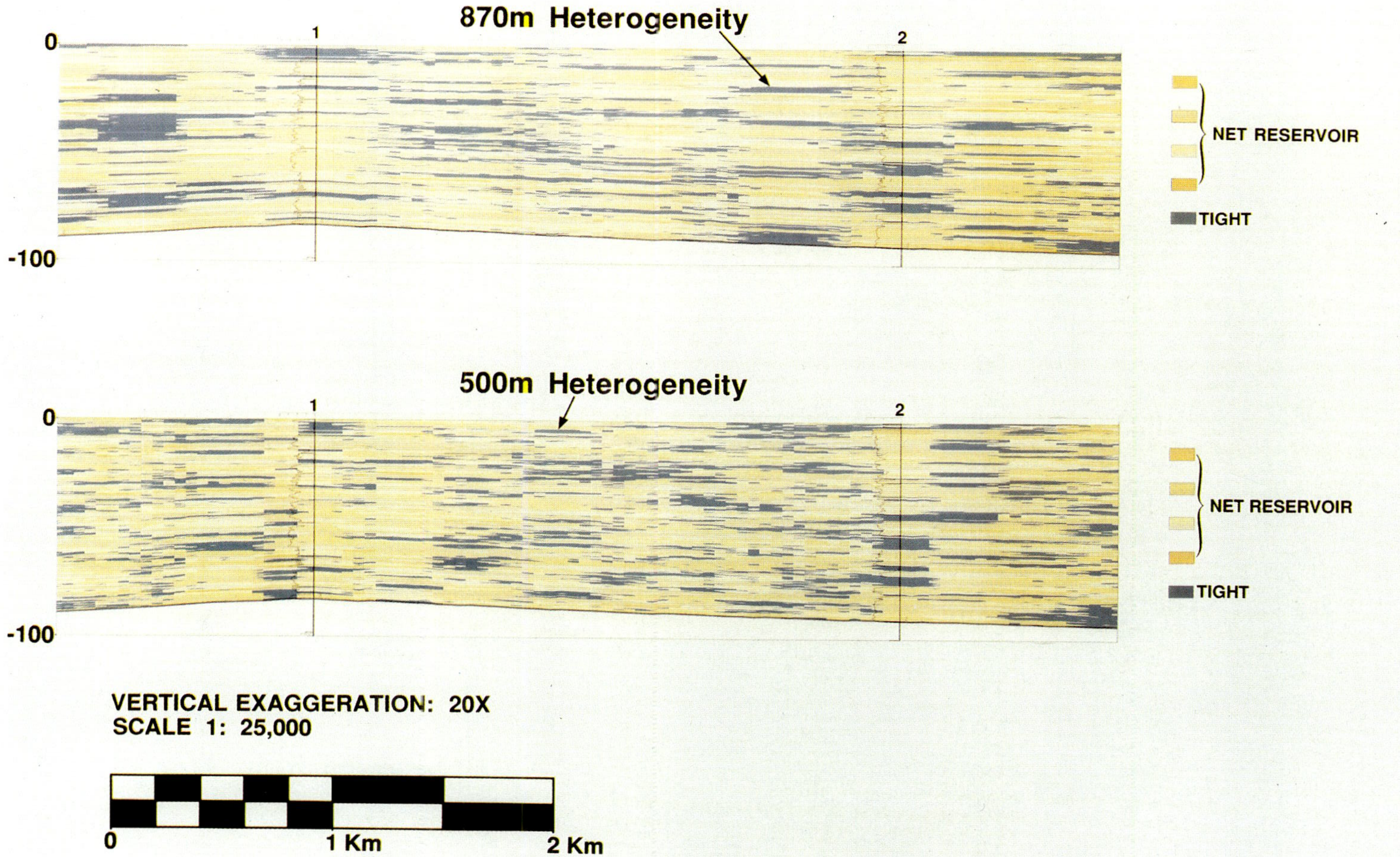

Plate 5. Brage Statfjord cross-section.

that should be incorporated in the 3-D geologic model.

ADVANTAGES AND CURRENT LIMITATIONS OF 3-D MODELLING

Advantages of 3-D geologic models in reservoir management

Most reservoir engineering simulations can benefit from the detail and convenience that a computerized 3-D model of reservoir geology provides. We have modelled more than one hundred reservoirs to assist in planning primary development, monitoring the drilling programme, assessing aquifer support, and assisting in-fill drilling programs.

The obvious advantages of 3-D geologic modelling are validation of geologic interpretation and simulator input array building; however, a 3-D model can have other uses, such as analysis of reservoir communication (continuity), developing alternate geologic interpretations, and maintaining interpretation archives.

Reservoir communication

A useful feature of a computerized 3-D reservoir description (a GEOSET model) is that it can be used to identify (to search for) individual, non-communicating reservoir units. The model can be searched for contiguous cells with specific attributes. This collection of cells, which we call 'geosets', are assembled into a new model, a geoset model, with the cells of each group individually identified.

The geoset model is used to identify the porous units that do not communicate with the major portion of the reservoir. If these stray units are small, they could be ignored by setting the porosity values in their cells equivalent to shale. Should the smaller units be large enough to be economic, they can be mapped and simulated as separate reservoirs. By mapping the individual geosets, we map continuity and fluid communication over the reservoir and estimate preferential flow patterns. We could estimate the number of wells needed to effectively drain the reservoir for a screening level economic analysis.

Alternative geologic interpretations

A reservoir description based on traditional procedures usually falls short of what the reservoir engineer needs most of all from the geologist: the alternative interpretations and maps and cross-sections for each alternative. Even if adequate maps are available from a traditional analysis, they can quickly become out-of-date. The computer 3-D reservoir description allows the ready availability of many different types of reservoir maps and cross-sections from the current or any alternative interpretation. Sometimes our engineers put together '3-D fences' of cross-sections from the geologic model, or cover the walls of workrooms with sections, or suspend them from ceilings; all of this to get a feel for the most typical areas for simulation pilots or problem spots in the field. This detailed look at the reservoir prior to the actual simulation work pays off in better design of the reservoir block model.

Because the re-modelling process generally takes only a day, alternate geologic interpretations can be easily developed to establish the economic 'sensitivity' of a specific, but perhaps uncertain, interpretation. Correlation of interbedded impermeable units, relative size of internal heterogeneities, aquifer quality, and even the exact porosity cutoff itself, are but a few of the reservoir features that may be difficult to establish firmly. With an automated reservoir description, any one or a combination of these uncertainties can be built into scenarios for reservoir simulation.

Archiving the geologic interpretation

Oil and gas fields have a long productive life as we all know. Putting the reservoir description in a 3-D computer model is one way to retain the geologic description; microfilmed maps and sections are much less effective. Off-line storage allows us to retain a complete reservoir description for our fields. This description can be retrieved and rapidly updated when the need for new reservoir simulations require it.

Limitations of 3-D geologic models in reservoir management

After years of experience in working with many field models, we are aware of the limitations of our program and the difficulties in applying it in any specific field. Below are some problems which we recognize.

1 Training. Our 3-D modelling system is a system of more than one hundred separate programs, all interfaced to an internal database and with each other. It takes extensive training and practice for a geologist to use it properly. A panel-driven system can help, but it still takes skill and experience to operate it effectively.

2 Hard technical decisions. A geologist may hesitate in making modelling decisions for engineering purposes unless he feels he has a reliable answer; not many geologists are comfortable with a 'best guess' model. The multiplicity of geologic options within the bounds of the interpretation requires the geologist to make hard decisions on reservoir description that he has not previously had to consider. The size and distribution of the clay plugs in Brage Statfjord is an example of such a decision. The careful construction of a database of reservoir heterogeneities and their geometries developed from outcrops and modern analogues would be particularly useful at this stage of the modelling process.

3 Data. Getting data into the modelling program is time consuming if there are a large number of wells. Retrieving it from a data storage and retrieval system is easy enough, but too often the log analyses are incomplete and produced by different analysts using different logs with different programs. Stratigraphic tops seem always to require new correlation. Old data may be detailed only on paper logs and maps which require laborious data entry.

SUMMARY

The building of a reservoir description into a 3-D computer matrix is a powerful asset in reservoir simulation work. All facets of the geologist's interpretation can be incorporated into such a model and readily interfaced to the production simulator. In addition, the availability of reservoir maps and cross-sections in large numbers permits the engineer to become thoroughly familiar with the geology prior to designing his reservoir block model.

The models have additional advantages for geologists. They provide a platform to update and display interpretations with new data or revised interpretations. Alternate interpretations can be modelled easily and are useful for examining sensitivity to specific geologic inferences. The model can be easily archived and can provide a source for reservoir information for many years after it is built.

ACKNOWLEDGEMENTS

We are members of a user support team at Exxon Production Research Company in Houston. Our team does user support for data management, mapping, and modelling software. We wish to acknowledge that we do not act alone. Our own support group included Tom Jones, Doug Palkowsky, Glenn Hebert and Mike Ryan. Without their work, assistance, and support, we would not have modelling software and no reason to write this paper.

The authors thank Exxon Production Research Company, Esso Norge, Statoil, and the Brage partners in general for permission to publish this paper.

REFERENCES

Haldorsen, H.H., Chang, D.M. & Begg, S.H. (1987) Discontinuous vertical permeability barriers: a challenge to engineers and geologists. In: *North Sea Oil and Gas Reservoirs—I* (Ed. by J. Kleppe, E. Berg, A.T. Buller, O. Hjelmeland & O. Torsæter), pp. 127–152. Graham and Trotman, Boston.

Jones, T.A. (1988) Modeling geology in three dimensions. *Geobyte*, **3**, 14–20.

Jones, T.A., Hamilton, D.E. & Johnson, C.R. (1986) *Contouring Geologic Surfaces with the Computer*. Van Nostrand Reinhold, New York, 314 pp.

Index

Page numbers in *italic* refer to figures, those in **bold** refer to tables

Abrasion shale smears 117–18, 121
Accommodation,
  and base-level changes 23, 49
  changes in 44, 46, 50
  rates 45, 47
  space 12, 15, 37, 39
Acoustic impedance 181, 182–3, 184, 185, 186
Aeolian deposits,
  Cambro-Ordovician *130*, 131
  geometry of 11
  permeability of 131
  sheetflood 132
Aggradation rate *see* Fluvial
Ainsa Basin, Spain 60, *61*, 62–4, **71**, 76
Ainsa, Spain 101, 102
Akhdar Group, Oman 138–9, *140*, 141
Alberta, Canada 144
Al Khlata Formation, Oman 128, 129, 131–2, *133*, *134*, 141
  sedimentology 132
Alluvial,
  architecture 12, 24, 25, 45–6
    3-D model for 234, 235
    process-based models 214–15
  deposits,
    avulsion controlled 215
    classification of 26–7
    Permo-Triassic, Australia 82, 84, 85
    sequence boundaries 43–4
    stacking patterns 23, 43
    -transgressive deposits 44
  fan deposits 64, 160
    braided stream sands 76
    Cambro-Ordovician *130*, 131
    distributaries 73
  plain deposits 25, 42, 44, 49, 82
    distal 135
    see also Braided stream
  sandbodies 22
    architecture 22, 50
    channel-fill 8, 10
    interconnectedness 12
    oil reserves in 22
    ribbon 26
    sequence stratigraphy and 22–4, 25, 42–5, 46, 47, 49
    sheet 26
  stratigraphy model,
    cross-sections for *223*
    input parameters and procedure 215–17, 219–20, *221*, 222
    mean sandstone body width 222, *223*, *224*, 226
    simulating geometry and distribution 234
Amal Eastern High Field, Oman 129
Amin Formation, Oman 129, *130*, 131, *134*
Ammonites 25
Amplitude mapping 182, *184*, 184, 185–6
Analogue modelling 7, 10, 24, 171, 241
  for flow unit definition 170, 193–4, 202
  reservoir architecture 179–80
  turbidite channels 102
  vertical dip profiling 143
  see also Outcrop data
Anhydrite 138
Anticlines,
  and palaeodrainage 60, 76–7
  roll-over 172
  and sediment accommodation 62
Aquifers 22
Arabian Gulf 128
  see also Oman
Ara Formation, Oman 128
Ara salt basins, Oman 125
Arbuckle Field, California 109
Arc volcanism 84
Åre Formation 59
Arenicola 41
Aspect ratio 211
Australia,
  Permo-Triassic 82, 84
    see also Bowen Basin
Autocorrelation functions 214
Avulsion(s) 215
  frequency 214
  period 219
    and sandbody thickness 229
  rates 46
  time between 226

Back flow ripples 164
Bankfull,
  channel depth 214
    from well data 232
  channel width **217**, 219
  depth 233
    measurements *216*, 219
  width vs. bankfull depth 217
Baralaba Coal Measures, Australia 96
Bar deposits 32, 37
Barents Sea 59
Barform, shale drapes 237
Barrier-bar sandbodies 152, 187
Barrier-spit sandbodies *88*, 94
Base-level,
  changes 23, 46, *48*, 49, 50, 76
    and sediment preservation 47
    see also Sea-level
  defined 45
Baylines 46, 47
Bedload deposits 27
Bell Ridge Field, USA 179
Belsue Formation, Spain 60
Biodegradation, oil 128
Bioturbation 63, 85
  turbidite deposits 105, 106–7, 108, 109
Blackhawk Formation, USA 13, 145, 147, 152, *153*, 154
Bleached sandstones 33–4, 64
Blue Cove, Utah *35*, *36*, *38*
Boltaña Anticline, Spain *61*, 76, *101*, 102, 108, 110
Book Cliffs, Utah, USA 13, 145, 147, 152, *153*, 154
Boolean methods 8
Borehole,
  imaging 202
    see also Formation microscanner (FMS)
  sandstone thickness 242
  washouts 135
Bottomsets 163
  permeability of **164**, 164, *166*, 167
Bouma sequences 104
Bounding surface, hierarchy 67, *68*, 69, 70
Bowen Basin, Australia 82, 96
  cross-section *84*, *85*
  geology 84–6
  map of *82*, *83*
Brage Fensfjord Field, Norway 253, 255
Braidbelts 241
Braided channels 219

Braided stream deposits *34*, 59, 60, 71, 76, 77, 161
braid bars 33
Cambro-Ordovician 129, *130*
permeability of 129
Breccias 117
Brent Field, UKCS *5*, *6*
Brent Group, UKCS, reservoir simulations 194–8
Burrows 106–7, 109
escape 105

Calcite cementation *see* Cementation
Calclithites 64
Calcrete horizons 135
Caliche nodules 63
Calico bed, Utah 27, *30*, 30–1, 32–4, 43, 44, 49
California, USA 109, 179
Cambro-Ordovician, Oman 128, 129, *130*, 131
Campodarbe Group 60
Canada,
channel deposits, Alberta 144
Cretaceous 42
Capillary pressure 158, 167
Carbonate(s),
cement 13, 102, 135
from FMS 129, 131
ooidal, Oman 128, 138, 139, *140*
reefs 181
reservoir,
fractured 128, 129
modelling 10
Carboniferous,
fluvial sands, Ireland 144–5, *146*
Kentucky *9*
Limestone, UK *208*, 209–10
tectonics 116, 141
Westphalian, UK 115–16
Carlson reservoir, USA *172*, 172–3, 174, *176*, 178–9
transition probabilities,
lateral *178*
upward *177*
Carlston's equations 216
Cayce reservoir, USA 172–3, *174*, 175, *176*, 177–9
CDP *see* Channel deposit proportion
Cell-based modelling 5, 8
Cementation,
carbonate,
calcite 102, 135
from FMS 129, 131
and permeability 194
and flow barriers 13, 14–15
mud drape 162
and parasequence boundaries 12–13
and permeability 131, 194
point bar sands 164
Channel bars 217, 219
Channel-belt(s),
aggradation 215
rate 219–20
ancient vs. modern 214
bifurcation 244
compartmentalization 241
connectedness ratio 215
deposits 24, 45, 86, 95
coarse-grained 215, 241
down-valley variation 234
orientation 244–5
and geometry 238–9
parameters defining variability 248
sandstones *38*, 50, **74**
aggraded 214
geometry and distribution 213, 214–15
step length 243
thickness 214
measuring 215–16
width,
vs. bankfull channel width **216**, 217
vs. bankfull depth 215
vs. channel planform geometry **217**
empirical expressions for 232
estimating 216–17, 219
vs. floodplain width 219
measuring 215–16
normalized 233
predicted vs. observed **218**
theoretical expressions for 231–2
vs. thickness 10, 214
Channel complexes, low-sinuosity 145
Channel deposit proportion (CDP) 215, 220, *234*, 234
estimates for 232, 233
lateral variation in 231
parameters affecting 222, *226*, 226, 227
Channel deposits 39, 69, *93*, 96, 137
abandonment 7, 39, 63, 167, 256
alluvial 49
channel-fill 13, 27, **28**, 32, 37, 44
cross-bedding *see* Cross-bedding
deposition times of 10
dimensions of 96, 178–9
geometries of 10
heterolithic 42
channel mouth 86
dip data for 144, 145
heterogeneous 96
inclined heterolithic strata (IHS) 95
interconnectivity of 95
laterally-accreted *see* Lateral accretion
low-sinuosity *146*, *149*
Permo-Triassic, Australia 89, *90*, *91*
sandbodies 5, *6*, 8, *9*, 50, 66
amalgamated *35*, 49
amplitude mapping of *184*, 184
architecture of 70
connectivity of 241
correlation of 77
Cretaceous, Utah *see* Cretaceous
isolated 37–9
orientation of 67, 137–8
ribbon 26
on seismic data 185
stacked 145, 152
submarine fan 99
vertically-accreted 89
see also Point bar
Channel plugs 39
Channel(s),
bankfull depth 215
vs. channel belts 241
dimensions of, from channel-belt width 214
geometry, controlling variables and 219
hydraulics of 27
instability of 241
patterns of 26–7
reaches, length of 243
reactivation of 161, 167
wavelength and sinuosity 217, 219
width of, and discharge 216
Channel scours 43
Chronostratigraphic,
correlation 12, 24
frameworks 22
markers 13
Chronostratum 254
Clastic wedges 85
Clay plugs 256
Clay smears *see* Shale smears
Climatic changes 76
and fluvial architecture 47
Clinoforms 183
Clontarf submarine fan, Ireland 183
Coal 15, 22, 26, 30, 42, 45, 95, 186
sealing potential of 94
seams 85, 86, 90, 94, 147
splitting 86, 91
Coal-bearing strata 26, 31, *92*, 94, 119
Permo-Triassic, Australia *85*, 85–6
reservoir characteristics 82, 84, *93*, 94–5

Coal Canyon, Utah, USA *152*, *153*
Coal Measures, UK 116
Coastal complexes 84–5
Coastal deposits, geometry of 10
Coastal plain deposits 12, 59
Collegats Formation, Spain 62
Colorado Plateau, USA 24
Colorado River, USA 24
Columbia River, USA 42
Compaction 170, 171, 215
  differential 245
  effect of, on sandbody width/thickness 232
Compartmentalization *see under* Reservoir
Compressional tectonics 84, 160
Condensed sections 23, 26, 43, 44, 49
Conditional simulation methods 8
Conditioned random modelling 170
Conglomerates 33
  lag 104
  lenses 30–1, 32
  mud-clast 32, 33, 34, 37, 39, 131
  mudflake 131
Coning 129
  see also Water
Connectedness 233, *234*, 234
  ratio (CR),
    defined 220
    parameters affecting 222, *226*, 226, 227
Connectivity 11
  sandbody 121–2, 238, 241
    parameters controlling 143
Continuous interpolation 255
Continuous variation modelling 8
Convolute lamination 37, 38
Cordilleran fold and thrust belt 24
Core data,
  and FMS calibration 129, 138
  and probe permeametry 198–9, 200–1
  problems with data collection 199
  and wireline log calibration 7
Core plug,
  measurements 192
  permeability 195, 196, 197, 199, 202
    compared with probe permeameter data 198, *199*
  sampling rate 196–7, 198
Cormorant Field, UKCS 114–15, 120–1
Cornelius reservoir, USA 172–3, 174–5, *176*, 177, 178–9
  transition probabilities *177*, *178*
Corvus interseam, Australia *88*
County Clare, Ireland *144*, 144–5
Crati Fan, Mediterranean 102
Cratonic basins 160
Cretaceous,
  Alberta 42
  California, USA 109
  Interior Seaway, USA 24
  New Mexico, USA 152
  shoreface sands, Utah 144, 145, 154
    dip profile data 147, 152, *153*, *154*
  tectonism, USA 24
  -Tertiary orogeny 128
  Utah 13, 22, 24–6
    sedimentary facies 26–7, **28–9**, 30–4, *35*, 36–42
Crevasse channel(s) **87**, 241, 246–8
  deposits 90
  described 245–6
Crevasse splay(s) 238, 241, 246
  deposits 26, 42, 45, 63, **87**, 90, 95
Crocker Formation, Australia 85
Cross-bedding 164
  compound **29**
  and dipmeter processing 12
  and FMS 131, 138, 139
  and permeability 143, *165*
  planar/tabular **28**, 31, 32, 34, 36, 37, 40, 64, 152
  point bar deposits 162–3
  trough 40
    channel-fill deposits **28–9**, 31, 32, 34, 36, 37, 38, 39, 64
    lateral accretion unit 162–3
    lunate 145, *151*
    permeability patterns of *165*
    shoreface sands 147, 152, 154
  see also Cross-stratification, Foresets
Cross-sections, 2-D outcrop 64–7, 68 70
Cross-strata, hummocky 152, 154
Cross-stratification 40, 158
  epsilon 95
  planar/tabular 33
  trough 33
  wind ripple 131
  see also Cross-bedding
Crustal extension 84

Dam-engineering 47
Darcy's Law 158
Debris flow deposits 107
Decompaction, in modelling 219
Deep marine sandbodies, geometry of 10–11
Deltaic deposits
  coal-bearing 82
  delta front 85
    distal **87**, 88, 94
  delta plain 59, 86
    lower 85, 88
    upper 86, 90, 91, 96
  fluvially-dominated 144–5, *146*
  glacio-lacustrine 132
  Permo-Triassic, Australia 84–5
  Utah 147
  wedges 181
  see also Fluvio-deltaic deposits
Deltaic systems, mixed-influence 88–9, 96
Densely-drilled fields 170–1
Density logs 139, 196
Depositional sequences, defined 24
Depositional systems tracts 23
Deposition rates, compacted 232
Deterministic modelling 13, 15, 170
Dhofar, Oman 125, 135
Diagenesis 113, 129, 170, 171
  carbonates 138
  turbidite deposits 102
  see also Cementation
Diamictites 132
Diplocraterion 41
Dipmeter,
  accretion channels and 144
  data,
    filtering 143, 144, 152
    and sandbody orientation 143
  High Resolution Dipmeter tool (SHDT) 138, 143, 144
  interpretation 12
  outcrop analogues for 143
  processing 12
    parameters 143
  profiles 7
Dip(s),
  azimuth 12, 152
    foreset 143
  data,
    filtering of 145, *151*, 152, 154
    shoreface deposits 152, *153*
  from cross-bed-foresets 145
  profiles 143–4
    Carboniferous, Ireland *146*
    Cretaceous, Utah 147, 152, *153*, *154*
    outcrop 144–5, *146*, 147, *148–51*, 152
    statistical manipulation of 145
  sedimentary 7, 12, 143
Discrete object modelling 8
Dish structures 106
Distributary,
  channel deposits 86, **87**, 91, *92*, 94–5, 94, 96
    dips *150*, *151*
    heterogeneity of 91, 94
    low-sinuosity 152
    sandbody architecture *93*
  channels, on seismic data 186

deposits, and seismic data 181
mouth bars 86, *88*
Dolomite 138, 139, 141
Draugen Field, Norway 183
Drip Tank Member, Utah 27, 30, 31–3, *33*, *34*, 43, 49
Dropstones 132
Drowned river valleys 49
Dunes, fluvial 33
Dunlin Field, UKCS 196
Dynamic normalization (FMS) 128–9

Earthquakes 37
Eastern Flank, Oman 128, 129, 135
Element simulation models 194, *195*
Enhanced oil recovery (EOR) 57, 58, 128, 131, 141, 158, 196
braided stream deposits 129
fluvial reservoirs 167
Oman 132
Eocene,
fluvial deposits, Spain *see* Escanilla
Ireland 182–3
turbidite deposits, Spain 101–2
Epicontinental seas 24
Epsilon cross-strata 95
Equilibrium point 46
Escanilla Formation, Spain 65, *66*, 67
architectural elements **74–5**
braided stream deposits 71, *72*
channel sandbodies of 77
geology 60, 62–4
meandering stream deposits 70, *72*, 76
palaeocurrents *69*, 76–7
palaeogeography *65*
SAFARI data 67–70, 70–1, 73, 76–7
valley-fills *72*
Escape burrows 105
Estuarine deposits 41
Etive Formation, UKCS 194, 196
Eustatic changes 45, 46, 47, 49
see also Base-level
Evaporites 128
see also Salt
Extensional tectonics 84

Facies,
analysis, vertical 4, 12, 15
architecture, and sequence stratigraphy 26
dislocations 47, 49
models 82
Facies-tract,
dislocation 43
offsets 26
Fault(s),
dilation 118
displacement 114–15, 118, *119*, 121
drag on 117, 118
growth 113, 172
hydraulic effect of 115
and permeability 115
quantitative modelling of 11
and reservoir heterogeneity 120
on seismic data 114, 115
and shale smears 113
sub-seismic 114
syn-sedimentary 116
throw 120
Fault scaling 114–15, 120–1, 122
Fault seals 11, 114, 115, 120, 121, 122
Feldspar,
alteration 33–4, 64
in log analysis 135
Ferron Sandstone Member, Utah 44, 49
Floodbasin deposits 38, 39, 42, 45, 50, 90
Flooding surfaces 7, 13, 15, 24
maximum **29**, 41, 43, 44, 49
and well correlation 10
Floodplain,
aggradation rate 227
deposition rate 219
deposits 49, 63, 69, 95, 135, 185, 238, 241
widths 232, 233
Flow barriers 11, 13–15, 100, 109, 129
Flow equations 192, 193
Flow simulation 158
Flow units 8, 13, 67, 77, 100
defining 200
and permeability choice 196
described 191–2
distribution modelling 70
permeability of 195, 197–200
sedimentologically-defined 191
Fluid communication, channel sands 70
Fluid escape pipes 106
Fluid flow 8, 70, 167
and anisotropic reservoirs 91, 94
baffles 100, 170
barriers 11, 13, 15, 100, 109, 122
faults/fractures 129
in braided stream deposits 129
characteristics 191, 193
and compartmentalization *see under* Reservoir
equations 169
and faults 120
pathways 77
pseudofunctions 194
simulations 5, 169
Fluidization,
and permeability 106
sediment 109
Flute casts/marks 37, *104*
Fluvial aggradation **29**, 34, 46–7, 49
rate 37, 43, 44
Fluvial deposits,
amalgamated 43, *48*, 50
sheet sand 44
architecture 23, 24, 45–6, *48*
and accommodation space 33
Escanilla Fm, Spain **74–5**
predicting 49–50
and sequence stratigraphy 42–5
braided *see* Braided
Carboniferous, Ireland 144–5, *146*
channel-belt *see* Channel-belt
channel *see* Channel deposits
communication in 68
Eocene (Spain) 60, 62–4
fan sediments 160
distal 158, 161
sedimentary facies 161–4
heterogeneity 69, 70
high-sinuosity *36*, 37, 39, *48*
hydrocarbon recovery 157, 158
low-sinuosity 32
meanderbelt *see* Meanderbelt
meandering stream *see* Meandering stream
multi-channel systems 32, 33
Oman 128
permeability 158
heterogeneity 164, 167
point bar *see* Point bar
and seismic data 181
sheet-braided 71
stacked sands 158
tidally-influenced **29**, 41, 42, 43, 44–5, *48*, 49
see also Alluvial deposits
see also Fluvial sandbodies *below*
Fluvial profiles 45, 46
Fluvial reservoirs,
enhanced oil recovery 167
Norway **58**, 59–60
recovery rates 158
Fluvial sandbodies,
amalgamated *33*
channel-fill 8, *9*
deposition times of 10
crevasse channels 245–6
and cross-sectional parameters 241–2
geometry of 10, **239**, *240*, 240–1
levees/overbank deposits 245
modelling geometry of *8*
parameters describing 241–6
plan geometry 242–5
thickness distribution 241–2
three-dimensional parameters 245
Fluvial stacking patterns, controls

on 45–7, *48*, 49
Fluvial systems,
bedload-dominated rivers 237
mixed-load 237–8, 248
Fluvial terrace deposits **28**
Fluvio-deltaic deposits 116, 143
heterogeneous 237–8
FMS *see* Formation micro-scanner
Footwall traps 122
Forbes Formation, USA 109
Foreland basins 24
Bowen, Australia 84
Pyrenean *100*, 100–10
Foreset(s),
dip *151*
azimuth 143
and palaeocurrent directions 145
oil entrapment in 164, 167
permeability **164**, 164, *166*, 167
point bar deposits 163
resistivity, from FMS 138
Foreshore deposits 86, **87**, *89*, 91, *93*, 94, 96, 177
Formation micro-scanner (FMS),
applications 132
carbonate reservoirs 138–9, 141
Oman 128–9, 137–9, 141
and channel orientation 141
and core data integration 129, 138
and cross-bedding 131, 138, 139
data processing 131
and fault/fracture analysis 129, 132, 139, 141
image(s),
mottled 138
processing 128–9
and lithofacies analysis 135
and pay identification 131, 132, 141
and permeability barriers 129, 131
and porosity 138
sedimentological interpretation from 129, 138–9, 141
static normalization 128–9, 131
tool described 128
Forties Field, UKCS 109, 194
Forward modelling 10–11, 170
dip 12
seismic response 186
Fractal dimension, defined 114
Fractures 117, 118, 138
in carbonate reservoirs 128, 129
extensional 139, 141
and FMS data 129, 132, 139, 141
and permeability 113
shear 141
Fracturing, thermal *196*
France, Holocene 42
Frigg Field, Norway 109, *182*, 182
Frio Formation, USA 172–3
Fucosopsis 105
Full-field simulations 194

Gas,
cuts 138
initially in place (GIIP) 247–8
Oman 125, 138
reserves 59
seismic detection of 184, 186
Gasfields 109
Leman 11
Genetic units,
modelling 174, 177–8
scale of variation in 170–1
Geological variation, quantifying 170
Geomathematical modelling 58, 70
Geostatistical models 8, 248–9
data collection for 173–5, 177–9
object-based 170, 238
sequence-based 170
German Creek Formation, Australia 84–5, 86, *89*, *91*, *92*
coal seams in 85
facies assemblages 86, **87**, *88*, 88–9
palaeogeography of *94*
reservoir,
characteristics 91, *92*, 94
facies 95–6
sandbody architecture *93*
Ghaba Salt Basin, Oman *136*
Gharif Formation, Oman 128, 141
FMS applications in 137–8
palaeogeography 135, *137*
sedimentology 135, *136*, 137
Gironde River, France 42
Glacial deposits 128, 131–2, *133*, 135
see also Al Khlata
Glacio-fluvial deposits 132, *136*
Glacio-lacustrine deposits 132
Gondwana glaciation 135
Goonyella mine, Australia *90*
Gouge 117
Grain size,
measurements of 201
and permeability 164
Grain sorting 163
Granularia 105, 107
Gravel deposits 73, **74**, 76
Gravity flows 108
see also Turbidite
Gregory mine, Australia 86, *88*
Groove marks *104*
Groundwater 22
Growth faults 113, 172
Gullfaks Field, Norway 60, *195*
Gutter casts 64, 67
Gypsum 163–4
alabastrine nodular 63

Haima Group, Oman 128, 129, *130*, 131, 141
Halokinetic structures 125
Haltenbanken, Norway 59
Haradh Formation, Oman 129, *130*, 131
Hartzog Draw Field, Wyoming, USA 138, 191, *192*
Hasirah Field, Oman 135, *136*, 137
Haushi Group, Oman 128, 131–2, 135, 137–8, 141
palaeogeography *137*
stratigraphy *133*, *134*, *136*
Hecho Group, Spain 101
Hemipelagic deposits 108
Heterogeneities,
fluvio-deltaic sequences 237–8
meandering channel sandbody 70
reservoir *see under* Reservoir
Heterolithic deposits 49
fluvial 39–42
inclined strata 40
sandstones **29**
palaeocurrent data for *32*
tidally-influenced 44
Hierarchical bedding surfaces 241
High Resolution Dipmeter tool (SHDT) *see under* Dipmeter
Highstand deposits 44–5
Highstand systems tracts 26, **29**, *39*, 47, 49, 50
Highwall mapping 82, 84
Holocene, France 42
Horizon amplitude analysis *183*
Horizontal drilling 4, 128, 129, 131
Horizon tracking 187
Howick, Northumberland *208*, 209–10
Hummocky clinoforms 183
Huqf, Oman 125, 135
Hydraulic barriers 122
Hydrocarbon,
production,
primary 128
rates 138
recovery *see* Recovery
reserves 59, 60
turbidite reservoirs 99
saturations, residual 15
traps 99, 114, 122, 125, 128, 172
see also Gas, Oil

Iberian Peninsula *61*, 159–60
Iberian Plate 160
Ichnofossils 86
see also Trace fossils
Incised valley fill 43
Inclined strata,

heterolithic 40
laterally- 34, 36
Infracambrian,
Huqf Group, Oman 128
salt basins, Oman 125
Injection programmes 158
Injection shale smears 117, 118–19, 121
Inoceramids 26, 41
Interconnectedness ratio 233
Interdistributary bay deposits 86, **87**, *88*, *89*, *92*, 94, 95
Interdistributary lakes 90–1
Interdune deposits 131
Interwell distance 194
Intraclast lags 237
Intra-sandstone slurries 107
Ireland,
Carboniferous 144–5
Eocene 182–3
Isochron maps 183

Jawdah Field, Oman 135
John Henry Member, Utah 25, 34, 36–42, 43, 45, 49
Jurassic,
fluvial reservoirs 59
see also Brent, Statfjord

Kaiparowits Plateau, Utah, USA 22, 24, 31, 42–3, 44, 50
alluvial strata of *27*
channel sandstones of **28–9**
depositional sequences 26
map of *22*, *23*
marine strata 25–6
sequence stratigraphy 26, 47, 49
stratigraphy 24–6
Kaolinite 33, 64
Karim West Field, Oman 131
Kentucky, USA *9*
Khuff Formation, Oman 128, 138–9, *140*, 141
sedimentology 138
Killard, Ireland 145, *149*, *150*, *151*
Knowledge base, developing geological 170–9

Labyrinth reservoirs 158
Lacustrine deposits 42, 135, *136*, 137, 160, 185, 186
see also Playa
Lags,
granule **29**, 34
intraclast 237
sandstone 34, 37, 39
transgressive 41
Laminated sandstones **29**
permeability of 199, *201*
Lancashire, UK 115–16
Lateral accretion **74**, 76
bedding 64
deposits 27, 37, 39, 89, *90*, 95, 96, 144
point bar 161, 162, *163*
tidally-influenced rivers *40*, *41*
surfaces **28**, 41, 70, 91
Laterally inclined strata 34, 36
Leaching 138
Left Hand Collet Canyon, Utah 40
Leman gasfield 11
Levees 238
width of 245
Limestones,
fractured 128, 141
micritic 63
oolitic grainstone 138, 139, *140*
Lithofacies, core descriptions 15
Lithohydraulic units *see* Flow units
Lithostratigraphic correlation 12
Longshore drift 94
Loranca Basin, Spain 158, 159, 160
cross-section *160*
sedimentary facies 161–4
Los Angeles Basin, USA 109
Lower Ainsa Channel, Spain 101
lithofacies of 102, *103*, 104–9
Low net/gross sequences, modelling 238, **239**
Lowstand deposits 44, 47
Lowstand wedges *48*
Lunde Formation, Norway 59, 60

Macmillan Formation, Australia 85
McMurray Formation, Canada 144
Mahwis Formation, Oman 129, 131, 141
Mancos Shale, Utah 44, 145
Maria Formation, Australia 85
Markov chains 12
Markov transition matrix 214
Marmul Field, Oman 132
Mathematical modelling 57–8
Maximum flooding surfaces *see* Flooding surfaces
Meanderbelt(s) 43, 49, 219, 241
width of 216
Meandering stream deposits *41*, 59, 70, 76, 77, 161, 164
Mediano anticline, Spain *61*, 76
Mediterranean, Crati Fan 102
Mesozoic,
Oman 128, 138
Spain 160
Micropalaeontology 7
Micro-resistivity tools *see* Formation micro-scanner (FMS)
Midland Basin, USA 109
Minipermeameters 101
described *158*, 158–9
see also Probe permeameters
Miocene, California, USA 109
Mire deposits 25
see also Peat
Mississippi River 241, 243
Mobility ratio 167
Modern environments *see* Analogue modelling
Molasse 160
Mondot Member, Spain 64, 69, 77
Monte Carlo sampling 226
Moranbah Coal Measures, Australia 84, 85–6, *93*, 96
channel subfacies 89, *90*, 90, *91*
facies assemblages **87**, 89–91
Mouth bar(s),
complexes 86, **87**, *88*
deposits *93*, 94
lacustrine 186
sands *93*, 94, 96
amalgamated *89*
proximal 91
Mud drapes 40, 41, 161–2, 237
Mudstone plugs 242
Multi-storey,
channel belts 242
sandstones 29, 32, 33, 34, 37, 38, 71, 73
Mutti & Ricci Lucchi facies scheme 105

Namurian, Ireland 144–5, *146*
Navier-Stokes equation 192
Navy Fan, Pacific Ocean 102
Ness Formation 59
Brent *5*, *6*
Cormorant Field 120–1
Net/gross ratio 120
and fault seals 115
Net pay zones 198
Net reservoir volume 238, 247–8
Net sand, definition of 201
Neutron logs 139
New Mexico, USA 152
Nickpoints 47
Nigeria 120
Nimr Field, Oman 131, 132, *134*
North Markham-North Bay City Field, USA *171*, 172–5, *176*, 177–9
North Sea,
Brent Group, reservoir simulations 194–8
Cormorant Field 114–15, 120–1
fluvial reservoirs 59
Forties Field 109, 194

Frigg Field 100, 109
sandbody imaging for 182
Pelican Field 121
see also Norway
Northumberland, Carboniferous Limestone *208*, 209–10
Norway,
continental shelf 57, 58
Draugen Field 183
Haltenbanken 59
offshore, submarine fan channels 106
Snorre Field 194
Statfjord Field 59, 60, 253
see also North Sea
Norwegian Petroleum Directorate 58
Nun River Field, Nigeria 120

Oaky Creek mine, Australia *83*, 86
Object-based modelling 6, 13, 238
Oil,
bypassing of 196
entrapment 158
heavy 128, 129
recovery 57
enhanced (EOR) *see* Enhanced
simulations 191
remigration of 128
reserves 22, 59
Oman 125, 128
reservoirs, Texas Gulf Coast 172–3
residual 167
saturation 132
residual 143, 202
see also hydrocarbons
Oilfields 109
Oligocene,
Frio Formation, USA 172–3
Spain 62
Olson Member, Spain 64, *65*, 69, 70, 71, 73, 76, 77
Olson, Spain 64–5, *66*
Oman,
Cambro-Ordovician 128, 129, *130*, 131
gasfields 125, 128, 138
North Oman nappes 141
oilfields of *126*, *127*, 128, 138
Palaeozoic 128, 138
Permo-Carboniferous 128, 131–2, *133*, *134*
Permo-Triassic 138–9, *140*, 141
petroleum habitat in 125, 128
Rub al Khali Basin 135
stratigraphy *127*
structural elements 125, *126*
Oman Mountains 138, 139
Onlap 34, 44, 99
Organic matter 36, 105
Outcrop,
data 4–5, 7, 11, 82, 84, 193
dip profiles 143, 144–5, *146*, 147, *148–51*, 152
permeability studies 15
SAFARI project 64–7
standard sedimentological 100
see also Analogue models
dipmeter profiles 7
studies 23, 170, 171
alluvial strata 45–6
channel-belt dimensions 215–16
and shale length 206–10
distribution 212
and shale smears 121–2
Overbank,
deposits 37, 45, 63, **75**, 215, 220, 238, 239
coarse 248
deposition rate 219
interdistributary lake 90
sedimentation, rate 42, 227
'wings' 95

Pacific Ocean, Navy Fan 102
Palaeocurrents 27
in amalgamated sandstones 36, 37
bimodal 86
channel, sandbodies 31, *32*, 32, 33, 38, 89, *91*
fluvial deposits 62
tidally-influenced 41, 42
from cross-bed foreset dips 145
inclined strata 40
indicators 137–8
multi-storey sandbodies 73
and palaeodrainage 76–7
turbidite deposits *102*, 102, *104*, 108
Palaeodrainage, fluvial sequences 76–7
Palaeogeographic maps 171–2, 178
Palaeontology 7, 13
Palaeosols 135, 164
see also Caliche, Pedogenic
Palaeovalley infills 77
Palaeozoic, Oman 128, 138
Parallel lamination 64
Parasequence(s) 12, 13
aggradational shoreface 45, 50
boundaries 7, 12–13, 15
correlation of 13
shoreface 47
retrogradational 44, 49, 50
stacking 22, 26, 43
Partial occurrence probability (Pp) 207, 208
Pay zones, from FMS images 131, 132, 141
Peat mire deposits 25, 86, **87**, 90–1
Pedogenic horizons 237
see also Caliche, Palaeosols
Pelican Field, UKCS 121
Percolation theory 222
Permeability,
anisotropy 15, 131, 158, 164
baffles 167
barriers 7, 13, 107, 108, 109, 135, 194, *208*, 210, 256
fluvial systems 237
from FMS 129, 131
see also Fluid flow barriers
and bioturbation 108
core plug 195, 196, 197, 198, 199, 202
cross-bedding 143, *165*
directional 154
distribution 15
fluvial deposits 158
effective *212*
estimation of 210, 211
factors controlling 113
vs. flow velocity *159*, 159
and fluidization 106
FMS 141
glacio-fluvial sands 132, *136*
heterogeneity 167, 196–7, 201–2
point bar deposits 158
sampling 198
small-scale **164**, 164
high resolution data 198–200
homogeneous reservoirs 194
horizontal (Kh) 194, 195, 210
anisotropy 131
imaging 202
log-derived (KLOGH) 196–7, 198, 199–200
mapping 67
minipermeameter *see* Probe permeameter *below*
modelling 8
point bar sands 158, **164**, 164, *165*, *166*, 167
probe-permeametry 158–9, 198, *199*, 199–200
ratio of Kv/Kh 195
and shale smears 113–14
of shaly sandstones 131, 199–200
turbidite deposits 99, 102, 105, 106, **107**, 107–8, 109
variations, and scale 15
vertical (Kv) 129, 170, 193, 195, 200, 210, 237
anisotropy 131
and carbonate horizons 194
and shale occurrence 100
see also Probe permeameter
Permeability-feet 201

Permian, Texas, USA 109
Permo-Carboniferous, Oman 128, 131–2, *133*, *134*
Permo-Triassic,
  Australia 82, 84
    see also German Creek Fm.
  Oman 138–9, *140*, 141
Petroleum reservoirs, *see* Gas, Hydrocarbons, Oil
Photomosaics 82, 84
Placer deposits 22
Plastic clays 119
Playa lake deposits 160
Pocahontos Basin, Kentucky *9*
Point bar deposits 29, 37, 76, 158, 159
  abandonment facies 164
  bounding surfaces 161
  cementation 162
  cross-bedding in 162–3
  permeability of **164**, 164, *165*, *166*, 167
  porosity 163–4
  sedimentary facies 161–4
  and seismic data 181, 185, 186
  stacked *161*, *162*, 167
  Tertiary, Spain 161–4
  see also Channel deposits, Fluvial
Polygonal interpolation 255
Polymer adsorption 131
Polymer injection 132
Porcupine Basin, Ireland 183
Porosity,
  -depth curves 219
  in dolomites 138
  log-derived 196
  micro- 138
  modelling 8
  point bar sands 163–4
  profiles 132
  secondary,
    dissolution 163–4
    mouldic 138
  types of, from FMS 138
Pressure data 11, 13
Pressure seals 7, 13
Probabilistic modelling 24
Probability density functions 8
Probe permeameters,
  advantages of 202
  data,
    compared with core plug values 198, *199*
    and core data integration 198–9, 200–1
  sampling rate 198
Probe-permeametry 198
  for Kh 201
  and net sand definition 199–200
  and shaly sand interpretation 199–200
Process-based modelling 214–15, 234
Pseudofunctions 194
Pseudowells (dummy wells) 253–4, 255
Puente Formation, USA 109
Pyrenean Foreland Basin 60, *65*, 73, *100*, 100–10
  cross-section *101*
  map of *61*
  turbidite channels in *102*, *103*
Pyrenees 76
  see also Escanilla Formation

Quantitative modelling,
  3-D, 12
  methodology 5, 7
Queensland, Australia 82

Rahab Shale, Oman 135
Random Interval Selection Algorithm (RISA) 255–6
Rangal Coal Measures, Australia 84–5, 96
Rannoch Formation, UKCS 194, 196
Ravinement surfaces 26
Reactivation surfaces **29**, 40, 41–2
Recovery, *see also* Hydrocarbons
  efficiency 99, 157, 194, 200
  enhanced oil *see* Enhanced oil
  factors 60
  mechanisms, secondary and tertiary 128
  methods, secondary and tertiary 50
  rates, fluvial reservoirs 158
Red beds *136*
  Oman 135, 138
  'wings' 245
Reef facies, on seismic sections 181
Regression equations 217, 219
Representative elementary volume (REV) 115, 192–3
Repulsion function 245
Reservoir,
  anisotropy 91, 94, 95
  architecture 7, 11, 12–13, 82, 169–70, 213
    3-D modelling 12
    and FMS 135
    interwell area 170
  characterization 191–2, 194
    scales of investigation *193*, 193
  communication 68, 76, 77, *78*, 257
  compartmentalization 11, 100, 114, 158, 164, 167, 241
    fault 121, 122
  connectivity,
    fluvial systems 246
    parameters controlling 143
  continuity 100, 120
    and faults 115, 131
  depletion 77
  drainage 12, 13, 15
    units 135
  engineers, geological data for 192–4
  facies associations 7
  flow,
    barriers 11, 13–15, 100, 109, 129
    units *see* Flow units
  fluid flow *see* Fluid flow
  forward modelling 10–11
  geometry,
    forward modelling of 12
    and orientation 8, *9*, 10–11, 10–12
  heterogeneity 13, 157, 158, 202
    and faults 120
    and modelling 58, 192, 193, 194, 201, 202, 248
  injection trends 100
  interconnectedness 45, 47, 49, 95, 135
  lateral variability 132
  production, and depositional architecture 99–100
  property, modelling 13, **14**, 14–15
  sandstone/shale ratio 115
  seals 94, 135
  shaly 131, 199–200
  simulation 169, 192, 193, 194
    3-D models 194, *195*
    models 3
    procedure for 194
    scales of analysis 194
  simulators 256–7
  space occupied by 7
  three-dimensional model,
    building of 253–7
    interfacing with reservoir simulator 256–7
    interwell interpolation 255, 256
    mapping systems 253, 254–7
  unit recognition 7–8
Residual hydrocarbon saturations 15
REV *see* Representative elementary volume (REV)
Rhône, deep-sea fan 183
Ribbon sandbodies 37, *38*
Ridge deposits 177, 178–9
Rio Grande 243
Rip current deposits 147
Ripple-laminated sandstones **28**, 31, 33, 36, 38, 39, 40, 162
Rock House Cove, Utah *33*, *39*, *40*, *41*
Rock properties,

interwell 169–70
quantifying 170
volume for averaging (REV) 192–3
Rocky Mountains, USA 24
Roll-over anticlines 172
Round O Quarry, Lancashire, UK 115–16, *121*
Rub Al Khali Basin, Oman 135

Sacramento Basin, USA 109
SAFARI project,
database **58**, *59*
data collection 67–70
data interpretation 70–1, 73, 76–7
described 57–9
field data acquisition 64–7
Salt,
basins 125, 135
deposits, Oman 128
tectonics 132
withdrawal 135
structures 125, 129
Sandbody,
depositional thickness 242
dimensions,
effects of compaction 231
effects of tilting 231
from field data 233
simulation of *232*
generation, in stochastic models 239
interconnectedness 45, 47, 49, 70, 76, 77, *78*, 95, 135
mean thickness,
vs. CDP and connectedness *226*, 226–7
vs. channel-belt aggradation rate *225*
frequency distribution of 229, *230*, 231
mean width 222, *223*, *224*, 226, 227
orientation 145
and connectivity 143
from dipmeter 152
stacked 242
storm-deposited 14
tabular **29**, 41
thickness 220, 222
vs. CDP and connectedness **233**
and channel-belt width/floodplain width 234
distribution 241–2
from well data 242
volume 77
width 220
vs. CDP and connectedness *224*, **233**, 233
and channel depth 242
vs. channel width 222, 226
frequency distribution of 220, 222, 227, 229, *230*
vs. normalized channel belt width 233
/thickness *228*, *229*, *231*, 231
decompaction of 231
see also Width/thickness
Sand-shale interfaces, modelling 255
Sandstone,
amalgamation 34, 36–7, 43, 45–6, 47, 49
connectivity 121–2
continuity *see* Reservoir
San Joaquin Basin, USA 109
Sansinena Field, USA 109
Santa Maria de Buil syncline (Ainsa) 60
San Vicente Formation, Spain 101
Schmidt plots *150*
Scolicia 105
Scour surfaces 30, 31, 33, 37, 40
Scroll bars 37
Sea-level changes,
from seismic data 181
low stands 12
see also Base-level, Eustatic
Seals 11, 13
diamictite 132, 135
intra-reservoir 121, 122
shale smears as 114
Seatearths 116, 119
Sedimentary,
architecture 67
of field analogues for reservoir information (SAFARI) *see* SAFARI
architecture *see* Reservoir architecture
dips *see* Dip(s)
facies, vs. lithohydraulic units 100
structures 7, 12, 13, 38
in fluvial deposits 63, 64
in interdistributary bay 86
permeability of 15, 198
secondary 63
see also Cross-bedding
see also Ripple
Sediment liquefaction 37
Seiscrop mapping 182, *186*, *187*
Seismic,
amplitude mapping 182, *184*, 184, 185–6
frequency 181, 187
resolution 4, 181
surveys 10
3-D, *4*
Seismic data,
3-D,
enhanced resolution 181
for reservoir modelling 183–4
and sandbody imaging 181, 182–3, 184, *185*, 185, 186–7
and estimating channel dimensions 219
and faults 114, 115
and genetic unit delineation 8
reef identification on 181
reservoir architecture from 183–6
and sandbody orientation 12
sedimentary feature imaging 181, 182–3
and shale dimensions 11
tying to log data 7, 10
Seismic stratigraphy 12, 182, 186
Sequence boundaries 23, 26, 47, 49, 254
alluvial strata 43–4
shoreface strata 45
Sequence stratigraphy 7, 10, 13
in alluvial realm 22–4, *25*, 25, 42–5, 46, 47, 49
application of 12–13
Sevier orogenic belt, USA 24
Shale,
characterization 7
continuity 7, 12, 13, *100*, 100
deterministic modelling of 13
dimensions 11, 100
modelling 13–14
distribution 100
drape 237
see also Mud drapes
flow 117, 118–19
barriers 13
geometry 11, 194
types, and seismic interpretation 184–5
veneers 117, 121
Shale length 13
characteristic 208–9
distribution *109*, 205, *206*, 206
1-D 210
3-D geometry from 2-D data 211, 212
curve *100*, 100
dimensionality 211–12
and incomplete outcrop data 206–9
normalizing 208, *209*, 212
orientation and shape 210–11
partial lengths 207–8, *209*
streamline method 210
volume (3-D) 209–10
vs. sample dimensions 205–6
frequencies 205–6
turbidite deposits 108–9
Shale smears,
abrasion 117–18, 121
application of studies 121–2
at outcrop 115–16
'clay smear potential' (CSP) 120
composite 119, 120

continuity of 122
described 115
factor (SSF) *119*, 119–20, 122
and faults 113
injection 117, 118–19
mechanisms 117–19
and permeability 113–14
shear zone 117, 118, 121
thickness 118, 119
types of 116–17
Shallow marine sandbodies 138
deposition times of 10
geometry of 10, *11*
sand bars 183
stacking patterns 23
storm/wave influenced 209
Shaly sandstones 131, 199–200
SHDT *see under* Dipmeter
Shear,
lateral 37
strength 118
Shear zone(s) 118, 121
geometry of *118*
smears 117, 118, 121
Sheetflood deposits 63, 131
Sheet sandstones 32, 37, *38*, *39*
Shelf sandbodies 12
Shoal deposits 147
Shoreface,
aggradation 49
architecture *48*
deposits 22–3, 25, 49
architecture of 24, 48
Brent Group, UKCS 194–8
Cretaceous, Utah 145, 147, 152, *153*, 154
dip data for *153*, *155*
directional trend of 152, 154
highstand *48*
lowstand *48*, 49
sedimentary structures *155*
stacking patterns 23
transgressive *48*, 49
see also Shallow marine deposits
Shrinkage cracks **29**, 40, 42
Siderite 42
Sierra de Altomira, Spain 160–1
Sigmoidal bedding 41
Siltation projects 47
Simulation models 194
grid cells 194, 195
numerical *195*, *196*
Sinuosity values 32
'SIRCH' software 238, 241, 245
applications of 246–8
shape generation 239
Sleipner Formation, Norway 59
Slickensides 42
Slip surfaces 117, 121
Slump(s) 107, 108
axes *104*
Smectites 131, 135
Smoky Hollow Member, Utah 27, 30, 37–9, 42, 43
Snadd Formation, Barents Sea 59
Snorre Field, Norway 60, 194
Soil horizons 42
see also Calcrete, Caliche
Sorting, grain 163
Source rocks 128
Southern Pyrenees *see* Pyrenean
Spain,
Ainsa Basin *see* Ainsa Basin
Loranca Basin 158, 159, 160, 161–4
submarine fan channel study 100–10
Special core analysis 202
Sporomorphs 132
Spraberry Trend, USA 109
Statfjord Field, Norway 59, 60, 253
Statfjord Formation 59, 60
Static normalization (FMS) 128–9, 131
Steam injection 131
Stevens Sandstone, USA 109
Stochastic modelling 11, 193, 234
channel scale heterogeneity, interactive (SIRCH) 238, 241, 245, 246–8, 249
described 238
and faults 246
sequence based 214
Straight Cliffs escarpment 43, 45
Straight Cliffs Formation, Utah 24–6, 34, *35*, 36–7, 42, 44, 49
Strain softening 118
Strandplain deposits *172*, *173*, 173, *174*, 174, 178
Stratal patterns 26
Stratigraphic traps 99
Stream equilibrium profiles 46–7
Striae 117
Stromatolites 138
Subaqueous fan sands 132
Submarine fan deposits,
channel,
mapping 183
sands 99
distributary systems, seismic imaging of 182–3
sand lobe(s) 99
identification 183
shale length in 100
see also Turbidites
Subsidence,
differential 76
rates 12
and alluvial architecture 45, **46**
tectonic 45, 46
thermal 84
Surfactant flooding 167
Sweep efficiency 158, 164, 201
Synaeresis cracks 42
Systems tract analysis 7

Tabular sandbodies **29**, 41
Tasman Orogenic Phase 84
Teredolites 41
Terrace deposits 34, 47, *48*, 49
Tertiary,
Ireland 182–3
Spain 101, 159–60
fluvial deposits 158
tectonics 159–60
Middle East 139
Texas, USA 109, 152, *171*, 172–5, *176*, 177–9
Thallassinoides burrows 105, 107
Thief zones 195, 197, 201
Three-dimensional models,
advantages of 257
building,
and geologic heterogeneity 253, 255, 256
stratigraphic framework 253, 254
and interfacing with reservoir simulator 256–7
limitations 257–8
reservoir 12
Thrust,
belts 161
faulting 24, 60, 84
tectonics 141
Tibbet Canyon, Utah *30–1*, *34*, *38*, 44
Tidal channel deposits 44
Tool marks 37, 117
Tórtola Fan, Spain 160
sedimentary facies 161–4
Trace fossils 42
of channel deposits 41
of turbidite deposits 105, 107
Transgressive systems tracts 26, **28**, *36*
alluvial 44, 49, 50
Transition probabilities, lateral and vertical 174–5, *177*, *178*
Transport directions 12, 31, 69
channel sandstones 38
from palaeocurrents 36
Traps,
footwall 122
hydrocarbon 99, 114, 125, 128, 172
Tremp-Graus Basin, Spain 60, *61*, 63, 101
Triassic, fluvial reservoirs 59
Trusklieve, Ireland 145, *146*, *148*, *151*
Tullig Cycle, Ireland 145, *146*, *148*
Turbidite channels,

diagenesis of 102
dimension of 102
mapping 185–6
Turbidite deposits, *see also* Submarine fans
basin plain 101
bioturbation in 105, 106–7, 108, 109
channel-fill 101–2, *103*, 109–10
burrows 105
lithofacies of 102, 104–9
permeabilities **107**, 107
sedimentary log for *104*
as exploration targets 99
fan,
fringe 107
sediments *186*
on seismic data 181, 182
Frigg Field, Norway 100, *182*
glacio-lacustrine 132
Ireland 183
meandering channel 108
palaeocurrents from *102*, 102, *104*, 108
sand lobes 99, 100, 101, 107, 109
seismic mapping of *182*, 185–6
and shale length 100
South-Central Pyrenean Basin *102*
stratigraphic traps in 99
Turbidity currents 105, 107

Unconformities 7, 13, 43, 44
intraformational 132
sequence boundary 44, 50
and sequence stratigraphy 24
United Kingdom
Northumberland *208*, 209–10
Round O Quarry, Lancs 115–16
see also North Sea
United States,
California 109
Bell Ridge Field 179
Hartzog Draw Field 138, 191, *192*
Holocene 42
Kentucky *9*
New Mexico 152
Rocky Mountains 24
Texas 152
Tertiary field example *171*, 172–5, *176*, 177–9
Utah *see* Utah
Wyoming 191, *192*
Upturned laminae 106
Utah 40
Blue Cove *35*, *36*, *38*
Book Cliffs 25
Cretaceous 13, 22, 144
shoreface sands 145, 147, 152, *153*, 154
thrust faulting 24

Valley,
fill deposits **28**, *36*, 44
incision deposits *48*, 49
Variograms 8, *9*, 13
Villalba de la Sierra Fan System, Spain *161*

Wafra Field, Oman 135
Walther's Law 50
Waste disposal 22
Water,
breakthrough 194–5, 196, *197*, 197–8
cut 129, 138, 195
fingering 167
flood 167
prediction 141, 195, *196*, 202
imbibition 164, 167
injection 128, 132, 135, 201
tonguing 167
Weibull distribution 226
Well data,
correlation 4–5, 13
high resolution 4–5
Western Interior Seaway, USA *152*
Westphalian,
Oman 132
UK 115–16
Width/thickness ratios *8*, 8, *9*, *11*, 11–12, 227, *228–9*, *231*, 231, 242
alluvial stratigraphy model 219–20
distributary channel deposits *96*, 96
fluvial channel deposits 64, *73*
high sinuosity rivers 36–7
low sinuosity rivers 32
turbidite channels 102
Wilcoxon test 164
Wireline logs 3–4
resolution 7
Workstations 4, 181–2, 184
Wyoming, USA 191, *192*

Yibal Field, Oman 138
Yoluwne Field, California, USA 109

Zauliyah Field, Oman 135